The Absolute, Ultimate Guide to
Lehninger Principles of Biochemistry

The Absolute, Ultimate Guide to
Lehninger Principles of Biochemistry
THIRD EDITION

Study Guide and Solutions Manual

Marcy Osgood
University of New Mexico School of Medicine,
Department of Biochemistry and Molecular Biology

Karen Ocorr
University of Michigan, Ann Arbor

Solutions Manual based on
the previous edition by

Frederick Wedler
Robert Bernlohr
Ross Hardison
Teh-Hui Kao
Ming Tien

Pennsylvania State University

WORTH PUBLISHERS

The Absolute, Ultimate Guide to
Lehninger Principles of Biochemistry, Third Edition
Study Guide and Solutions Manual

Printed in the United States of America

ISBN: 1-57259-167-6

Fourth printing, 2003

Cover images by Jean-Yves Sgro

Worth Publishers
41 Madison Avenue
New York, NY 10010
www.worthpublishers.com

Study Guide

Contents

Solutions Manual

Contents

Preface

Learning a complex subject, such as biochemistry, is very much like learning a foreign language.

In the study of a foreign language there are several distinct components that must be mastered: the vocabulary, the grammatical rules, and the integration of these words and rules so they can be used to communicate ideas. Similarly, in the study of biochemistry there is a very large (some would say vast) number of new terms and concepts, as well as a complex set of "rules" governing biochemical reactions that must all be memorized. All of this information must be integrated into an interrelated whole that describes biological systems.

Memorizing vocabulary and grammatical rules will not make you fluent in a foreign language; neither will such memorization make you fluent in biochemistry.

Similarly, listening to someone speak a foreign language will not, by itself, make you more capable of producing those same sounds, words, and sentences. The key to mastery of any new subject area, whether it is a foreign language or biochemistry, is the interaction of memorization, practice, and application until the information fits together into a coherent whole.

In this workbook we attempt to guide you through the material presented in Lehninger Principles of Biochemistry, Third Edition *by Nelson and Cox. The* **Step-by-Step Guide** *to each chapter includes three parts:*

- A one to two page summary of the **Major Concepts** helps you to see the "big picture" for each chapter.
- **What to Review** helps you make direct connections between the current material and related information presented elsewhere in the text.
- **Topics for Discussion** for each section and subsection in the chapter focus your attention on the main points being presented and help you internalize the information by *using* it.

Each chapter also includes a **Self-Test** *for you to assess your progress in mastering biochemical terminology and facts, and learning to integrate and apply that information.*

- **Do You Know the Terms?** asks you to complete a crossword puzzle using the new vocabulary introduced in the chapter.
- **Do You Know the Facts?** tests how well you have learned the "rules" of biochemistry.
- **Applying What You Know** tests how well you "speak the language" of biochemistry, often in an experimental or metabolically relevant context.

There are two features new to this edition of the Absolute, Ultimate Guide.

- The **Biochemistry on the Internet** problems will expose you to just a few of the analytical resources that are available to scientists on the Internet. The molecular models are fun to play with, and many of the questions provide you with an opportunity to analyze "data" as you might in an actual laboratory setting.
- The new **Cell Map** is based on many semesters of use by our students. It is designed to help you place the biochemical pathways that you are learning about into their proper cellular perspective. The Cell Map questions (designated with a icon) tell you what to include in your Map but you can be creative!

Students have told us it is a great study aid, that really helped them to make the connections between the various pathways.

In our experience, this material can best be assimilated when it is discussed in a study group.

A study group is nothing more than two to four people who get together on a regular basis (weekly) to "speak biochemistry." This type of interaction is critical to fluency in a foreign language and is no less critical in the successful assimilation of biochemistry. We designed our Step-by-Step Guide to each chapter with study groups in mind. The questions posed for each section can be used as springboards for study group discussions. We purposefully have *not* supplied answers to this section to force you to wrestle with the concepts. It is the struggle that will make you learn the material. In addition, since most of the answers can be readily worked out by a careful reading of the section of the text, the questions will focus your attention on the more important aspects of the material.

Detailed **Solutions** *to all the end-of-chapter textbook problems are included as a separate section in the* Absolute, Ultimate Guide.

We have taken great care to ensure that the solutions are correct, complete, and informative. The final answer to numerical problems has been rounded off to reflect the number of significant figures in the data.

We thank each and every one of our students for their invaluable feedback and input which has helped to make this study guide its absolute and ultimate best.

About the Authors

Karen Ocorr has taught Introductory Biochemistry for seven years in the Biology Department at the University of Michigan, Ann Arbor. She has nine years of experience teaching the department's Animal Physiology course and has also taught Cell Biology, Seminars for Biology Majors, and Biology for Non-Majors. Dr. Ocorr received her Ph.D. from Wesleyan University where she studied the physiology and neurochemistry of the lobster cardiac ganglion. As an NIH postdoctoral research fellow at the University of Texas in Houston she examined the role of enzymes and second messengers in modifying neuronal activity in the sea snail, *Aplysia californica*. She continued her investigations of the biochemical basis of neuronal plasticity in the vertebrate hippocampus at the California Institute of Technology and Stanford University. At the University of Michigan her research efforts have shifted to an investigation of the roles of protein phosphorylation in developmental plasticity.

Marcy Osgood received her Ph.D. from Rensselaer Polytechnic Institute. After two postdoctoral positions, one that involved the investigation of mammalian mitochondrial Complex III, and the other that involved an attempt to construct a mercury-detecting microbial biosensor, she became an Assistant Professor in the Department of Physical Sciences at Albany College of Pharmacy, Union University. She then took a position as Lecturer in the Biology Department at the University of Michigan, Ann Arbor. She taught Introductory Biochemistry, a Seminar in Teaching Biochemistry, as well as Biology for Non-Majors, for eight years there. She won numerous teaching awards and organized and participated in teaching workshops at national meetings. Her current position is as Assistant Professor of Biochemistry Education in the University of New Mexico's School of Medicine, Department of Biochemistry and Molecular Biology.

Osgood and Ocorr have worked together for six years to refine this study guide. It includes thousands of discussion, quiz, and exam questions from their Biochemistry courses, and also incorporates their teaching philosophy, which has been developed by years of instruction at many levels and in many areas of biology. Between them, they have been responsible for teaching Biochemistry to over 10,000 undergraduates.

Study Guide

The Molecular Logic of Life

STEP-BY-STEP GUIDE

Major Concepts

This chapter is the first introduction to underlying themes in biochemistry that will be reiterated throughout the text.

At this point, the student may not completely understand each concept or term. However, the overview this chapter provides may be referred to later as the material becomes more detailed.

This chapter should be largely a review for those students with a general biology and chemistry background.

Most, if not all, of the information presented will be very familiar. Because this chapter is intended as introduction and review, this study guide presents an abbreviated treatment of the material.

Macromolecules found in biological systems are formed from a relatively small number of simple compounds.

A set of 20 different amino acids is used to make all proteins. Four different deoxyribonucleotides are used to construct DNA, and four ribonucleotides are used to construct RNA. Polysaccharides are constructed from monosaccharides.

Living organisms are interdependent; they exchange energy and matter with each other and the environment.

The sun is the ultimate source of (almost) all energy used by organisms. The **anabolic** reactions that use energy to form biological macromolecules and the **catabolic** reactions that liberate energy constitute the **metabolic pathways.** Most of these reactions require **enzymes** in order to proceed at useful rates. All metabolic reactions, anabolic and catabolic, are linked through **ATP,** which can capture or release stored energy as needed by cells.

Information needed for the formation and function of all living organisms is contained in their genetic material, which is usually DNA.

The linear sequence of nucleotides in a strand of DNA provides information specifying the linear sequence of amino acids in all the proteins made by an organism. In turn, the complement of proteins determines the repertoire of metabolic reactions and functions that an organism can perform. (In some viruses, the genetic material is RNA.)

Topics for Discussion

Answering each of the following questions, especially in the context of study group discussions, should help you understand the important points of the chapter.

The Chemical Unity of Diverse Living Organisms

1. What properties of biological macromolecules generate such diversity in the types of cells and organisms?

Energy Production and Consumption in Catabolism

2. In terms of **potential energy,** energy transductions, and entropy, explain the following normal human daily activities: eating, moving, excreting. Where do the sun and ATP fit into this scheme?

3. Why are enzymes essential in biochemical reactions? How do enzymes overcome activation barriers? Why is the option of increasing temperature to overcome these barriers *not* possible in living cells?

Biological Information Transfer

4. How are the instructions contained in the linear sequence of DNA translated into a three-dimensional enzyme or structural protein? What forces contribute to the three-dimensional structure? How can a change in the DNA instructions affect this final form?

SELF-TEST

Do You Know the Terms?

ACROSS

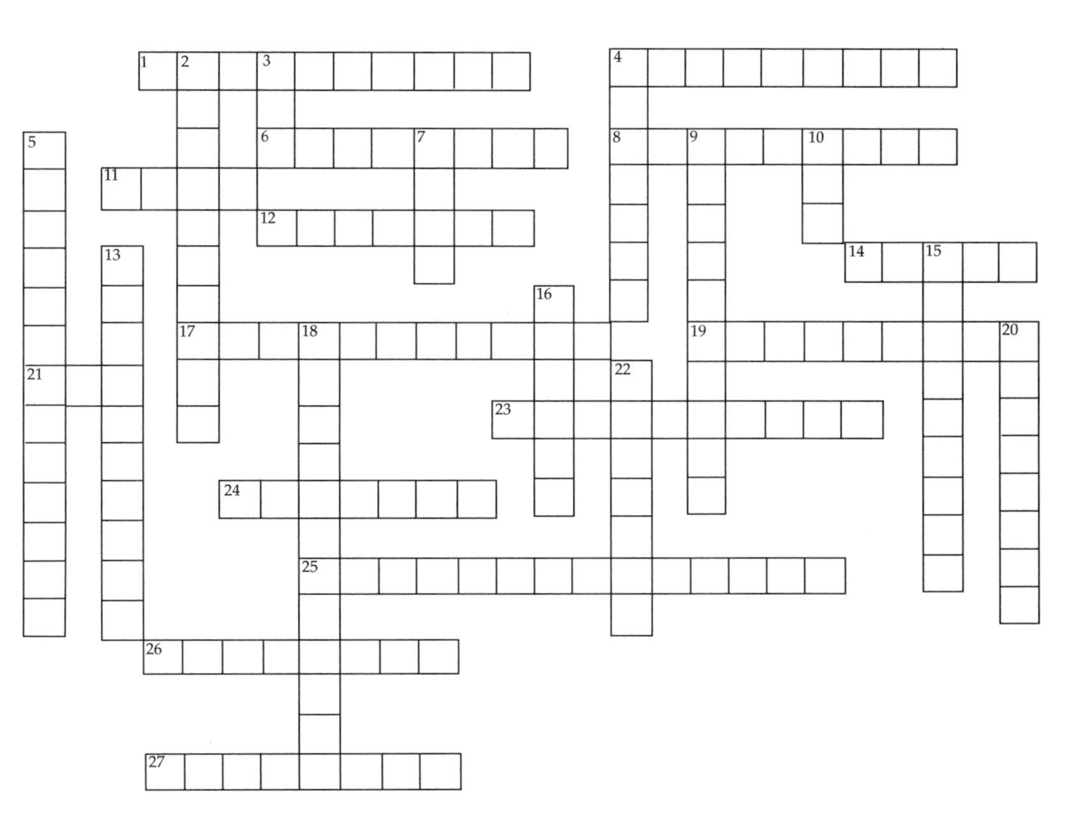

1. The network of interrelated catabolic and anabolic pathways in cells is referred to as cellular ________.
4. Reactions for which ΔG is negative are ________ reactions.
6. Consecutive reactions that convert initial reactants into final products.
8. Interconvert energy from one form to another.
11. A system that exchanges both energy and material with its surroundings is said to be ________.
12. Biological reactions that share reactants and/or products are ________ reactions.
14. ________-synthetic organisms convert solar energy into ATP.
17. A type of weak interaction that stabilizes the native conformation of a biomolecule or supramolecular complex.
19. Synthesis of large, complex molecules from smaller precursor molecules.
21. The monomeric subunits of this molecule are ribonucleotides.
23. The stretching and breaking of bonds that occurs during the conversion of a reactant to a product creates a ________ state.
24. A macromolecule composed of a chain of amino acids.
25. Proteins, nucleic acids, and polysaccharides.
26. Many reactions in biological systems are kept in balance by ________ feedback.
27. Complex macromolecules are comprised of a small, common set of monomeric ________.

DOWN

2. Reactions for which the products have more free energy than do the reactants.
3. A chemical form of energy.
4. A measure of randomness.
5. The principles of ________ describe energy transformations and exchanges.
7. A genome without mutations that affect phenotype would produce a ________-type organism.
9. Enzymes enhance the rate of chemical reactions by lowering the ________ energy that constitutes an energy barrier between reactants and products.
10. The monomeric subunits of this molecule are deoxyribonucleotides.
13. Degradation of organic compounds into simple end products to extract chemical energy.
15. ________-reduction reactions involve the flow of electrons.
16. G; free ________.
18. The three-dimensional structure of a protein; native ________.
20. A change in the nucleotide sequence of a chromosome.
22. Chemical reactions in cells occur at rates up to 10^{10}- to 10^{14}-fold greater than the same reactions in test tubes due to the presence of cellular ________.

ANSWERS

Do You Know the Terms?

ACROSS

1. metabolism
4. exergonic
6. pathways
8. transduce
11. open
12. coupled
14. photo
17. noncovalent
19. anabolism
21. RNA
23. transition
24. protein
25. macromolecules
26. negative
27. subunits

DOWN

2. endergonic
3. ATP
4. entropy
5. bioenergetics
7. wild
9. activation
10. DNA
13. catabolism
15. oxidation
16. energy
18. conformation
20. mutation
22. enzymes

chapter

2

Cells

STEP-BY-STEP GUIDE

Major Concepts

All cells have common structural elements.

All cells are defined by a **plasma membrane,** which separates the contents of a cell from its surroundings and is a barrier to diffusion. All cells are divided into two major internal regions, the cytoplasm and a nuclear region. The **cytoplasm** contains soluble enzymes, metabolites, and cellular organelles; the **nuclear region** contains primarily DNA and associated proteins.

Cells are classified according to the complement of cellular membranes and the complexity of the nuclear region.

Prokaryotes are relatively small cells that have only a plasma membrane. Their nuclear region, the **nucleoid,** has no membrane to separate it from the rest of the cytoplasm. In addition, there are no other internal membranes and no internal **organelles. Eukaryotes** generally are larger than prokaryotes and include all protists, fungi, plants, and animals. Eukaryotic cells have plasma membranes as well as many membrane-bounded intracellular organelles and a membrane-bounded nucleus.

The chart below provides a comparison of the sizes and cellular components of a "typical" prokaryotic and eukaryotic cell.

Prokaryotic Cell	Size (μm)	Eukaryotic Cell	Approximate Size (μm)
Cell diameter	1–10	Cell diameter	5–100
Nucleoid	1	Nucleus	10
		Chloroplast	2 × 8
		Mitochondrion	0.5 × 3
		Lysosome	1
Ribosome	0.025	Ribosome	0.03
Flagellum, diameter	0.01–0.02	Flagellum, diameter	0.015
Chromosome, condensed	<1	Chromosome, condensed	4
Chromosome, extended	100–1000	Chromosome, extended	4 cm

Topics for Discussion

Answering each of the following questions, especially in the context of study group discussions, should help you understand the important points of the chapter.

Cellular Dimensions

1. Cellular dimensions are difficult to imagine on a human scale. To put cells into perspective it is helpful to magnify them. The following table shows the relative sizes of cells and their components magnified 10,000-fold (from micrometer to millimeter scale). A typical eukaryotic cell, when magnified 10,000-fold, is about the size of a large platter. Complete the table below with familiar objects that are approximately the sizes indicated and assemble them on a large serving platter.

Cell or Cellular Component	Size at 10,000× (mm)	Equivalent Size Object
Ribosome	0.3	
Lysosome	10	
Flagellum, diameter	0.15	
Nucleoid	10	
Mitochondrion	5 × 30	
Prokaryotic cell	10–100	
Chloroplast	20 × 80	
Nucleus	100	
Eukaryotic cell	500 (50 cm)	Large serving platter

Cells and Tissues Used in Biochemical Studies

2. What are some of the features of experimental or model systems that are especially critical for biochemical studies?

Evolution and Structure of Prokaryotic Cells

3. What are the two groups of extant prokaryotes?

4. What feature(s) allow discrimination between different types of eubacteria and archaebacteria?

5. How would you classify mushrooms with respect to their energy sources?

6. Why is it likely that autotrophs evolved only after the appearance of heterotrophs?

7. What are structural features of prokaryotes that distinguish them from eukaryotes, which are discussed in the next section of the text?

Evolution of Eukaryotic Cells

8. As cells became larger, the ratio of cell surface area to cell volume decreased. What changes occurred that facilitated transmembrane exchange of nutrients and wastes despite a reduction in the surface area of cells?

9. Which structures of modern eukaryotic cells probably did *not* arise as a result of endosymbiotic associations with a common ancestral prokaryote?

Major Structural Features of Eukaryotic Cells

10. What are the different cellular functions of the plasma membrane and the associated membrane proteins in eukaryotes?

11. Why do some cells require a cell wall in addition to a plasma membrane?

12. List at least three functions of the endoplasmic reticulum.

13. What is the orientation of the Golgi complex in cells, and what is its primary function?

14. Which organelle is responsible for the degradation of used proteins in animal cells? What is the corresponding organelle in plants? Why is it sensible to have a specialized compartment for this process?

15. What specialized role do peroxisomes play in catabolic processes in cells?

16. What is the role of the nucleolus?

17. What role do nucleosomes play in the nucleus?

18. Compare the structure and function of mitochondria and chloroplasts.

19. What are the three classes of cytoskeletal proteins?

20. What are three functions of the cytoskeleton?

Study of Cellular Components

21. How can researchers study cellular organelles? Describe the procedures that researchers use to separate and isolate the different organelles of a cell.

Evolution of Multicellular Organisms and Cellular Differentiation

22. How do cells in multicellular organisms associate with each other? What types of molecules are involved in these associations?

Viruses: Parasites of Cells

23. What aspects of viruses make them unable to be classified as cells?

SELF-TEST

Do You Know the Terms?

ACROSS

2. mRNA molecules with two or more attached ribosomes.

4. Component of eukaryotic cells consisting of microtubules, actin filaments, and intermediate filaments.

6. ________ cytosis involves the invagination of a portion of the plasma membrane to produce an intracellular vesicle that contains extracellular particles.

7. A small, circular segment of DNA located outside the nucleoid is called a ________.

8. Small molecules such as Na^+ and K^+, which are charged, cannot diffuse across plasma membranes. These molecules require ion ________ to cross the membrane.

10. Large, cytoplasmic organelles that produce ATP, which provides the energy for cellular work.

14. The internal components of cells and the aqueous solution in which they are suspended.

15. Archaebacteria and eubacteria.

16. Essential, organic participant in many enzyme-catalyzed reactions.

18. Organisms that obtain energy by transferring electrons to substances other than oxygen.

22. A ________ is a supramolecular complex that can replicate only within a host cell.

23. During cell replication, the genetic material must be faithfully duplicated and partitioned between the two daughter cells. The process of division of the genetic material is called ________.

24. To make an omelet, one has to break a few eggs; to study cells, one often must crack them open by ________.

25. Complex that modifies newly synthesized proteins. Here, proteins are provided with an address ensuring that they reach their proper destinations in the cell.

27. Solution surrounding the intracellular contents of cells.

28. Region of the nucleus that contains a high concentration of RNA (ribosomal RNA).

30. Fusion of intracellular, membrane-bounded vesicles with the cell membrane, resulting in the release of cytoplasmic contents into the extracellular space.

32. Membrane proteins that bind extracellular molecules and relay information about this binding to the interior of the cell.

34. Invagination of a portion of the plasma membrane, which then buds off to produce a fluid-filled vesicle called an endosome.

36. Membrane-bounded compartment, present only in eukaryotes, that contains chromosomes.

37. ________ organisms use light as their primary source of energy.

39. Centrifugation can be used to fractionate the particulate components of cells (for example, membranes) into a pellet at the bottom of the centrifuge tube, away from the soluble components (cytosol), which remain in the ________ fraction.

40. Organisms that produce energy from fuel molecules by transferring electrons to oxygen use ________ metabolism.

41. Tight junctions and ________ are points of intercellular attachment that hold multicellular organisms together.

DOWN

1. The cell ________ provides some structural rigidity to plant cells.

3. Mitochondria are thought to have evolved from bacteria that formed ________ associations with the ancestors of modern eukaryotes.

4. Plastid containing the pigment chlorophyll.

5. ________ proteins in the plasma membrane circumvent permeability barriers to the diffusion of nutrients and wastes into and out of cells.

6. Viruses that infect bacterial cells.

9. Intracellular organelle containing enzymes that work best at pH 5 and are responsible for most of the degradation and recycling of cellular components.

10. Intermediates in biosynthetic and degradative pathways.

11. Organisms that cannot synthesize all of the molecules they need for growth; the missing molecules are obtained from nutrients in their environment.

12. Organisms that can synthesize most of the molecules necessary for their growth from simple compounds such as CO_2 and NH_3.

13. The final step of cell division, in which two daughter cells are formed.

14. Molecular complexes of DNA plus associated histone and nonhistone proteins.

17. Intracellular organelle in plant cells that degrades and recycles cellular components.

19. Protists, yeast, fungi, plants, and animals.

20. DNA-binding proteins.

21. A complex of RNA and protein.

26. The complete set of genetic material containing all the genes required for the growth and replication of a cell.

29. Structural units of genetic material in eukaryotes.

31. The internal membranes of chloroplasts in which the chlorophyll pigment is localized.

33. An organism that obtains its energy from chemical fuel molecules.

35. Gamete is to haploid as somatic cell is to ________.

38. Membrane that defines the periphery of cells and prevents their contents from diffusing away.

ANSWERS

Do You Know the Terms?

ACROSS

2. polysomes
4. cytoskeleton
6. phago
7. plasmids
8. channels
10. mitochondria
14. cytoplasm
15. prokaryotes
16. coenzyme
18. anaerobic
22. virus
23. mitosis
24. homogenization
25. Golgi
27. cytosol
28. nucleolus
30. exocytosis
32. receptors
34. endocytosis
36. nucleus
37. phototrophic
39. supernatant
40. aerobic
41. desmosomes

DOWN

1. wall
3. symbiotic
4. chloroplast
5. transporter
6. phages
9. lysosome
10. metabolites
11. heterotrophs
12. autotrophs
13. cytokinesis
14. chromatin
17. vacuole
19. eukaryotes
20. histones
21. ribosome
26. genome
29. chromosomes
31. thylakoids
33. chemotroph
35. diploid
38. plasma

chapter

3

Biomolecules

STEP-BY-STEP GUIDE

Major Concepts

Living matter is composed of low atomic weight elements.

Hydrogen, oxygen, nitrogen, and carbon are the most abundant elements in biomolecules. The bonding versatility of carbon makes it the most important and defining element in biochemical compounds. Most biomolecules are derivatives of hydrocarbons with a variety of attached functional groups. These functional groups determine the chemical behavior of biomolecules.

Biochemistry is three-dimensional.

In addition to the functional groups, the overall shape of a biomolecule greatly affects the types of interactions in which it can participate. Most biomolecules are asymmetric. Typically, only one of the possible (**chiral**) forms is found in living organisms. Interactions between many biomolecules (for example, enzymes and their substrates) depend upon their ability to differentiate between **stereoisomers.**

Biochemical reactions are fundamentally similar to other chemical reactions.

The strength of chemical bonds, expressed as bond energies, depends on the relative electronegativity of the elements involved. As bonds are formed or broken, energy is either gained from or lost to the surroundings. Five general categories of chemical transformations occur in living cells: **oxidation-reduction** reactions, breakage or formation of carbon–carbon bonds by **nucleophilic substitution, internal rearrangements** of the bonds around carbon atoms, **group transfers,** and **condensation** reactions. Many biochemical reactions involve the giving up of electrons by nucleophilic functional groups and the acceptance of electrons by electrophilic groups.

Biological systems have a hierarchy of structure.

A relatively small number of monomeric subunits are the building blocks used to construct macromolecules. Macromolecules can then be arranged into supramolecular complexes. Different classes of macromolecules have different roles in cells. For example, membranes (supramolecular complexes of lipids, proteins, and carbohydrates) define and compartmentalize cells, whereas chromosomes (supramolecular complexes of DNA and proteins) encode and store genetic information.

Biomolecules first arose by chemical evolution.

The conditions under which life arose on Earth cannot be absolutely determined, but chemical evolution can be simulated in the laboratory. Considering the simplicity of the chemical reactions that probably led to the formation of small biomolecules, and eventually to macromolecules, the evolution of life seems to be possible on other planets.

What to Review

Most of this chapter is a review of organic and inorganic chemical principles. You may find it helpful to consult a good inorganic or organic chemistry text for additional background information. Using this and other texts as sources, make sure that you thoroughly understand the following definitions and concepts.

- Covalent bonds and their relationship to the number of unpaired electrons (Fig. 3–2; Table 3–3).
- The significance of the atomic numbers.
- Stereoisomerism; chirality (pp. 57–61; Fig. 3–9).
- Electronegativity (p. 64; Table 3–2).
- Nucleophilic and electrophilic centers in reactions (p. 66; Table 3–4).

Topics for Discussion

Answering the following questions, especially in the context of a study group discussion, should help you understand the important points of this chapter.

Chemical Composition and Bonding

1. What are the four most common elements in living organisms?

2. What is the importance of trace elements to animal life?

Biomolecules Are Compounds of Carbon

3. What is meant by the "bonding versatility" of carbon?

4. What factors influence the strength, length, and rotation of carbon bonds?

Functional Groups Determine Chemical Properties

5. List ten to fifteen common families of functional groups or organic compounds that are encountered in biomolecules.

Three-Dimensional Structure: Configuration and Conformation

6. What are the three types of models that are used to represent the three-dimensional structure of molecules? What different kinds of information are provided by each type of model?

The Configuration of a Molecule Is Changed Only by Breaking a Bond

7. What is the difference between a diastereomer and an enantiomer?

Molecular Conformation Is Changed by Rotation about Single Bonds

8. How does the high potential energy of the eclipsed conformation effect the dynamics of ethane?

Louis Pasteur and Optical Activity: In Vino, Veritas

9. Use the structural formulas of Pasteur's two forms of tartaric acid to follow the description of RS nomenclature in the text.

Configuration and Conformation Define Biomolecular Structures

10. What are the advantages and disadvantages of the techniques of x-ray crystallography and NMR spectroscopy?

Interactions between Biomolecules Are Stereospecific

11. Why do cells produce only one form of a chiral compound rather than a racemic mixture?

Chemical Reactivity

Bond Strength Is Related to the Properties of the Bonded Atoms

12. How do the relative electronegativities of atoms in biomolecules influence the polarity and strength of their bonds?

13. What is the difference between the bond energies and the enthalpy of a reaction?

Five General Types of Chemical Transformations Occur in Cells

Committing these five general types of reactions to memory *now* will make understanding later chapters much easier.

All Oxidation-Reduction Reactions Involve Electron Transfer

14. Based on the discussion of oxidation-reduction reactions, does it make intuitive sense that CO_2 and H_2O are the final end products of ATP-yielding catabolism?

Carbon–Carbon Bonds Are Cleaved and Formed by Nucleophilic Substitution Reactions

15. Why are the different types of nucleophilic substitutions good examples of the importance of the "bonding versatility" of carbon to biological reactions?

Electron Transfers within a Molecule Produce Internal Rearrangements

16. Are any atoms lost or gained in the phosphohexose isomerase reaction?

Group Transfer Reactions Activate Metabolic Intermediates

17. Why does it make sense that "activator" groups such as phosphoryl groups and thioalcohols are common participants in many anabolic and catabolic reactions?

Biopolymers Are Formed by Condensations

18. Why is the condensation of two amino acids into a dipeptide not at all a "simple" reaction?

Macromolecules and Their Monomeric Subunits

Macromolecules Are the Major Constituents of Cells

19. What are macromolecules and how are they categorized?

Macromolecules Are Composed of Monomeric Subunits

20. Proteins and nucleic acids are considered **informational macromolecules** while polysaccharides are not; why is this not an absolute difference between these types of macromolecules?

Monomeric Subunits Have Simple Structures

21. What are the functional groups that all amino acids have in common?

Subunit Condensation Creates Order and Requires Energy

22. What are the qualities that determine the free energy (G) in a system?

23. How are cells able to synthesize polymers if such reactions are thermodynamically unfavorable?

Cells Have a Structural Hierarchy

24. What are the bonds and/or interactions that are important at each of the three levels of cell structure?

Prebiotic Evolution

Biomolecules First Arose by Chemical Evolution

25. Under what environmental conditions are the first biomolecules thought to have been formed?

Chemical Evolution Can Be Simulated in the Laboratory

26. What energy sources may have been available to drive prebiotic evolution?

RNA or Related Precursors May Have Been the First Genes and Catalysts

27. What are the lines of evidence suggesting that RNA or a similar molecule was both the first gene and the first catalyst?

28. Why would the development of impermeable layers of lipid-like compounds have favored continued chemical evolution?

Biological Evolution Began More Than Three and a Half Billion Years Ago

29. The fossil record indicates that organisms evolved mechanisms to adapt to dramatic changes in the composition of Earth's atmosphere. How can we use this information in evaluating the consequences of modern global warming?

SELF-TEST

Do You Know the Terms?

ACROSS

1. Compounds having different arrangements of substituent groups around nonrotating double bonds are ________–________ isomers.

3. Compound comprising multiple subunits.

6. Compounds having electron-deficient functional groups; they tend to bond to electron-rich (negatively charged) sites.

8. Describes molecules containing covalently bonded carbon backbones.

9. Conversion of glucose-6-phosphate to fructose-6-phosphate is an internal ________ reaction.

11. Amino acids are the ________ subunits of proteins.

12. Molecules having the same chemical formula but different arrangements of atoms.

16. Spatial arrangement defined by the specific location of double bonds and/or chiral centers in a molecule.

17. Structural components of membranes; energy storage molecules.

18. Compounds having electron-rich functional groups, for example, H_2O, —NH_2, or —SH; they tend to bond to electron-deficient (positively charged) sites.

20. In ________–________ reactions, electrons are transferred from a more reduced to a more oxidized molecule.

21. Glycine is the only amino acid lacking an asymmetric, or ________ carbon.

22. Stereoisomers that cannot be superimposed.

25. The arrangement in space of a molecule that is free to assume different positions.

26. Reaction in which two smaller molecules form a larger molecule with the production of H_2O.

27. A heterolytic ________ will produce a carbanion and a carbocation.

28. Shifting a phosphoryl group from a molecule of ATP to a molecule of glucose is a group-________ reaction.

DOWN

2. Molecules having the same composition and order of atomic connections, but different spatial arrangements among the atoms.

3. Amino acid is to ________ as monomer is to polymer.

4. An equimolar mixture of the D and L isomers of an optically active compound is a ________ mixture.

5. The randomness of the components of a chemical system; *S*.

6. Reactions requiring an input of energy from the surroundings are ________ thermic reactions.

7. ________ groups determine the chemical properties of biomolecules.

10. The energy or heat content of a system; *H*.

11. Proteins and nucleic acids are informational ________.

13. An ________ thermic reaction releases energy to the surroundings (ΔH is negative).

14. Polymers of nucleotides. (2 words)

15. One of a pair of stereoisomers that are *not* mirror images of each other.

19. Polymers of simple sugars.

22. Describes a reaction for which the free-energy change (ΔG) is positive.

23. Describes a reaction for which the free-energy change (ΔG) is negative.

24. ________ activity is the ability of a molecule to rotate the plane of polarized light.

Do You Know the Facts?

1. Add the appropriate elements to make each of the following functional groups:

Aldehyde R—

Amido R—

Amino R—

Carboxyl R—

Disulfide R—

Ester R—

Ether R—

Ethyl R—

Guanidino R—

Hydroxyl R—

Imidazole R—

Ketone R—

Methyl R—

Phenyl R—

Phosphoryl R—

Sulfhydryl R—

2. Identify the following functional groups and categorize them as either electrophilic or nucleophilic.

(a) R—OH

(b) R—SH

(c)

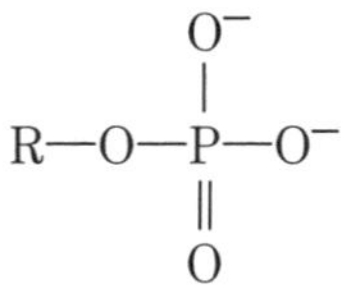

```
         O⁻
         |
R—O—P—O⁻
         ||
         O
```

(d)

```
R—C=O
   |
   O⁻
```

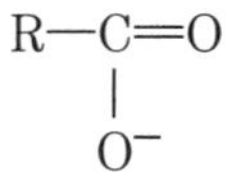

3. In the chart below, list the four classes of macromolecules in cells, their monomeric subunits, and general functions.

Macromolecule	Subunits	General Functions

4. The following reactions are involved in carbohydrate metabolism. Each is representative of one of the five general types of chemical transformations discussed in this chapter; classify each according to its reaction type.

(a)

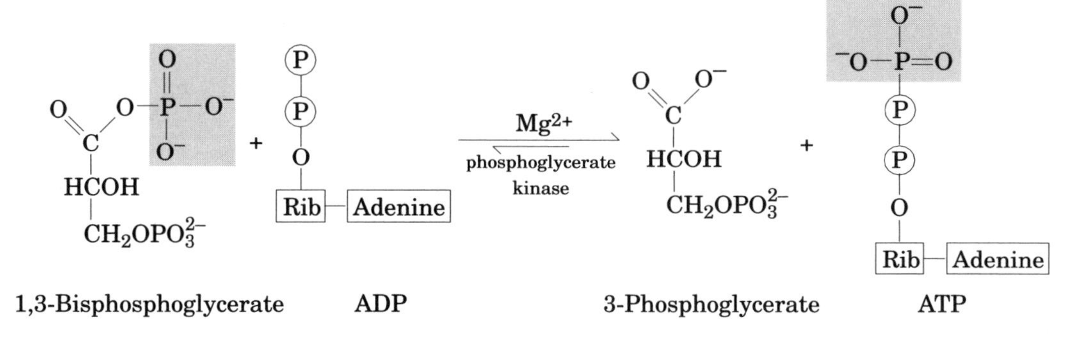

$\Delta G'^{\circ} = -18.5$ kJ/mol

(b)

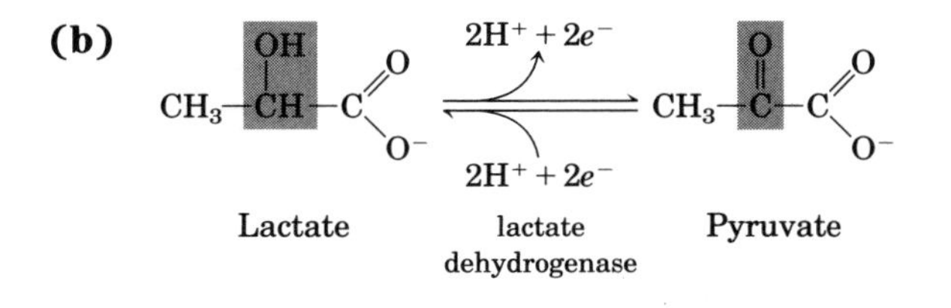

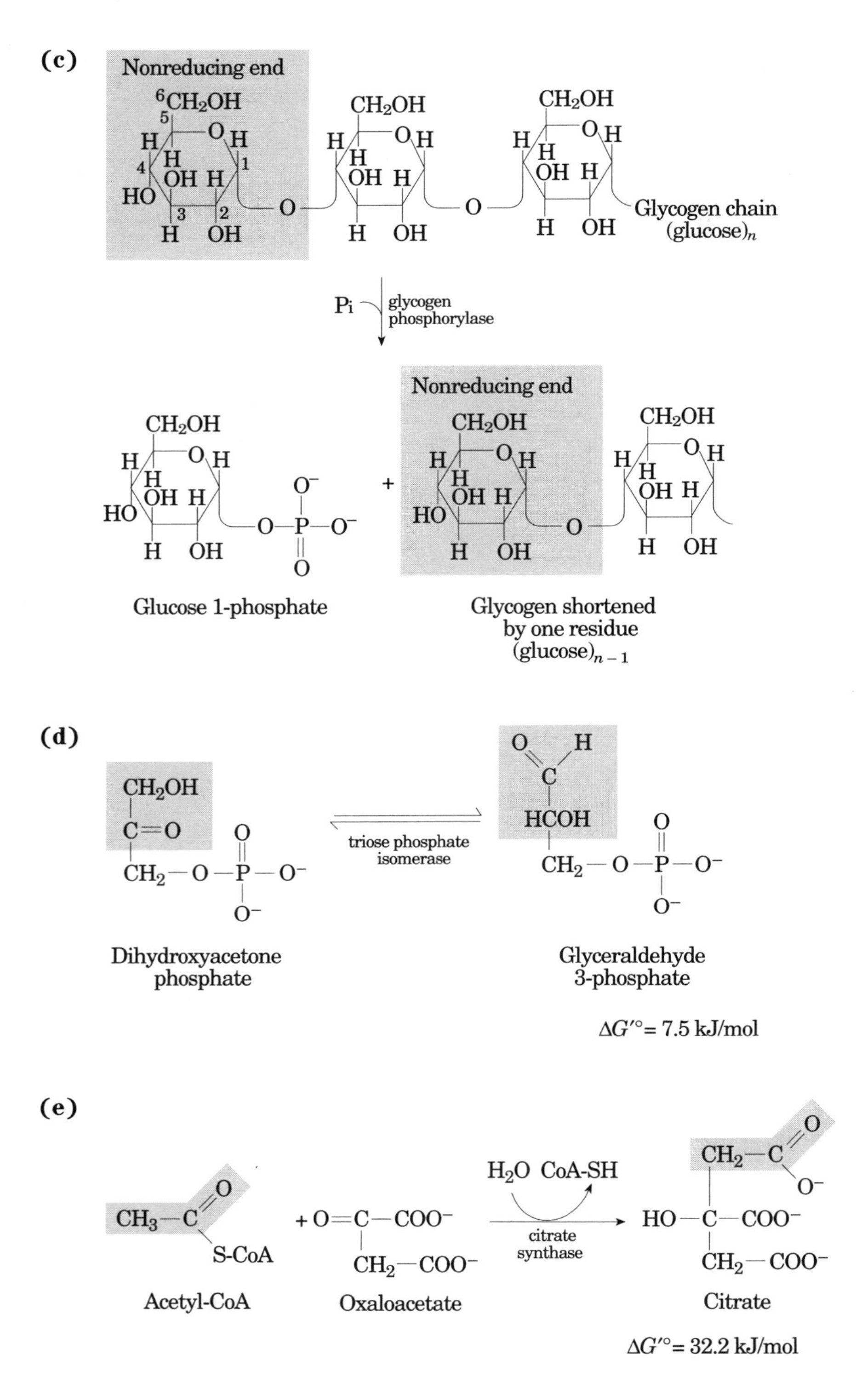
(c)
Nonreducing end
Glycogen chain
(glucose)$_n$
Pi
glycogen phosphorylase
Glucose 1-phosphate
Nonreducing end
Glycogen shortened by one residue (glucose)$_{n-1}$
(d)
triose phosphate isomerase
Dihydroxyacetone phosphate
Glyceraldehyde 3-phosphate
$\Delta G'^{\circ}$ = 7.5 kJ/mol
(e)
H_2O CoA-SH
citrate synthase
Acetyl-CoA
Oxaloacetate
Citrate
$\Delta G'^{\circ}$ = 32.2 kJ/mol

ANSWERS

Do You Know the Terms?

ACROSS

1. cis-trans
3. polymer
6. electrophiles
8. organic
9. rearrangement
11. monomeric
12. isomers
16. configuration
17. lipids
18. nucleophiles
20. oxidation-reduction
21. chiral
22. enantiomers
25. conformation
26. condensation
27. cleavage
28. transfer

DOWN

2. stereoisomers
3. protein
4. racemic
5. entropy
6. endo
7. functional
10. enthalpy
11. macromolecules
13. exo
14. nucleic acids
15. diastereomer
19. polysaccharides
22. endergonic
23. exergonic
24. optical

Do You Know the Facts?

1. Refer to Figure 3–5.

2. **(a)** hydroxyl, nucleophilic
(b) sulfhydryl, nucleophilic
(c) phosphoryl, electrophilic
(d) carboxylate, nucleophilic

3.

Macromolecule	Subunits	General Functions
Proteins	Amino acids	Structure, catalysis, signaling, transport
Nucleic acids	Nucleotides	Store, transmit, translate genetic information
Polysaccharides	Simple sugars	Fuel storage, structure, signaling
Lipids	Diverse, mainly hydrocarbon chains	Fuel storage, structure, signaling

4. **(a)** group transfer
(b) oxidation-reduction
(c) cleavage by nucleophilic substitution
(d) internal rearrangement
(e) condensation

chapter

4 Water

STEP-BY-STEP GUIDE

Major Concepts

The special properties of water, all of which are derived from its polarity and hydrogen-bonding capability, are central to the behavior of biomolecules dissolved in it.

The shape and polarity of the water molecule are responsible for its hydrogen-bonding capability.

This capability is what gives water its internal cohesion and its versatility as a solvent. An increase in entropy (randomness) of the ions dissolving from a crystal lattice drives the dissolution of crystalline substances, and an increase in entropy of surrounding water molecules drives the behavior of hydrophobic and amphipathic molecules in water. Individual hydrogen bonds, ionic interactions, hydrophobic interactions, van der Waals interactions are weak relative to covalent bonds. However, cumulatively, these small binding forces result in strong associations that are critical to the function of a biomolecule.

The ionization of water is an experimentally measurable quantity, which is expressed as K_w, the ion product of water.

This K_w is the basis for the **pH scale,** which designates the concentration of H^+ in an aqueous solution. The pH of the environment and the pK_a (the negative log of the equilibrium constant) of each of the participating biomolecules are extremely influential in the reactions of biological systems.

Buffers are mixtures of a weak acid (proton donor) and its conjugate base (proton acceptor).

Buffers will resist changes in pH most effectively when the concentrations of the proton donor and proton acceptor are equal. Under these circumstances, the pH of the solution is equal to the pK_a of the weak acid, as illustrated by the **Henderson-Hasselbalch equation:**

$$\text{pH} = \text{p}K_a + \log \frac{[\text{proton acceptor}]}{[\text{proton donor}]}$$

The phosphate and bicarbonate systems are critical buffer systems in cytoplasm and blood.

Water acts as a reactant as well as a solvent in many biochemical reactions.

A molecule of water is eliminated in condensation reactions, added to bonds in hydrolytic reactions, and split in the process of photosynthesis.

What to Review

Answering the following questions and reviewing the relevant concepts, which you have already studied, should make this chapter more understandable.

- The electronegativity of different elements affects the types of covalent and noncovalent interactions in which they can participate. What are the relative electronegativities of the elements O, N, C, and H (Table 3–2)?
- A number of functional groups are commonly found in biomolecules. Review the structures of the following groups: hydroxyl, carbonyl, carboxyl, amino (Fig. 3–5).
- Provide an example of a condensation reaction and a hydrolysis reaction.

Topics for Discussion

Answering each of the following questions, especially in the context of a study group discussion, should help you understand the important points of this chapter.

Weak Interactions in Aqueous Systems

Hydrogen Bonding Gives Water Its Unusual Properties

1. What is the shape of the water molecule? How is its unequal charge distributed over that shape?

2. How do hydrogen bonds contribute to the high melting and boiling points of water?

3. Why is ice less dense than liquid water?

Water Forms Hydrogen Bonds with Polar Solutes

4. What other functional groups of biomolecules can form hydrogen bonds? Why *don't* —CH groups participate in H bonds?

Water Interacts Electrostatically with Charged Solutes

5. What are some of the functional groups of biomolecules that will interact with water electrostatically?

6. What is the force or strength of the ionic attraction between a Na^+ and Cl^- (10 nm apart) in water? In benzene?

Entropy Increases as Crystalline Substances Dissolve

7. Describe what occurs when a crystalline salt dissolves in water in terms of the enthalpy (H) and the entropy (S) of the system.

Nonpolar Gases Are Poorly Soluble in Water

8. What physical property of oxygen could contribute to its concentration being a limiting factor for aquatic animals in deep water?

Nonpolar Compounds Force Energetically Unfavorable Changes in the Structure of Water

9. What is the driving force behind the formation of micelles?

10. Amphipathicity is important to the structure (and therefore the function) of which biomolecules?

Van der Waals Interactions Are Weak Interatomic Attractions

11. Van der Waals interactions are a weak, transient subcategory of which type of noncovalent interaction?

Weak Interactions Are Crucial to Macromolecular Structure and Function

12. How do the four types of weak interactions among biomolecules compare in strength to each other and to covalent bonds?

13. Why would a zipper or a Velcro® strip be an appropriate analogy to weak interactions in biochemical reactions?

Solutes Affect the Colligative Properties of Aqueous Solutions

14. Which of the following would have the greatest effect on the freezing point of a liter of water: the addition of 2 mol of NaCl or 2 mol of glucose?

Ionization of Water, Weak Acids, and Weak Bases

Pure Water Is Slightly Ionized

15. What specific information does the equilibrium constant of a reaction provide?

The Ionization of Water Is Expressed by an Equilibrium Constant

16. Why is the ion product of water (K_w) at 25 °C always equal to 1×10^{-14}?

The pH Scale Designates the H^+ and OH^- Concentrations

17. How does K_w relate to the pH scale?

18. Consider two sets of solutions: one set is composed of a solution of pH 1 and a solution of pH 2; the second set is composed of a solution of pH 11 and a solution of pH 12. Both sets thus comprise solutions differing from one another in pH by one pH unit. Is the difference in H^+ concentration between the solutions in each set equal? Why or why not?

Weak Acids and Bases Have Characteristic Dissociation Constants

19. Does a strong acid have a greater or lesser tendency to lose its proton than does a weak acid? Does the strong acid have a higher or lower K_a? A higher or lower pK_a?

Titration Curves Reveal the pK_a of Weak Acids

20. At a pH equal to the pK_a of a weak acid, what can be said about the concentrations of the acid and its conjugate base? What point on a titration curve indicates the pK_a of that weak acid?

Buffering against pH Changes in Biological Systems

Buffers Are Mixtures of Weak Acids and Their Conjugate Bases

21. What are the two equilibrium reactions that are simultaneously adjusting during an experimental titration of a weak acid?

22. What does the relatively flat zone of a titration curve tell you about the pH changes within that zone?

A Simple Expression Relates pH, pK_a, and Buffer Concentration

23. How does the Henderson-Hasselbalch equation prove that the pK_a of a weak acid is equal to the pH of the solution at the midpoint of its titration?

Weak Acids or Bases Buffer Cells and Tissues against pH Changes

24. What is the importance of the functional groups of proteins that act as weak acids or bases in biological systems?

25. What is the extracellular buffering system that is used by animals with lungs? What is the primary intracellular buffering system?

26. Where and how does the phosphate buffer system function?

Blood, Lungs, and Buffer: the Bicarbonate Buffer System

27. What are the three reversible equilibria involved in the bicarbonate buffer system in animals with lungs?

28. What is the most sensitive aspect of cell function (mentioned many times in this chapter) in relationship to changes in pH?

Water as a Reactant

29. What general types of cellular reactions form water, and which consume water? Which of these reactions are endergonic and which exergonic?

The Fitness of the Aqueous Environment for Living Organisms

30. What are some of the ways that organisms on earth exploit the special properties of water in terms of temperature regulation?

SELF-TEST

Do You Know the Terms?

ACROSS

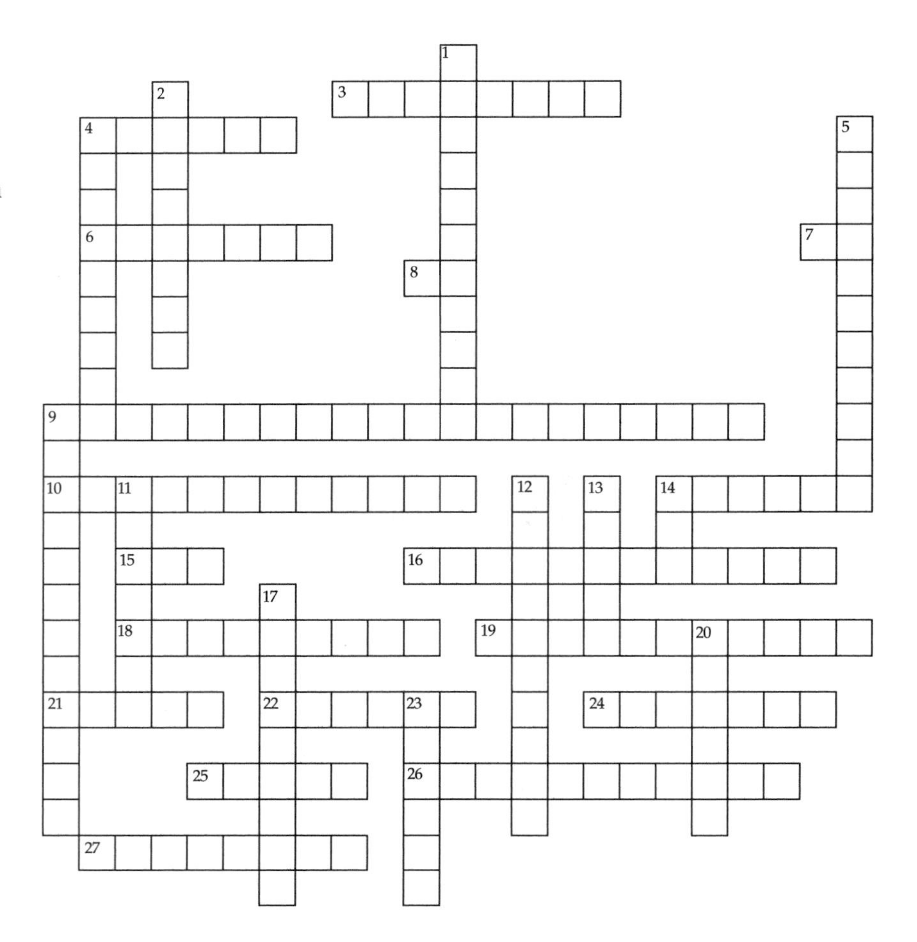

3. Describes a solution with a $[H^+]$ of 1×10^{-8}.
4. Hydro________ molecules can form energetically favorable interactions with water molecules.
6. Water is often referred to as the "universal ________" because of its ability to hydrate molecules and screen charges.
7. Describes the concentration of H^+ (and therefore of OH^-) in an aqueous solution.
8. The ion product of water; it is 1×10^{-14} M in aqueous solutions at 25 °C.
9. The ________–________ equation; describes the relationship between pH and the pK_a of a buffer.
10. The equilibrium constant for the reaction $HA \rightleftharpoons H^+ + A^-$ is also called the ________ constant, K_a.
14. Hydro________ molecules decrease the entropy of an aqueous system by causing water molecules to become more ordered.
15. The numbers 1, 10, 100, 1000 are placed at equal intervals on a ________ scale.
16. Reaction in which two reactants combine to form a single product with the elimination water.
18. A plot of pH vs. OH^- equivalents added is a ________ curve.
19. Weak interactions that are crucial to the structure and function of macromolecules.
21. Describes a solution in which $[OH^-]$ is greater than $[H^+]$.
22. A mixture of a weak acid and its conjugate base.
24. Enzymes show maximum activity at a characteristic pH ________.
25. HA is a proton ________.
26. The point in a reversible chemical reaction at which the rate of product formation equals the rate of product breakdown to the starting reactants.
27. A^- is a proton ________.

DOWN

1. The ________ radius is approximately twice the distance of a covalent radius for a single bond. (3 words)
2. Stable structures formed by lipids in water, which are held together by hydrophobic interactions.
4. $H_2PO_4^- \rightleftharpoons H^+ + PO_4^{2-}$ describes a ________ buffer system.
5. Compound containing both polar and nonpolar regions.
9. The electrostatic interactions between the hydrogen and oxygen atoms on adjacent H_2O molecules constitute a ________. (2 words)
11. Dissolved molecules.
12. Covalent bond breakage by the addition of water.
13. Water molecules readily dissolve compounds such as NaCl because they screen ________ interactions between Na^+ and Cl^-.
14. pH at which $[HAc] = [Ac^-]$.
17. $H_2CO_3 \rightleftharpoons H^+ + HCO_3^-$ describes a ________ buffer system.
20. Describes a solution in which $[H^+]$ is greater than $[OH^-]$.
23. Noncovalent bonds have weaker bond ________ than covalent bonds.

Do You Know the Facts?

In questions 1–4, decide whether the statement is true or false, then explain your answer.

1. The oxygen atom in water has a partial positive charge.

2. The hydrogen atoms of water each bear a partial positive charge.

3. H bonds can form only between water molecules.

4. H bonds are relatively weak compared to covalent bonds.

In questions 5–9, choose the one best response.

5. You are running your first marathon on a very warm day. You start to sweat heavily and realize that you may be in danger of dehydration. Why is severe dehydration potentially life-threatening?

A. Water is a solvent for many biomolecules.
B. Water is a chemical participant in many biological reactions.
C. Water is necessary for buffering action in the body.
D. Water's attraction to itself drives hydrophobic interactions.
E. All of the above are true.

6. Which of the following is true of hydrogen bonds?

A. The attraction between the oxygen atom of a water molecule and the hydrogen atom of another molecule constitutes a hydrogen bond.
B. Hydrogen bonds form as covalent bonds between positively and negatively charged ions.
C. Hydrogen bonds form between nonpolar portions of biomolecules.
D. A and B are true.
E. A, B, and C are true.

7. Which of the following is true of pH?

A. pH is the negative logarithm of $[OH^-]$.
B. Lemon juice, with a pH of 2.0, is sixty times more acidic than ammonia, with a pH of 12.0.
C. Varying the pH of a solution will alter the pK_a of an ionizable group in that solution.
D. Varying the pH of a solution will alter the degree of ionization of an ionizable group in that solution.
E. All of the above are true.

8. Consider a weak acid in a solution with a pH of 5.0. Which of the following statements is true?

A. The weak acid is a proton acceptor.
B. The weak acid has a lower affinity for its proton than does a strong acid.
C. At its pK_a, the weak acid will be totally dissociated.
D. The $[H^+]$ is 10^{-5} M.
E. All of the above are true.

9. Water derives all its special properties from its

A. cohesiveness and adhesiveness.
B. high boiling point and melting point.
C. small degree of ionization.
D. polarity and hydrogen-bonding capacity.
E. high dielectric constant.

10. The pK_a's for the three ionizable groups on tyrosine are: pK_a (—COOH) = 2.2, pK_a (—NH_3^+) = 9.11, and pK_a (—R) = 10.07. In which pH range will this amino acid have the greatest buffering capacity?

A. at all pH's between 2.2 and 10.07
B. at pH's near 7.1
C. at pH's between 9 and 10
D. at pH's near 5.7
E. Amino acids cannot act as buffers.

11. In a typical eukaryotic cell the pH is usually around 7.4. What is the [H^+] in a typical eukaryotic cell?

A. 0.00000074 M
B. 6.6 μM
C. 4×10^{-8}
D. 2.3 nM
E. 7.4×10^{-5} M

12. Formic acid has a pK_a of 1.78×10^{-4} and acetic acid has a K_a of 1.74×10^{-5}. Which of the following is true?

A. Formic acid has the higher pK_a of the two acids and would be the best buffer at pH 6.
B. The solution with the larger K_a is a good proton acceptor.
C. Formic acid is the stronger acid with a lesser tendency to lose its proton compared to acetic acid.
D. Neither formic acid nor acetic acid can be effective buffers.
E. Acetic acid is a weaker acid and has a lesser tendency to lose its proton compared to formic acid.

13. As climbers approach the summit of a mountain they usually increase their rate of breathing to compensate for the "thinner air" due to the lower oxygen pressures at higher elevations. This increased ventilation rate results in a reduction in the levels of CO_2 dissolved in the blood. Which of the following accurately describes the effect of lowering the $[CO_2]_{dissolved}$ on blood pH?

A. It will result in an increase in the dissociation of $H_2CO_3 \rightarrow H^+ + HCO_3^-$ and a drop in pH.
B. It will result in an increase in the dissociation of $H_2CO_3 \rightarrow H_2O + CO_2$ and a drop in pH.
C. It will result in an increase in the association of $H^+ + HCO_3^- \rightarrow H_2CO_3$ and an increase in pH.
D. It will result in a decrease in the association of $H^+ + HCO_3^- \rightarrow H_2CO_3$ and a decrease in pH.
E. Lowering $[CO_2]_{dissolved}$ has no effect on blood pH.

14. You mix 100 ml of solution of pH 1 with 100 ml of a solution of pH 3. The pH of the new 200 ml solution will be:

A. 1.0.
B. 2.0.
C. 3.0.
D. between pH 1.0 and pH 2.0.
E. between pH 2.0 and pH 3.0.

15. Which of the following are true?

A. Amino acids can act as proton donors and acceptors.
B. The stronger the acid, the higher the K_a.
C. The stronger the acid, the lower the pH.
D. Amino acids are amphoteric.
E. All of the above are true.

16. The pH of a sample of blood is 7.4; the pH of a sample of gastric juice is 1.4. The blood sample has an [H^+]:

A. 5.29 times lower than that of the gastric juice.
B. a million times higher than that of the gastric juice.
C. 6,000 times lower than that of the gastric juice.
D. a million times lower than that of the gastric juice.
E. 0.189 times that of the gastric juice.

Applying What You Know

1. Certain insects can "skate" along the top surface of water in ponds and streams. What property of water allows this feat, and what bonds or interactions are involved?

2. When you are very warm, because of high environmental temperature or physical exertion, you perspire. What property(ies) of water is your body exploiting when it sweats?

3. What is the absolute difference in $[H^+]$ between two aqueous solutions, one of pH 2.0 and one of pH 3.0? What is the $[OH^-]$ of the solution of pH 2.0?

4. Formic acid has a pK_a of 3.75; acetic acid has a pK_a of 4.76. Which is the stronger acid? Does the stronger acid have a greater or lesser tendency to lose its proton than the weaker acid?

5. In a solution of pH 4.76, containing both acetic acid and acetate, what can you say about the concentrations of acetic acid (CH_3COOH) and acetate (CH_3COO^-) present?

6. The Henderson-Hasselbalch equation is $\text{pH} = \text{p}K_a + \log \frac{[\text{proton acceptor}]}{[\text{proton donor}]}$. Show how it proves that the pK_a of a weak acid is equal to the pH of the solution at the midpoint of its titration.

7. Define pH and pK_a and explain how they are different.

ANSWERS

Do You Know the Terms?

ACROSS

3. alkaline
4. philic
6. solvent
7. pH
8. K_w
9. Henderson-Hasselbalch
10. dissociation
14. phobic
15. log
16. condensation
18. titration
19. noncovalent
21. basic
22. buffer
24. optimum
25. donor
26. equilibrium
27. acceptor

DOWN

1. van der Waals
2. micelles
4. phosphate
5. amphipathic
9. hydrogen bond
11. solutes
12. hydrolysis
13. ionic
14. pK_a
17. carbonate
20. acidic
23. energy

Do You Know the Facts?

1. False; the oxygen atom has a partial negative charge.
2. True; because the oxygen atom is more electronegative than the two hydrogen atoms, the electrons are more often in the vicinity of the oxygen, giving each of the hydrogen atoms a partial positive charge.
3. False; they can form between any electronegative atom (usually oxygen or nitrogen) and a hydrogen atom covalently bonded to another electronegative atom in the same or another molecule.
4. True; hydrogen bonds have a stabilization energy of 8–21 kJ/mol, whereas covalent single bonds have a stabilization energy of approximately 200–460 kJ/mol.

5. E
6. A
7. D
8. D
9. D
10. C
11. C
12. E
13. C
14. D
15. E
16. D

Applying What You Know

1. Extensive hydrogen binding among water molecules accounts for the surface tension that allows some insects to walk on water.
2. The high heat of vaporization of water, which is a measure of the energy required to overcome attractive forces between molecules, allows your body to dissipate excess heat through the evaporation of the water that is perspired.
3. The solution with a pH of 2.0 has a $[H^+]$ of 10^{-2} M, and an $[OH^-]$ of 10^{-12} M; the solution with a pH of 3.0 has a $[H^+]$ of 10^{-3} M. The difference in $[H^+]$ is 10^{-2} M $-$ 10^{-3} M $=$ 0.009 M.
4. Formic acid is the stronger acid. It has a greater tendency to lose its proton than does acetic acid.
5. At pH 4.76, the concentrations of acetic acid and acetate in the solution will be equal.
6. $$\text{pH} = \text{p}K_a + \log \frac{[\text{proton acceptor}]}{[\text{proton donor}]}$$
   At the midpoint of the titration, [proton acceptor] = [proton donor]. The log of 1 = 0, so pH = pK_a + 0; pH = pK_a.
7. pH is the negative logarithm of $[H^+]$ in an aqueous solution. It provides a standard way to measure the H^+ concentration in an aqueous solution. pK_a is the negative logarithm of an equilibrium constant. It is equal to the pH at which a weak acid is one-half dissociated; i.e., the pH at which there are equal concentrations of a weak acid and its conjugate base. The pK_a occurs at the midpoint of the titration curve of a weak acid, the center of the range providing the maximum buffering capacity of the conjugate acid-base pair. The pK_a is an integral, unchanging property of an ionizable group. It is the extent of ionization of an ionizable group that varies with the pH of the solution.

Amino Acids, Peptides, and Proteins

STEP-BY-STEP GUIDE

Major Concepts

Amino acids share a common structure.

There are 20 standard amino acids found in proteins and they are usually referred to by either a three-letter or a one-letter code. Each amino acid contains a **carboxyl group,** an **amino group,** and a hydrogen atom bonded to a central carbon atom, the α **carbon.** The α carbon is **chiral** when the R group is anything other than a hydrogen. Therefore, with the exception of glycine, there are two **enantiomers** of each amino acid, a D-form and an L-form. Proteins contain only the L isomers of amino acids.

The different chemical properties of amino acids are the result of the different properties of their R groups, which are the basis for categorizing amino acids as nonpolar, aromatic, polar, positively charged, or negatively charged.

Amino acids ionize in aqueous solution.

Both the carboxyl and the amino groups can ionize, and amino acids are capable of acting as both weak acids (proton donors) and weak bases (proton acceptors). The R groups of some amino acids can also ionize, affecting overall acid–base behavior. All of these ionizable groups affect the net charge on each amino acid. Consequently, amino acids are often characterized by the pH at which they have no net charge; this characteristic is their **isoelectric point,** or **pI.**

Polymers of amino acids are polypeptides or proteins.

Amino acids in these polymers are covalently linked through a peptide bond formed by a condensation reaction between the carboxyl group of one amino acid and the amino group of a second amino acid. In general, proteins are very large molecules, generally containing more than 100 amino acid residues. Peptides generally contain less than 100 amino acid residues. Small peptides perform a number of biological functions such as intercellular signaling. Some proteins contain non-amino acid prosthetic groups, which usually play an important role in the protein's biological function.

Proteins can be studied using a variety of techniques.

Proteins can be purified on the basis of **solubility, size** (size-exclusion chromatography, gel electrophoresis), **shape** or **binding characteristics** (affinity chromatography), or **charge** (anion- or cation-exchange chromatography, isoelectric focusing). Often, purification requires multiple steps, each step taking advantage of a different characteristic of the protein being purified.

A protein's function depends on its primary structure, which can be determined experimentally.

Protein structure is described in terms of four levels: primary (1°), secondary (2°), tertiary (3°), and quaternary (4°). Primary structure is the specific number and sequence of amino acids in a specific protein. The location of disulfide bonds is also a part of the primary structure of a protein. The primary structure determines how a protein will fold into its three-dimensional, functional form. Altering the primary structure in critical areas can therefore affect the protein's function. Primary structure is specified by DNA coding regions, therefore DNA sequence information provides a fast and accurate method for determining the amino acid sequence of a protein. Alternatively, the amino acid sequence can be determined directly by first breaking disulfide bonds, cleaving the protein into small fragments, sequencing the fragments by the Edman degradation procedure, and then reconstructing the amino acid sequence of the intact protein.

Proteins that have similar functions in different species have similar amino acid sequences.

Often stretches of amino acid sequences are invariant in the same proteins from different species. These invariant residues are usually found in functionally critical regions.

What to Review

Answering the following questions and reviewing the relevant concepts, which you have already studied, should make this chapter more understandable.

- Review the chemical properties of the biologically important functional groups (Fig. 3–5), especially $—NH_3^+$, and $—COO^-$. How do these functional groups behave in aqueous solution (p. 87)? How will they influence the behavior of molecules (amino acids and peptides) containing them?
- Be sure you understand the difference between a strong acid or base and a weak acid or base (pp. 98–99). Into which category do $—COO^-$ and $—NH_3^+$ groups fit?
- What are buffers? How do they work? How can a titration curve tell you at what pH a molecule will act as an effective buffer (Fig. 4–16)? Use this information to interpret the titration curve of glycine (Fig. 5–10).
- Using models if necessary, satisfy yourself that enantiomers are *not* superimposable and cannot function as identical molecules (Fig. 3–10).
- Review condensation reactions (Fig. 3–22). The formation of a peptide bond is just one of many examples of monomers joining to form polymers through condensation reactions.
- Review what a positive free-energy change means in terms of the spontaneity of a reaction (pp. 65, 72, 84). Does this mean that an endergonic reaction cannot occur? If not, what circumstances must prevail in order for such a reaction to occur?

Topics for Discussion

Answering each of the following questions, especially in the context of a study group discussion, should help you understand the important points of this chapter.

Amino Acids

Amino Acids Have Common Structural Features

1. Why is glycine *not* optically active?

2. Both the three-letter and one-letter abbreviations for the 20 standard amino acids are very commonly used in the biochemical literature and should be committed to memory. Use the Worksheet in the Self-Test to help you memorize the various amino acid structures, names, and abbreviations.

The Amino Acid Residues in Proteins Are L-Stereoisomers

3. Could a life form that evolved on a planet where all proteins were made up of D-amino acids survive on Earth? Why or why not? (More on this topic in Chapter 8.)

Amino Acids Can Be Classified by R Group

4. The structures and R-group classifications of the 20 standard amino acids are essential parts of a working vocabulary for biochemistry, and should be committed to memory.

5. Which amino acids are capable of forming hydrogen bonds? Which will promote hydrophobic interactions? Which amino acid fits into both of these categories?

6. Which amino acid allows the most structural flexibility when found in a protein? Which allows the least? Which is the only amino acid that forms disulfide bridges?

7. Does the ability of some amino acids to absorb ultraviolet light translate into a useful technique for the detection of *all* proteins? Why or why not?

8. What are the pK_a values of R groups on the charged amino acids (see Table 5–1)? Which amino acid has an R group with a pK_a value near the pH of most living systems?

Absorption of Light by Molecules: The Lambert-Beer Law

9. The measurement of protein concentration is a common laboratory procedure. Measurements of light absorption by unknown samples are compared to a standard curve that can be generated by measuring the absorption of solutions with known amounts of protein. Why must these standard solutions be made in the same solvent and at the same pH as that of the unknown protein solutions?

Nonstandard Amino Acids Also Have Important Functions

10. Proteins can contain nonstandard amino acids which are modified versions of the 20 amino acids discussed in this chapter. Do these modifications occur before or after the amino acids are incorporated into the protein molecule?

Amino Acids Can Act as Acids and Bases

11. How does the zwitterionic nature of amino acids in solution affect their solubility in water? Why *must* individual amino acids be water soluble? (Note that although R groups vary in water solubility, *all* amino acids having a single amino group and a single carboxyl group are zwitterionic in neutral aqueous solution.)

12. Are amino acids weak or strong acids and bases?

Amino Acids Have Characteristic Titration Curves

13. Over what pH range(s) is glycine an effective buffer? Over what pH range is glycine not at all an effective buffer?

14. Refer to Table 5–1. Note the pK_a values for the —COOH on the R group of aspartate and glutamate. Why are these pK_a values higher than those of —COOH groups bonded to the α carbon of amino acids?

The Titration Curve Predicts the Electric Charge of Amino Acids

15. At approximately what pH would you expect glycine to have a net charge of -1?

Amino Acids Differ in Their Acid-Base Properties

16. Note that the pI values for glutamate and histidine are *not* the average of the pK_a values for each of their three ionizable groups. Can these pI values be calculated as the average of some set of pK_a values? Does this calculation hold for pI values of the other charged amino acids?

Peptides and Proteins

Peptides Are Chains of Amino Acids

17. What is lost from amino acids in the formation of a peptide bond?

Peptides Can Be Distinguished by Their Ionization Behavior

18. Is the tetrapeptide Ala–Glu–Gly–Lys (see Fig. 5–15) a good buffer at pH 7.0? Will it move in an electric field at pH 7.0?

Biologically Active Peptides and Polypeptides Occur in a Vast Range of Sizes

19. List some functions of small bioactive peptides.

20. Provide an example of a very large protein and a very small protein.

Polypeptides Have Characteristic Amino Acid Compositions

21. How might the number, sequence, and properties of amino acids in a protein affect its structure and function?

Some Proteins Contain Chemical Groups Other Than Amino Acids

22. Name five or six different types of prosthetic groups that can be attached to proteins.

There Are Several Levels of Protein Structure

23. Which of the proteins listed in Table 5–2 exhibit quaternary structure?

Working with Proteins

Proteins Can Be Separated and Purified

As you read the rest of this chapter you will be able to answer the following questions about protein purification:

24. Which protein purification and separation techniques separate proteins on the basis of size?

25. What technique might you use to separate two proteins that comigrate on an SDS polyacrylamide gel?

26. Which techniques separate proteins on the basis of charge?

27. Which techniques separate proteins on the basis of solubility?

28. Which techniques separate proteins on the basis of binding specificity?

Proteins Can Be Separated and Characterized by Electrophoresis

29. What does a single band on a protein gel represent?

30. What is SDS, and why is it used in electrophoretic procedures?

31. Explain why two-dimensional gel electrophoresis is a more sensitive analytical method than one-dimensional gel electrophoresis.

Unseparated Proteins Can Be Quantified

32. When you finish this chapter, make a list of different methods for quantifying the amount of a protein in a complex solution, such as a blood sample or tissue homogenate.

33. Why is it important to be able to measure the amount of a specific protein in a sample?

34. Why is it important to be able to measure the activity of a protein?

The Covalent Structure of Proteins

The Function of a Protein Depends on Its Amino Acid Sequence

35. How does information about primary structure contribute to an understanding of a protein's function?

Protein Homology among Species

36. What is the biological significance of the variable residues of homologous proteins?

37. What can be inferred about stretches of invariant residues of homologous proteins?

38. What constitutes a conservative substitution for an amino acid residue in a protein?

The Amino Acid Sequence of Polypeptide Chains Can Be Determined

39. Use the amino acid sequence of insulin (Fig. 5–24) to quiz yourself: What are the one-letter abbreviations and R-group classifications for each amino acid listed?

Short Polypeptides Are Sequenced Using Automated Procedures

40. Outline the individual steps required to sequence a protein.

Large Proteins Must Be Sequenced in Smaller Segments

41. What general steps must be taken before larger proteins can be sequenced using automated procedures?

42. Placing peptide fragments in their proper order is like piecing together a puzzle. Make sure to understand the example in Fig. 5–27; there is a similar puzzle in the Self-Test.

Amino Acid Sequences Can Be Deduced by Other Methods

43. What information concerning the primary structure of a protein would *not* be available from the DNA sequence?

Investigating Proteins with Mass Spectrometry

44. Why is the partial sequencing of a protein by mass spectrometry such a useful tool for modern proteome research?

Amino Acid Sequences Provide Important Biochemical Information

45. How does information about primary structure contribute to an understanding of a protein's function?

Small Peptides and Proteins Can Be Chemically Synthesized

46. Explain how the use of a solid support for chemical synthesis of polypeptides eliminates a step of traditional organic synthesis.

SELF-TEST

Worksheet

The following worksheet helps you memorize the structures, one- and three-letter abbreviations, and R-group category of each amino acid. The first example is filled in as a guide. (Note: Many students find it useful to make "flash-cards" of the amino acid structures in order to facilitate memorization.)

$$\begin{array}{c} COO^- \\ | \\ H_3\overset{+}{N}-C-H \\ | \\ CH_2OH \end{array}$$

Name	Serine
1 letter	S
3 letter	Ser
Group	Polar, uncharged

$$\begin{array}{c} COO^- \\ | \\ H_3\overset{+}{N}-C-H \\ | \\ H \end{array}$$

Name	
1 letter	
3 letter	
Group	

$$\begin{array}{c} COO^- \\ | \\ H_3\overset{+}{N}-C-H \\ | \\ H-C-OH \\ | \\ CH_3 \end{array}$$

Name	
1 letter	
3 letter	
Group	

$$\begin{array}{c} COO^- \\ | \\ H_3\overset{+}{N}-C-H \\ | \\ CH_2 \\ | \\ COO^- \end{array}$$

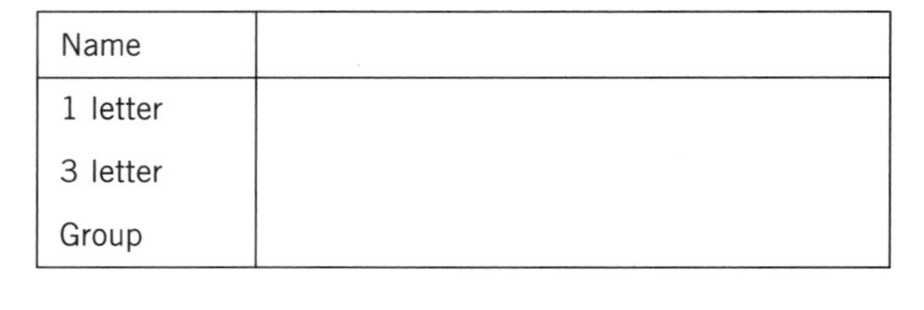

Name	
1 letter	
3 letter	
Group	

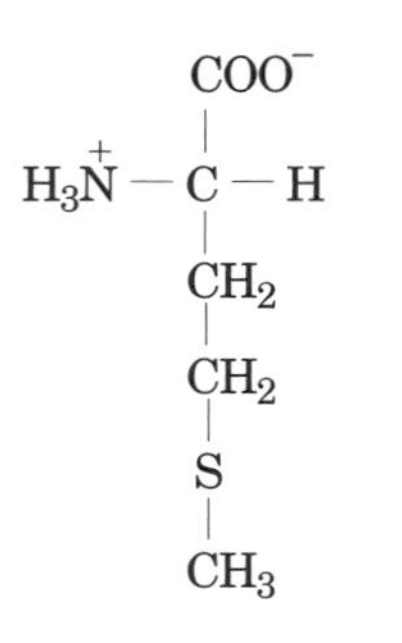

Name	
1 letter	
3 letter	
Group	

$$\begin{array}{c} COO^- \\ | \\ H_3\overset{+}{N}-C-H \\ | \\ CH_2 \\ | \\ CH_2 \\ | \\ CH_2 \\ | \\ CH_2 \\ | \\ {}^+NH_3 \end{array}$$

Name	
1 letter	
3 letter	
Group	

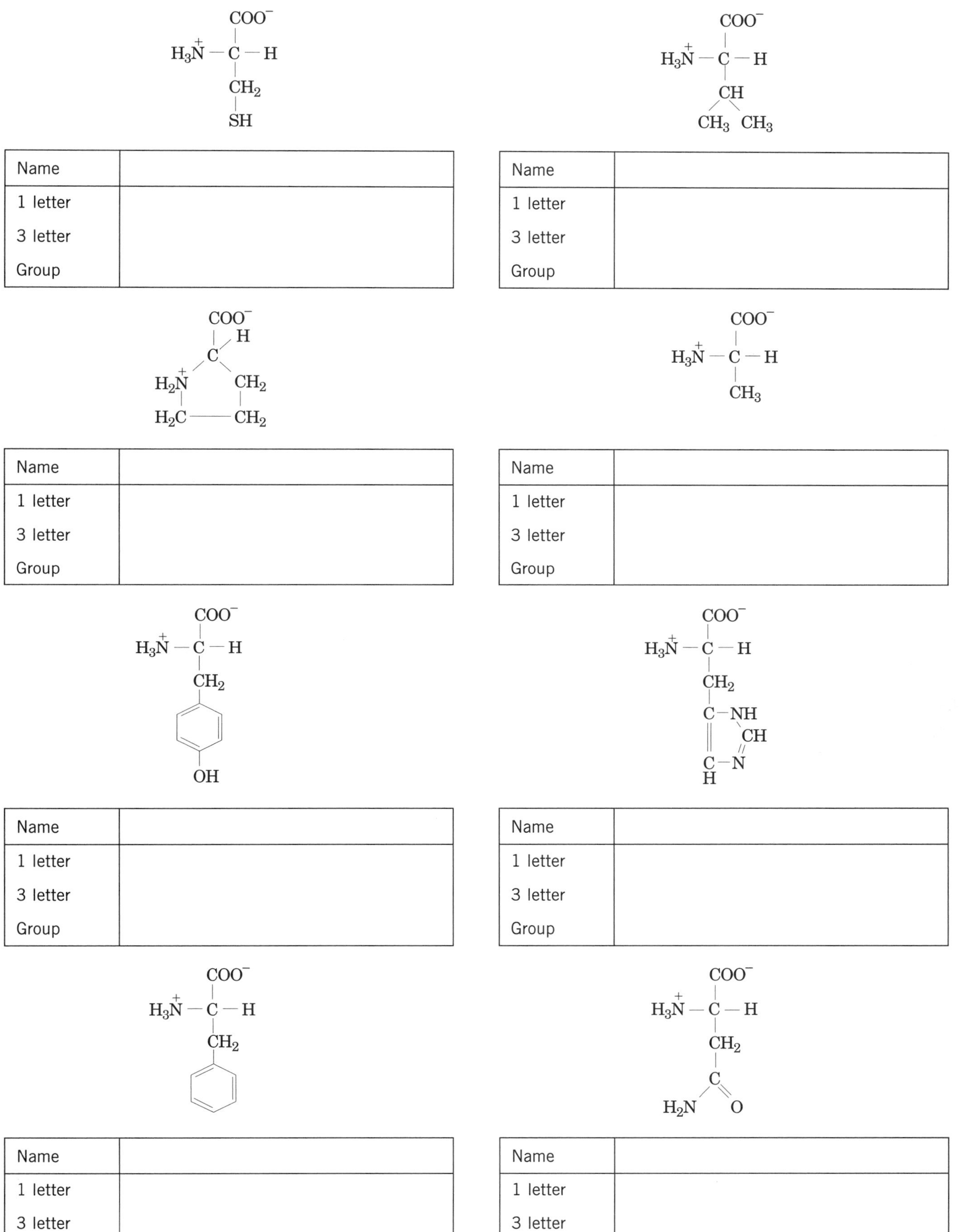

Name	
1 letter 3 letter Group	

Name	
1 letter 3 letter Group	

Name	
1 letter 3 letter Group	

Name	
1 letter 3 letter Group	

Name	
1 letter 3 letter Group	

Name	
1 letter 3 letter Group	

Name	
1 letter 3 letter Group	

Name	
1 letter 3 letter Group	

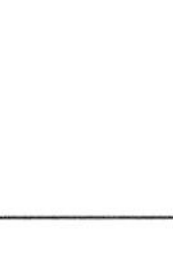

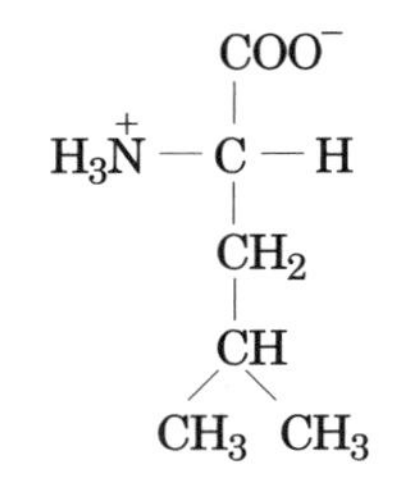

Name	
1 letter	
3 letter	
Group	

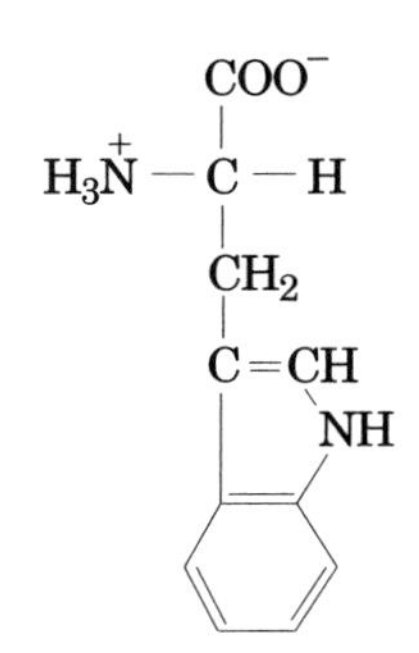

Name	
1 letter	
3 letter	
Group	

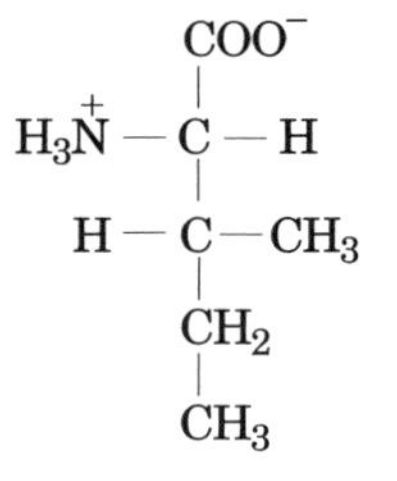

Name	
1 letter	
3 letter	
Group	

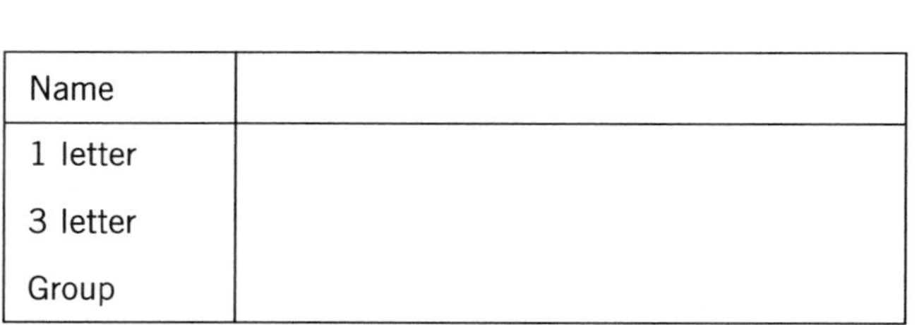

Name	
1 letter	
3 letter	
Group	

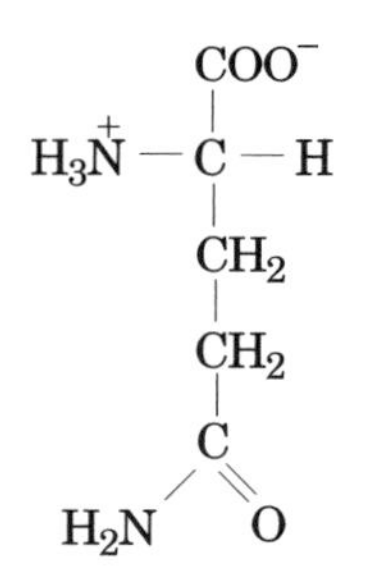

Name	
1 letter	
3 letter	
Group	

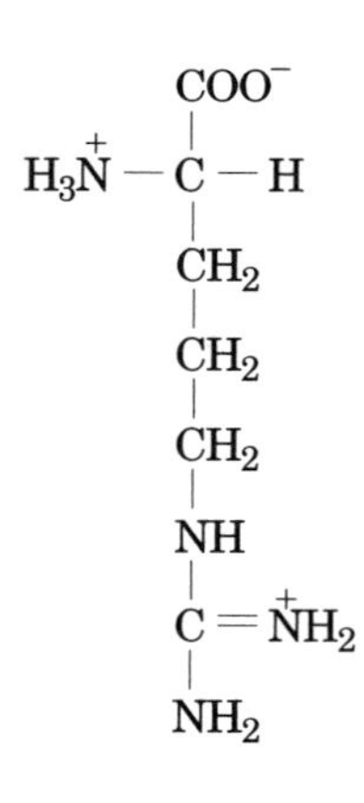

Name	
1 letter	
3 letter	
Group	

SELF-TEST

Do You Know the Terms?

ACROSS

1. The lipid portion of a lipoprotein is known as a ________ group.

4. A covalent bond between two nonadjacent cysteines in a polypeptide chain is a ________ bond.

6. An example of a(n) ________ amino acid is histidine, which can either accept protons or donate them at a pH that is close to physiological pH values.

9. All stereoisomers must have at least one ________ center.

11. ________ out is a technique that selectively precipitates some proteins, while others remain in solution. Ammonium sulfate ($(NH_4)_2SO_4$) is often used for this purpose.

12. The bond type that forms the primary structure.

13. The pH at which the numbers of positive and negative charges on an amino acid are equal is referred to as the ________ point or pI.

15. A single unit within a polymer: for example, lysine in a protein molecule.

16. At pH 7, any amino acid with an uncharged R group is a ________.

DOWN

2. A reagent used in electrophoresis to separate polypeptides on the basis of mass. (abbr.)

3. Many types of separation can be done using this chromatographic technique; its advantage lies in the reduction of transit time on the column, limiting diffusional spreading of protein bands and improving resolution. (abbr.)

4. After "salting out" proteins, removal of excess ammonium sulfate can be accomplished by ________ of the protein-salt solution overnight against large volumes of buffer.

5. The whole assortment of proteins in an organism; analogous to the genome.

7. A linear chain of amino acid residues that usually has a molecular weight less than 10,000 daltons.

8. The ________ degradation procedure provides information about a protein's primary structure.

10. Insulin obtained from sheep can be used to treat human diabetics because sheep and human insulin are ________ proteins.

14. Hemoglobin, which contains two sets of identical subunits, is often referred to as a(n) ________.

Do You Know the Facts?

1. Draw the chemical structures of the following amino acids. Give their one- and three-letter abbreviations, and R-group classification.

   (a) Alanine

   (b) Proline

   (c) Cysteine

   (d) Phenylalanine

   (e) Lysine

   (f) Aspartate

2. Name the following structures. Give their one- and three-letter abbreviations, and R-group classifications.

   (a)

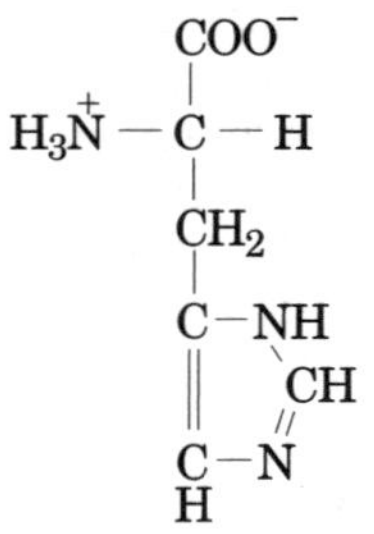

   COO^-
   $H_3\overset{+}{N}-C-H$
   CH_2
   $C-NH$
   $\| \quad CH$
   $C-N$
   H

   (b)

   COO^-
   $H_3\overset{+}{N}-C-H$
   $H-C-OH$
   CH_3

   (c)

   COO^-
   $H_3\overset{+}{N}-C-H$
   CH_2
   CH_2
   CH_2
   NH
   $C=\overset{+}{N}H_2$
   NH_2

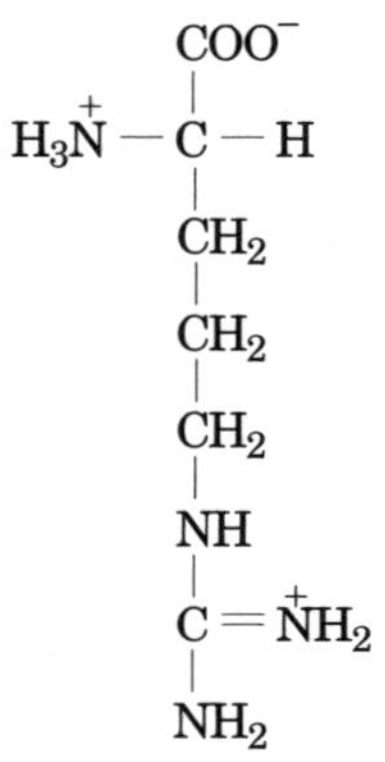

   (d)

   COO^-
   $H_3\overset{+}{N}-C-H$
   CH_2
   CH_2
   COO^-

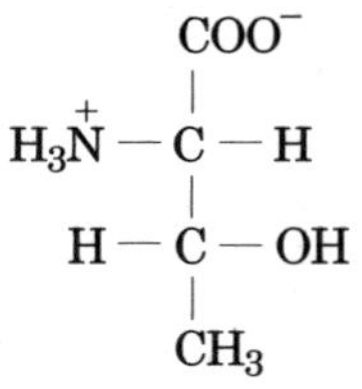

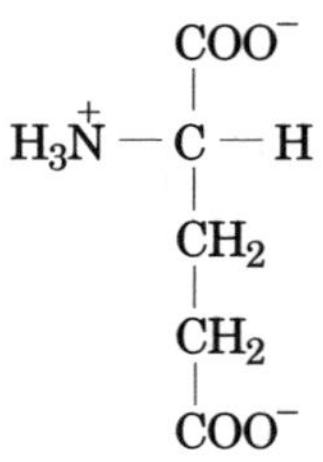

3. Name the amino acid(s) described.

(a) Provides the least amount of steric hindrance in proteins.

(b) Positively charged at physiological pH.

(c) Aromatic R group; hydrophobic and neutral at any pH.

(d) Saturated hydrocarbon R group; important in hydrophobic interactions.

(e) The only amino acid having an R group with a pK_a near 7; important in the active site of some enzymes.

(f) The only amino acid with a substituted α-amino group; influences protein folding by forcing a bend in the chain.

4. What is the amino acid sequence of the peptide abbreviated KQNY?

5. What is the amino acid sequence of the peptide abbreviated RLWEQ?

6. What is the one-letter abbreviation for the amino acid that has an aromatic R group capable of forming hydrogen bonds?

7. What is the one-letter abbreviation for the amino acid that has a sulfur-containing R group and is able to form a disulfide bridge?

8. Which of the following statements describe(s) the peptide bond? (More than one answer may be correct.)

A. It is the only covalent bond between amino acids in polypeptides.
B. It is a substituted amide linkage.
C. It is formed in a condensation reaction.
D. It is formed in an exergonic reaction.
E. It is unstable under physiological conditions.

9. Which of the following explains why all individual amino acids are soluble in water but not all peptides are soluble?

A. Individual amino acids are zwitterions at physiological pHs.
B. All peptides are insoluble in water.
C. The R groups of the amino acid residues in the peptide are charged at physiological pHs.
D. All the amino acid residues in the peptide are zwitterions at physiological pHs.
E. The R groups on all amino acids can interact noncovalently with water at pH 7.4.

10. Name the techniques described for separating cellular proteins.

(a) Taking advantage of unique structural or functional properties of a protein, this technique specifically removes the protein of interest from a solution.

(b) Proteins leave the mobile phase, associating with a negatively charged immobile substrate such as a bead or resin.

(c) Proteins are separated on the basis of their ability to migrate in an electric field, an indicator of relative size.

(d) Proteins are chromatographically separated solely on the basis of size.

Refer to the table below for questions 11 and 12.

	Molecular Weight	Number of Residues	Number of Polypeptide Chains
Cytochrome *c* (human)	13,000	104	1
Myoglobin (equine heart)	16,890	153	1
Serum albumin (human)	68,500	~550	1
Apolipoprotein B (human)	513,000	4,536	1

11. Which protein is likely to be retained the longest on (i.e., will elute last from) a size-exclusion chromatographic column? Why?

12. Which protein would show up as the band at the top of an SDS polyacrylamide gel after electrophoresis? Why?

Applying What You Know

1. What do the amino acids threonine and tyrosine have in common?

2. What makes the amino acid cysteine important? Can methionine perform the same function?

3. What structural property of amino acids permits the measurement of protein concentration by UV light absorption? Which amino acids have this property?

4. Indicate whether the following statements concerning histidine are true or false. Refer to the following titration curve.

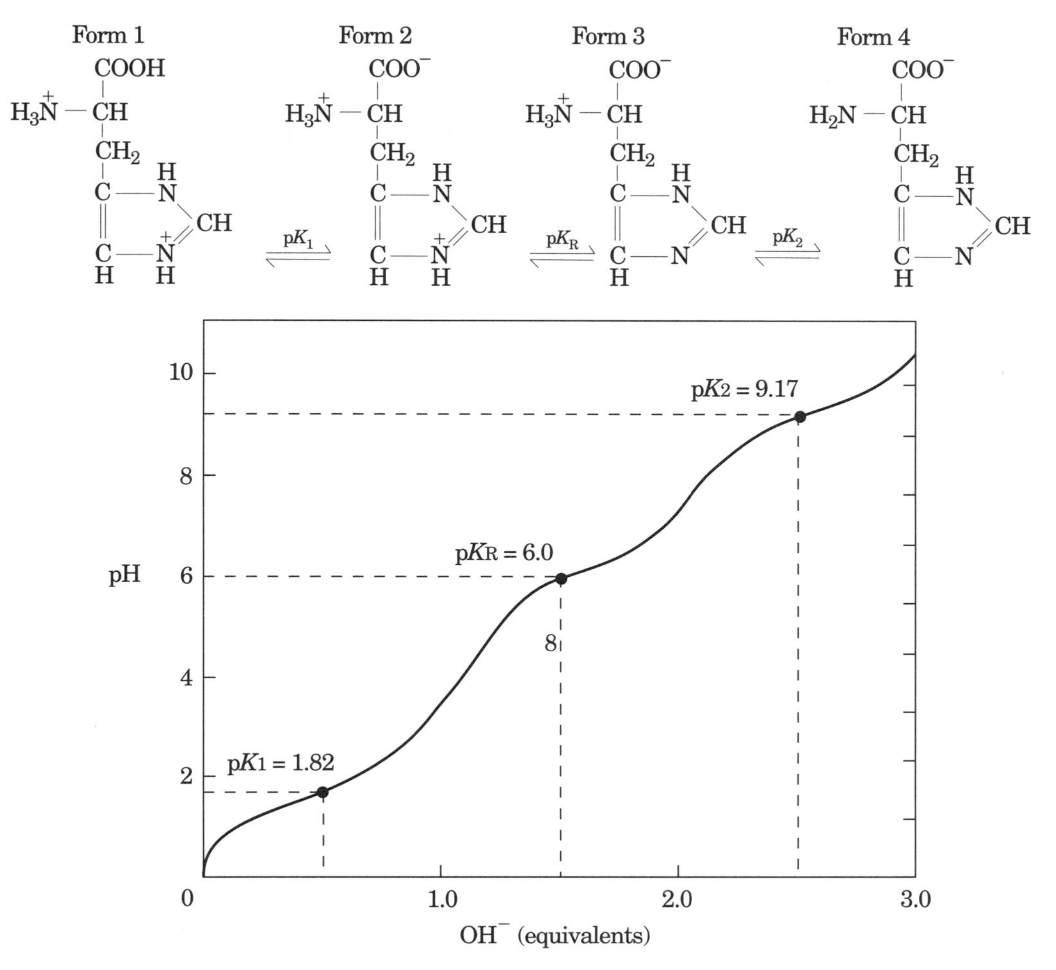

______ **(a)** At pH 1.82 there are equal amounts of form 1 and form 2.

______ **(b)** The α-carboxyl group is half dissociated at pH 9.17.

______ **(c)** Histidine would be a good buffer at pH values near 1.8 and 9.2.

______ **(d)** Histidine has biological importance because the pK_a of its side chain is close to physiological pH.

______ **(e)** Histidine's pI is between pH 1.82 and 6.0.

______ **(f)** Histidine's pI is between pH 6.0 and 9.17.

______ **(g)** Histidine is an aromatic amino acid.

5. Indicate whether the following statements concerning glutamate are true or false. Refer to the following titration curve.

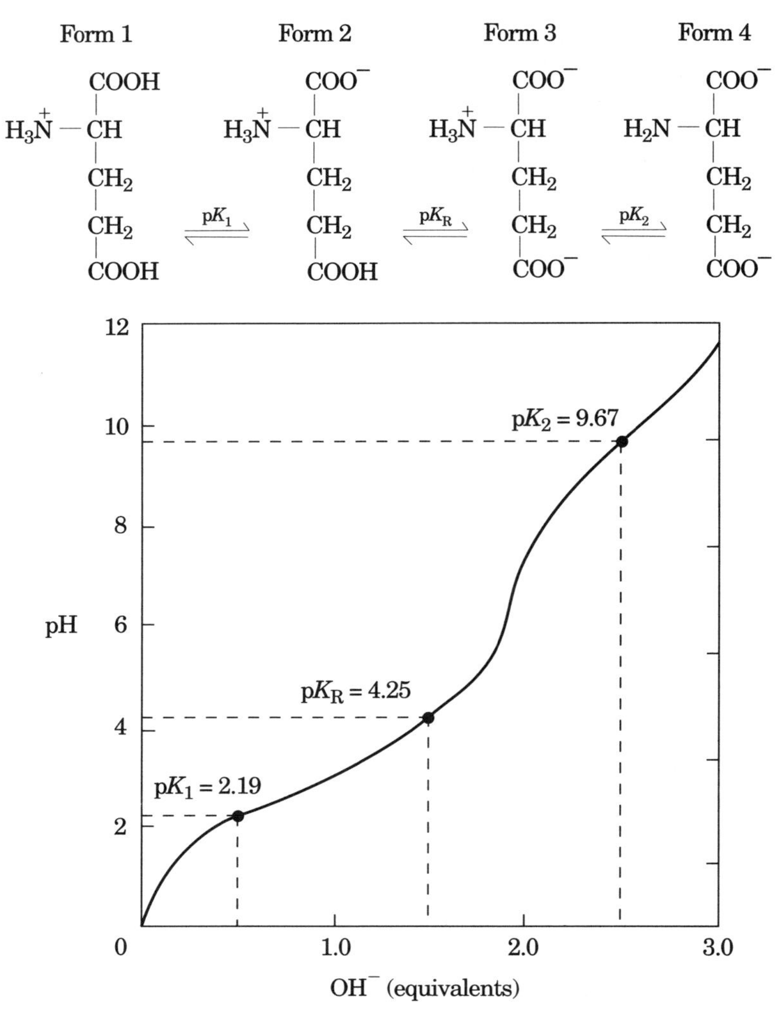

______**(a)** At pH 2.19 there are equal amounts of form 1 and form 2.

______**(b)** The α-carboxyl group is half dissociated at pH 9.67.

______**(c)** Glutamate would be a good buffer at pH values near 2.2 and 9.7.

______**(d)** Glutamate has biological importance because the pK_a of its side chain is close to physiological pH.

______**(e)** Glutamate's pI is between pH 2.19 and 4.25.

______**(f)** Glutamate's pI is between pH 4.25 and 9.67.

______**(g)** Glutamate is an aromatic amino acid.

6. Refer to the following table and titration curve to answer the questions below.

Compound	pK_1	pK_2	pK_R
Lactate	3.86		
Acetate	4.76		
Glycine	2.34	9.60	
Glutamate	2.19	9.67	4.25
Arginine	2.17	9.04	12.48

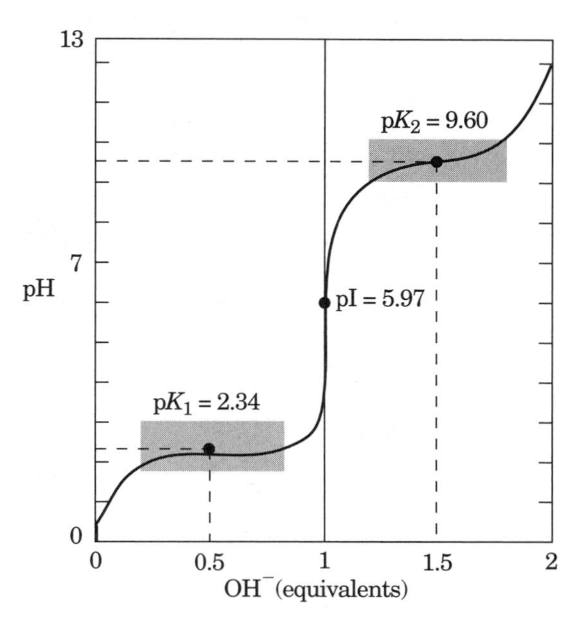

(a) Which compound from the table is illustrated by the titration curve?

(b) Which compound from the table has a net charge of zero nearest to physiological pH?

7. (a) Name the peptide shown below.

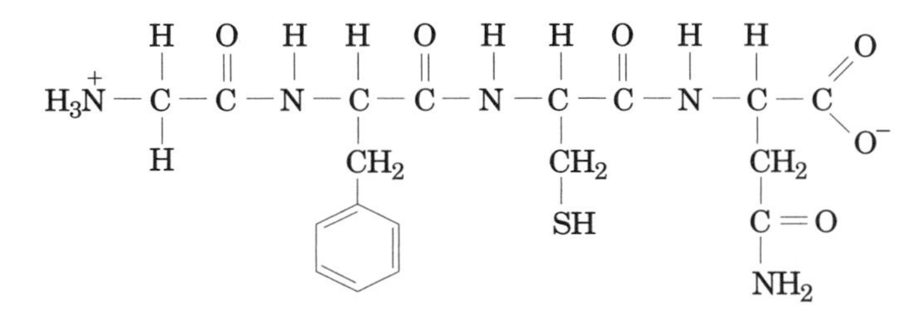

(b) How many peptide bonds are present in this peptide?

(c) How many disulfide bonds?

8. Calculate the net charge on the following tripeptides at pH 7.0:

(a) Asp–Glu–Ser **(b)** Ser–Gly–Thr **(c)** Gly–Lys–Arg

(d) Which tripeptide will be retained the longest on a cation-exchange chromatographic column in a pH 7.0 buffer?

9. Although gel electrophoresis allows researchers to identify specific protein bands on a gel and provides approximate molecular weights for proteins, it does not provide any information about amino acid sequence. Sequence information is extremely valuable in that it can be used to predict protein structure and function. Assume you have isolated a relatively abundant protein, and you want to obtain the amino acid sequence. You perform the following experiments:

(a) Addition of dithiothreitol to the protein sample followed by gel electrophoresis results in the protein gel shown on the next page. What can you conclude?

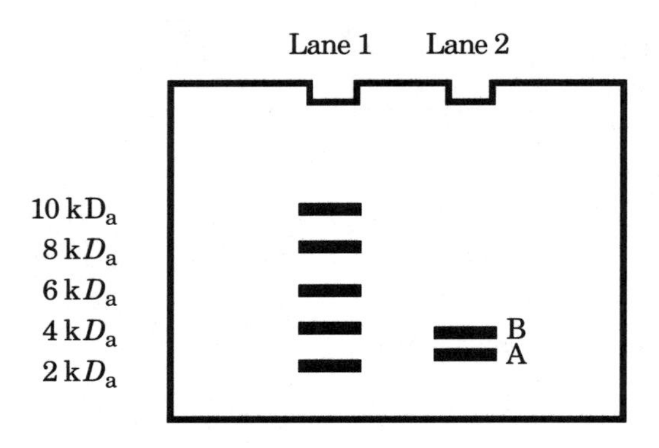

Lane 1 contains molecular weight markers as indicated to the left of the figure.

Lane 2 contains the DTT-treated, purified protein.

(b) Cleavage with chymotrypsin produces the following fragments:

Band A	CN NLQNY GIVEQCCHKRCSEY
Band B	F Y DPTKM IACGVRGF RTTGHLCGKDLVNALY

Cleavage with *Staphylococcus aureus* V8 protease produces the following fragments:

Band A	GIVE YNLQNYCN QCCHKRCSE
Band B	PTKM RTTGHLCGKD LVNALYIACGVRGFFYD

What is the amino acid sequence of your isolated protein?

(c) The protein you have sequenced is actually found in the pancreas of hagfish. An analysis of protein databases indicates a high degree of similarity with bovine insulin protein shown in Figure 5–24 of the textbook.

What residues are invariant or conserved?

What residues are variable or nonconserved?

Why do you think the conserved regions are important to the function of this protein?

Residues tend to be conserved because of their functional importance. What aspects of the conserved regions in these two proteins might contribute to their overall conformation and function?

ANSWERS

Worksheet

1. Refer to Fig. 5–5 for the identification and classification of each amino acid structure. Refer to Table 5–1 for the one- and three-letter codes.

Do You Know the Terms?

ACROSS

1. prosthetic
4. disulfide
6. amphoteric
9. chiral
11. salting
12. peptide
13. isoelectric
14. residue
16. zwitterion

DOWN

2. SDS
3. HPLC
4. dialysis
5. proteome
7. polypeptide
8. Edman
10. homologous
14. oligomer

Do You Know the Facts?

1. **(a)**

COO^-
$H_3\overset{+}{N}-C-H$
CH_3

Alanine, Ala, A, nonpolar

(b)

COO^-
$C-H$
$H_2\overset{+}{N}$ CH_2
H_2C-CH_2

Proline, Pro, P, nonpolar

(c)

COO^-
$H_3\overset{+}{N}-C-H$
CH_2
SH

Cysteine, Cys, C, polar, uncharged

(d)

COO^-
$H_3\overset{+}{N}-C-H$
CH_2
(benzene ring)

Phenylalanine, Phe, F, aromatic

(e)

COO^-
$H_3\overset{+}{N}-C-H$
CH_2
CH_2
CH_2
CH_2
$^+NH_3$

Lysine, Lys, K, positively charged

(f)

COO^-
$H_3\overset{+}{N}-C-H$
CH_2
COO^-

Aspartate, Asp, D, negatively charged

2. **(a)** Histidine, His, H, positively charged
   (b) threonine, Thr, T, polar, uncharged
   (c) arginine, Arg, R, positively charged
   (d) glutamate, Glu, E, negatively charged

3. **(a)** Glycine; **(b)** lysine, arginine, histidine (possible); **(c)** phenylalanine, tryptophan; **(d)** alanine, valine, leucine, isoleucine; **(e)** histidine; **(f)** proline; **(g)** tyrosine

4. Lysine, glutamine, asparagine, tyrosine

5. Arginine, leucine, tryptophan, glutamate, glutamine

6. Y

7. C

8. B, C

9. A

10. **(a)** affinity chromatography; **(b)** cation exchange chromatography; **(c)** electrophoresis; **(d)** gel filtration.

11. Cytochrome *c*. Size-exclusion chromatography separates molecules on the basis of size. Larger molecules are unable to fit into the tiny pores in the chromatographic beads, and so are excluded based on their size. These larger molecules take a direct route through the column, eluting before smaller molecules, which are detoured into the bead pores. The smallest protein, cytochrome *c*, would be retained the longest, and elute last. (This result assumes that all of proteins in the mixture are globular.)

12. Apolipoprotein B. SDS polyacrylamide gel electrophoresis is used to separate proteins on the basis of molecular weight. SDS molecules, which are negatively charged, bind to proteins uniformly along their length, rendering the intrinsic charge of the protein insignificant. Proteins in SDS thus assume a similar shape and have a net negative charge roughly proportional to their mass. Electrophoresis in the presence of SDS separates proteins almost exclusively on the basis of mass, with smaller polypeptides migrating more rapidly to the bottom of the gel when an electric field is applied, and larger proteins moving through the gel much more slowly.

Applying What You Know

1. Both have an —OH group, which contributes polarity and hydrogen-bonding capability.

2. Cysteine's sulfhydryl group readily oxidizes to form a covalently linked disulfide bridge, a dimeric amino acid called cystine. Cystine residues stabilize the structures of proteins in which they occur. Methionine cannot form disulfide bridges.

3. The aromatic rings of the side chains of Trp and Tyr, and to a much lesser extent Phe, absorb light strongly around 280 nm.

4. **(a)** T; **(b)** F; **(c)** T; **(d)** T; **(e)** F; **(f)** T; **(g)** F

5. **(a)** T; **(b)** F; **(c)** T; **(d)** F; **(e)** T; **(f)** F; **(g)** F

6. **(a)** Glycine; (b) glycine

7. **(a)** Glycylphenylalanylcysteinylasparagine; **(b)** 3; **(c)** 0

8. **(a)** −2; **(b)** 0; **(c)** +2; **(d)** Gly–Lys–Arg

9. **(a)** From the gel results, one can conclude that the protein is composed of two subunits corresponding to the two bands in lane 2. The protein in band A has a molecular mass of approximately 3 kD, and the protein in band B has a molecular mass of approximately 4 kD.

(b) Protease treatment provides the information that the protein in band A has the sequence

GIVEQCC**HKR**CS**E**Y**N**L**Q**NYCN

The protein in band B has the sequence

RTTGHLCG**KD**LV**N**ALY**IA**CGVRGFFY**D**P**TKM**

(c) It appears that you have isolated and sequenced hagfish insulin. (A glutamate residue, E, has replaced an isoleucine residue for didactic purposes.) The amino acid residues in hagfish insulin that differ from the bovine protein are shown in bold type. The protein in band A is 71% similar to the A chain of bovine insulin and the protein in band B is 58% similar to the B chain of bovine insulin.

The structural features of insulin that appear to be highly conserved across species include the position of the cysteine residues (and the three disulfide bonds). In addition, both the amino and carboxyl termini of the A chain are highly conserved. The hydrophobic residues near the carboxyl terminus of the B chain are also conserved, although this is not obvious based on the present analysis.

chapter

The Three-Dimensional Structure of Proteins

STEP-BY-STEP GUIDE

Major Concepts

A protein's conformation, or three-dimensional structure, is described by its secondary, tertiary, and quaternary structure.

All of these are dependent on the primary structure. Weak, noncovalent interactions stabilize these levels of architecture, which determine a protein's overall shape.

Peptide bonds connect amino acid residues in proteins.

These C—N bonds have some double-bond character, which restricts rotation, limiting the possible conformations of a polypeptide chain. The N—C_α and C_α—C bonds can rotate, but their possible conformations are limited by steric hindrance. A **Ramachandran plot** depicts the conformations theoretically permitted for peptides.

Secondary structure refers to the arrangement of adjacent amino acids in regular, recurring patterns.

Some common secondary structures are the α **helix,** β **conformation,** β **turn,** and **collagen helix.** These structures are formed, in part, in response to noncovalent interactions between neighboring amino acids. They are stable because hydrogen-bond formation is maximized and steric repulsion is minimized in these conformations. Specific amino acids are more likely to be found in some secondary structures than in others; consequently, information about the amino acid sequence of a protein can be used to predict where these structures are likely to occur.

Tertiary structure refers to the actual three-dimensional arrangement of a single chain of amino acids.

The linear amino acid sequence is a critical determinant in the final shape of a protein. Secondary structure, bend-producing amino acids, and hydrophobic interactions also contribute to the folding of a protein molecule and therefore to its overall shape. Protein folding is a complex process that is still not well understood. In some cases, protein folding occurs spontaneously, determined solely by the primary sequence. Weak, noncovalent interactions between individual amino acid residues play a role in stabilizing a protein's tertiary structure. Most soluble proteins are compact, globular structures that are somewhat flexible. The ability to change conformation is critical to protein function; enzymes, for example, often change shape when they bind substrate. **X-ray diffraction** techniques have provided direct information on the tertiary structure of many proteins.

Quaternary structure is a level of protein architecture found only in proteins having more than one polypeptide chain (subunit).

The three-dimensional arrangement of the separate subunits in the protein is referred to as its quaternary structure. Interactions between the individual subunits are stabilized by the same weak, noncovalent interactions that help stabilize the secondary and tertiary structure. One well-characterized protein with quaternary structure is hemoglobin, which consists of four distinct polypeptide chains. The association of these subunits into an oligomer contributes to the ability of hemoglobin to both bind and release oxygen under the appropriate physiological conditions (see Chapter 7 for more information). The quaternary structure of fibrous proteins such as collagen contributes to the tensile strength of these proteins. Very large assemblies of polypeptides have also been described, for example, ribosomes which contain over 80 proteins.

What to Review

Answering the following questions and reviewing the relevant concepts, which you have already studied, should make this chapter more understandable.

- Protein chemistry takes place in aqueous solution. Renew your appreciation of the role water plays in all biochemistry discussions (pp. 85–89).
- Weak, noncovalent interactions are extremely important in protein architecture. What are the various types of weak interactions common in biomolecules, and what are their relative strengths (pp. 90–91)?
- How are peptide bonds formed? Does a polypeptide have direction (p. 126)?
- Recall that proteins are made using only L-amino acids. Draw a representation of an L- and a D-amino acid (p. 117).
- The amino acid sequence, or primary structure, of a protein influences all other levels of protein architecture. What bonds and interactions are most important in this level of protein structure? How can the sequence of amino acids in a polypeptide be determined (pp. 126, 141–145)?

Topics for Discussion

Answering each of the following questions, especially in the context of a study group discussion, should help you understand the important points of this chapter.

Overview of Protein Structure

A Protein's Conformation Is Stabilized Largely by Weak Interactions

1. Would the "simple rules" stated at the end of this section of the text (p. 161) be the same if protein chemistry occurred in a nonpolar solvent?

The Peptide Bond Is Rigid and Planar

2. What makes peptide bonds planar?

3. Why can't psi (ψ) and phi (ϕ) both be zero?

Protein Secondary Structure

The α Helix Is a Common Protein Secondary Structure

4. Find the ψ and ϕ angles for an α helix on the Ramachandran plot shown in Figure 6–3. Are these angles theoretically allowed?

5. Are the side chains of amino acids in an α helix on the outside or inside of the helix?

Knowing the Right Hand from the Left

6. Attempt to draw for yourself an α helix containing a *mixture* of L- and D-amino acids. What does this attempt tell you about α helices?

Amino Acid Sequence Affects α Helix Stability

7. The properties of the various amino acid side chains dictate their interactions in secondary structure elements. In α helices, the properties of the side chains place constraints on the stability of the helix.

The β Conformation Organizes Polypeptide Chains into Sheets

8. Why is the α helix often referred to as a condensed secondary structure, whereas the β conformation is often called extended?

9. In silk fibroin, which amino acid residues have functional groups participating in interchain H bonds? Which have groups extending above and below the plane of the sheet?

10. What are the limitations on the kinds of amino acids found in β structures? Compare these to the constraints discussed for the α helix.

β Turns Are Common in Proteins

11. What is unusual about peptide bonds involving the imino nitrogen of proline in β turns?

12. How many amino acid residues are typically found in a β turn and how is this structure stabilized?

Common Secondary Structures Have Characteristic Bond Angles and Amino Acid Content

13. Note where each of the secondary structures falls on the Ramachandran plot in Figure 6–9.

Protein Tertiary and Quaternary Structures

14. What distinguishes tertiary from quaternary structure in proteins? Do all proteins have quaternary structure?

15. How could you determine if a protein has quaternary structure?

Fibrous Proteins Are Adapted for a Structural Function

16. What contributes to the insolubility of fibrous proteins such as α-keratin and how does this insolubility contribute to the function of these proteins?

17. What covalent bonds contribute strength in each of these fibrous proteins: α-keratin, collagen, elastin?

18. Wool and silk are both composed of fibrous proteins; wool can stretch and shrink but silk cannot. What is the molecular basis for the different characteristics of these two fibers?

Permanent Waving Is Biochemical Engineering

19. Why can a badly done perm be so damaging to hair?

Structural Diversity Reflects Functional Diversity in Globular Proteins

20. Try to envision how protein structure contributes to the functions of the various globular proteins mentioned in this section.

Myoglobin Provided Early Clues about the Complexity of Globular Protein Structure

21. Why does it make functional sense that a globular protein, such as myoglobin, is soluble in water?

22. What weak interactions contribute to the close packing and stability of the structure of myoglobin?

Methods for Determining the Three-Dimensional Structure of a Protein

23. Why can't you see proteins under a high-power light microscope?

24. Why must x rays be used to generate "images" of proteins?

25. Why is it important to know about protein structure in aqueous solutions?

Globular Proteins Have a Variety of Tertiary Structures

26. Lysozyme and ribonuclease are enzymes; cytochrome *c* is a component of the respiratory chain. What three-dimensional characteristic is visible in the space-filling models of lysozyme and ribonuclease that is *not* seen in the cytochrome *c* model? (See Fig. 6–18.)

Analysis of Many Globular Proteins Reveals Common Structural Patterns

27. What is the difference between a domain and a motif?

28. Which amino acids will usually be found in the interior of a globular protein? Which will be found on the exterior?

29. Provide some examples of common protein motifs.

Protein Motifs Are the Basis for Protein Structural Classification

30. Why are the proteins belonging to a structural "family" likely to be related with respect to their evolution?

Protein Quaternary Structures Range from Simple Dimers to Large Complexes

31. After reading this section, list some supramolecular complexes consisting only of proteins and some consisting of proteins associated with other types of biomolecules.

32. What kinds of bonds or interactions would be important in the assembly of large protein complexes?

There Are Limits to the Size of Proteins

33. What two factors limit the size of proteins?

Protein Denaturation and Folding

Loss of Protein Structure Results in Loss of Function

34. Under what conditions are proteins denatured?

35. What bonds or interactions are disrupted during denaturation?

Amino Acid Sequence Determines Tertiary Structure

36. When denatured proteins refold in the presence of low levels of detergent or denaturant, they form many incorrect disulfide bonds. What does this tell you about the process of protein folding?

Polypeptides Fold Rapidly by a Stepwise Process

37. What are the two models of protein folding presented in this section?

38. What are the noncovalent interactions that drive folding in each of these models?

Death by Misfolding: The Prion Diseases

39. The conversion of the normal PrP protein to the PrP^{Sc} form is the result of an alteration that changes the structure of the prion protein. Based on information from this chapter, what other changes in a protein's amino acid composition would be expected to change a protein's structure and therefore its function?

Some Proteins Undergo Assisted Folding

40. Yeast mutants that are deficient in heat shock proteins (some of which are of molecular chaperones) have proteins that do not fold correctly. One role of molecular chaperones is to bind to nascent protein chains (proteins that are being synthesized) and *prevent* associations between newly synthesized regions. Why would this be necessary?

SELF-TEST

Do You Know the Terms?

ACROSS

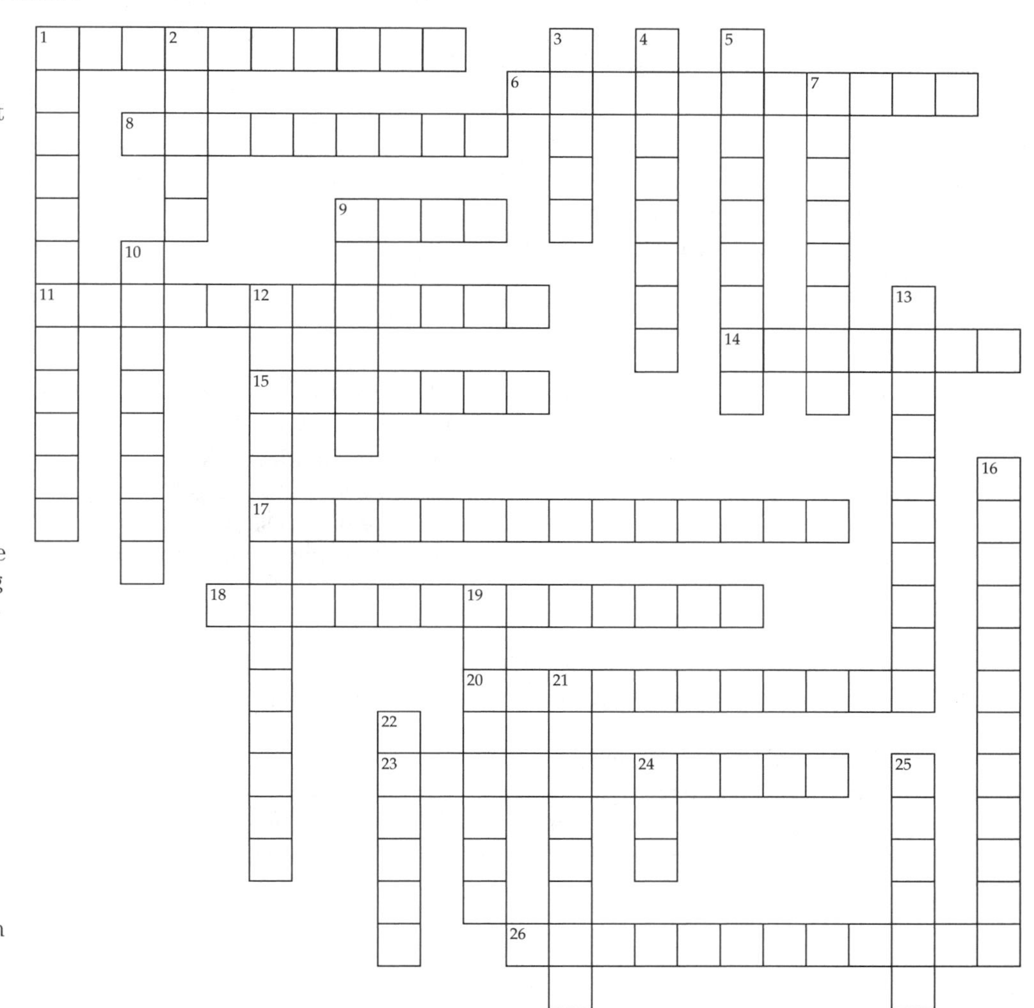

1. Cellular agents that assist in protein folding at elevated temperatures.
6. Covalently linked amino acids with a single amino terminus and a single carboxyl terminus is called a(n) ________.
8. Bonds that occur between cysteine residues in proteins.
9. Also called a "motif."
11. Hemoglobin is a(n) ________ protein because it has two or more polypeptide chains.
14. They protrude in opposite directions from the zigzag structure of the β conformation. (2 words)
15. GCKKGGLVCAH (S—S linking the two C residues) for example; ________ structure.
17. Muscle fibers are an example of a(n) ________ complex.
18. Protein secondary structure that extends 0.35 nm per amino acid residue.
20. Though unrelated based on their amino acid sequences, proteins that belong to a(n) ________ have related structural features.
23. The noncovalent interactions that are thought to be the driving force behind the formation of a "molten globule."
26. An example is the re-formation of disulfide bonds during permanent waving.

DOWN

1. A native protein is in its functional ________.
2. An example of protein misfolding that has lethal consequences.
3. A stable arrangement of a few secondary structures.
4. α helices are stabilized by ________ bonds between the carbonyl oxygen and the amino hydrogen.
5. A β turn is an example of ________ structure.
7. Disrupting the hydrophobic interactions of a single-subunit protein would have the greatest effect on the ________ structure of that protein.
9. An example of a supramolecular assembly is the collagen ________.
10. α-keratin is referred to as a supramolecular complex of protein subunits; hemoglobin with only four subunits is referred to as a(n) ________.
12. The saddle conformation is a(n) ________ structure.
13. Myoglobin is to tertiary as hemoglobin is to ________.
16. Roasting a chicken results in the permanent ________ of myosin and actin proteins in the muscle cells.
19. Individual amino acids when polymerized in a protein.
21. The αβ subunits in hemoglobin comprise a single ________; the intact hemoglobin tetramer contains two of these.
22. Protein secondary structure that extends 0.15 nm per amino acid residue.
24. This class of proteins binds to and shields hydrophobic portions of unfolded polypeptides in cells. These proteins also are denatured by elevated temperatures.
25. Refers to the portion of a protein that is often composed of noncontiguous amino acid sequences and is usually defined on the basis of its contribution to protein function.

Do You Know the Facts?

1. Which of the following statements is/are true concerning peptide bonds?
   A. They are the only covalent bond formed between amino acids in polypeptide structures.
   B. The angles between the participating C and N atoms are described by the values psi (ψ) and phi (ϕ).
   C. They have partial double-bond character.
   D. A and C.
   E. All of the above are true.

2. In an α helix, the R groups on the amino acid residues:
   A. are found on the outside of the helix spiral.
   B. participate in the hydrogen bonds that stabilize the helix.
   C. allow only right-handed helices to form.
   D. A and B are true.
   E. A, B, and C are true.

3. Which of the following bonds or interactions is/are possible contributors to the stability of the tertiary structure of a globular protein? (Hint: remember the amino acid categories.)
   A. peptide bonds between a metal ion cofactor and a histidine residue
   B. hydrophobic interactions between histidine and tryptophan R groups
   C. covalent disulfide cross-links between two methionine residues
   D. hydrogen bonds between serine residues and the aqueous surroundings
   E. All of the above contribute.

Refer to the following numbered statements to answer questions 4–7.

(1) Found in the same percentage in all proteins.
(2) Stabilized by H bonds between —NH and —CO groups.
(3) Found in globular proteins.
(4) Affected by amino acid sequence.
(5) An extended conformation of the polypeptide chain.
(6) Includes all 20 standard amino acids in equal frequencies.
(7) Hydrophobic interactions are responsible for the primary structure.

4. Which statements are true of α helices?

5. Which statements are true of β sheets?

6. Which statements are true of both α helices and β sheets?

7. Which statements are true of neither α helices nor β sheets?

8. Describe and compare the positions of the R groups of amino acids that are in an α helix and in a β conformation.

9. Fibrous proteins, such as α-keratin, collagen, and elastin, have evolved to be strong and/or elastic. They are also insoluble in water. What would you guess this insolubility means about the spatial positioning of the various groups of amino acids that comprise these proteins?

10. What does denaturation mean? Under what conditions are proteins denatured?

11. From the types of bonds and interactions below, identify which is *most* responsible for the structures described in (a)–(j).

ionic interactions	covalent bonds
hydrophobic interactions	hydrogen bonds
van der Waals interactions	

(a) the association of hemoglobin subunits in its quaternary structure

(b) primary structure of proteins

(c) secondary structure of proteins

(d) the interaction between hemoglobin subunits upon binding oxygen

(e) the binding of iron within the heme group

(f) an α helix

(g) a β sheet

(h) association of the heme group within the myoglobin polypeptide

(i) the hardness of rhinoceros horn

(j) the compactness of the interior of myoglobin

Refer to Table 5–2 in the textbook for questions 12 and 13.

12. How many proteins in the table exhibit quaternary structure? How can you tell?

A. all
B. none
C. half
D. five
E. One can not tell from the data provided.

13. How many proteins in the table exhibit α helix secondary structure? How can you tell?

A. all
B. none
C. half
D. two
E. One can not tell from the data provided.

Applying What You Know

1. Answer the following questions concerning the polypeptides whose amino acid sequences are shown below. It may help to refer to Table 5–1 for the three-letter abbreviations and to Figure 5–5 for amino acid structures.

(1) Gly–Ile–Trp–Leu–Ile–Ile–Phe–Gly–Val–Val–Ala–Gly–Val–Ile–Gly–Trp–Ile–Leu–Leu–Ile
(2) Gly–Pro–Hyp–Gly–Pro–Met–Gly–Pro–Ser–Gly–Pro–Arg–Gly–Pro–Hyp–Gly–Pro–Hyp–Gly
(3) Gly–Met–Trp–Pro–Glu–Met–Cys–Gly–Glu–Pro–Ala–His–Val–Arg–Asp–Tyr–Pro–Leu–Leu
(4) Gly–Met–Trp–Pro–Glu–Met–Cys–Gly–Glu–Pro–Ala–His–Val–Arg–Asp–Tyr–Cys–Leu–Leu

(a) Which would be *most likely* to form an α-helical structure?

(b) Which could have at least one disulfide bond?

(c) Which would be *most likely* to be part of a fibrous protein found in cartilage?

(d) Suppose that the first amino acid residue in peptide (1) were altered from Gly to Pro. Would you expect this change to have an effect on the secondary structure of this peptide? Why or why not?

2. Myoglobin and the individual subunits of hemoglobin are similar, though certainly not identical, in size, overall shape, and function. Would you expect a molecule of myoglobin or a subunit of hemoglobin to have a greater ratio of nonpolar to polar amino acids? Why?

3. How would the following agents and/or procedures interfere with or disrupt the different levels of protein architecture? Why?

(a) Addition of SDS (see p. 134).

(b) Oxidation of cystine with performic acid to cleave disulfide bonds (Fig. 5–26).

(c) Addition of proteases such as trypsin (p. 143).

Biochemistry on the Internet

In Chapters 5 and 6 you have been introduced to the structural hierarchy of cellular proteins. The extent to which the linear sequence of amino acids, the primary structure, determines the final three-dimensional shape, or tertiary structure, of proteins is quite remarkable.

As our understanding of the mechanisms underlying protein folding has expanded, it has become possible to identify unknown proteins and predict their shape, and therefore their function, based only on amino acid sequence information. This type of protein analysis relies on information that is available from sequence databases around the world. The key to unlocking this information is some knowledge of the DNA or amino acid sequence of the "mystery" protein.

The problem outlined below provides a key (an amino acid sequence) and directs you to some of the doors (database access sites) to the ever expanding amount of scientific information available on the Internet. The list of databases presented here is by no means exhaustive; this information is intended only to provide a limited and somewhat structured exposure to protein databases and molecular modeling.

As you proceed through this exercise you will be able to obtain additional information by following any of the links that appear during your search. Also, once a specific protein has been identified, you can use the protein's name as a keyword on any Internet search engine (such as Google, Lycos, Yahoo, or HotBot). It will be up to you to sift through all this information to find what is relevant to your research, so don't stray too far in your initial forays and be sure to be able to find your way back.

The Problem: Recently a doctor in southeast Algeria examined a 6-year-old patient who had bronchitis. In the course of evaluating the patient, a systematic hematological study of the child's blood was performed. An abnormal protein was detected and additional tests on the patient's blood sample were performed. These tests, which included isoelectric focusing, electrophoresis, and cation-exchange high performance liquid chromatography, confirmed the existence of a novel blood protein. The protein was also observed in blood samples from the child's father and four other blood relatives, indicating a genetic basis for the abnormality. The abnormal protein was isolated in a pure form and the following amino acid sequence was determined:

VLSPADKTNV KAAWGKVGAH AGEYGAEALE RMFLSFPTTK TYFPHFDLSH
GKKVADALTN AVAHVDDMPN ALSALSDLHA HKLRVDPVNF KLLSHCLLVT
LAAHLPAEFT PAVHASLDKF LASVSTVLTS KYR

(a) Your job is to try to identify this novel blood protein. First locate a protein database on the Internet. A number of excellent sites are listed below. For this initial analysis try using the SCOP database.

- Structural Classification of Proteins or SCOP at http://scop.mrc-lmb.cam.ac.uk/scop/
- PIR—International Protein Sequence Database at http://pir.georgetown.edu/pirww/search/textpsd.shtml
- ProDom at http://prodes.toulouse.inra.fr/prodom/doc/prodom.html
- BCM Search Launcher at http://searchlauncher.bcm.tmc.edu
- Prosite at http://us.expasy.org/prosite

(A list of current urls for these sites is also provided at the *Principles of Biochemistry* Web site at http://www.worthpublishers.com/lehninger.)

For any of these sites, follow the internal links to locate a sequence comparison engine. For example, the SCOP site has a function called **Sequence Similarity Search.** Copy (or type) a segment of your sequence (20 amino acid residues should be enough) into the appropriate search field and submit your sequence for analysis. What does this analysis tell you about the identity of your protein?

Try using a different set of 20 amino acids residues for your analysis. How certain are you that you have correctly identified the abnormal protein?

(b) Your research indicates that the abnormal protein has a primary sequence that is similar to that of another protein that is normally present in the blood. To what "Class," "Superfamily," and "Family" does the abnormal protein probably belong?

(c) The molecular mass of the abnormal blood protein is approximately 764 daltons less than the mass of the "normal" protein. What does this suggest about the abnormal protein and what might you do to verify your hypothesis?

(d) What information can you find on the secondary, tertiary, and quaternary structure of the "normal" version of this protein? You can start your structural analysis at a variety of Web sites that provide information on the three-dimensional structure of proteins. For this initial analysis try using the RCSB Protein Data Bank at http://www.rcsb.org/pdb/. (For a list of other sites that provide information on the three-dimensional structure of proteins, see the *Principles of Biochemistry* Web site.)

ANSWERS

Do You Know the Terms?

ACROSS

1. chaperones
6. polypeptide
8. disulfide
9. fold
11. multisubunit
14. R groups
15. primary
17. supramolecular
18. β conformation
20. superfamily
23. hydrophobic
26. renaturation

DOWN

1. conformation
2. prion
3. motif
4. hydrogen
5. secondary
7. tertiary
9. fibril
10. oligomer
12. supersecondary
13. quaternary
16. denaturation
19. residues
21. protomer
22. α helix
24. hsp
25. domain

Do You Know the Facts?

1. C. Disulfide bonds between cysteine residues are also covalent bonds in polypeptides. Psi (ψ) and phi (ϕ) describe the C_α—C and the N—C_α bonds, not those involved in the peptide bond.

2. A. H bonds form between —NH and —CO groups of the polypeptide backbone, not between R groups. Both left- and right-handed helices can occur (though extended left-handed helices have not been observed in proteins).

3. D. The hydroxyl group on the serine side chain can form H bonds with surrounding water molecules. Peptide bonds occur only between amino acids, not between an amino acid and a cofactor. Histidine has no hydrophobic R group. Only cysteines can form disulfide linkages.

4. 2, 3, 4

5. 2, 3, 4, 5

6. 2, 3, 4

7. 1, 6, 7

8. In an α helix, the polypeptide backbone is tightly wound along the long axis of the molecule; the R groups extend outward from the rod, like spokes, in a helical array. In a β conformation, the backbone of the polypeptide chain is extended into a zigzag rather than a helical rod; the R groups of adjacent amino acids protrude in opposite directions from the zigzag structure, creating an alternating pattern when viewed from the side.

9. These fibrous proteins are an exception to the rule that hydrophobic R groups of amino acid residues must be buried in the interior of a protein; they have a high concentration of hydrophobic amino acids both in the interior and on the surface of the protein.

10. Denaturation is the total loss or randomization of three-dimensional structure. It can be reversible or irreversible. Extremes of heat and pH, exposure to some organic solvents such as alcohol or acetone, or some other substances such as urea or some detergents, will denature proteins. The covalent peptide bonds of proteins are not broken, but denaturation disrupts the many weak, noncovalent interactions that are most important in maintaining the native three-dimensional conformation.

11. **(a)** Hydrophobic interactions; **(b)** covalent bonds; **(c)** hydrogen bonds; **(d)** ionic interactions; **(e)** covalent bonds; **(f)** hydrogen bonds; **(g)** hydrogen bonds; **(h)** covalent bonds; **(i)** covalent bonds; **(j)** hydrophobic interactions.

12. D. A protein has quaternary structure only if it has more than one polypeptide chain, or subunit. Five of the proteins listed have at least two subunits, and so have quaternary structure.

13. E. None of the data in the table provides any information regarding secondary structure.

Applying What You Know

1. **(a)** 1; **(b)** 4; **(c)** 2; **(d)** No; peptide (1) could adopt an α-helical structure before the change of the amino-terminal residue to proline. Though proline residues are rarely found in α helices, the amino terminus is one location where the proline R group would have relatively little effect. The amino (imino) group would not interfere with the helical structure.

2. The hemoglobin subunit would have a higher ratio of nonpolar to polar amino acids. In hemoglobin's native conformation, each subunit is in contact with the other subunits; hydrophobic interactions between the subunits are important in the stability of the tetramer, and nonpolar amino acids can be on the surface of the subunit. The external surface of the myoglobin molecule would have more polar amino acid residues, to increase hydrogen bonding with the aqueous surroundings.

3. (a) SDS, a detergent, would interfere with hydrophobic and ionic interactions, and would therefore interfere in secondary, tertiary, and quaternary structure; primary structure would not be affected. (b) For proteins with disulfide bridges, oxidation of cystine with performic acid to cleave disulfide bonds would alter all levels of protein architecture. Disulfide bonds are part of primary structure, and a loss of these stabilizing covalent cross-links would very likely disrupt a protein's native conformation. Proteins without disulfide bridges would be unaffected. (c) Proteases cleave peptide bonds, destroying primary structure. All other levels depend on the amino acid sequence, and so would be affected as well.

Biochemistry on the Internet

(a) The following is one example of a sequence similarity search using the SCOP database. If you used the first 20 amino acid residues, the output shows that the abnormal protein bears a strong sequence similarity to the alpha subunit of hemoglobin. Using other portions of the amino acid sequence almost always results in the identification of the alpha chain of hemoglobin.

(b) Click on the SCOP ID number to the left of the hemoglobin entry (1a3na) to see the classification hierarchy. If the abnormal protein is a hemoglobin variant, it belongs to the

Class: All α (helix) proteins
Fold: Globin-like, 6 helices, folded leaf, partly opened
Superfamily: Globin-like
Family: Globins, heme-binding protein
Protein: Hemoglobin, α chain
Species: human (*Homo sapiens*)

(c) The lower molecular mass coupled with the sequence similarity to the α chain of hemoglobin suggests that the protein is a mutant form of hemoglobin that has fewer amino acids than the normal form. Since the average weight of a single amino acid is around 110, one could predict that there would be about 7 amino acids missing from the abnormal hemoglobin.

This can be verified by directly comparing the amino acid sequence of the abnormal protein and the human hemoglobin α-chain sequence. Type in the entire sequence for the abnormal protein and rerun the **Blast Query.** This time scroll down the resulting data screen until you see the listing for 1a3na, where the amino acid sequences are compared. Amino acids 50–109 of the abnormal protein match with residues 58–117 of alpha hemoglobin suggesting that a deletion has occurred in the abnormal hemoglobin around residue 50. A direct comparison of the complete sequences for both the abnormal protein and the α chain of hemoglobin (which you can obtain by clicking on **Full Sequences**) will confirm that 8 residues somewhere between residues 50 and 59 of the normal hemoglobin (depending upon the reading frame) are absent in the abnormal protein.

In fact, the amino acid sequence of the abnormal protein provided in the original question is the complete sequence for an α-chain variant called Hb-J Briska, named after the town in southeast Algeria where it was first discovered. For the complete reference see Wajcman *et al.*, Haemoglobin J-Briska: a new mildly unstable α 1 gene variant with a deletion of eight residues (α 50–57, α 51–58 or α 52–59) including the distal histidine. *British Journal of Haematology,* (1998) 100: 401–406.

(d) This example of an Internet-based analysis of protein structure uses the Protein Data Bank. If you type in the PDB ID 1A3N, you will be presented with a page of information and links related to this particular alpha chain of hemoglobin. The **View Structure** link will allow you to access three-dimensional models of the hemoglobin tetramer. You will most likely need to download a plug in program to allow you to view the molecular structure of your protein. Two popular methods for viewing three-dimensional structures are Chime and Rasmol. Instructions on how to download these programs are available at the *Principles of Biochemistry* Web site.

In Chime or Rasmol you can view the protein's secondary structure by setting the **Display** option to **Ribbons.** This will show the protein's beta conformation and alpha-helical structure. You can also manipulate the movement of the proteins by clicking on the structure and then dragging it with your mouse.

Note the heme molecules bound to each of the four subunits. You can view the entire hemoglobin tetramer by using the **Display** option **Spacefill** in either the Chime or Rasmol viewers.

If you click on **Other Sources** in the PDB page, you will be presented with a list of links to additional information on this protein. For example, the **PDBsum** link provides a summary of the locations of secondary structures. It gives the total number and position in the linear amino acid sequence of the various secondary structures in a protein.

chapter

7

Protein Function

STEP-BY-STEP GUIDE

Major Concepts

Several key principles of protein function can be illustrated by the behavior of oxygen-binding proteins, immune system proteins, and muscle proteins: proteins are not rigid but, in fact, must be flexible; proteins interact with other molecules called ligands; and the interaction of ligands with proteins can affect the binding of other ligands.

Myoglobin is a protein consisting of a single polypeptide chain with a heme prosthetic group that binds oxygen reversibly.

The binding properties of the heme group *within* the globin protein are different from those of the free heme group. The binding of oxygen (the ligand, L) to myoglobin can be described quantitatively; for a monomeric protein, the fraction of binding sites occupied by the ligand is a hyperbolic function of [L].

Hemoglobin is a multisubunit, heme-containing, oxygen-binding protein.

This protein is a tetramer with two α chains and two β chains. Each subunit contains a heme prosthetic group; consequently each is capable of binding one molecule of oxygen. Hemoglobin displays **positive cooperativity** with respect to its oxygen binding; the binding of oxygen to hemoglobin follows a sigmoid curve. This cooperativity is attributed to alterations in noncovalent interactions between amino acids in adjacent subunits. These alterations, which are the result of subtle changes in the porphyrin ring structure of the heme upon oxygen binding, result in a conformational shift in the entire hemoglobin tetramer.

Hemoglobin also binds the excess hydrogen ions and carbon dioxide produced by respiring tissues.

H^+ can bind to a number of amino acid residues in the hemoglobin molecule, which stabilizes the deoxy form of hemoglobin. CO_2 can bind to amino-terminal residues to form carbaminohemoglobin; the binding of CO_2 forms additional ion pairs, which stabilize the deoxy form of hemoglobin. Binding of both H^+ and CO_2 promotes the release of oxygen from hemoglobin to the tissues where it is needed. In the lungs, H^+ and CO_2 are released, increasing the affinity of hemoglobin for oxygen.

2,3-Bisphosphoglycerate (BPG) also regulates oxygen binding by hemoglobin.

A single BPG molecule binds to hemoglobin in the central cavity between the four subunits. Binding is due to interactions between negatively charged groups on BPG and positively charged amino acid residues that line the central cavity. BPG stabilizes the deoxy form of hemoglobin.

Sickle-cell hemoglobin (HbS) illustrates the importance of protein conformation to protein function.

The hemoglobin molecules of individuals with **sickle cell anemia** have different amino acid sequences than hemoglobin from normal individuals. A hydrophobic residue on the outer surface of the β subunit promotes clumping of hemoglobin molecules, producing abnormally shaped erythrocytes with decreased lifespan.

The vertebrate immune system protects the organism by utilizing incredibly specific interactions between white blood cells and their associated proteins, and non-self molecules.

T-lymphocytes produce receptor proteins and B-lymphocytes produce immunoglobulins. MHC proteins are found on all cells. Recognition and binding by these proteins allow the organism to determine self from non-self. When a virus, bacterium, or other invader is detected, an immune response is mounted that involves the coordinated control of proliferation of specific B- and T-cells and their proteins for attack against the invading foreign proteins.

Immunoglobulins are the proteins produced by B-cells.

Their shape, flexibility, and their combination of conserved and variable regions make them capable of recognizing an almost limitless variety of antigens.

The principles of protein flexibility, and protein-protein interaction through non-covalent interactions, are illustrated at an extreme level by the cyclic conformational changes that cause muscles to contract.

The proteins myosin and actin are organized into thick and thin filaments, which are themselves arranged in sarcomeres, the repeating units of myofibrils. Many myofibrils make up a muscle fiber; skeletal muscle consists of bundles of muscle fibers. The sliding of the myosin thick filaments along the actin thin filaments, coordinated by nerve impulses, produces the contraction of the muscle. Many other accessory proteins are involved in the structure and temporal control of these assemblies.

What to Review

Answering the following questions and reviewing the relevant concepts, which you have already studied, should make this chapter more understandable.

- Altering a single amino acid in a protein's sequence can have enormous consequences to the secondary, tertiary, and quaternary structure, and to the binding of ligands to proteins. Review the properties of the different R groups of amino acids (Fig. 5–5).
- Ionic interactions are critical to the cooperativity of oxygen binding in hemoglobin. How strong are ionic interactions compared to hydrogen bonds? To covalent bonds? (Table 3–3, pp. 84, 90).
- From the amino acid sequences given in Figure 7–7, what portions of hemoglobin and myoglobin contain invariant residues (Box 5–2)? Which reagents and/or techniques of protein purification denature proteins? Which leave them in their native conformation (pp. 130–137)?

Topics for Discussion

Answering each of the following questions, especially in the context of a study group discussion, should help you understand the important points of this chapter.

Reversible Binding of a Protein to a Ligand: Oxygen-Binding Proteins

Oxygen Can Be Bound to a Heme Prosthetic Group

1. Why is it important that iron is incorporated into the heme group, rather than free in the cell?

2. What is the shape of the heme group?

Myoglobin Has a Single Binding Site for Oxygen

3. How is the heme group attached to and positioned within the myoglobin molecule?

Protein-Ligand Interactions Can Be Described Quantitatively

4. Making a real effort to understand the quantitative relationships described in this section will help you to understand critical concepts in enzyme kinetics (Ch. 8).

5. In oxygen-binding curves for myoglobin, why is the term pO_2 rather than $[O_2]$ used?

Protein Structure Affects How Ligands Bind

6. Given that free heme binds CO 20,000 times better than it binds O_2, why don't we all succumb to carbon monoxide poisoning?

Oxygen Is Transported in Blood by Hemoglobin

7. Why is it important that erythrocytes are *small* cells?

Hemoglobin Subunits Are Structurally Similar to Myoglobin

8. Could a myoglobin molecule substitute for one of the subunits in the hemoglobin tetramer? Why or why not?

9. Why are the histidine residues E7 and F8 invariant between myoglobin and the α and β chains of hemoglobin? (See Fig. 7–5 and Fig. 7–7).

10. What weak bonds or interactions, *other than* ion pairs, are important to hemoglobin's quaternary contacts?

Hemoglobin Undergoes a Structural Change on Binding to Oxygen

11. How does oxygen binding cause rearrangement of ion pairs? (It may help to imagine tugging on a stiffly coiled garden hose. Moving one part of the hose shifts the rest of it.)

Hemoglobin Binds Oxygen Cooperatively

12. How does a sigmoid binding curve illustrate the presence of a low-affinity state and a high-affinity state?

13. Why does hemoglobin have a lower affinity for the first oxygen molecule it binds than it does for binding additional oxygen molecules? One way to think of this is by analogy to a block of four postage stamps, arranged in a two-by-two array:

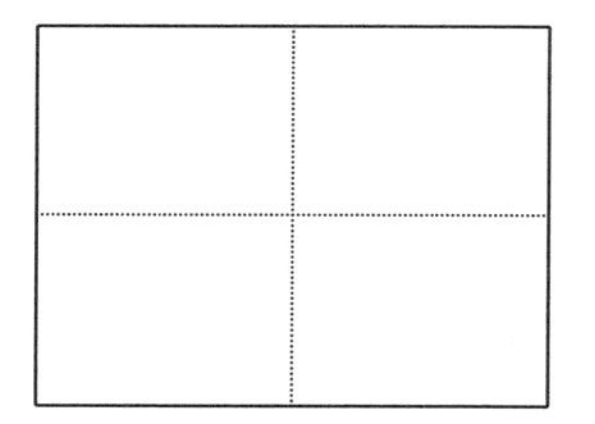

How many perforations (bonds) is it necessary to tear to free the first stamp from the other three? How many to free the second?

14. Oxygen acts as what type of modulator for hemoglobin?

Cooperative Ligand Binding Can be Described Quantitatively

15. Under what circumstances is the Hill coefficient (n_H) of a multisubunit protein equal to 1? When is n_H greater than 1? When would it be less than 1?

Two Models Suggest Mechanisms for Cooperative Binding

16. Is the concerted model or the sequential model more consistent with current knowledge of hemoglobin's T→R transition?

Hemoglobin also Transports H^+ and CO_2

17. The equation $HHb^+ + O_2 \rightleftharpoons HbO_2 + H^+$ indicates that both O_2 and H^+ can bind to hemoglobin. Do these two ligands bind at the same site? Do they both cause the same changes in conformation?

18. Carbon dioxide also binds to hemoglobin. Do CO_2 and O_2 bind at the same site? Do they both cause the same changes in conformation?

19. How are the binding of H^+, CO_2, and O_2 interdependent? Why does this make physiological sense?

Oxygen Binding to Hemoglobin Is Regulated by 2,3-Bisphosphoglycerate

20. The equation $HbBPG + O_2 \rightleftharpoons HbO_2 + BPG$ indicates that both oxygen and BPG can bind to hemoglobin. Do these two ligands bind at the same site? Do they both cause the same changes in conformation?

21. How would the oxygen-binding curve of fetal hemoglobin compare to that of maternal hemoglobin? Does this make physiological sense?

Sickle-Cell Anemia Is a Molecular Disease of Hemoglobin

22. What protein purification techniques might be useful for separating HbS from HbA?

23. Would it be better for patients with sickle-cell disease to live at sea level or at high altitude?

Complementary Interactions between Proteins and Ligands: The Immune System and Immunoglobulins

The Immune Response Features a Specialized Array of Cells and Proteins

24. What are the defense responsibilities of the humoral immune system? Of the cellular immune system?

25. Which type of white blood cell produces antibodies? What are the analogous recognition proteins produced by the cells of the cellular immune system?

26. Would a single amino acid be an effective antigen? A dipeptide? Hemoglobin?

Distinguishing Self from Nonself Involves the Display of Peptides on Cell Surfaces

27. How is a person's specific set of MHC proteins similar to a fingerprint?

28. Which class of MHC proteins and which type of white blood cell are responsible for the rejection of transplanted organs?

29. Why is the seemingly wasteful selection process that weeds out so many immature T_C and T_H cells so crucial to the proper functioning of the immune system?

Molecular Interactions at Cell Surfaces Trigger the Immune Response

30. How are viral proteins made "visible" to the immune system cells?

31. How do interleukins stimulate the reproduction of *only* those immune system cells needed in a particular viral attack on an organism?

32. What is the role of T_H cells in the protection of the organism under viral attack? List two ways that they participate.

33. What is the role of B cells in the defensive strategy?

34. Why is a secondary immune response so specific?

Antibodies Have Two Identical Antigen-Binding Sites

35. What type of bonds hold the portions of the heavy and light chains together in the Fab?

36. Antigen binds between the *variable* domains of the heavy and light chains; why does this make sense for the specificity of the antigen/antibody interaction?

37. How are basophils and mast cells conscripted into the army of defensive cells of the immune response?

Antibodies Bind Tightly and Specifically to Antigen

38. Which amino acid residues are negatively charged? If these were found in an epitope of a particular antigen, which amino acid residues in the corresponding antibody would interact with them? Take a moment and review the categories of amino acid side chains.

39. How does induced fit increase the strength of the interaction between antibody and antigen?

The Antibody-Antigen Interaction Is the Basis for a Variety of Important Analytical Procedures

40. What type of chromatography is shown in Fig. 5–18c?

41. The techniques illustrated in Fig. 7–28 that use antibodies to detect proteins all require that the proteins be in native conformation. What reagents or conditions can therefore *not* be used in these procedures?

Protein Interactions Modulated by Chemical Energy: Actin, Myosin, and Molecular Motors

42. What are the important protein-protein interactions among motor proteins? (This is the major theme of this chapter.)

The Major Proteins of Muscle Are Myosin and Actin

43. Do thick filaments show directionality? What makes up the different "ends" of the thick filament?

44. What is the role of ATP in actin filament assembly?

Additional Proteins Organize the Thin and Thick Filaments into Ordered Structures

45. The Z-disk anchors which type of filament? The M-disk anchors which type?

46. The size hierarchy of the architectural levels of the muscle is a useful thing to commit to memory, because it helps in understanding the interactions between all the proteins involved in contraction. Order the components, starting with smallest (actin and myosin) and continuing up to the largest level—a muscle itself.

47. Calculate the (approximate) molecular weight of titin.

Myosin Thick Filaments Slide along Actin Thin Filaments

48. In terms of the protein-protein interactions that make muscle contraction possible, why is it critical to prevent overextension of the sarcomere?

49. What types of interactions are involved in the binding of the myosin head to actin? Are there any *covalent bonds* made or broken in the cycle of reactions that moves the thick filament relative to the thin filament?

50. Does Ca^+ act to *directly* or *indirectly* expose the attachment site for the myosin head groups?

SELF-TEST

Do You Know the Terms?

ACROSS

4. ________ occurs when the binding of one ligand increases or decreases the binding of additional ligands.
7. The ________ immune system protects against bacterial infections.
9. This protein can exist in a globular or filamentous form; hydrolysis of ATP is necessary to convert one to the other.
10. Composed of many sarcomeres, many of these make up a muscle fiber.
13. The covalent binding of CO_2 to the amino termini of hemoglobin subunits favors the ________ form.
14. This protein has a hyperbolic O_2 binding curve and no quaternary structure; it serves as an O_2 "reservoir" in muscle cells.
15. The metabolic intermediate 2,3- ________ binds to hemoglobin molecules with a stoichiometry of 1:1 and promotes the release of O_2.
16. Red blood cell.
20. Programmed cell death.
22. Immunoglobulin.
23. Types of white blood cells; T- and B-cells.
24. A helper T cell can signal nearby lymphocytes by secretion of a signal protein called a(n) ________.

DOWN

1. The production of lactic acid in muscle tissue contributes to the ________ effect, which explains the link between lactate production and an increased release of O_2 from hemoglobin.
2. The iron in this prosthetic group can bind either CO or O_2 at its sixth coordination position.
3. A molecule reversibly bound by a protein.
5. Antibodies are produced by the immune system as part of a defense against invasion by a foreign particle known as a(n) ________.
6. Red blood cells transport CO_2 produced by respiring tissues in two forms: as bicarbonate ions and as ________.
7. Oxygen transport protein that binds O_2 with a stoichiometry of 4 O_2:1 molecule transport protein.
8. Individual molecules of ________ aggregate to form thick filaments.
11. All vertebrates have an immune system capable of distinguishing ________ from invader.
12. Cleavage of an IgG with the protease papain separates the basal fragment from the "branches," called ________.
17. Small molecules covalently attached to large proteins in the laboratory in order to elicit an immune response.
18. This collection of about 20 proteins produces holes in the cell walls of bacteria.
19. A particular molecular structure within an antigen that binds an individual antibody.
21. Heme groups are covalently bound to globin through the ________ histidine residue.

Do You Know the Facts?

1. Indicate whether each of the following statements about the heme-binding site in myoglobin is true or false. If you think a statement is false, explain why.

(a) The proximal histidine covalently binds iron.

(b) The distal histidine covalently binds oxygen.

(c) The distal histidine binds iron.

(d) Free heme binds CO with the Fe, C, and O atoms in a linear array.

(e) The iron in heme binds the oxygen atom of CO.

(f) A molecule of carbon monoxide can bind to myoglobin at the same time as a molecule of oxygen.

2. Which of the following statements is/are true concerning the structures of myoglobin and hemoglobin?

(a) The tertiary structure of myoglobin is similar to that of a subunit of hemoglobin.
(b) The quaternary structure of myoglobin is similar to that of a subunit of hemoglobin.
(c) Myoglobin could substitute for one of the subunits in hemoglobin in erythrocytes.
(d) Myoglobin contains one binding site for oxygen per molecule.
(e) Myoglobin contains one binding site for oxygen per heme.
(f) Hemoglobin contains one binding site for oxygen per molecule.
(g) Hemoglobin contains one binding site for oxygen per heme.

3. Which of the following statements is/are true of *both* hemoglobin and myoglobin?

(a) Acidic conditions lower the affinity for oxygen.
(b) The iron atom of the heme prosthetic group is bound at five of its six coordination sites to nitrogen atoms.
(c) The Hill coefficient is equal to the number of subunits in the molecule.
(d) O_2 binding occurs in the cleft where the polypeptides come into contact.
(e) None of the above is true of both molecules.

4. Which of the following explain(s) how the reaction of carbon dioxide with water helps contribute to the Bohr effect? Refer to the figure below.

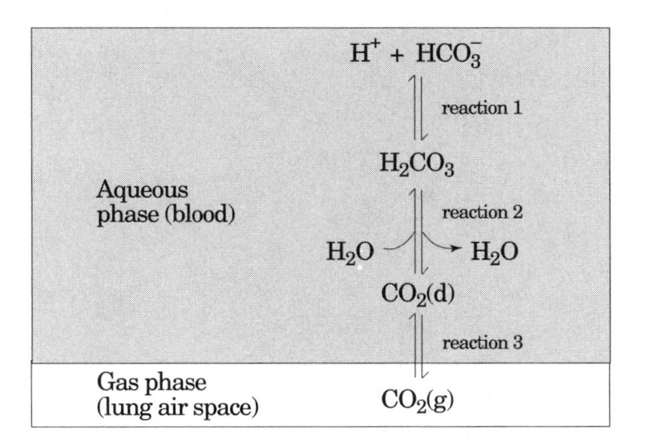

(a) The H^+ generated in this reaction decreases the pH of the blood, stabilizing the deoxy form of hemoglobin.
(b) The H_2CO_3 generated in this reaction stabilizes the deoxy form of hemoglobin.
(c) The H^+ generated by this reaction results in a lower affinity of hemoglobin for O_2, causing more O_2 to be released to the surrounding tissue.
(d) The H^+ forms a carbamate group with the amino termini of the four Hb chains.
(e) The increased concentration of HCO_3^- decreases the pH of the blood, stabilizing the deoxy form of hemoglobin.
(f) The CO_2 (dissolved) can form carbaminohemoglobin, which promotes the release of O_2.
(g) The H^+ generated interferes with the ionic interactions between the subunits, increasing the affinity for O_2.

5. High-altitude adaptation is a complex physiological process that involves an increase in both the number of hemoglobin molecules per erythrocyte and the total number of erythrocytes. It normally requires several weeks to complete. However, even after one day at high altitude, there is a significant degree of adaptation. This effect results from a rapid increase in the erythrocyte BPG concentration. (a) If oxygen-binding curves for both high-altitude-adapted hemoglobin and normal, unadapted hemoglobin were plotted together, would the curve for high-altitude-adapted hemoglobin be to the left of, to the right of, or the same as the curve for unadapted hemoglobin? (b) Is the O_2-binding affinity of high-altitude-adapted hemoglobin higher, lower, or the same as that of unadapted hemoglobin?

6. What is the relationship among θ (theta), K_d, and K_a?

7. What types of interactions are responsible for the close association of the nonidentical subunits (α and β) in the quaternary structure of hemoglobin?

8. Where does 2,3-bisphosphoglycerate interact with hemoglobin? What bonds or interactions are responsible for this interaction?

9. Which of the following describe the humoral immune system? The cellular immune system?

(a) It destroys host cells infected with viruses.
(b) It destroys some parasites.
(c) It responds to bacterial infections.
(d) It responds to extracellular viruses.
(e) The recognition proteins are immunoglobulins.
(f) The recognition proteins are cell surface receptors.
(g) The major cell type is B-cell.
(h) The major cell types are T_C and T_H.

10. (a) Differentiate between antigen, epitope, and hapten. (b) Suppose a molecule of hemoglobin were bound by several different antibodies: one at each BPG binding site, one at each amino terminus of the α subunits, and one at each carboxyl terminus of the β subunits. How many total antibody molecules would be bound? How many antigens, epitopes, and haptens are involved in this scenario?

11. What is the role of T_H cells in the protection of the organism under viral attack? Why is the depletion of these cells (as in HIV infection) so devastating to the immune system?

12. Generate a hierarchical list for the components in muscles. Start with the individual actin and myosin molecules, and work up in complexity to full muscle tissue.

13. Place the following states or processes that occur during muscle contraction in their proper order. Use any step as a starting point.

(a) "Power stroke"; the thick filament moves with respect to the thin filament.
(b) ATP is hydrolyzed; the heads of myosin are displaced along the thin filament.
(c) ATP binds to the myosin head; reduces the affinity of the myosin head for actin.
(d) The myosin head rebinds to actin closer to the Z-disk.

14. Consider the following events associated with the regulation of the contraction of skeletal muscle. Place the following events in the order in which they occur.

(a) A nerve impulse leads to the release of Ca^{2+} from the sarcoplasmic reticulum.
(b) Tropomyosin changes position on the thin filament.
(c) The formation of the complex between troponin and Ca^{2+} causes the tropomyosin-troponin complex to undergo a conformational change.
(d) The heads of myosin bind to actin.
(e) Ca^{2+} binds to troponin.

Applying What You Know

1. Though carbon monoxide is hindered in its binding to hemoglobin by the distal histidine, hemoglobin's affinity for CO is nonetheless about 200 times that for hemoglobin compared to that for O_2. Like oxygen, carbon monoxide shows **positive cooperativity** in binding to the heme sites of hemoglobin. A molecule of carbon monoxide and a molecule of oxygen can bind to the hemoglobin tetramer at the same time, though at different heme-binding sites. Use this information to analyze the following clinical situation.

Two people are present in a hospital emergency room. John's blood hemoglobin concentration is one-half the normal value (only one-half the normal number hemoglobin molecules are present). David has been poisoned with carbon monoxide by car exhaust, such that the number of oxygen-binding sites available to bind oxygen is only one-half the normal value (because, on average, every hemoglobin molecule has two carbon monoxides bound to two of its four hemes). Assuming that John and David are otherwise similar, which patient is in more serious danger? Explain why.

2. Why is hemoglobin better suited for its job as the oxygen-carrying protein in the bloodstream than myoglobin? (Include a discussion of oxygen pressure in the lungs and peripheral tissues, and the O_2-binding curves of the two proteins.)

3. How do carbon dioxide levels both *indirectly* and *directly* affect the ability of hemoglobin to bind oxygen?

4. Antibodies against an injected protein can be generated in a laboratory animal. The antibodies can then be purified, and these antibodies themselves can be used as antigens to generate new antibodies in a different lab animal. For example, antibodies can be raised in a rabbit against the muscle protein α-actinin. A goat can then be injected with the rabbit immunoglobulin molecules to generate goat anti-rabbit immunoglobulins. The goat antibodies can then be tagged with a fluorescent molecule that will be visible under UV light. The rabbit anti-α-actinin antibodies will react with α-actinin, and the goat anti-rabbit antibodies will bind to the bound rabbit antibodies. The complex is then visualized under UV light. This technique is called indirect immunofluorescence. What is the advantage of using such a technique over *directly* tagging the rabbit antibodies themselves with a fluorescent tag in order to visualize the presence of α-actinin?

Biochemistry on the Internet

In Chapter 6 you were presented with the case of a sick child with an abnormal form of hemoglobin. The Hb variant, Hb-J Briska, is the result of a deletion of 8 amino acids in the alpha chain. The positions and identities of the amino acid residues in a normal alpha chain and in the Hb-J Briska variant are shown below:

Normal Alpha Chain

1 VLSPADKTNV	11 KAAWGKVGAH	21 AGEYGAEALE	31 RMFLSFPTTK
41 TYFPHFDLSH	51 GSAQVKGHGK	61 KVADALTNAV	71 AHVDDMPNAL
81 SALSDLHAHK	91 LRVDPVNFKL	101 LSHCLLVTLA	111 AHLPAEFTPA
121 VHASLDKFLA	131 SVSTVLTSKY	141 R	

J-Briska Alpha Chain

1 VLSPADKTNV	11 KAAWGKVGAH	21 AGEYGAEALE	31 RMFLSFPTTK
41 TYFPHFDLSH	51 GKKVADALTN	61 AVAHVDDMPN	71 ALSALSDLHA
81 HKLRVDPVNF	91 KLLSHCLLVT	101 LAAHLPAEFT	111 PAVHASLDKF
121 LASVSTVLTS	131 KYR		

(a) The residues that are deleted in the mutant alpha chain are underlined in the normal alpha chain sequence. Using this sequence information and the Protein Data Bank, see if you can determine where in the three-dimensional structure of the protein this deletion occurred. The PDB ID for human hemoglobin is 1A3N.

(b) Based on the information presented in Chapter 7 concerning oxygen binding in hemoglobin, predict how the altered amino acid sequence of the Hb-J Briska alpha chain mutant will affect the function of the resulting hemoglobin $\alpha_2\beta_2$ tetramer.

(c) Can you find a three-dimensional model that shows the conformation of hemoglobin with one of its ligands (oxygen or carbon monoxide) bound?

ANSWERS

Do You Know the Terms?

ACROSS

4. cooperativity
7. humoral
9. actin
10. myofibrils
13. deoxy
14. myoglobin
15. bisphosphoglycerate
16. erythrocyte
20. apoptosis
22. antibody
23. lymphocytes
24. interleukin

DOWN

1. Bohr
2. heme
3. ligand
5. antigen
6. carbaminohemoglobin
7. hemoglobin
8. myosin
11. self
12. Fab
17. hapten
18. complement
19. epitope
21. proximal

Do You Know the Facts?

1. **(a)** T; **(b)** F; the distal histidine does bind oxygen, but it is a hydrogen bond, not a covalent one. **(c)** F; His[93] (the proximal histidine) binds iron; **(d)** T; **(e)** F; it binds the carbon atom of CO. The details of the chemistry here are very important to remember. In carbon monoxide, carbon is *triple-bonded* to O. Carbon can make four bonds to other atoms; oxygen cannot. **(f)** F; there is only one binding site, which can be occupied by either O_2 or CO, but not both at the same time.
2. a, d, e, g
3. b
4. a, c, f
5. **(a)** right; **(b)** lower
6. K_a is the association constant of the equilibrium expression $P + L \rightleftharpoons PL$, where P is a protein L is a ligand, and PL represents protein bound to ligand.

$$K_a = \frac{[PL]}{[P][L]}$$

K_d, the dissociation constant, is the reciprocal of K_a.

$$K_d = \frac{[P][L]}{[PL]}$$

θ is the fraction of ligand-binding sites on the protein that are occupied by ligand.

$$\theta = \frac{[PL]}{[PL] + [P]}$$

The [L] at which half of the available ligand-binding sites are occupied ($\theta = 0.5$) corresponds to $1/K_a$.

7. Hydrophobic interactions, hydrogen bonds, and a few ion pairs. *All* of these are important, even though ionic interactions are given the most attention.
8. A single BPG molecule binds to hemoglobin in the central cavity formed by the four subunits. Negatively charged groups on BPG interact with positively charged amino acid residues lining the central cavity. BPG binding stabilizes the deoxy form of hemoglobin.
9. Humoral: (c), (d), (e), (g)
   Cellular: (a), (b), (f), (h)
10. **(a)** Antigen: any molecule or pathogen capable of eliciting an immune response. A large antigen may have several epitopes. Epitope: a particular molecular structure within the antigen that binds an individual antibody or T-cell receptor. Hapten: a small molecule (generally under M_r 5,000) that can be attached to larger molecules in order to elicit an immune response.
    (b) Five total antibodies bound, of three different binding specificities: two (identical) at the amino termini, two at the carboxy-termini, one at the single BPG site. Hb is one antigen; it has three epitopes; haptens are not involved.
11. Specific T_H cell receptors recognize and bind to the viral peptides complexed to Class II MHC proteins displayed on the surface of macrophages and B-lymphocytes. T_H cells have another receptor, CD4, that enhances the binding interaction. The two types of receptor binding activate the T_H cells. Activated T_H1 cells secrete interleukin-2, which stimulates the reproduction of nearby T_C and T_H cells with receptors for that particular interleukin. Activated T_H2 cells secrete interleukin-4, stimulating the reproduction of appropriate B-cells that are nearby.

    The T_H cells therefore act to signal the nearby cells, which will then actively defend the host. However, the T_H cells do not directly destroy the invader; they are helpers. Without their signals to cause proliferation of the other immune system cells and proteins, the entire immune system is incapacitated.
12. Individual myosin molecules aggregate into thick filaments. Individual actin (G-actin) molecules aggregate into F-actin which, with troponin and tropomyosin, make up the thin filaments. Nebulin is also part of the thin filament.

    Thick and thin filaments align and overlap in sarcomeres, which are the individual unit of contraction. Sarcomeres are bounded by Z-disks, which are made up of the proteins α-actinin, desmin, and vimentin. The Z-disks are the anchors for the two sets of thin filaments found in each sarcomere. The M-disk, which is seen in the center of each sarcomere, anchors the thick filaments with the help of paramyosin, C-protein, and M-protein. Titins connect the thick filaments to the Z-disk.

    Many sarcomeres in linear array make up the myofibril. Many (~1000) myofibrils in lateral alignment make up one muscle fiber. Long parallel bundles of muscle fibers (large, single, multinucleated cells) make up a muscle.
13. (c), (b), (d), (a)
14. (a), (e), (c), (b), (d)

Applying What You Know

1. Remember that both the number and affinity of the binding sites must be considered. Although John has half the normal number of hemoglobin molecules available, O_2 can bind to all four sites, some of which will be low-affinity sites, readily releasing oxygen in the peripheral tissues. David is in more serious danger. Although he and John have the same number of available binding sites for O_2, David's are all high-affinity sites because CO already occupies the low-affinity sites. David's bound O_2 will not be given up readily in the tissues.
2. An efficient O_2-carrying protein in the bloodstream must be able to bind oxygen efficiently in the lungs, where the pO_2 is about 13.3 kPa, and release oxygen in the peripheral tissues, where the pO_2 is about 4 kPa. Myoglobin, with its hyperbolic binding curve, is ill-suited to this function. Mb binds O_2 with high affinity, but will not release it in the tissues because of that same high affinity. Hemoglobin is a much better candidate because it undergoes a transition from a low-affinity state to a high-affinity state as more O_2 molecules are bound, as evidenced by its sigmoid O_2-binding curve. The transition from low- to high-affinity occurs cooperatively through the interaction of multiple subunits. The first O_2 that binds to (deoxy)hemoglobin binds weakly, because it must bind to a heme in the T state. Its binding leads to changes in conformation that are communicated to adjacent subunits, making it easier for additional molecules of O_2 to bind. A single-subunit protein, like myoglobin, cannot produce a sigmoid binding curve.
3. Carbon dioxide levels indirectly affect the ability of hemoglobin to bind oxygen: respiring tissues produce a large amount of CO_2 which diffuses into erythrocytes. There, CO_2 is converted to H_2CO_3 by the enzyme, carbonic anhydrase. H_2CO_3 dissociates into H^+ and HCO_3; the net result is a lowering of pH and a decrease in the affinity of hemoglobin for oxygen. Carbon dioxide levels directly affect the ability of hemoglobin to bind oxygen: CO_2 can bind to amino terminal residues to form carbaminohemoglobin; the binding of CO_2 also forms additional salt bridges, which stabilize the deoxy form of hemoglobin. Binding of CO_2 promotes the release of oxygen in the tissues where it is needed. In the lungs, CO_2 is released, increasing the affinity of hemoglobin for oxygen.
4. This approach increases the sensitivity of the method because several fluorescently-labeled goat antibodies can bind to each rabbit immunoglobulin; it is a large enough protein to have several epitopes.

Biochemistry on the Internet

(a) There are a number of ways you can accomplish this task. Two are outlined below.

Using your Web browser, go to the Protein Data Bank at http://www.rcsb.org/pdb. Type 1A3N in the search window to retrieve the data page for human hemoglobin. On the resulting page click on **Download/Display Structures** and save the file to your hard drive. Then use your Chime or RasMol viewers to display the structure and manipulate the image. Using the mouse menu in Chime, click on **Display → Ribbon** to show the hemoglobin in ribbon format. To display the heme ligands, click **Select → Hetero → Ligand** and then **Display → Ball and Stick.**

You are looking for residues in positions 50–58. If you click on portions of the 3-D structure you will see the name and position of the corresponding residue displayed in the window. The deleted amino acid residues in Hb-J Briska alpha chains include the "distal" histidine (His^{58} in the normal alpha chain). See if you can locate this residue by clicking on the protein's backbone in the area where you would expect to find this amino acid. Click on **Select → Residue → His** and then **Display → Spacefill** to see how the histidine residue projects into the oxygen binding crevice.

An alternative method makes use of a Sting analysis. In the original PDB data page, click on **Other Sources** and then click on **Sting.** The resulting image can be manipulated in the same ways as the one in the Chime viewer; however, in Sting the complete sequence for the protein is displayed in a window separate from the 3-D structure. You can click on any of the amino acid residues to see that residue highlighted in the 3-D image.

Using either one of these methods you should be able to identify the deleted portion as one belonging to the alpha helix that is adjacent to one face of the heme moiety. The deleted region includes a key histidine residue that is thought to stabilize the heme interactions with the globin protein.

(b) The absence of the distal histidine in the region of the heme binding site for oxygen might change the affinity of the hemoglobin molecule for both carbon monoxide and oxygen. The absence of the positively charged distal His eliminates the steric hindrance that normally precludes the linear binding of CO to heme. Removing the His residue also affects the ability of oxygen to access the heme pocket and therefore would affect oxygen binding. In fact, the oxygen affinity of this variant is greater than that of normal hemoglobin, and the cooperativity of oxygen binding is decreased. However, the authors note that no abnormal red blood cell parameters were found in the patients with this Hb variant and that the distal histidine may not be crucial to the function of Hb since histidine is not present in the Hb of some animal species such as the opossum.

(c) Go to the PDB SearchLite page and enter the appropriate keywords to locate the database entries for ligand-bound hemoglobin.

One example of a hemoglobin molecule bound to a ligand is the entry with the PDB ID 2HCO (hemoglobin bound to carbon monoxide). You can view the molecule in three dimensions as you have previously for the 1A3N entry. Display the model as a ribbon structure (click on **Display → Ribbon**) and then give each chain a different color (**Color → Chain**). To view the bound ligands click on **Select → Hetero → Ligand** and then **Display → Ball and Stick.** The heme and bound CO are then displayed. To identify the individual atoms within the selected heme-CO complex click on **Color → CPK.**

One example of a hemoglobin molecule bound to oxygen is 1HBI (hemoglobin dimer from clam with oxygen bound).

chapter

8 Enzymes

STEP-BY-STEP GUIDE

Major Concepts

Enzymes are biological catalysts.

Most enzymes are proteins. Many have multiple subunits and require additional components—cofactors and/or coenzymes—for activity. Enzymes are formally named according to the class of reaction they catalyze, but usually have a common or trivial name as well. Enzymes bind the substrate within an **active site,** forming an enzyme-substrate (ES) complex. The transition from reactant to product is limited by an energetic barrier. Enzymes increase the rate of a reaction by lowering this **activation energy.** The binding energy, derived from numerous weak interactions between enzyme and substrate, is the source of this rate increase. The binding energy also contributes to an enzyme's specificity since it allows an enzyme to discriminate between a substrate and a competing molecule. The **equilibrium** of a reaction is dependent on the difference in the free energy of the ground states of the reactants and the products. The equilibrium position is not influenced by the presence of an enzyme.

Understanding the kinetics of enzyme-catalyzed reactions can be very helpful for grasping the mechanisms of catalysis.

Through a series of simplifying assumptions and definitions, the initial reaction equation $\mathbf{E + S \rightleftharpoons ES \rightarrow E + P}$ can be manipulated to give the **Michaelis-Menten equation:**

$$V_0 = \frac{V_{max}\,[S]}{K_m + [S]}$$

This equation describes the relationship among the initial velocity (V_0) of an enzymatic reaction, the substrate concentration ([S]), and the maximal velocity (V_{max}), via the Michaelis is constant K_m. K_m and V_{max} are measurable and are characteristic for each enzyme and each substrate. The turnover number, k_{cat}, is another useful parameter. It describes the number of substrate molecules that can be converted to product when the enzyme is fully saturated and functioning at its maximal velocity. K_m, V_{max}, and k_{cat}/K_m are all useful parameters for comparing the activities of different enzymes. Enzyme activity can be altered by environmental pH and by the presence of several distinct types of reversible inhibitors: competitive, uncompetitive, and mixed.

Although complete elucidation of enzymatic reaction mechanisms is not currently possible, certain enzymes can be used to illustrate key concepts.

Chymotrypsin provides an example of transition-state stabilization, hexokinase illustrates induced fit, and enolase illustrates one type of metal ion catalysis.

Regulatory enzymes control the overall rates of sequences of reactions.

Regulatory enzymes are either activated or inhibited by two major mechanisms: reversible, noncovalent allosteric interactions (e.g., the binding of an end product to the regulatory enzyme in a multienzyme pathway), or reversible, covalent modifications (e.g., addition of phosphoryl groups). Allosteric enzymes do not conform to Michaelis-Menten kinetics. Irreversible modification of enzymes also occurs, as in zymogen activation.

What to Review

Answering the following questions and reviewing the relevant concepts, which you have already studied, should make this chapter more understandable.

- Many enzymes are regulated by group-transfer reactions and oxidation-reduction reactions (review Fig. 3–16 and 3–21).
- You must be familiar with the ways energy is utilized and transferred in biological systems if you are to understand how enzymes facilitate reactions (review p. 72).
- The local pH and degree of ionization influence the types of interactions in which enzymes and their substrate molecules can participate. For this reason, you should remind yourself what the pK_a of an amino acid indicates (Table 5–1).
- Chymotrypsin is used as a model for enzymatic function in this chapter. Look up the proteolytic specificity of chymotrypsin (Table 5–7).
- A few amino acids have particularly critical roles in regulating the conformation of an enzyme and the types of noncovalent associations it can form. Review the structures and biochemical behaviors of three of these key amino acids: histidine, aspartate, and serine (Fig. 5–5).
- Antibodies are now being engineered to function as enzymes. Review the structure and functions of antibodies (Chapter 7).
- Hemoglobin exhibits cooperativity in the binding of oxygen by each of its subunits (pp. 214–215). How is this cooperativity similar to allosteric effects observed for regulatory enzymes?

Topics for Discussion

Answering each of the following questions, especially in the context of a study group discussion, should help you understand the important points of this chapter.

An Introduction to Enzymes

Most Enzymes Are Proteins

1. What components besides amino acid residues are sometimes necessary for an enzyme to be active?

2. What are the functions of coenzymes? Note that many are derivatives of vitamins.

Enzymes Are Classified by the Reactions They Catalyze

3. By international agreement, every enzyme is classified according to the type of reaction it catalyzes. Refer back to Figs. 3–16 and 3–20 in the textbook for examples.

How Enzymes Work

4. Why are catalysts necessary for reactions in living systems to proceed at a useful rate?

Enzymes Affect Reaction Rates, Not Equilibria

5. What is the difference between a transition state and a reaction intermediate?

6. What affects the reaction equilibrium between S and P? What affects the reaction rate of the conversion from S to P? Which aspect of a reaction can an enzyme alter?

Reaction Rates and Equilibria Have Precise Thermodynamic Definitions

7. What does a large, positive K'_{eq} (for example, $K'_{eq} = 1{,}000$) for a reaction mean in terms of the final relative concentrations of product and reactants? What does a very small K'_{eq} (for example, $K'_{eq} = 0.001$) mean?

8. What does K'_{eq} mean in terms of the standard free-energy change of the reaction? How does it relate to the speed at which equilibrium is reached?

9. In qualitative terms, what is the relationship between the rate constant K and the activation energy for an enzymatic reaction?

A Few Principles Explain the Catalytic Power and Specificity of Enzymes

10. What is the specific source of energy for lowering the activation energy barriers in enzyme-catalyzed reactions?

Weak Interactions between Enzyme and Substrate Are Optimized in the Transition State

11. Why is it important for an enzyme to be complementary to the reaction transition state rather than to the substrate?

12. What is one reason that some enzymes are very *large* molecules?

Binding Energy Contributes to Reaction Specificity and Catalysis

13. How does binding energy contribute to the high degree of specificity shown by enzymes?

14. Describe four physical and thermodynamic barriers to reaction, and explain how enzymatic catalysis overcomes them.

15. Explain the difference between the "lock and key" hypothesis and the "induced fit" mechanism.

Specific Catalytic Groups Contribute to Catalysis

16. When is general (as opposed to specific) acid-base catalysis observed?

17. Why must the covalent bond formed between enzyme and substrate in covalent catalysis be transient?

18. In what ways can metal ions participate in catalysis?

Enzyme Kinetics as an Approach to Understanding Mechanism

Substrate Concentration Affects the Rate of Enzyme-Catalyzed Reactions

19. What assumption concerning substrate concentration is made in the discussion of enzyme kinetics? Why is this important?

20. Explain the effect of saturating levels of substrate on enzyme-catalyzed reactions.

The Relationship between Substrate Concentration and Reaction Rate Can Be Expressed Quantitatively

21. In common terminology, what do the terms [S], V_0, V_{max}, E, ES, E_t, k_1, k_{-1}, and k_2 mean?

22. Why is there no k_{-2} in the equation describing the reaction from E+S to E+P?

23. What is the steady-state assumption?

24. What does the Michaelis-Menten equation describe?

25. When enzyme concentration is held constant, what is the relationship between V_0 and [S] at low [S]? At high [S]? How does the Michaelis-Menten equation illustrate these relationships mathematically?

26. What are the two definitions of K_m? What are its units?

Transformations of the Michaelis-Menten Equation: The Double-Reciprocal Plot

27. What makes transformation of the Michaelis-Menten equation to the Lineweaver-Burk equation useful? (Understanding the different ways to plot experimental data on enzyme activity is critical to grasping the material in this chapter. Working through the short-answer questions in the Self-Test section will help you to answer this question and to understand V_0 vs. [S] and double-reciprocal plots.)

Kinetic Parameters Are Used to Compare Enzyme Activities

28. Under what conditions does K_m (in this case called K_d) represent a measure of affinity of the enzyme for the substrate?

29. What does a high k_{cat} value mean? What does a high K_m value mean?

30. Why is k_{cat}/K_m, the specificity constant, the most useful parameter for discussing catalytic efficiency?

Many Enzymes Catalyze Reactions Involving Two or More Substrates

31. Which of the general reaction types discussed in Chapter 3 would you expect to follow Michaelis-Menten kinetics? In other words, which are simple enzyme reactions involving a single substrate molecule? Which of these reactions might use a double-displacement mechanism? Which of these reactions might involve formation of a ternary complex?

Pre-Steady State Kinetics Can Provide Evidence for Specific Reaction Steps

32. What kinds of information can be obtained by the study of pre-steady state kinetics that cannot be obtained by steady state kinetics alone?

Enzymes Are Subject to Inhibition

33. Be sure you can explain how the different types of enzyme inhibition affect the apparent K_m and V_{max}.

34. Would you expect to find irreversible inhibitors as normal enzymatic control mechanisms in biological systems?

Kinetic Tests for Determining Inhibition Mechanisms

35. How would it be possible to distinguish experimentally between a competitive enzyme inhibitor and a mixed inhibitor?

Enzyme Activity Is Affected by pH

36. Explain how alterations in the surrounding pH can affect enzyme activity.

Examples of Enzymatic Reactions

37. What types of experiments would be necessary to provide the information on enzyme mechanisms listed in statements 1–5 of the text (p. 272)?

Evidence for Enzyme–Transition State Complementarity

38. What techniques are used to study enzyme–transition state complementarity?

Reaction Mechanism Illustrates Principles

39. How do the serine, histidine, and aspartate residues in chymotrypsin act together to stabilize the transition state during catalysis?

40. How does Fig. 8–20 show that the deacylation phase is rate-limiting? (Note that the x-axis of this graph is *not* [S].)

41. What is the basis for the specificity of hexokinase for its substrate?

42. Which residues are important in the mechanism of enolase?

Regulatory Enzymes

43. Why is a regulatory enzyme very often the first enzyme in a multiple reaction sequence?

44. What is the most significant difference in how the two major classes of regulatory enzymes are controlled?

Allosteric Enzymes Undergo Conformational Changes in Response to Modulator Binding

45. How are allosteric enzymes different from other enzymes?

The Regulatory Step in Many Pathways Is Catalyzed by an Allosteric Enzyme

46. What are the general characteristics of feedback inhibition, as illustrated by the regulation of the conversion of L-threonine to L-isoleucine?

The Kinetic Properties of Allosteric Enzymes Diverge from Michaelis-Menten Behavior

47. Why is the term $K_{0.5}$ (or $[S]_{0.5}$), rather than K_m, used to describe the substrate concentration that produces a half-maximal velocity in an allosteric enzyme-catalyzed reaction?

Some Regulatory Enzymes Undergo Reversible Covalent Modification

48. What functional groups other than $—PO_4^{2-}$ can be added to or removed from enzymes to turn them on or off?

Phosphoryl Groups Affect the Structure and Catalytic Activity of Proteins

49. In what ways can phosphoryl groups interact with other groups on an enzyme? How can this affect catalysis?

50. Is glycogen phosphorylase made *more* or *less* active by the addition of phosphoryl groups? (Note: This may be different in other enzymes that are modified in this manner.)

Multiple Phosphorylations Allow Exquisite Regulatory Control

51. Table 8–9 lists myosin light chain kinase as one that recognizes a consensus sequence. What is the function of myosin light chain kinase? (Refer back to Chapter 7.)

52. Note that phosphorylation of glycogen synthase *inactivates* the enzyme; this is the reverse of the effect that phosphorylation has on glycogen phosphorylase. (There is a very good reason for this: see Chapter 20, Carbohydrate Biosynthesis.)

53. Why is it important in terms of regulatory control that the dephosphorylation reaction is not simply the reverse of the phosphorylation reaction?

Some Types of Regulation Require Proteolytic Cleavage of an Enzyme Precursor

54. Compare enzymes regulated by allosteric modulation, by reversible covalent modification, and by proteolytic cleavage with respect to how they are activated and inactivated.

Some Regulatory Enzymes Use Multiple Regulatory Mechanisms

55. Why is control of catalysis just as important as the fact of catalysis itself?

SELF-TEST

Enzymes and the study of enzyme reaction rates (enzyme kinetics) are among the most difficult areas of biochemistry for students to assimilate. Consequently, more problems have been included in this chapter's Self-Test. If you work through these problems carefully, your understanding of this material will be greatly enhanced. However, these problems will only be beneficial if you work through them to completion without looking at the answers.

Do You Know the Terms?

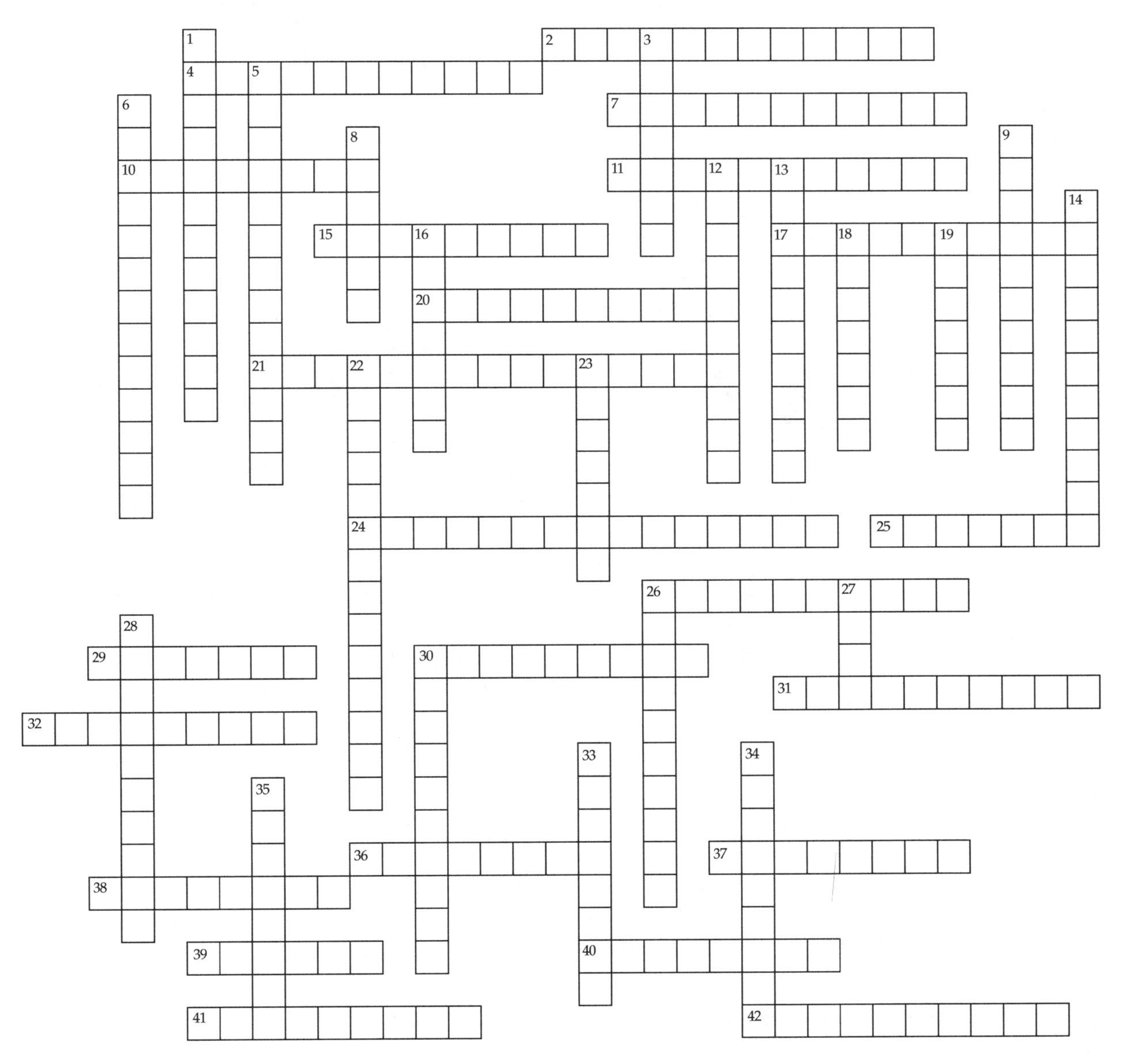

ACROSS

2. The slowest reaction in a sequence is the ________-________ step.
4. State of a system in which no further net change is occurring.
7. The assumption that the rate of formation of ES is exactly equal to the rate of breakdown of ES is called the ________ ________ assumption. (2 words)
10. k_{cat} is known as the ________ number. At saturating substrate concentrations $k_{cat} = V_{max}/[E_t]$.
11. Type of inhibitor that alters the K_m of an enzyme without altering V_{max}.
15. Molecule that binds to the active site of an enzyme.
17. Relatively small portion of an enzyme that is involved in substrate binding. (2 words)

20. Describes changes in the conformation of an enzyme upon substrate binding. (2 words)
21. A ________-________ analog binds more tightly to the active site than does the substrate molecule.
24. ________-________ kinetics describes the enzymatic activity of an idealized enzyme.
25. Trypsin is to trypsinogen as active enzyme is to ________.
26. Complete enzyme complex including all the protein subunits and prosthetic groups.
29. Type of energy derived from enzyme-substrate interactions that lowers the activation energy for a reaction.
30. An enzyme without its prosthetic group.
31. Hemoglobin without the heme, for example.
32. Common structural motifs recognized by specific protein kinases are known as ________ sequences.
36. Organic cofactor required for certain enzymes to be active.
37. An enzyme that acquires a phosphate from ATP, releases the ATP, binds a second substrate and transfers the phosphate group to the second substrate uses a ________-________ mechanism.
38. Type of inhibition in which increasing concentrations of a reaction pathway's product decrease the activity of an allosteric enzyme early in that pathway.
39. A specific enzyme that contains Ni^{2+} as a cofactor; it enhances the rate of its reaction by 10^{14}; Sumner's first enzyme crystallized.
40. A molecule essential to the functioning of an enzyme, but not part of the enzyme protein itself (e.g., a divalent cation).
41. Modulator that increases the overall rate of an enzyme-catalyzed reaction.
42. Enzymes that catalyze the rate-limiting step in a reaction sequence usually are ________ enzymes.

DOWN

1. An enzyme whose activity is regulated by a modulator other than its substrate.
3. Binding of a substrate molecule to an enzyme promotes catalysis by reducing the relative motions of the participants, and so reducing ________.
5. Inhibitor that binds only to the ES complex, and therefore cannot bind to the substrate-binding site.
6. EP and ES are examples of reaction ________.
8. Energetic state of a substrate or product molecule in its most energetically stable form.
9. Enzymes catalyze biological reactions by lowering the ________ energy needed for the reaction to proceed.
12. Some enzymes have a covalently linked ________ group essential to its activity.
13. The top of the energy hill is the ________ state.
14. ________ of a substrate occurs when hydrogen bonds between a substrate molecule and water are replaced by noncovalent interactions between the substrate molecule and an enzyme.
16. Inhibitors that irreversibly bind to an enzyme are known as ________ inactivators.
18. ES_1S_2 is a ________ complex.
19. "Ferments" according to Louis Pasteur; the names of many end in "ase."
22. A specific, rare type of mixed inhibitor that alters V_{max} without affecting K_m
23. Some allosteric enzymes show ________ kinetic behavior, which reflects cooperativity.
26. Allosteric enzyme whose substrate is also a modulator of activity.
27. When the K'_{eq} = one, $\Delta G'^{0}$ = ________.
28. ________-________ reactions involve a single substrate molecule, and the rate depends only on [S].
30. Class of regulatory enzymes that change their conformation when bound to a modulator.
33. The study of reaction rates in biological systems is referred to as enzyme ________.
34. Agent that reduces the overall rate of an enzyme-catalyzed reaction.
35. Regulation of enzyme activity by the reversible binding of a phosphoryl group is an example of regulation by ________ modification.

Do You Know the Facts?

1. In some enzymes, components other than amino acid residues are necessary for activity. These components are called ________________.
In such enzymes, the complete, active form is called a(n) ________________ and the enzyme without its additional component is called a(n) ________________.

2. Are enzymes always proteins, consisting only of amino acids? Be sure to explain your answer.

3. Why are enzymes necessary for the catalysis of reactions in living systems?

4. Why must the covalent bond formed between enzyme and substrate in covalent catalysis be transient?

5. Why are regulatory enzymes very often the first enzyme in a multireaction sequence?

6. How are allosteric enzymes different from other enzymes?

7. What is the difference between reversible covalent modification and proteolytic cleavage?

8. Label the double-reciprocal plot below with the letters corresponding to items (a)–(g).

(a) A typical Michaelis-Menten enzyme in the absence of inhibitors

(b) Enzyme activity in the presence of a noncompetitive inhibitor

(c) Enzyme activity in the presence of a competitive inhibitor

(d) 1/[S]

(e) $1/V_0$

(f) $-1/K_m$

(g) $1/V_{max}$

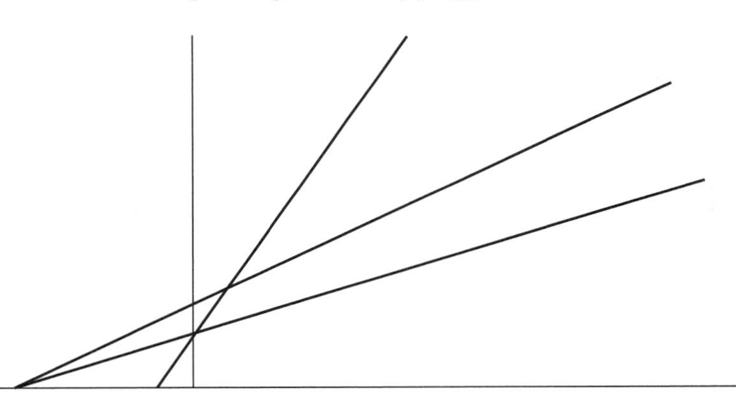

For the situations described in questions 9–12, indicate whether K_m, V_{max}, both parameters, or neither parameter would be altered.

9. In the presence of a mixed inhibitor:

10. After [S] has been doubled:

11. In the presence of a competitive inhibitor:

12. After the [E] has been doubled:

13. An enzyme facilitates chemical reactions by:

A. increasing the free-energy difference between reactants and products.
B. decreasing the free-energy difference between reactants and products.
C. lowering the activation energy of the reaction.
D. raising the activation energy of the reaction.
E. none of the above.

14. Which of the following is not a cellular mechanism for regulating enzyme activity?

A. binding of regulatory peptides via disulfide bonds
B. proteolysis
C. covalent modification
D. induced changes in conformation

15. Using site-directed mutagenesis, you have created a mutant form of chymotrypsin that has alanine substituted for the usual serine at position 195. Which of the following effects would you expect to observe?

A. no effect or a slight increase in affinity for substrate coupled to a complete loss of enzyme activity
B. a decrease in the affinity for substrate coupled to a decrease in enzyme activity
C. an increase in the rate of peptide-bond cleavage due to an increase in the rate of acid-base catalysis
D. an increase in the rate of peptide-bond cleavage due to an increase in the rate of covalent catalysis
E. a complete loss of enzyme activity due to the inability to bind substrate

In questions 16–19, calculate the indicated parameters using the information below.
You wish to characterize a new enzyme that you have isolated. You set up a series of test tubes, each containing 0.1 μM enzyme and various substrate concentrations. You measure the activity of the enzyme in each tube and plot the enzyme activity (V_0) vs. [S]. The plot is shown below.

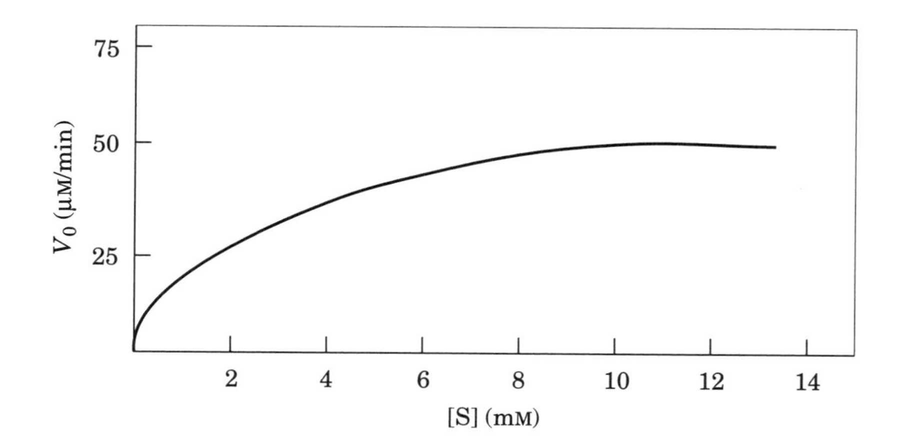

16. What is V_{max} for your enzyme with this substrate?

A. 10 μM/min
B. 30 μM/min
C. 0.8 μM/min
D. 0.05 mM/min
E. 50 mM/min

17. What is the K_m of your enzyme for this substrate?

A. 1.5 μM
B. 10 μM
C. 0.5 mM
D. 1.0 mM
E. 1.5 mM

18. What is the k_{cat} (turnover number) for your enzyme?

A. 5×10^2/s
B. 1.0/min
C. 5×10^2/min
D. 5×10^3/min
E. 1×10^2/s

19. Why does your plot eventually plateau?

A. the enzyme becomes inhibited at high substrate concentrations
B. the enzyme affinity for substrate changes
C. the enzyme is being degraded by proteases
D. the active site is saturated with substrate
E. the substrate is being degraded

20. For the following enzyme-catalyzed reaction, why is the term k_{-2} not included in the Michaelis-Menten equation?

$$E + S \underset{k_{-1}}{\overset{k_1}{\rightleftharpoons}} ES \overset{k_2}{\rightleftharpoons} E + P$$

A. This reaction never occurs.
B. It simplifies the math.
C. The Michaelis-Menten equation only describes initial reaction rates.
D. It is part of the term K_m.
E. The Michaelis-Menten equation only describes the formation of the ES complex.

21. Indicate whether each of the following is true or false regarding Michaelis-Menten kinetics.

(a) K_m has the dimensions of rate per second.
(b) $K_m = k_2 + k_{-1}/k_1$.
(c) K_m is equal to the substrate concentration that will bring the reaction to maximal velocity.
(d) At very high substrate concentration, the velocity of the reaction is independent of substrate concentration.
(e) The concentration of the enzyme-substrate complex stays constant throughout the reaction.
(f) Enzyme concentration is much lower than substrate concentration.
(g) The velocity of the reaction is equal to the (mathematical) product of (concentration of enzyme-substrate complex) and (the rate constant), or $V_0 = [ES] \times k_2$.

22. Indicate whether each of the following statements about enzymes is true or false.

(a) To be effective, they must be present at the same concentration as their substrate.
(b) They increase the equilibrium constant for the reaction, thus favoring product formation.
(c) They increase the rate at which substrate is converted to product.
(d) They ensure that all substrate is converted to product.
(e) They ensure that the product is more thermodynamically stable than the substrate.
(f) They lower the activation energy for conversion of substrate to product.
(g) They are consumed in the reactions that they catalyze.

Applying What You Know

1. In your studies of the enzyme ribonuclease A, you obtain activity data for a wild-type enzyme and a mutant ribonuclease A. The two enzymes differ at one amino acid position in the protein. From the activity data you calculate the following kinetic parameters:

	Maximum Velocity	K_m
Wild Type	100 μmol/min	10 mM
Mutant	1 μmol/min	0.1 mM

(a) Which enzyme has a higher affinity for substrate? (Assume a two-step reaction, with k_2 the rate-limiting step.)

(b) What is the initial velocity of the reaction catalyzed by the wild-type enzyme at a substrate concentration of 10 mM?

(c) Which enzyme shifts K_{eq}, the equilibrium constant, more in the direction of product?

2. The sensitivity of individuals of certain ethnic backgrounds to alcoholic beverages has a biochemical basis. In such individuals, much less ethanol is required to produce the vasodilation that results in facial flushing than is required to achieve the same effect in those without such sensitivity. This physiological effect arises from the acetaldehyde generated by liver **alcohol dehydrogenase (AD),** which catalyzes the following reaction:

$$\underset{\text{Ethanol}}{CH_3CH_2OH} + NAD^+ \underset{AD}{\rightleftharpoons} \underset{\text{Acetaldehyde}}{CH_3CHO} + H^+ + NADH$$

In sensitive individuals, it is the next step that confers sensitivity. Sensitive individuals have a different isozyme of the enzyme aldehyde dehydrogenase (ADD), which converts acetaldehyde to acetate.

$$\underset{\text{Acetaldehyde}}{CH_3CHO} + NAD^+ \underset{ADD}{\rightleftharpoons} \underset{\text{Acetate}}{CH_3COO^-} + NADH + H^+$$

This isozyme allows a higher concentration of acetaldehyde to accumulate in the blood after alcohol consumption.

(a) Which ADD isozyme has a higher K_m for acetaldehyde?

(b) You are investigating the effects of several agents on the activity of alcohol dehydrogenase. The enzyme activity data are shown in the table below. (Hint: If you put these data into a spreadsheet it will be much easier to work with. You can use the spreadsheet tools to do curve fitting and to perform a linear regression analysis.)

[Alcohol] (mM)	AD Activity (V_0, mM/min)	AD Activity + Agent A (V_0, mM/min)	AD Activity + Agent B (V_0, mM/min)
0.1	14	2	5
0.5	50	7	8
1.0	65	10	30
2.0	72	12	45
4.0	80	14	62
8.0	85	15	75
32.0	90	16	90

Construct a **[substrate] vs. activity plot** and a **double-reciprocal plot** for this enzyme. Be sure to label all axes.

(c) Determine the V_{max} and K_m for AD from the graphs in each type of plot.

(d) Using the same plots, graph the data for AD activity in the presence of agent A and agent B. What information does this provide about the effects of these compounds on the activity of alcohol dehydrogenase?

ANSWERS

Do You Know the Terms?

ACROSS

2. rate-limiting
4. equilibrium
7. steady state
10. turnover
11. competitive
15. substrate
17. active site
20. induced fit
21. transition-state
24. Michaelis-Menten
25. zymogen
26. holoenzyme
29. binding
30. apoenzyme
31. apoprotein
32. consensus
36. coenzyme
37. Ping-Pong
38. feedback
39. urease
40. cofactor
41. activator
42. regulatory

DOWN

1. heterotropic
3. entropy
5. uncompetitive
6. intermediates
8. ground
9. activation
12. prosthetic
13. transition
14. desolvation
16. suicide
18. ternary
19. enzymes
22. noncompetitive
23. sigmoid
26. homotropic
27. zero
28. first-order
30. allosteric
33. kinetics
34. inhibitor
35. covalent

Do You Know the Facts?

1. Cofactors, coenzymes, and, if tightly bound, prosthetic groups; holoenzyme; apoenzyme

2. Enzymes are usually proteins, but some nucleic acids (RNA) show enzymatic activity. Many enzymes, though not all, also have non-amino acid portions, called cofactors or coenzymes, that are necessary for catalytic activity.

3. At the relatively mild temperature and pH of cells, most biomolecules are quite stable. Many common reactions in biochemistry do not occur at an appreciable rate. Enzymes circumvent these problems by providing a specific environment (the active site) within which a given reaction will occur at a significantly increased rate.

4. The enzyme must be regenerated in free form (the covalent bond to the substrate/product must be undone) to be able to catalyze the reaction again. Enzymes are catalysts and must be recycled many times.

5. In cell metabolism, groups of enzymes work together in sequential pathways to carry out a given metabolic process. Catalyzing even the first few reactions of a pathway leading to an unneeded product is a waste of energy and materials. Therefore, the first enzyme in the pathway is often the ideal place to turn on or off the whole pathway.

6. Allosteric enzymes are generally larger and more complex than simple enzymes. Most have two or more polypeptide subunits. In addition to active sites (catalytic sites), allosteric enzymes have at least one regulatory (allosteric) site that binds a modulator(s). Allosteric enzymes do not follow Michaelis-Menten kinetics.

7. Reversible covalent modification is just that—reversible. Phosphorylation, methylation, etc., are reversible modifications used to regulate the activity of enzymes and other proteins. Proteolytic cleavage is also a covalent modification of an enzyme, but it is irreversible. This process is also referred to as zymogen or proprotein activation.

8.

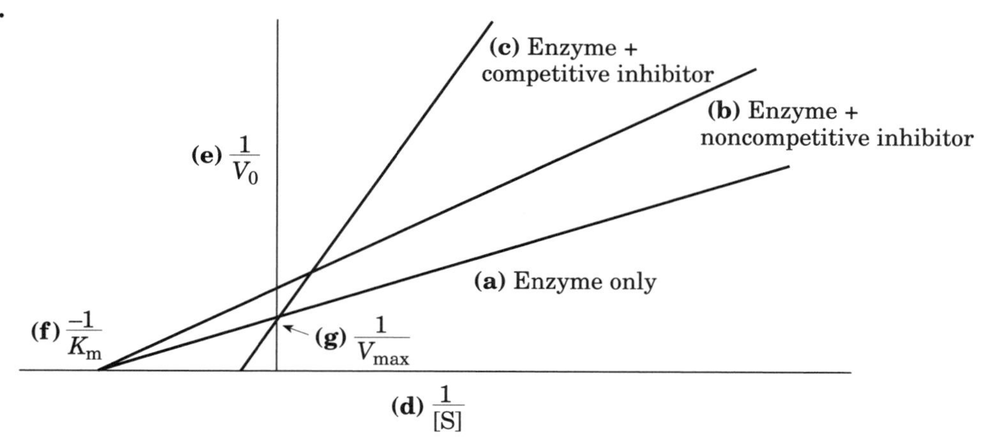

9. V_{max} and K_m both (usually). In the rare case of $\alpha = \alpha'$, which is called noncompetitive inhibition, K_m is not affected.

10. Neither

11. K_m

12. V_{max}

13. C

14. A

15. A

16. D

17. E

18. C $k_{cat} = V_{max}/E_t$ (the total enzyme present is 0.1 μM)
$= (50\ \mu M/\text{min})/0.1\ \mu M$
$= 500$ (molecules of substrate transformed per molecule of enzyme)/min

19. D

20. C

21. **(a)** F; **(b)** T; **(c)** F; **(d)** T; **(e)** T; **(f)** T; **(g)** T

22. **(a)** F; **(b)** F; **(c)** T; **(d)** F; **(e)** F; **(f)** T **(g)** F

Applying What You Know

1. (a) The mutant enzyme binds substrate more strongly than does the wild-type enzyme because the amount of substrate that results in a half-maximal velocity is much lower for the mutant enzyme (0.1 mM) than for the wild-type enzyme (10 mM).

(b) The velocity of the enzyme can be calculated for any substrate concentration if you know the values of K_m and V_{max}.

$$V_0 = \frac{V_{max}[S]}{K_m + [S]}$$

$$= 100\ \frac{\mu\text{mol/min} \times 10\ \text{mM}}{10\ \text{mM} + 10\ \text{mM}}$$

$$= 50\ \mu\text{mol/min.}$$

Be sure to use the same units when substituting values (i.e., either mM or μM).

(c) Neither enzyme alters the equilibrium position for this reaction; they only alter the rate at which that equilibrium position is reached.

2. (a) The K_m for the ADD isozyme in alcohol sensitive individuals is higher than that in nonsensitive individuals.

(b) See graph at right. From the V_0 vs. [S] plot, one can estimate the V_{max} for AD to be 90 mM/min. V_{max} in the presence of agent A can be estimated to be approximately 16 mM/min. V_{max} in the presence of agent B is unclear because it is not obvious that the curve has reached a plateau. K_m values can be obtained by estimated [S] at 1/2 V_{max}.

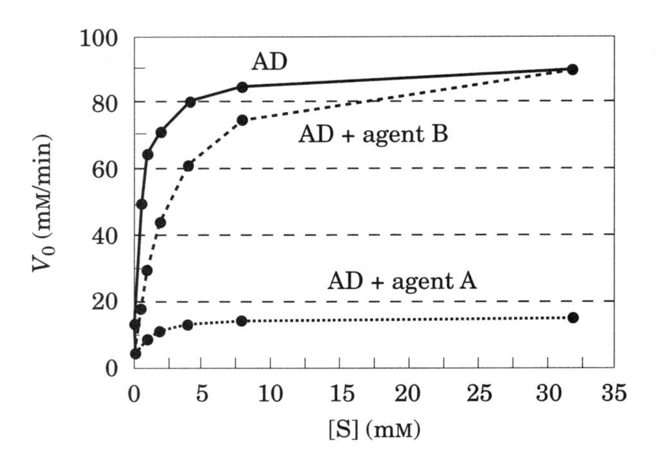

	AD Alone	AD + Agent A	AD + Agent B
V_{max} (1/*b*)	90 mM/min	16 mM/min	?
K_m ($m \times V_{max}$)	<1 mM	approx. 1 mM	?

The V_{max} can be calculated accurately for AD in all three samples. To do this, calculate the values for 1/[S] and $1/V_0$ for the enzyme and enzyme plus each inhibitor. Graph $1/V_0$ vs. 1/[S], and perform a linear regression on each set of data. This analysis will fit the data to a straight line using the formula ($y = mx + b$). The value for b will be the y-intercept, which is equal to $1/V_{max}$. The constant m is the slope of the line, which is equal to K_m/V_{max}. The value for R^2 tells you how well the data fit the formula; with a perfect fit, R^2 would be 1.0. Such analysis yields the following information:

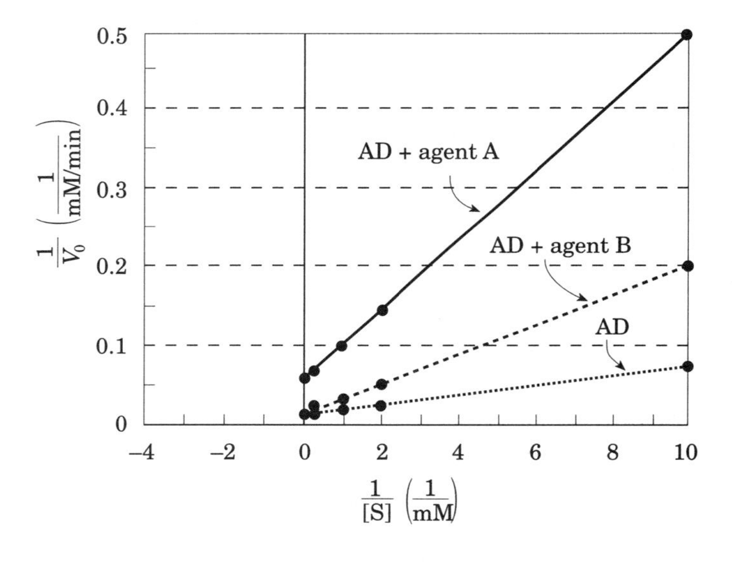

	AD Alone	AD + Agent A	AD + Agent B
m	0.006	0.044	0.019
b	0.010	0.060	0.013
V_{max} (1/*b*)	98 mM/min	18 mM/min	78 mM/min
K_m ($m \times V_{max}$)	0.59 mM	0.78 mM	1.5 mM
R^2	1.0	1.0	1.0

An examination of the values in the previous table provides the following insights as to the actions of agents A and B:

Agent A decreased the V_{max} of AD by 82%, while increasing K_m 34%. Agent A acts primarily as a mixed inhibitor of AD.

Agent B, on the other hand, does not affect the V_{max} as significantly as Agent A. (The maximal velocity decreased by 20%.) However, the K_m increased by more than 100% in the presence of agent B. This suggests that agent B may be a competitive inhibitor of AD or a mixed inhibitor. Further analysis is needed to determine which type of inhibitor it is.

chapter

9 Carbohydrates and Glycobiology

STEP-BY-STEP GUIDE

Major Concepts

Monosaccharides are the monomeric subunits from which di-, oligo-, and polysaccharides are formed.

They are classified according to the *position* of the carbonyl group (aldose or ketose), the total *number* of carbon atoms in the chain (triose to heptose), the *isomeric form* (L or D, **epimers, anomers**), whether or not they occur in *cyclic* form, and whether or not they have another **chemical substituent.**

Polysaccharides vary in monomeric composition, type of glycosidic bond connecting the monosaccharide units, chain length and degree of branching, and biological function.

Starch and **glycogen** are **homopolysaccharides** of glucose that function as fuel stores in plants and animals, respectively. Both starch and glycogen are chains of glucose monomers connected by linear linkages, (α**1→4**) glycosidic bonds, but they have different degrees of branching through (α**1→6**) glycosidic bonds. Branching allows faster degradation of a polysaccharide to individual monosaccharide units, which can then be used to generate biologically useful energy. **Cellulose** is also a homopolymer of glucose, but its (β**1→4**) linkages result in rigid extended polymers that are used by cells for support rather than as energy storage molecules. More complex **heteropolysaccharides,** composed of more than one type of monosaccharide, can be covalently cross-linked by peptides or proteins to form **peptidoglycans** or **proteoglycans.** These carbohydrate–protein aggregates function to provide structural support (bacterial cell wall), connectivity (extracellular matrix), and lubrication (synovial fluid).

Glycoconjugates include proteoglycans, glycoproteins, and glycolipids. These hybrid molecules have carbohydrate portions that alter or augment the functional properties of their protein and lipid components.

Proteoglycans are important components of the extracellular matrix of multicellular animals. They are composed primarily (by mass) of **glycosaminoglycan** chains and are important for structure and resilience. In **glycoproteins,** which are primarily proteins, oligosaccharides are linked to amino acid residues of polypeptides via ***O***- or ***N***-glycosidic bonds; these information-rich portions commonly act as recognition sites for enzymes or receptors, or for targeting a protein to a specific cellular location. **Lectins,** oligosaccharide-binding proteins, are central in these recognition processes. Membrane **glycolipids,** such as the gangliosides and bacterial lipopolysaccharides, are lipids that are covalently bound to oligosaccharide moieties.

Chemical analysis of carbohydrates makes use of some of the same techniques used for protein analysis.

Full determination of the structure of complex heteropolysaccharides involves many steps not only because there are a variety of monosaccharide subunits, but also because there are a variety of linkages possible. Mass spectrometry and NMR spectroscopy are two important procedures used to characterize carbohydrates.

What to Review

Answering the following questions and reviewing the relevant concepts, which you have already studied, should make this chapter more understandable.

- Review the structure of aldehydes, ketones, and hydroxyl groups (Fig. 3–5); all of these groups are found in carbohydrates.

- Simple sugars exist in a variety of isomeric forms. Review what is meant by enantiomers (Fig. 3–9) and the various conformations of biological molecules (pp. 58–63).
- Many sugars can be oxidized and are termed reducing sugars. Be sure you understand the overall concept of oxidation–reduction reactions (Fig. 3–16).
- Some polysaccharides form long fibers and are functionally similar to fibrous proteins. Review the structural properties of these proteins and review how structure relates to their function (Table 6–1, pp. 170–175).
- Review the various analytical techniques used in the study of amino acids and proteins (Figs. 2–20, 5–17 through 5–22). Which ones can be applied to the study of carbohydrates?

Topics for Discussion

Answering each of the following questions, especially in the context of a study group discussion, should help you understand the important points of this chapter.

Monosaccharides and Disaccharides

1. Why are monosaccharides appropriately named *"carbo hydrates"*?

The Two Families of Monosaccharides Are Aldoses and Ketoses

2. Why are monosaccharides soluble in water but not in nonpolar solvents?

3. What are the naming conventions for monosaccharides?

Monosaccharides Have Asymmetric Centers

4. The word "chiral" is derived from the Greek word meaning "hand." How does this relate to the structure of carbohydrates?

5. How many D isomers would an aldopentose have? How many total isomers?

6. What is a possible explanation for the observation that most of the hexoses found in living organisms are D isomers? (Hint: see Chapter 8.)

7. Is galactose an epimer of mannose?

The Common Monosaccharides Have Cyclic Structures

8. In an aqueous solution of D-glucose, why will there always be a very small amount of the linear form of the monosaccharide?

9. Does the discussion on conformations of cyclic forms of monosaccharides suggest an explanation for why the more common anomer of D-fructofuranose is the β anomer? (Hint: see Fig. 9–8.)

10. How many D isomers would a cyclized aldopentose have? How many total isomers?

Organisms Contain a Variety of Hexose Derivatives

11. What are some of the biologically more important monosaccharide derivatives?

12. Why is phosphorylation of sugars beneficial to a cell?

Monosaccharides Are Reducing Agents

13. What is reduced and what is oxidized in the reaction between a monosaccharide and a ferric ion?

Disaccharides Contain a Glycosidic Bond

14. Be able to draw and give the complete name of the disaccharides commonly known as **maltose, lactose,** and **sucrose.** Which of these is/are reducing sugars?

Polysaccharides

15. What are the structural and functional differences between homopolysaccharides and heteropolysaccharides?

16. Why are *homo*polysaccharides not useful as informational molecules?

Starch and Glycogen Are Stored Fuels

17. Suppose you had three polysaccharides, amylose, amylopectin, and glycogen, each with the same number of monosaccharide subunits. Which would be degraded the fastest, assuming that the enzymes acting to degrade the polysaccharides all worked at the same rate?

Cellulose and Chitin Are Structural Homopolysaccharides

18. How can such a simple difference—the difference between α- and β-glycosidic bonds—produce differences in function and three-dimensional structure as drastic as those seen between amylose and cellulose?

19. What are the most important noncovalent bonds or interactions in cellulose?

Bacterial Cell Walls Contain Peptidoglycans

20. How does lysozyme act as a first defense against bacterial infection? How does penicillin combat bacterial infections?

Glycosaminoglycans Are Components of the Extracellular Matrix

21. What chemical property of glycosaminoglycans contributes to their role as lubricants?

22. How does heparin work to inhibit blood coagulation?

Glycoconjugates: Proteoglycans, Glycoproteins, and Glycolipids

23. The distinction between proteoglycans and glycoproteins is entirely relative. The second half of the name indicates the predominant species. Proteo*glycans* are primarily glycans (or polysaccharides), whereas glyco*proteins* contain more protein (by weight).

Proteoglycans Are Glycosaminoglycan-Containing Macromolecules of the Cell Surface and Extracellular Matrix

24. How does the R-group of Ser make it a good amino acid for connecting to a carbohydrate moiety?

25. If a proteoglycan aggregate is compared to an evergreen tree, with hyaluronate as the "trunk," what compound would be the "needles" on this tree? What role do the core proteins play?

26. What is the relationship among glycosaminoglycans, proteoglycans, fibrous proteins, integrins, adhesion proteins, and the extracellular matrix? It may be helpful to diagram the cellular locations of each of these components.

Glycoproteins Are Information-Rich Conjugates Containing Oligosaccharides

27. How are the functions of glycoproteins different from those of glycosaminoglycans?

28. What types of linkages connect oligosaccharides to proteins?

29. What kinds of biologically important information are encoded by the oligosaccharide portions of glycoproteins?

30. How does the addition of carbohydrate moieties alter the chemistry of glycoproteins and glycolipids?

31. How can so much distinguishing information be packed into an oligosaccharide of comparatively few monosaccharide units?

Glycolipids and Lipopolysaccharides Are Membrane Components

32. How do glycolipids and lipopolysaccharides differ?

Oligosaccharide-Lectin Interactions Mediate Many Biological Processes

33. Why is it important that there are signals for removal and destruction of "old" cells and hormones?

34. Why do people of blood type O tend to have gastric ulcers more often than do people of blood type A or B?

Analysis of Carbohydrates

35. What makes analysis of carbohydrates so much more difficult than analysis of proteins?

36. How can the specificity of lectins and glycosidases be used to identify and analyze carbohydrates?

SELF-TEST

Do You Know the Terms?

ACROSS

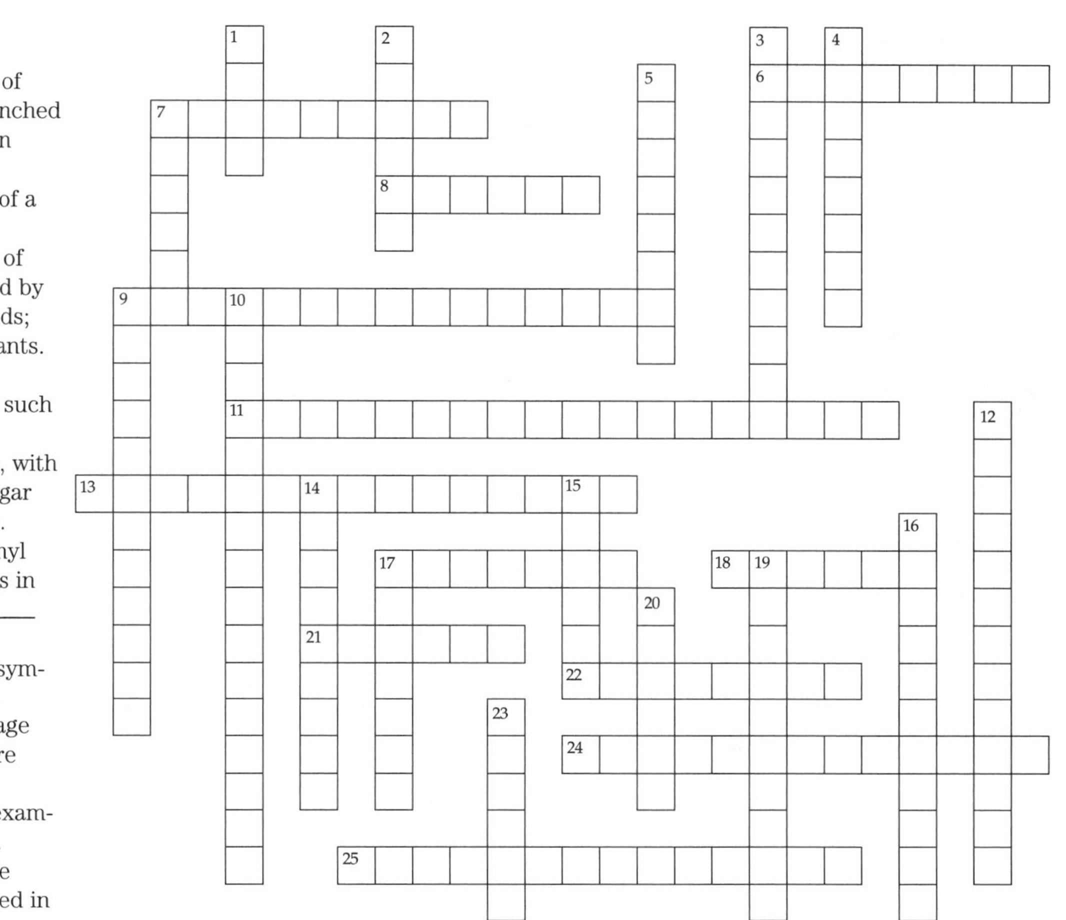

6. A homopolysaccharide of glucose; it is highly branched and found exclusively in animal cells.
7. Formed by cyclization of a ketose sugar.
8. A homopolysaccharide of glucose units connected by ($\alpha 1 \rightarrow 4$) glycosidic bonds; found exclusively in plants.
9. Simple sugars.
11. Heteropolysaccharides such as hyaluronate.
13. Glycogen and cellulose, with thousands of simple sugar subunits, are examples.
17. Oxidation of the carbonyl carbon of sugars results in the formation of ________ acids.
18. A compound with an asymmetrical atom allowing formation of mirror-image isomers has one or more ________ centers.
21. β-D-glucuronate is an example of a ________ acid.
22. End of a polysaccharide chain that is not involved in a glycosidic bond and has a free carbonyl carbon.
24. Lactose and sucrose are examples.
25. Polysaccharides cross-linked by peptides and found in bacterial cell walls.

DOWN

1. A ________ polysaccharide is a linear polymer of repeating monosaccharides.
2. A sugar with a carbonyl group at C-2 (or any position other than C-1).
3. Carbohydrate moieties are attached to glycoproteins through either *N*- or ________-________ bonds.
4. Six-membered ring form of sugars.
5. Five-membered ring form of sugars.
7. A polysaccharide containing more than one type of sugar is a ________ polysaccharide; an example is chondroitin sulfate.
9. Process that interconverts isomers of pyranoses.
10. Lectins are proteins that bind to specific ________; one such ligand is Gm1 pentasaccharide.
12. Glycoconjugates containing protein and oligosaccharide portions: for example, immunoglobulins.
14. A homopolysaccharide of glucose units connected by ($\beta 1 \rightarrow 4$) glycosidic bonds; it is found exclusively in plants.
15. An isomer that differs at only one of two or more chiral centers.
16. Gangliosides, for example.
17. The α and β forms of a pyranose, for example.
19. In the formation of pyranoses, linkage between the aldehyde on C-1 and the alcohol on C-5.
20. A sugar with the carbonyl group at C-1.
23. Animal tissues have an extracellular ________ composed of glycoconjugates and fibrous proteins.

Do You Know the Facts?

1. Which of the following is *not* a characteristic of carbohydrates in cells?
A. They serve as energy stores in both plants and animals.
B. They are major structural components of plant tissues.
C. They act as binding sites for proteins.
D. They are organic catalysts.
E. They play a role in cell–cell recognition.

2. Which of the following contributes to the structural rigidity of cellulose?
A. Adjacent glucose polymers are stabilized by hydrogen bonding.
B. Glucose residues are joined by (α1→4) linkages.
C. Cellulose is a highly branched molecule.
D. The conformation of the glucose polymer is a coiled structure.
E. Adjacent polymers are covalently linked by short peptides.

3. Which of the following is an epimer of glucose?
A. talose
B. idose
C. gulose
D. altrose
E. allose

4. Why are sugars usually found as phosphorylated derivatives in cells?
A. Phosphorylated sugars are important in regulating cellular pH.
B. Unphosphorylated sugars can be transported across cell membranes.
C. Unphosphorylated sugars are rapidly degraded by cellular enzymes.
D. Phosphorylated sugars encode genetic information.
E. None of the above is a correct explanation.

5. Which of the following disaccharides could be extended to form a cellulose polymer?
A. sucrose
B. trehalose
C. maltose
D. lactose
E. none of the above

6. Which of the following is a heteropolysaccharide? What is its function?
A. glycogen
B. hyaluronate
C. starch
D. cellulose
E. chitin

7. (a) Draw the pyranose form of β-D-glucose.

(b) How many isomers of glucose are possible?

(c) Describe this monosaccharide using at least four chemical terms.

8. **(a)** Draw the furanose form of α-D-fructose.

(b) How many isomers of fructose are possible?

(c) Describe this monosaccharide using at least four chemical terms.

9. In addition to simple sugars, what other compounds have you encountered in the text that exist as enantiomers?

10. Cellulose is a homopolysaccharide of glucose units. Why can't this molecule be used by most organisms to generate energy?

11. Draw both straight-chain (Fischer projection) and cyclic (Haworth perspective) representations of the following molecules. Identify your Fischer projections as showing the L or the D form, and identify your Haworth perspectives as the α or β form.

(a) Glyceraldehyde

(b) Dihydroxyacetone

(c) Ribose

(d) Galactose

12. Draw the Haworth perspective formulas for disaccharides containing two glucose molecules:

(a) in an (α1→4) linkage.

(b) in a (β1→4) linkage.

(c) in an (α1→6) linkage.

Applying What You Know

1. Why are many membrane proteins glycoproteins?

2. Describe the relationship between glycosaminoglycans and a proteoglycans aggregate. How is collagen involved in this arrangement?

3. What structural differences between glycogen and cellulose explain their functional differences?

4. Amylopectin is a homopolysaccharide containing 300–6,000 glucose residues. Because a single glucose residue is approximately 0.001 μm long, the largest amylopectin molecules would be 6 μm long, if all the residues were arranged in a single chain. Plant cells store amylopectin in granules whose diameters are 3–100 μm. How do plant cells maximize the storage capacity of these (relatively) small structures? What other factor(s) probably have influenced the way cells store glucose molecules?

5. The enzyme hyaluronidase is found as a component of some snake and insect toxins. How does it increase a toxin's effectiveness?

6. Consider the following experimental observations: (1) In many glycoproteins, the oligosaccharide portion attaches to the protein portion at sequences that form β bends. (2) The three-dimensional protein architecture of glycoproteins is *not* affected by removal of the associated oligosaccharide(s). What do these observations tell you about the spatial placement of the oligosaccharides in glycoproteins?

7. The bacterium *Helicobacter pylori* adheres to the surface of the stomach by interactions between bacterial membrane lectins and specific oligosaccharides of membrane glycoproteins of gastric epithelial cells. One of the binding sites recognized by *H. pylori* is the oligosaccharide called Le^b. Chemically synthesized analogs of the Le^b oligosaccharide can be administered orally to inhibit bacterial attachment.

Refer back to Fig. 7–4a; draw a ligand binding curve to illustrate the *H. pylori* lectin-Le^b oligosaccharide interaction, with and without the presence of Le^b analog. What type of interaction is this similar to? (Hint: See Chapter 8).

8. The starch present in barley consists of polymers of glucose that are linked by two types of bonds: (α1→4) and (α1→6) linkages. The enzymes found in malt (the dried, germinated barley that is rich in hydrolytic enzymes) are used to hydrolyze some of these bonds in starch to produce free glucose, which can then be fermented to ethanol and CO_2. There are other methods for making fermented beverages; for example, "spit beer" is made by a person chewing corn briefly and then spitting it into a container. The action of salivary amylase is similar to that of the barley enzymes. Neither the barley nor salivary enzymes can degrade starch completely; the pieces of starch left unhydrolyzed are highly branched and are referred to as limit dextrins. In "light" beer production, however, a mold enzyme is used to split all the limit dextrins to individual glucose molecules. What bond(s) must this accessory mold enzyme be able to cleave?

ANSWERS

Do You Know the Terms?

ACROSS

6. glycogen
7. hemiketal
8. starch
9. monosaccharides
11. glycosaminoglycans
13. polysaccharides
17. aldonic
18. chiral
21. uronic
22. reducing
24. disaccharides
25. peptidoglycans

DOWN

1. homo
2. ketose
3. *O*-glycosidic
4. pyranose
5. furanose
7. hetero
9. mutarotation
10. oligosaccharides
12. glycoproteins
14. cellulose
15. epimer
16. glycolipids
17. anomers
19. hemiacetal
20. aldose
23. matrix

Do You Know the Facts?

1. D
2. A
3. E
4. B
5. E
6. B; Hyaluronate plays an important structural role in the extracellular matrix of animal tissues.
7. **(a)**

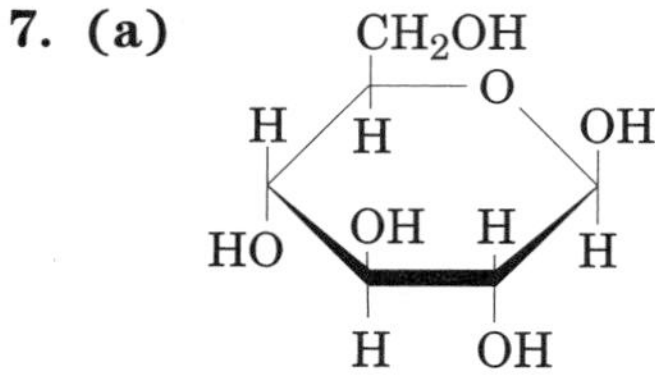

β-D-glucose

(b) 32 isomers (chiral carbons, $n = 5$; $2^n = 32$)

(c) hexose, aldose, pyranose, reducing sugar

8. **(a)**

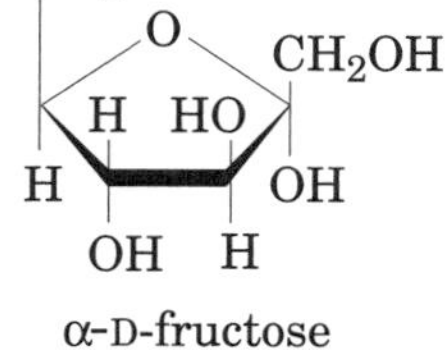

α-D-fructose

(b) 16 isomers (chiral carbons, $n = 4$; $2^n = 16$)

(c) hexose, ketose, furanose, reducing sugar

9. Amino acids other than glycine also have a chiral center—the α carbon.
10. The (β1→4) glycosidic bonds of cellulose can only be hydrolyzed by cellulase, which is produced by a few microorganisms such as bacteria and protists.
11. The different Fischer projection formulas are shown in Figure 9–3. The Haworth perspective formulas are shown below.

(a) not found as a ring structure

(b) not found as a ring structure

(c)

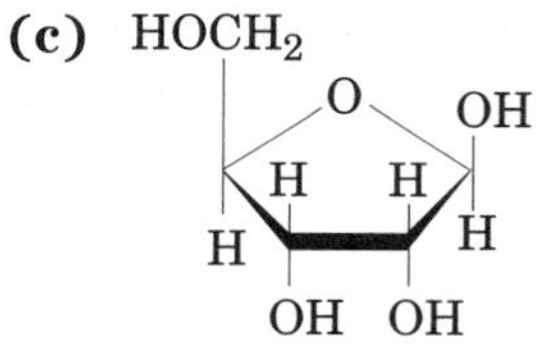

β-D-ribofuranose

(d)

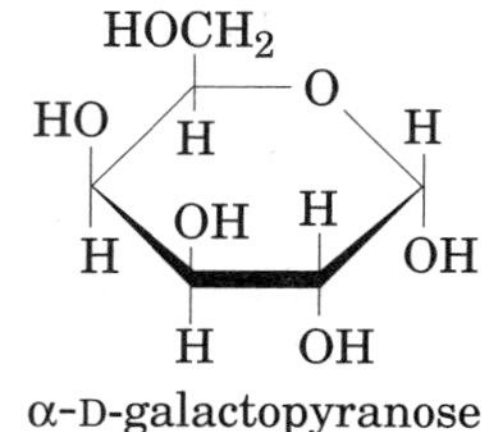

α-D-galactopyranose

12. **(a)** (α1→4) linkage

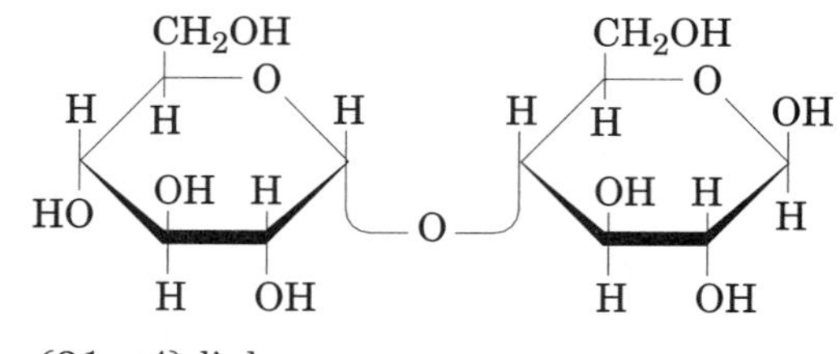

(b) (β1→4) linkage

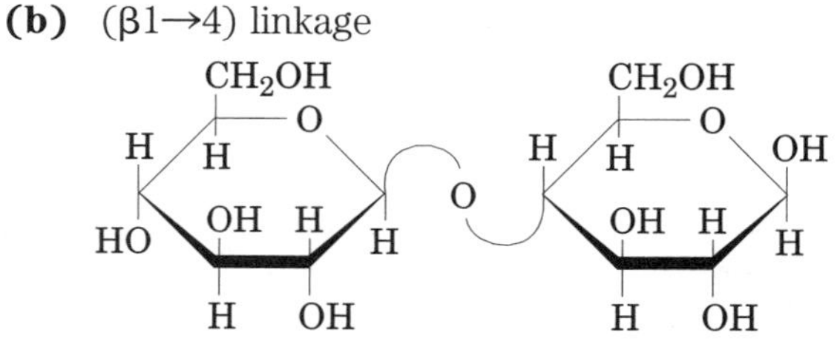

(c) (α1→6) linkage

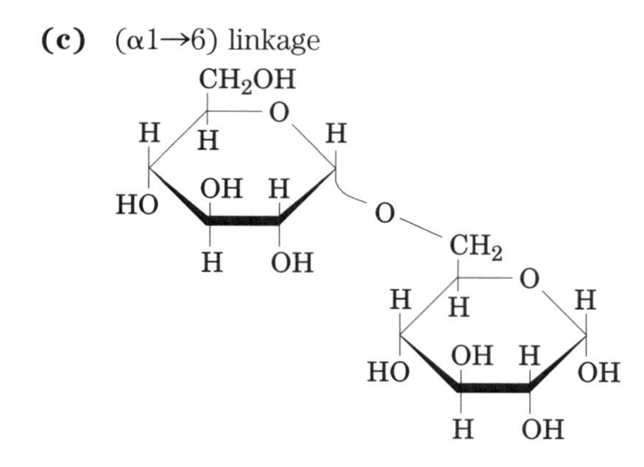

Applying What You Know

1. Many plasma membrane proteins are glycoproteins, with the oligosaccharide portion invariably located on the external surface of the membrane. Oligosaccharide moieties on glycoproteins are not monotonous, but rich in structural information. They act as biological markers, for example as "life-clocks" for individual proteins or cells, signaling whether they should be allowed to circulate or should be destroyed.

2. Glycosaminoglycans are heteropolysaccharides, a family of linear polymers composed of repeating disaccharide units. One of the two monosaccharides in the repeating disaccharide always has an amino group of some sort attached, hence glycos*amino*glycans. In a typical proteoglycan aggregate, a very long strand of a certain type of glycosaminoglycan (hyaluronate), which can be thought of as the central trunk of a tree, has numerous "branches" (core proteins) attached. To these core protein branches are attached many smaller glycosaminoglycan molecules of other types (such as keratan sulfate). In a typical proteoglycan aggregate in human cartilage, each hyaluronate trunk has about 100 core protein branches, and each core protein branch has about 150 smaller glycosaminoglycan branches. Interwoven with these enormous aggregates are fibrous proteins such as collagen, which give strength and resilience to the matrix.

3. Both polysaccharides are homopolysaccharides, meaning they contain only a single type of monomeric unit; in this case, glucose. Cellulose is a linear, unbranched homopolysaccharide, whereas glycogen is branched. Glycogen and cellulose each have variable numbers of subunits, but both are very large molecules.

The critical difference between the two is that glucose residues in cellulose have a β configuration, whereas glucose residues in glycogen are linked in the α configuration. Glycogen is a polymer of (α1→4)-linked subunits of glucose, with (α1→6)-linked branches every 8–12 residues. The most favorable three-dimensional conformation for glycogen is a tightly coiled helical structure stabilized by hydrogen bonds. Cellulose is a polysaccharide of (β1→4) glycosidic bonds. Each glucose unit is turned 180° relative to its neighbors, yielding a straight, extended chain. Multiple hydrogen bonds form between neighboring chains on all sides.

These structural differences correspond to the particular functions of the polysaccharides. Glycogen's role is energy storage. When glucose is needed, its branched structure allows for rapid degradation (from many branch ends) by the cell. Cellulose has a structural role; it gives rigidity and strength to plant cell walls. Extensive hydrogen bonding between linear molecules of cellulose forms sheets, and further hydrogen bonding between stacked sheets makes for very tough, rigid masses of material.

4. Glucose storage molecules are branched, with branches occurring every 2–30 residues in amylopectin and every 8–12 residues in glycogen. The most stable conformation for the adjacent and relatively rigid residues is a curved chain. The extensively branched molecules, therefore, pack more residues into a tighter space.

Other factors probably influenced the evolution of very large, branched molecules for storage. The enzymes that act on these polymers to mobilize glucose for metabolism act only on their nonreducing ends. With extensive branching, more such ends are available for enzymatic attack than in the same quantity of glucose stored as a linear polymer. In effect, branched polymers increase the substrate concentration for these enzymes. Another significant consideration is the effect of the same number of *individual* glucose molecules on cell osmolarity. The increase in intracellular osmolarity due to the presence of hundreds of thousands of glucose molecules probably would be sufficient to rupture the cell.

5. Hyaluronidase degrades hyaluronate, an important structural glycosaminoglycan in animal tissues. It hydrolyzes the (β1→4) linkage between the repeating disaccharide units. This allows invasion of the tissue by the other components of the toxins, such as those that act to disrupt cell membranes.

6. These two pieces of evidence, along with the highly hydrophilic character of oligosaccharides, indicate that the oligosaccharide portions of glycoproteins extend from the surface of the proteins, rather than being involved in the internal structure.

7. The presence of the analog of the Le^b oligosaccharide alters binding of the Le^b oligosaccharide to the *H. pylori* lectin in a manner similar to a competitive inhibitor of an enzyme. The analog competes with the Le^b oligosaccharide for the binding site of the lectin, much as a competitive inhibitor competes with substrate for the active site of an enzyme.

8. Barley and salivary enzymes can hydrolyze only the (α1→4) linkages; not only can they not hydrolyze the (α1→6) linkages at the branch points, usually the (α1→4) linkages just adjacent to the (α1→6) linkages are not cleaved. These short polymers remain in the final beer product, where they account for ~22% of the total starch (and therefore calories) in regular beer.

The mold enzyme used in light beer production can hydrolyze both the (α1→4) and (α1→6) linkages. Since it is added before the fermentation process, it hydrolyzes the starch completely, making all the glucose available to the yeast for fermentation. (This leads to higher alcohol content of the beer, which is then diluted, making light beer taste less than robust.)

chapter

10

Nucleotides and Nucleic Acids

STEP-BY-STEP GUIDE

Major Concepts

Nucleotides have a variety of roles in cells.

They serve as carriers of chemical energy, as components of enzyme cofactors, and as molecular messengers. Most important, they are the monomeric subunits of the polymers DNA and RNA. These nucleic acids carry the genetic information of every organism.

Nucleic acids are polymers of nucleotide residues.

Nucleotides contain three characteristic components: a nitrogenous base (purine or pyrimidine), a pentose (ribose or deoxyribose), and a phosphate group. The nucleotides in RNA contain ribose; in DNA, they contain 2′-deoxyribose. The common pyrimidine bases in RNA are uracil and cytosine, while in DNA they are thymine and cytosine. The common purines are adenine and guanine in both RNA and DNA. A **nucleoside** is a nitrogenous base covalently bound in an *N*-β-glycosidic linkage to the 1′ carbon of a pentose sugar. A **nucleotide** is a nucleoside with at least one phosphate group attached to the 5′ carbon of the sugar. Successive nucleotide residues within a polymer are linked through their phosphate groups in phosphodiester linkages between the 5′ hydroxyl of one pentose and the 3′ hydroxyl of the next. Just as individual amino acid residues influence the overall structure of the proteins they make up, the chemical properties of the individual nucleotide residues influence the structure and behavior of the nucleic acids they form.

Nucleic acids have levels of structure, similar to proteins.

Polymers of DNA or RNA exhibit polarity in that the two ends of the molecule are different. The 3′ end has no nucleotide attached at the 3′ carbon of the pentose, whereas the 5′ end has no nucleotide attached through the 5′ phosphate. Two polymers (strands) of DNA are associated in an **antiparallel** arrangement; the orientation of one strand (with respect to its 3′ and 5′ ends) is opposite to that of the other strand. In the Watson-Crick DNA structure, also called B-form DNA, the strands are stabilized in a right-handed double-helix conformation that is held together by hydrogen bonds and base-stacking interactions. The ability of nucleotides in one strand to form *specific* base pairs with nucleotides in the other (adenine pairs with thymine; cytosine pairs with guanine) is the result of hydrogen bonds. Consequently, each strand in a double helix **complements** the other with respect to nucleotide sequence. The overall content of the different bases can affect the form of the DNA. A-DNA is relatively shorter and thicker than B-DNA; Z-DNA is left-handed. The occurrence of structural variants may have biological significance.

RNA is transcribed as a complementary, single-stranded molecule using a DNA molecule as a template.

Three species of RNA in cells are **messenger RNA** (mRNA), **ribosomal RNA** (rRNA), and **transfer RNA** (tRNA). These different types of RNA have various secondary structures, which are related to their base sequence and, in part, determine their functions.

An understanding of the chemistry of nucleic acids facilitates understanding of their functions.

Disruption of the noncovalent hydrogen bonds and hydrophobic interactions between bases in the DNA double helix causes it to denature or "melt" into separate strands. Laboratory measurements of the temperature at which a specific DNA denatures can yield information on its base composition (G≡C pairs form more hydrogen bonds than do A=T pairs and therefore lead to higher melting temperatures). Denaturation of double-helical DNA is the first step in **hybridization** experiments that form the basis of many techniques used in molecular genetics. The sequence of a strand of DNA can be determined using automated procedures, and DNA polymers containing specific sequences can be chemically synthesized.

DNA is susceptible to chemical changes that cause mutations.

Deamination reactions, hydrolysis of *N*-β-glycosidic bonds, irradiation–induced formation of pyrimidine dimers, alkylating reactions, and oxidative damage all can cause mutations in DNA. Not all alterations to the DNA are deleterious; enzymatic methylation of certain bases is common and affects both the structure and function of DNA.

Nucleotides serve important cellular functions beyond their role as information molecules.

Nucleoside triphosphates, most importantly ATP, transfer energy in the cell. Many cofactors, including coenzyme A, contain nucleotides, as do **second messengers** such as cAMP.

What to Review

Answering the following questions and reviewing the relevant concepts, which you have already studied, should make this chapter more understandable.

- In a nucleotide, the pentose sugar is linked to the purine or pyrimidine base through an *N*-β-glycosidic bond. Review glycosidic bonds and remind yourself of other biologically important molecules that have glycosidic bonds (Fig. 9–11, Fig. 9–25).
- Nucleotides contain a carbohydrate residue in the furanose form. Review the structure and nomenclature of furanoses. Draw a furanose sugar, including the numbering of its carbon skeleton (Fig. 9–7).
- Like proteins, nucleic acids exhibit several levels of structural organization. Compare the levels of protein structure (Chapters 6 and 7) to the levels of nucleic acid structure discussed in this chapter. What bonds and/or interactions are important to each class of biomolecule?
- The nucleotide ATP affects both the reaction *equilibrium* and *rate* of many chemical reactions. Be sure you understand the distinction (pp. 249–250).

Topics for Discussion

Answering each of the following questions, especially in the context of a study group discussion, should help you understand the important points of this chapter.

Some Basics

1. What are the functions of DNA and of the various types of RNA?

2. What is the functional definition of a gene?

Nucleotides and Nucleic Acids Have Characteristic Bases and Pentoses

3. What are the general constituents of nucleosides and nucleotides, and what types of bonds are involved in each of these structures?

4. Which bases are commonly found in DNA? Which bases are found in RNA?

5. What minor bases are occasionally found in DNA? In RNA?

6. What are the possible phosphorylation sites in nucleotides?

Phosphodiester Bonds Link Successive Nucleotides in Nucleic Acids

7. What components form the backbone of DNA and RNA molecules?

8. Is DNA soluble in water? Why or why not?

9. Why is RNA sensitive to alkaline hydrolysis, whereas DNA is not?

The Properties of Nucleotide Bases Affect the Three-Dimensional Structure of Nucleic Acids

10. How does pH affect the structure of the purine and pyrimidine bases?

11. Are free bases soluble in water?

Nucleic Acid Structure

12. What are the hierarchical levels of nucleic acid structure? What are the analogous levels of structure in proteins?

DNA Stores Genetic Information

13. In the studies of *Streptococcus pneumoniae,* how was the specificity of enzymes used to provide evidence that strongly suggested DNA was the carrier of genetic information?

14. How did the use of radioactive phosphorus and sulfur in the Hershey-Chase experiment clarify that DNA was the genetic material?

DNA Molecules Have Distinctive Base Compositions

15. Which of "Chargaff's rules" was the most important clue leading to the model postulated by Watson and Crick?

DNA Is a Double Helix

16. Clearly describe the three-dimensional structure of double-helical DNA. Be sure you know where each of the nucleotide components is placed in relationship to the others, and the bonds and interactions involved. How does this structure suggest a mechanism for the transmission of genetic information?

17. What type of noncovalent interaction is critical for the specificity of base pairing between nucleotides of complementary strands of DNA? What type of interaction is the major contributor to the overall stability of the double helix?

DNA Can Occur in Different Three-Dimensional Forms

18. What are the similarities and differences among B-, A-, and Z-DNA?

19. What is the possible biological significance of the alternate forms of DNA?

Certain DNA Sequences Adopt Unusual Structures

20. What is the difference between a palindrome and a mirror repeat?

21. Why can't mirror repeats form hairpin structures?

22. What chemical conditions promote the formation of triplex DNAs? What nucleotide sequences promote triplex formation?

23. Why are these unusual DNA structures of interest?

Messenger RNAs Code for Polypeptide Chains

24. What is the minimum length of mRNA required to code for a polypeptide chain of 80 amino acid residues?

25. How are prokaryotic and eukaryotic mRNA molecules different?

26. What is one function of the noncoding regions of mRNA?

Many RNAs Have More Complex Three-Dimensional Structures

27. How does base-pairing differ in DNA–DNA interactions and RNA–RNA interactions?

28. What bonds and interactions are important in some RNA secondary structures?

29. What is it about RNA molecules that allows them to form more complex structures than do molecules of DNA?

Nucleic Acid Chemistry

Double-Helical DNA and RNA Can Be Denatured

30. Which bonds and interactions are affected during denaturation of DNA? How does this compare to protein denaturation?

31. What physical changes allow scientists to determine the melting points of various DNAs? How are these changes experimentally measured?

32. Remind yourself why guanine can form three hydrogen bonds with cytosine, whereas adenine–thymine pairs have only two.

Nucleic Acids from Different Species Can Form Hybrids

33. How does the degree of evolutionary closeness between two species affect the extent of hybridization between their nucleic acids?

Nucleotides and Nucleic Acids Undergo Nonenzymatic Transformations

34. Explain in evolutionary terms why it is important that DNA contains thymine rather than uracil.

35. Physicians caution against excessive suntanning. Explain the molecular basis for their concern.

36. Why is DNA so well-suited to the role of molecular repository of genetic information?

Some Bases of DNA Are Methylated

37. List some reasons for methylation of certain DNA bases and sequences.

The Sequences of Long DNA Strands Can Be Determined

38. What are the general principles involved in the two methods of sequencing DNA?

39. What is the role of the dideoxynucleoside triphosphate in the Sanger method?

40. What component is labeled in the sequencing reactions using the Sanger method?

41. Be able to "read" a sequencing gel.

The Chemical Synthesis of DNA Has Been Automated

42. The ability to synthesize DNA has proven to be an extremely useful tool for a wide variety of biochemical and biological studies. Think of a biochemical question that can be addressed using synthetic polynucleotides.

Other Functions of Nucleotides

Nucleotides Carry Chemical Energy in Cells

43. Theoretically, how do ATP and PP_i compare in their ability to affect reaction equilibria?

44. How do ATP and PP_i compare in their ability to affect reaction rates?

Adenine Nucleotides Are Components of Many Enzyme Cofactors

45. Explain the concept of evolutionary economy. How does it relate to the adenosine component of many cofactors?

Some Nucleotides Are Regulatory Molecules

46. Why is cAMP called a *second* messenger?

SELF-TEST

Do You Know the Terms?

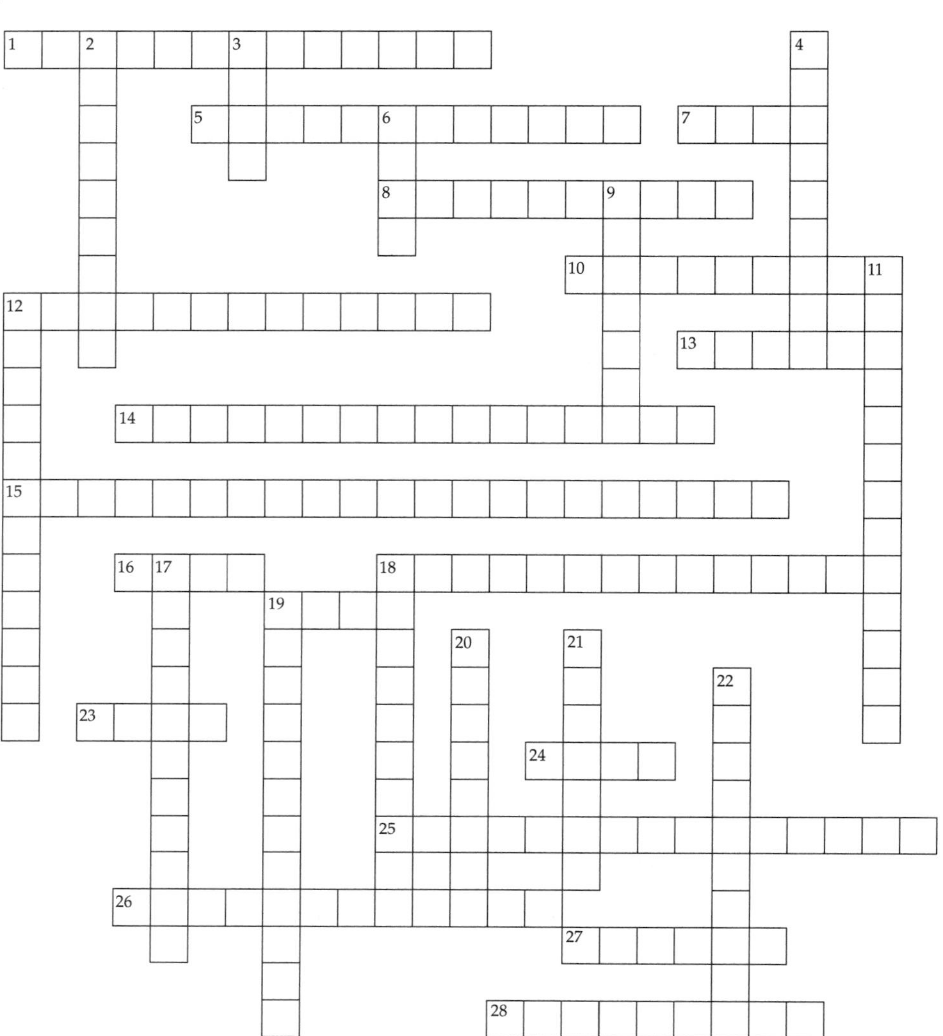

ACROSS

1. These two strands are ______ to each other.
   AATGCGGTCCTAT
   TTACGCCAGGATA
5. 3′ → 5′
   5′ ← 3′
7. A ribonucleic acid involved in protein synthesis; it binds amino acids.
8. Contains a phosphate group in an ester linkage to a ribose sugar and a nitrogenous base.
10. A common intracellular signalling molecule. (2 words)
12. Most eukaryotic mRNA codes for a single polypeptide and is ______.
13. What two complementary strands of DNA spontaneously do to form an intact duplex.
14. Thymidylate is a nucleotide found primarily in ______ acids.
15. A common protein domain found in proteins that bind ATP. (3 words)
16. A major structural component of the protein synthetic machinery of cells.
18. Covalent bonds that link the individual nucleotide residues in DNA and RNA.
19. Structure containing polypurine tracts and mirror repeats; forms a triple helix.
23. Left-handed double-helical structure.
24. Right-handed, Watson-Crick double helix.
25. Short polymers of nucleotides (50 or less), often used as complementary DNA "probes" for hybridization techniques.
26. AACCTTTTCCAA
    TTGGAAAAGGTT (2 words)
27. DNA duplex formed from DNA of different species.
28. Non–Watson-Crick, or ______ pairing; allows formation of triplex DNAs.

DOWN

2. Hydrogen peroxide, a byproduct of aerobic metabolism, is a common cause of these in DNA.
3. Carries genetic information from DNA to the ribosomes.
4. Contains a purine attached to a phosphorylated ribose; the base was first isolated from bird manure.
6. Dehydrated, compact form of DNA.
9. A major pyrimidine; has a methyl group at C-5.
11. Describes mRNA that is translated into more than one protein.
12. Determined for a solution of DNA by measuring UV light absorption as a function of temperature. (2 words)
17. Uracil is a nitrogenous base found predominantly in ______ acids.
18. AACCAATTGGTT
    TTGGTTAACCAA
19. The increase in UV light absorption when double-stranded DNA is denatured is referred to as the ______ effect.
20. A major pyrimidine; has an amino group at C-4.
21. Uracil attached through N-1 to ribose.
22. Purine or pyrimidine base covalently bound to furanose through an *N*-β-glycosidic bond.

Do You Know the Facts?

1. Draw the following structures.

(a) Adenylate **(b)** Deoxyguanylate **(c)** Deoxycytidylate

(d) Uridylate **(e)** Deoxythymidylate **(f)** Hypoxanthine

2. Compounds that contain a nitrogenous base, a pentose sugar, and a phosphate group are called **(a)** ________. Two purines found in DNA are **(b)** ________ and **(c)** ________. In DNA, the base pair **(d)** ________-________ is held together by three hydrogen bonds; the base pair **(e)** ________-________ has only two such bonds. In a solution of DNA, the purine and pyrimidine bases stack like coins because of their **(f)** hydro________ nature. When the solution is heated, the base stacking is disrupted, and the optical absorbance (at 260 nm) of the solution increases; this is called the **(g)** ________.

3. Indicate whether the following statements about DNA are true or false.

____ **(a)** A-form and B-form DNA are right-handed helices, but Z-form DNA is a left-handed helix found only in single-stranded DNA.

____ **(b)** Palindromic sequences can potentially form cruciform structures.

____ **(c)** The dideoxy method of sequencing DNA can be used on B-DNA, but not on Z-DNA.

____ **(d)** Deoxyribose is bound to the nitrogenous base at C-1.

____ **(e)** A pyrimidine in one strand of DNA always hydrogen bonds with a purine in the opposite strand.

____ **(f)** G≡C pairs share three glycosidic bonds.

____ **(g)** All the monosaccharide units lack an —OH at C-2.

____ **(h)** The phosphodiester bonds that link adjacent nucleotides join the 3′ hydroxyl of one nucleotide to the 5′ hydroxyl of the next.

____ **(i)** The two strands are aligned as parallel strands.

____ **(j)** The phosphodiester bonds that link adjacent nucleotides are uncharged under physiological conditions.

____ **(k)** B-form DNA predominates in aqueous solution; dehydration favors the A-form.

____ **(l)** A-form DNA is shorter and has a larger diameter than the B-form DNA.

____ **(m)** Nucleotide sequence has little or no effect on which form DNA takes.

4. ADP:
A. contains a furanose ring.
B. contains a ketose sugar.
C. contains two phosphoanhydride bonds.
D. contains a β-*O*-glycosidic bond.
E. contains a pyrimidine base.

5. Which of the following is true for DNA?
A. Phosphate groups project toward the middle of the double helix.
B. Deoxyribose units are connected by 3′,5′-phosphoanhydride bonds.
C. The 5′ ends of both strands are at the same end of the double helix.
D. G≡C pairs share three hydrogen bonds.
E. The ratio of A + T to G + C is constant for all naturally occurring DNA.

6. In the DNA sequencing by the Sanger (dideoxy) method,
A. the template strand of DNA is radioactive.
B. enzymes are used to cut the DNA into small pieces, which are then separated by electrophoresis.
C. ddTTP is added to each of four reaction mixtures prior to synthesis of complementary strands.
D. the role of ddATP is to occasionally terminate synthesis of DNA where dT occurs in the template strands.
E. the sequence is read from the top of the gel downward.

7. In living cells, various nucleotides and polynucleotides:
A. serve as structural components of enzyme cofactors.
B. alter the equilibria of chemical reactions.
C. carry metabolic energy.
D. serve as intracellular signaling molecules.
E. all of the above are true.

8. The compound that consists of ribose linked by an *N*-β-glycosidic bond to N-9 of adenine is:
A. a purine nucleotide.
B. a pyrimidine nucleotide.
C. adenosine.
D. AMP.
E. deoxyadenosine.

9. The phosphodiester bonds that link adjacent nucleotides in DNA:
A. are positively charged.
B. join the 3′ hydroxyl of one nucleotide to the 5′ hydroxyl of the next.
C. always link A with T and C with G.
D. are positively charged and always link A with T and C with G.
E. are positively charged and join the 3′ hydroxyl of one nucleotide to the 5′ hydroxyl of the next.

10. Nucleic acid samples have been isolated from three different organisms. The nucleic acids have the following base ratios (%):

	A	T	U	G	C	A + T/G + C	A + G/C + T
Sample 1	29	19	0	22	30	0.92	1
Sample 2	24	0	16	24	36	0.4	1.3
Sample 3	17	17	0	33	33	0.5	1

Which sample(s) are DNA? Which sample would you expect to have the highest t_m (melting point)?
A. 1 and 3; 3
B. 1, 2, and 3; 3
C. 1; 2
D. 1 and 3; 1
E. 2 and 3; 2

11. Two molecules of double-stranded DNA are the same length (1,000 base pairs), but differ in base composition. Molecule 1 contains 20% A + T; molecule 2 contains 60% A + T. Which molecule has a higher t_m (melting point)? How many C residues are there in the 60% A + T DNA molecule?
A. 2; 40
B. 1; 200
C. 2; 400
D. 1; 400
E. 2; 200

12. What crucial pieces of information from the Hershey-Chase experiment indicated that DNA is the genetic material?

13. Which of "Chargaff's rules" was the most important clue leading to the model postulated by Watson and Crick? Why?

14. Why does DNA contain thymine rather than uracil?

Applying What You Know

1. You wish to determine the relative degree of sequence similarity for DNA from different species of Galapagos finches (referred to as species A, B, and C). You make hybrid duplexes of the DNA between A and B, A and C, and B and C. You measure the increase in the absorption of UV light with increasing temperature for each of the duplex solutions and obtain the following data:

Hybrid Duplex	t_m
A + B	80°C
A + C	87°C
B + C	83°C

(a) Which two species have a greater degree of similarity in their DNA sequences? Why?

(b) What factors should be taken into account when interpreting these data?

2. How would you use sequence information for a gene in *Drosophila* to identify and localize a similar gene from a mouse?

3. Why would reducing the concentration of salts (lowering the ionic strength) in a solution of double-stranded DNA lower the melting point of the DNA?

Biochemistry on the Internet

Nucleic acids are a functionally and structurally diverse group of molecules. Their secondary structures range from the familiar Watson-Crick double helix to the exceedingly complex ribosomal structures formed by rRNA. The structures of the various forms of nucleic acids provide insights into their different functions. You can view them first-hand using 3-D viewers and the Internet.

(a) First look at the basic structure of a molecule of single-stranded DNA. Go to the Protein Data Bank Web site and use the PDB ID code 116D to find the file for Deoxyribonucleic Acid [DNA (5′-d(CpCpGpTpApCpGpTpApCpGpG)-3′)]. You can display this file using a number of viewers; both Chime and RasMol work very well. Identify the various components of the sugar-phosphate backbone and the individual bases. Which is the 5′ end of this single strand?

(b) To see how DNA forms a double-helical structure, find the file 1D92 and display the model in your viewer. This file contains a model of a synthetic double-stranded DNA molecule with the sequence (5′-d(GpGpGpGpCpTpCpC)-3′). How are these strands held together in a double-helical conformation? Two of the base pairs in this double helix are mismatched. Can you find them?

(c) To see the structure of a transfer RNA (tRNA) molecule, open the file 2TRA. How many strands of RNA make up a tRNA molecule? Where are the 5′ and the 3′ ends of this molecule?

(d) To see the structure of a ribosomal RNA (rRNA) molecule, open the file 1RRN. How many strands of RNA make up this rRNA molecule? Where are the 5′ and the 3′ ends of this molecule?

(e) To get a feel for the 3-D shape of each of the molecules above, display them in **Spacefill** format, which will allow you to visualize the surface of each molecule. There is often a **Shadows** option available, which enhances your perception of the molecule's surface.

Although this method for displaying molecules provides good insight into 3-D structure, the models are still primarily two-dimensional representations. Consequently, most viewers have a stereo viewing option that allows for a true 3-D view of the models. Turn on **Stereo Display,** which results in a display of two images of the chosen molecule. Sit in front of your computer so that your nose is approximately 10 inches from the monitor and focus on the tip of your nose. In the background you should see three images, two real and one "virtual." When the middle or "virtual" image is in focus, slowly shift your attention from the tip of your nose to this middle image. It should appear as a 3-D virtual image, which you can scan and study.

Tips for Successful 3-D Viewing:

1. Turn off or lower the room lighting.
2. Sit directly in front of the screen.
3. Use a ruler to make sure you are 10–11 inches from the screen.
4. Position your head so that when you focus on the tip of your nose, the screen images are on either side of the tip (that is, look down your nose at the structures).
5. Move your head slightly closer to or farther away from the screen to bring the middle image into focus. Don't look directly at the middle image as you try to bring it into focus.
6. If you find it uncomfortable to focus on the tip of your nose, try using the tip of a finger (positioned just beyond the tip of your nose) instead.
7. *Relax* as you attempt to view the 3-D image.
8. Rotating the image very slightly may help you to switch your focus from the tip of your nose to the 3-D virtual molecule.

   Note: You can view the model in 3-D using any of the display options (e.g., ball and stick, wireframe, or spacefill).

ANSWERS
Do You Know the Terms?

ACROSS

1. complementary
5. antiparallel
7. tRNA
8. nucleotide
10. cyclicAMP
12. monocistronic
13. anneal
14. deoxyribonucleic
15. nucleotide-binding fold
16. rRNA
18. phosphodiester
19. H-DNA
23. Z-DNA
24. B-DNA
25. oligonucleotide
26. mirror repeat
27. hybrid
28. Hoogsteen

DOWN

2. mutations
3. mRNA
4. guanylate
6. A-DNA
9. thymine
11. polycistronic
12. melting point
17. ribonucleic
18. palindrome
19. hyperchromic
20. cytosine
21. uridine
22. nucleoside

Do You Know the Facts?

1. Refer to Figures 10–4 and 10–5 for the correct structures.
2. **(a)** nucleotides; **(b, c)** adenine, guanine; **(d)** G–C; **(e)** A–T; **(f)** phobic; **(g)** hyperchromic
3. **(a)** F; **(b)** T; **(c)** F; **(d)** T; **(e)** T; **(f)** F; **(g)** T; **(h)** T; **(i)** F; **(j)** F; **(k)** T; **(l)** T; **(m)** F
4. A
5. D
6. D
7. E
8. C
9. B
10. A
11. D; remember, 1,000 base pairs = 2,000 bases.
12. Two populations of bacteriophage T2 (bacterial virus) were radioactively labeled. The DNA of one batch was labeled with ^{32}P, and the *proteins* of a second batch were labeled with ^{35}S. The bacteriophage particles were then allowed to infect separate suspensions of *E. coli*, and the infected cells were collected and analyzed for the presence of radioactivity. Only the *E. coli* that were exposed to ^{32}P-labeled bacteriophage had associated radioactivity indicating that the ^{32}P labeled DNA had been transferred from the viral particles. ^{35}S, associated only with protein molecules, was not found in the infected *E. coli* cells. New viral particles were made in *both* suspensions of *E. coli*, but only the DNA-associated label was transferred to the infected bacteria; this strongly suggested that DNA was the source of the genetic information.
13. The rule stating that all DNA, regardless of the species, contains equal numbers of adenosine and thymidine residues (A = T) and that the number of guanosine residues equals the number of cytidine residues (G = C), suggested specific pairing between the bases. Watson and Crick's double-stranded structure, with A always H-bonding with T, and C always H-bonding with G, provides the molecular basis for the rule.
14. Spontaneous deamination of cytosine to uracil is relatively common—about 100 events per mammalian cell per day. Uracil is recognized as incorrect in DNA, and is removed by a repair system. If DNA normally contained uracil instead of thymine (as does RNA), recognition of these deamination mistakes would be more difficult. Unrepaired uracils would pair with adenine during replication, leading to permanent base-sequence changes (mutations).

Applying What You Know

1. **(a)** The melting point of DNA provides information concerning the degree of hydrogen bonding between two strands of DNA; the higher the melting point, the greater the extent of noncovalent bonding. Therefore, for the hybrid duplexes, a higher melting point indicates a greater degree of base pairing and a greater degree of sequence similarity between the two strands. The data suggest that species A and C share the highest degree of similarity in their DNA sequences.

   (b) Because G≡C pairs share three hydrogen bonds, whereas A═T pairs share only two, DNA hybrids having a high proportion of G≡C pairs will have a higher melting temperature than hybrids having more A═T pairs. Consequently, a hybrid with fewer matching bases that were all G≡C pairs could theoretically have a higher melting temperature than a hybrid with more matches all occurring between A and T.
2. Molecular biology techniques allow ready detection of genes with significant sequence similarity across species. These techniques rely on the ability of DNA strands to form hybrid duplexes.

   Step 1. Digest mouse genomic DNA into small fragments using specific enzymes, and separate fragments according to size by electrophoretic techniques.

   Step 2. Use sequence information from *Drosophila* to synthesize oligonucleotides with sequences complementary to the gene under consideration. These complementary oligonucleotides are tagged, with ^{32}P, for example, and used as "probes."

   Step 3. Expose the electrophoretically separated mouse DNA to the Drosophila "probe" and allow hybrids to form. Wash away unhybridized probe. Remaining probe is the result of the formation of hybrid duplexes and can be detected (for example, using autoradiography).
3. The negative charges on the phosphate groups on each strand of DNA are shielded by positive ions in a high ionic strength solution. Lower ionic strength reduces the screening of the negative charges on the phosphate groups. The result is stronger charge–charge repulsion between the phosphate portions of the backbones of each strand, which favors strand separation.

Biochemistry on the Internet

(a) To identify the various structures, it may help to display the molecule as a ball-and-stick structure and to turn on CPK color in your viewer. In this coloring scheme, the phosphates will be colored orange and their associated oxygens will be red, highlighting the backbone of this structure relative to the individual (and unpaired) bases. Notice that the deoxyribose (sugar) moiety is oriented perpendicularly to the bases.

You can locate the 5′ end in one of two ways: either locate the end of the strand that has a free C5 on the ribose moiety or, given the nucleotide sequence in the PDB file, look for the end with a cytosine. The 3′ end is the one with a free C3 on the ribose or, alternatively, the end with a guanine.

(b) The strands are held together primarily by the numerous hydrogen bonds that form between adjacent bases. You can visualize these bases by turning on the hydrogen bond display (**Options → Display Hydrogen Bonds** in Chime or "hbonds" in the command line for RasMol). The hydrogen bonds between AT and GC pairs are indicated by dashed lines. Note that there are two hydrogen bonds for each AT pair and three hydrogen bonds for each GC pair.

The mismatched pairs are between G3 (A strand) and T14 (B strand) as well as T6 (A strand) and G11 (B strand). The mispairing is very evident with the hydrogen bond display turned on since the model shows no hydrogen bond formation between the mismatched bases. (The lack of hydrogen bonds in the model does not mean that it is impossible for hydrogen bonds to form between these pairs in real molecules, but rather that the rules used by Chime and RasMol to draw hydrogen bonds are somewhat restricted.)

(c) All transfer RNA molecules are composed of a single strand of RNA. There are two ways to view this molecule that allow you to visualize the backbone structure. You can select **Backbone** as your display option, which will show only the backbone in ribbon format. Alternatively you can display the entire tRNA molecule as a **Wireframe** model and then select the backbone (**Select → Nucleic → Backbone**) and display it in **Backbone** format. This second method will allow you to see the base-pairing interactions. (The actual hydrogen bonds in this model will not display using Chime or Rasmol.) If you change the **Backbone** display to **Ball and Stick,** you can click on the base at each of the free ends to identify them. The end that contains U1 is the 5′ end. Note the presence of a free phosphate group on C5 of the ribose portion of this nucleotide. C74 is at the 3′ end. Note the presence of a free C3 hydroxyl (represented by a red ball for the oxygen) on the ribose in this nucleotide. Also note the oxygen molecule on C2 of every ribose in this structure, which was not present on the ribose in the DNA models in (a) and (b); tRNA is a ribonucleic acid and not a *deoxy*ribonucleic acid after all.

(d) Ribosomal RNA (rRNA) provides the structural framework for the process of protein synthesis, which will be discussed in detail in Chapter 27. Most ribosomes are complexes composed of several different-sized pieces of RNA as well as numerous associated proteins. File 1RRN is a theoretical model of one small rRNA component of ribosomes from eukaryotes. This ribosomal component is a single strand of RNA, and the relatively complex secondary structure is due to intramolecular base pairing, similar to what is observed for tRNA. The 5′ end is the one containing a guanine residue (G1) and the 3′ end contains a cytosine (C118).

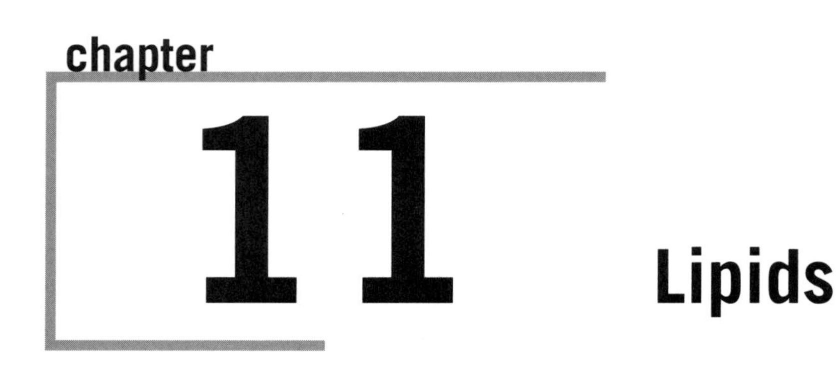

Lipids

STEP-BY-STEP GUIDE

Major Concepts

Biological lipids are chemically diverse and have a number of different functions in biological systems.

Lipids are poorly soluble in water. Consequently they are well suited to serve as the major component of cell membranes, forming a barrier between the cell interior and the extracellular environment. The hydrocarbon chains of lipids are highly reduced, making them excellent compounds for energy storage. A few select lipids also serve very specific biological functions, for example, as intracellular messengers and redox cofactors.

Fatty acids are hydrocarbon chains, generally 4 to 36 carbons long.

Saturated fatty acids contain *no* double bonds; fatty acids *with* double bonds (usually cis) are **unsaturated.** The physical properties of fatty acids and the lipids that contain them are determined largely by the chain length and number of double bonds. Saturated fatty acids have full rotation around each carbon bond and are very flexible; they also have a relatively high melting point. Unsaturated fatty acids cannot rotate around their cis double bonds and are less flexible. Unsaturated fatty acids have lower melting points than saturated fatty acids. Fatty acids with long hydrocarbon chains tend to be less soluble in water than those with short chains.

Triacylglycerols are lipids that are used for energy storage by biological systems.

In mammals, these lipids are found primarily in the large storage vacuoles of adipocytes (fat cells). Triacylglycerols have a glycerol backbone to which three fatty acid chains are attached via ester linkages.

Membrane lipids include phospholipids, sphingolipids, and sterols (cholesterol in animals).

These lipids contain one or two hydrophobic hydrocarbon chains (tails) attached to a glycerol, sphingosine, or sterol backbone. A hydrophilic moiety (head) is also attached to that backbone. Therefore, unlike triacylglycerols, membrane lipids are **amphipathic** molecules, with both hydrophobic and hydrophilic regions. This property is essential to their ability to spontaneously form bilayer structures in aqueous solutions. Some membrane lipids, such as gangliosides, contain carbohydrate moieties as well.

Biologically active lipids can serve both intra- and intercellular signaling functions.

Hydrolysis of the membrane lipid phosphatidylinositol produces two **intracellular** signaling molecules, diacylglycerol and inositol triphosphate. *Localized* intercellular messengers derived from arachidonic acid are called **eicosanoids,** and include prostaglandins, thromboxanes, and leukotrienes. Hormones such as testosterone and estrogen are derived from cholesterol; these steroid hormones are **intercellular** messengers. Vitamins D, A, E, and K are fat soluble isoprenoids with essential metabolic roles. Quinones function as redox cofactors.

Lipid analysis is difficult because lipids are insoluble in water.

The extraction of lipids from tissues requires the use of organic solvents. Chromatographic techniques are employed to further separate charged lipids from neutral lipids. Mass spectrometry is often required for complete structure determination.

What to Review

Answering the following questions and reviewing the relevant concepts, which you have already studied, should make this chapter more understandable.

- Carboxyl, phosphoryl, and hydroxyl groups, and ether and ester bonds, all play roles in the formation, structure, and function of lipids. Remind yourself again of these functional groups of biomolecules (Fig. 3–5).
- The presence of cis double bonds in unsaturated fatty acids has important consequences for the properties of lipids. Cis and trans configurations are shown in Figure 3–8 to emphasize the structural differences between these isomers.
- In some circumstances, structural formulas are as useful as space-filling models, but the shapes of lipids are only really well-illustrated by the latter. See Figure 3–7 to remind yourself of the rules behind each model, and look at Figures 11–1, 11–2, and 11–11.
- Hydrophobic interactions and the behavior of amphipathic compounds in water are critical in explaining the behavior of lipids in biological systems. These were previewed in Chapter 4. What factor is most important in explaining the strength of the forces that hold nonpolar regions of micelles together?
- Lipid extraction, separation, and identification involve the use of organic solvents and chromatography techniques that should be familiar from a course in organic chemistry. If you do not remember these techniques, it would be wise to review an organic chemistry (laboratory) text.
- Glycolipids contain carbohydrate moieties: review the various oligosaccharides and linkages that are commonly found in glycolipids (Fig. 9–25).

Topics for Discussion

Answering each of the following questions, especially in the context of a study group discussion, should help you understand the important points of this chapter.

Storage Lipids

1. Note that the categorization of lipids is not as neat and clear-cut as for proteins; the defining feature of lipids is their insolubility in water.

Fatty Acids Are Hydrocarbon Derivatives

2. Draw linolenic acid in the structural (not space-filling) style seen in Figure 9–1a. Would it be solid or liquid at room temperature?

Triacylglycerols Are Fatty Acid Esters of Glycerol

3. Why is olive oil (and other fats) so insoluble in water?

Triacylglycerols Provide Stored Energy and Insulation

4. Why are triacylglycerols such a good storage fuel, whereas carbohydrates are better as quick sources of energy?

Sperm Whales: Fatheads of the Deep

5. Why would solid spermaceti oil be more dense than liquid spermaceti oil? Contrast this property with the density change that occurs when going from solid water (ice) to liquid water.

Many Foods Contain Triacylglycerols

6. Vegetable oil and solid shortening are both mixtures of triacylglycerols. Why is shortening a solid and vegetable oil a liquid at room temperature?

Waxes Serve as Energy Stores and Water Repellents

7. Sheep shearers often have extremely soft, young-looking skin on their hands. What aspect of their job might account for this?

Structural Lipids in Membranes

8. Although storage lipids are quite hydrophobic, membrane lipids are amphipathic and can have noncovalent interactions with polar molecules. For both types of lipids, function is dictated by structure. Why can't storage lipids function as membrane lipids?

Glycerophospholipids Are Derivatives of Phosphatidic Acid

9. What is the nonpolar portion of phosphatidylcholine? This portion of the lipid would face away from the aqueous surroundings.

Some Phospholipids Have Ether-Linked Fatty Acids

10. What is the molecular basis for the suggestion that ether-linked fatty acids may confer resistance to phospholipases?

Sphingolipids Are Derivatives of Sphingosine

11. What is the polar portion of sphingomyelin? This portion of the lipid would face the aqueous surroundings.

12. Ceramides are structurally similar to diacylglycerols. Compare the attachment of head groups to the diacylglycerol unit in glycerophospholipids with their attachment to ceramide in sphingolipids.

Sphingolipids at Cell Surfaces Are Sites of Biological Recognition

13. What portions of glycosphingolipids confer the recognition function?

Phospholipids and Sphingolipids Are Degraded in Lysosomes

14. Why are lysosomes important to cellular metabolism?

Inherited Human Diseases Resulting from Abnormal Accumulations of Membrane Lipids

15. Why must there be several phospholipases to degrade phospholipids?

16. What is the enzymatic defect in individuals affected with Tay–Sachs disease?

Sterols Have Four Fused Carbon Rings

17. How are sterols similar to other membrane lipids? How are they different?

18. Do cholesterol esters have a polar portion?

Lipids as Signals, Cofactors, and Pigments

Phosphatidylinositols Act as Intracellular Signals

19. How is the location of phosphatidylinositol critical to its functions?

20. How does ibuprofen act to reduce inflammation and pain?

Steroid Hormones Carry Messages between Tissues

21. What characteristic of steroid hormones allow them to be soluble in, and transported by, the aqueous bloodstream?

22. How does prednisone act to reduce inflammation?

Eicosanoids Carry Messages to Nearby Cells

23. How are steroid hormones, phosphatidylinositol, and eicosanoids similar and different in the locality of their effects? In their location in the cell? In their structure?

Vitamins A and D Are Hormone Precursors

24. What is the structure of isoprene? Why is this structure biologically important?

25. What are the symptoms of deficiencies of vitamins A and D?

Vitamins E and K and the Lipid Quinones Are Oxidation-Reduction Cofactors

26. What are the symptoms of deficiencies of vitamins E and K?

27. Quinones function as electron carriers *within* the lipid bilayer. What is it about their structures that make them so lipophilic?

Dolichols Activate Sugar Precursors for Biosynthesis

28. What structural aspect of dolichols makes them interact so strongly with the hydrophobic portions of membrane lipids?

Separation and Analysis of Lipids

Lipid Extraction Requires Organic Solvents

29. What would happen to proteins in a mixture of biomolecules if you were trying to extract lipids? What would happen to the lipids in the mixture as you attempted to extract proteins?

Adsorption Chromatography Separates Lipids of Different Polarity

30. To what protein purification technique is adsorption chromatography most similar?

Gas-Liquid Chromatography Resolves Mixtures of Volatile Lipid Derivatives

31. Does gas-liquid chromatography leave the lipid in biologically functioning form?

Specific Hydrolysis Aids in Determination of Lipid Structure

32. Why are specific enzymes so useful in the structural analysis of lipids?

Mass Spectrometry Reveals Complete Lipid Structure

33. How would mass spectrometry detect the difference in two isoprenoids with different numbers of isoprenoid units? (Hint: See the legend to Fig. 11–21.)

SELF-TEST

Do You Know the Terms?

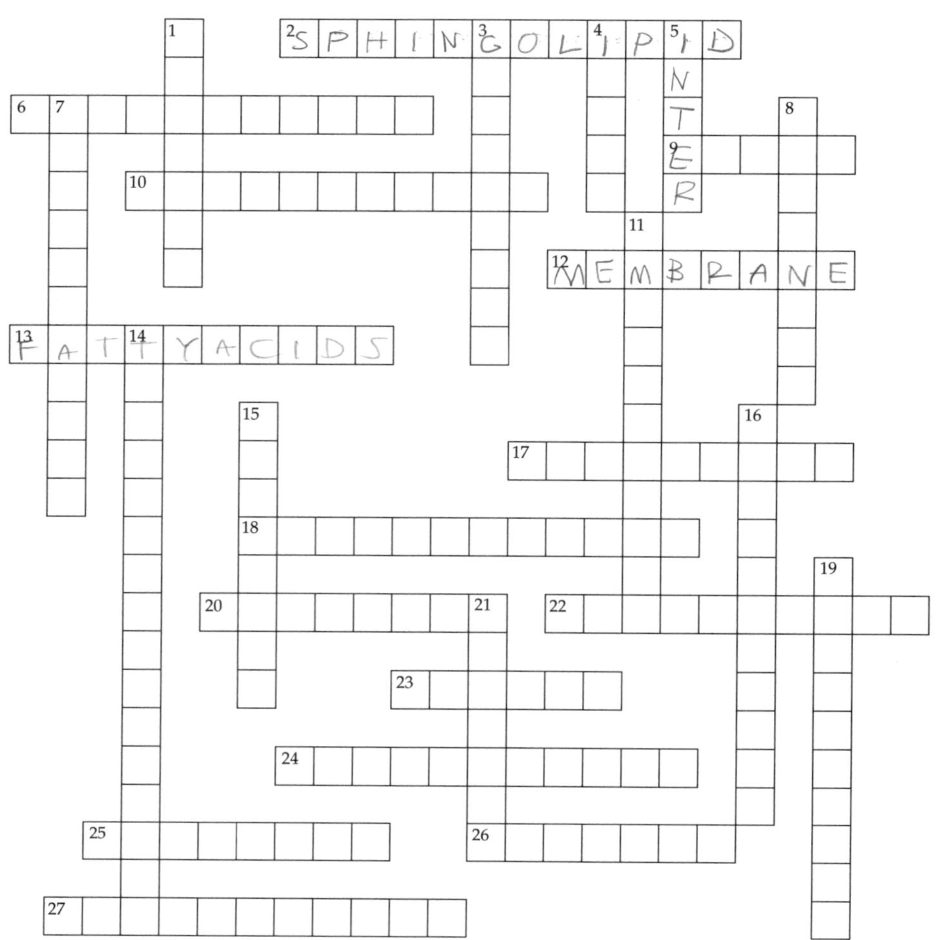

ACROSS

2. The possible head group of this class of lipids includes phosphocholine and sugars.

6. Sphingolipid with a very complex oligosaccharide head group.

9. Linkage joining the fatty acid tails of most glycerophospholipids to the glycerol backbone.

10. Describes the behavior of the lipid tail of a membrane lipid.

12. Intracellular organelle that compartmentalizes many degradative processes in cells.

13. Long hydrocarbon chains with carboxylic acid groups. (2 words)

17. Type of chromatography that separates lipids from a complex mixture on the basis of capillary action and differences in affinity for an immobile, polar matrix. (2 words)

18. Class of lipid whose hydrophilic moiety contains PO_4^{2-}.

20.

```
   H  OH  H
   |   |   |
H—C—C—C—H
   |   |   |
  OH  H  OH
```

22. Type of chromatography that uses a polar, immobile substrate to selectively remove polar and charged lipids from a complex lipid mixture.

23. Vertebrate heart tissue is uniquely enriched in ether ________, such as plasmalogens.

24. Animal lipid having a rigid sterol nucleus.

25. Extremely hydrophobic isoprenoid compound that anchors sugars to cell membranes.

26. Phosphatidylinositols act as intracellular ________.

27. Describes fatty acids having one or more double bonds.

DOWN

1. The arrangement of lipids in membranes.

3. ________-________ chromatography uses differences in the ability to partition between an inert column matrix and an inert gas to separate lipids from a complex mixture. (2 words)

4. Cleavage of phosphatidylinositol bisphosphate by a lipase (phospholipase C) produces two ________ cellular messengers.

5. Steroid hormones are ________ cellular messengers because they carry messages between tissues.

7. Describes phospholipids, sphingolipids, and cholesterol, but not triacylglycerols.

8. Phosphatidic acid is to glycerophospholipids as ________ is to sphingolipids.

11. Ubiquinone and dolichols are both biologically active ________.

14. Lipids stored in adipocytes (fat cells).

15. Compounds in which electrons are shared equally between the atoms are ________ and hydrophobic; they can form few hydrogen bonds with water.

16. Describes the behavior of the head group of a membrane lipid.

19. Prostaglandins belong to this class of paracrine hormones derived from the fatty acid arachidonic acid.

21. Class of enzymes contained in adipocytes that are used for the mobilization of storage lipids.

Do You Know the Facts?

1. For each class of lipids on the left, put a check mark in the column under the components found in these lipids. If only *some* of the lipids in a category contain a component, write "some" in the box. Is the compound a membrane lipid? The first one has been done for you as an example.

	Fatty Acid	Glycerol	Sphin-gosine	Phosphate	Carbo-hydrate	Membrane Lipid?
Triacylglycerol	✔	✔				
Phospholipid						
Sphingolipid						
Ceramide						
Ganglioside						
Cholesterol						

2. What are the polar and nonpolar portions of each of the following classes of lipids?

(a) Triacylglycerol

(b) Glycerophospholipid

(c) Sphingolipid

(d) Ganglioside

(e) Cholesterol

3. The term amphipathic means:

A. branched, with at least two branch points.
B. having one region that is positively charged and one region that is negatively charged.
C. different on the inside and outside of the lipid bilayer.
D. having one region that is polar and one that is nonpolar.
E. having two different types of bonds.

4. Coconut oil contains only a very small amount of unsaturated fatty acids. How can it still have a low melting point?

A. It contains a lot of long-chain fatty acids.
B. It contains mostly short-chain fatty acids
C. It has only a few hydrogen bonds per fatty acid chain.
D. A and C are true.
E. B and C are true.

5. The polar head group of cholesterol is:

A. the alkyl side chain.
B. glycerol.
C. the steroid nucleus.
D. the hydroxyl group.
E. choline.

For questions 6–8, refer to the structures labeled (a)–(d).

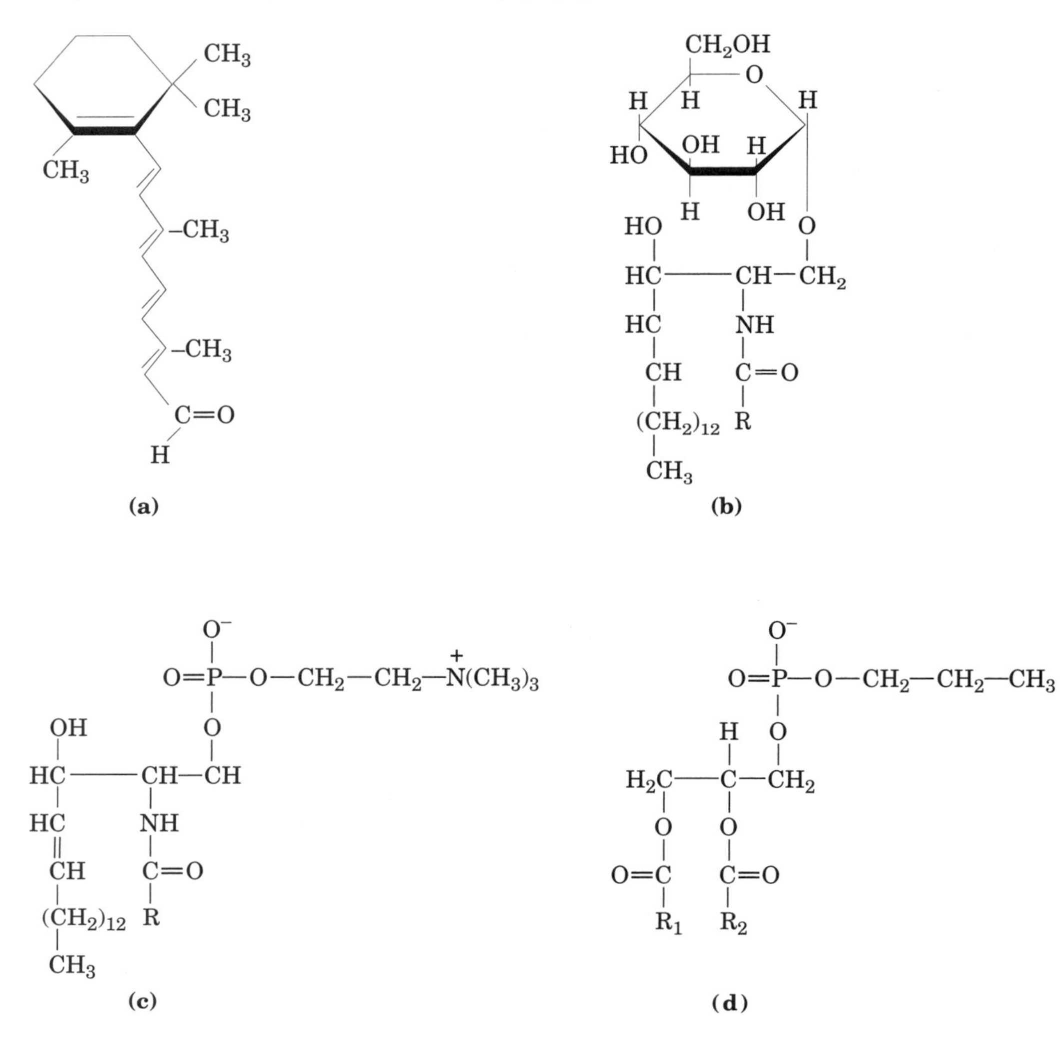

6. Which structure(s) contain(s) sphingosine?

7. Which structure(s) is a (are) glycolipid(s)?

8. Which structure(s) is a (are) glycerophospholipid(s)?

9. Match the following molecules to the descriptions:

(a) Testosterone and cortisol
(b) Phosphatidylinositol and its derivatives
(c) Eicosanoids (such as prostaglandins)
(d) Vitamin K and Vitamin E

________ Fatty acid derivatives that act on the tissue in which they are produced.
________ Isoprenoids that must be obtained from the diet.
________ Intracellular messengers that are components of the plasma membrane.
________ Steroid hormones that are produced in one tissue and carried through the bloodstream to target tissues.

Applying What You Know

1. How are the chemical properties of storage lipids and membrane lipids different? How does this relate to their functional properties? (i.e., why wouldn't triacylglycerols make good membranes?)

2. Mixtures of lipids can be separated by adsorption column chromatography. A mixture containing the lipids listed below was applied to a silica gel column, and progressively more-polar solvents were applied. What would be the order of elution of the individual lipids?

________ triacontanylpalmitate, a component of beeswax
________ lauric acid, a free fatty acid (12:0)
________ cholesterol
________ cholesterol palmitate, a cholesteryl ester (cholesterol esterified to a fatty acid at the —OH group)
________ lignoceric acid, a free fatty acid (24:0)
________ glucosylcerebroside, a sphingolipid with glucose as its "X" group

3. A naturally occurring fatty acid is systematically named 20:4 ($\Delta^{5,8,11,14}$) or cis-, cis-, cis-, cis- 5, 8, 11, 14-icosatetraenoic acid. What is the common name of this fatty acid? What is its shape? What class of compounds is this a precursor of?

4. Naturally occurring unsaturated fatty acids are nearly all found in the cis configuration. However, trans fatty acids are found in some foods, most often as a byproduct of the saturation process (hydrogenation) used to harden natural oils in the manufacture of margarine. A small number of trans fatty acids also are found in foods derived from ruminant animals, produced by the action of microorganisms in the rumen.

Oleic acid is a naturally occurring monounsaturated fatty acid, 18:1 (Δ^9). Elaidic acid is the trans form, also 18:1 (Δ^9), an unnatural isomer found in many margarine products. Would you expect a simple triaclyglycerol containing elaidic acid to have a higher or lower melting point relative to one containing oleic acid? Why?

5. Lipoproteins are lipid-protein complexes that transport lipids through the bloodstream from dietary and endogenous sources to peripheral tissues and to the liver. They include LDLs (the "bad" cholesterol"), HDLs (the "good" cholesterol), and others. They are essentially fat droplets, composed of triacylglycerols and cholesterol esters, enclosed by a monolayer of phospholipids, cholesterol, and protein. What is the functional significance of the monolayer of amphipathic lipids? How is it spatially arranged?

ANSWERS

Do You Know the Terms?

ACROSS

2. sphingolipid
6. ganglioside
9. ester
10. hydrophobic
12. lysosome
13. fatty acids
17. thin-layer
18. phospholipid
20. glycerol
22. adsorption
23. lipids
24. cholesterol
25. dolichol
26. signals
27. unsaturated

DOWN

1. bilayer
3. gas-liquid
4. intra
5. inter
7. amphipathic
8. ceramide
11. isoprenoids
14. triacylglycerols
15. nonpolar
16. hydrophilic
19. eicosanoid
21. lipases

Do You Know the Facts?

1.

	Fatty Acid	Glycerol	Sphin-gosine	Phosphate	Carbo-hydrate	Membrane Lipid?
Triacylglycerol	✔	✔				
Phospholipid	✔	✔ (some)	✔ (some)	✔	✔ (some)	✔
Sphingolipid	✔		✔	✔ (some)	✔ (some)	✔
Ceramide	✔		✔			✔
Ganglioside	✔		✔		✔	✔
Cholesterol	(has a short hydrocarbon side chain)					✔

2. (a) Fatty acid carboxylic acid groups are esterified to glycerol's —OH groups: the whole molecule is nonpolar.

(b) Phosphate groups and attached head groups are polar; fatty acid tails are nonpolar.

(c) The —OH groups on the alcohol sphingosine are polar; some sphingolipids have phosphate groups, some have sugar groups, both of which are polar; fatty acid tails are nonpolar.

(d) The carbohydrate moiety is polar; fatty acid tails are nonpolar.

(e) The —OH group is polar; the steroid nucleus and acyl chain are nonpolar.

3. D

4. B

5. D

6. b and c

7. b

8. d

9. c, d, b, a

Applying What You Know

1. Storage lipids include primarily fats and oils. These are highly reduced compounds, containing fatty acids that, when oxidized, provide a large amount of energy for the organism. Triacylglycerols are one type of storage lipid. These are made up of three fatty acids in ester linkage with a single glycerol molecule. Because the polar hydroxyls of glycerol and the polar carboxylates of the fatty acids are bound in ester linkages, triacylglycerols are nonpolar, hydrophobic molecules, essentially insoluble in water.

   Membrane lipids, on the other hand, are amphipathic. The orientation of their hydrophobic and hydrophilic regions directs their packing into membrane bilayers. The three general types of membrane lipids are glycerophospholipids, sphingolipids, and sterols. All have polar "heads" and nonpolar "tails," and all form bilayers, which constitute the central architectural feature of biological membranes. Hydrophobic triacylglycerols do not orient into such bilayers, having no hydrophilic portion to face toward the aqueous surroundings.

2. In order of first eluted to last eluted:

   - cholesterol palmitate and triacontanylpalmitate (most nonpolar)
   - cholesterol (still quite nonpolar because of the steroid nucleus, but with a polar —OH group)
   - glucosylcerebroside (a neutral but polar lipid, due to its —OH group on the sphingosine backbone and the —OH groups of the glucose portion)
   - lignoceric acid (a long, charged fatty acid)
   - lauric acid (a charged fatty acid because of the $—COO^-$ group, but shorter than lignoceric acid so less nonpolar)

3. This is arachidonic acid. It is bent into a U-shape, due to the four cis double bonds, and is a precursor of the eicosinoids.

4. The triacyclglycerol containing the trans isomer of the fatty acid would likely have a higher melting point. In trans fatty acids, the acyl chains originate from the opposite side of the double bond, making the whole molecule relatively "straight." The cis isomer has the acyl chains originating on the same side of the double bond, giving the fatty acid a bend or "kink." The "straight" trans fatty acids will pack more tightly, increasing the interactions between them, and raising the melting point of the fat.

5. The monolayer of lipids are arranged so that their hydrophobic portions (the fatty acid tails of the phospholipids and the steroid nucleus of the cholesterol) are facing inward, in contact with the nonpolar center lipids. The hydrophilic portions (the polar heads of the phospholipids and the —OH group of the cholesterol) face outwards in contact with the aqueous surroundings (the blood), increasing the solubility of the lipoproteins and therefore their transport in the bloodstream.

chapter

12 Biological Membranes and Transport

STEP-BY-STEP GUIDE

Major Concepts

The major constituents of biological membranes are lipids and proteins.

Membrane lipids are primarily, by weight percentage, phospholipids, but also include varying amounts of sphingolipids and sterols. Membrane lipids are the primary *structural* component of biological membranes, but some serve as signaling molecules, as anchors for other membrane components, and probably in other roles, that have yet to be determined.

Membrane proteins are the primary *functional* entities of biological membranes. Proteins in membranes transport solutes, transduce extracellular signals, act as enzymes and enzyme activators, anchor other membrane components and intracellular structures, and provide intercellular contacts. The lipid and protein components of membranes vary from one cell to another in both their relative proportions and in the specific types of lipids and proteins.

Membrane structure is best described by the fluid mosaic model.

Membranes are structurally organized as a lipid **bilayer;** proteins "float" within and on this fluid bilayer. Membranes are asymmetric with respect to both the protein and lipid composition of the inner and outer layers. This indicates that movement between the two faces of the bilayer must be restricted, although lateral movement within a single side of the bilayer is allowed and often is extremely rapid. The lipid bilayer is a thermodynamically stable structure that forms spontaneously, driven by hydrophobic interactions among the fatty acid tails of lipids and hydrophilic interactions among the polar head groups and surrounding water molecules.

Membrane proteins are associated with the lipid bilayer in a number of ways.

Peripheral membrane proteins are loosely associated with the membrane, usually through hydrogen bonds and ionic interactions. Some peripheral membrane proteins are attached through a covalent linkage to a membrane lipid "anchor." **Integral membrane proteins** span the lipid bilayer and contain regions that are relatively hydrophobic. These regions are buried in the core of the bilayer where their association with the membrane is stabilized by hydrophobic interactions with membrane lipids. Integral proteins can contain one, or several bilayer-spanning regions. These transmembrane portions are commonly α-helices or multistranded β-barrels, and can often be predicted from primary structure (sequence) data.

Membrane fluidity is critical for many biological processes.

Exocytosis, endocytosis, movement of many single-celled organisms, cell movement during development, cell division and many other biological processes require a cell membrane that can change its shape. Membrane fluidity and the ability to reorganize spontaneously permit the fusion of biological membranes without disrupting the integrity of cells. Fusion is assisted by specific proteins called SNAREs.

The movement of many molecules across biological membranes is restricted.

The hydrophobic core of the lipid bilayer restricts the movement of polar or charged molecules. Water is an exception; it can diffuse slowly across the bilayer.

However, water is also transported via **aquaporins** in cells when large fluxes of water are needed.

Transport can be categorized based on the need for energy, kinetic behavior, or the direction of movement.

Metabolites and ions cannot cross the bilayer without specific **transporters,** which are specialized transmembrane proteins. Transporters are similar to enzymes in that they exhibit saturation and substrate specificity.

Passive transport moves solutes across membranes and down their respective concentration gradients, and therefore requires no direct input of energy. An example of passive transport is the movement of glucose into erythrocytes through the GluT1 transporter. **Active transport** moves solutes across membranes but against their respective electrochemical gradients, thus requiring the input of some form of energy. Primary active transport uses the energy of hydrolysis of ATP in a direct way. There are at least four types of ATPases in membranes. The electrochemical gradients generated by active transport often serve as potential energy sources to drive (via symport or antiport) the movement of other solutes against their concentration gradient. This process is **secondary active transport.**

Ion channels also provide routes for the transmembrane movement of specific ions. These do not show saturation behavior. Most ion channels are "gated" by either voltage or a particular ligand. The K^+ channel is an example of a membrane protein that forms a transmembrane pore or channel, which provides a route through the lipid bilayer. This channel is also an example of a multi-α-helical integral protein. **Porins** are β-barrel integral proteins that transiently open to allow passage of specific solutes.

What to Review

Answering the following questions and reviewing the relevant concepts, which you have already studied, should make this chapter more understandable.

- Hydrogen bonds, van der Waals interactions, and hydrophobic interactions are extremely important forces in biological systems. Review how they apply to lipid (Ch. 11) and protein structure (Chs. 5 and 6).
- When describing cell membranes, a shorthand "cartoon" of membrane lipids is usually used instead of drawing the complete structures. Be sure you understand what the components of that cartoon really represent (Fig. 11–1).
- Lipids are the primary structural component of cell membrane, but not all lipids are found in cell membranes. Be sure you know what types of lipids are found in membranes and know what properties they have in common (Fig. 11–6).
- Many proteins are associated with cell membranes. One of the more common conformations of membrane-associated proteins is the α helix. Review the α helix secondary structure, and note how it contributes to the structural stability of proteins (Ch. 6, p. 164–166).

Topics for Discussion

Answering each of the following questions, especially in the context of a study group discussion, should help you understand the important points of this chapter.

The Molecular Constituents of Membranes

Each Type of Membrane Has Characteristic Lipid Compositions and Proteins

1. What are the major classes of lipids found in biological membranes?

2. What do proteins contribute to the properties of biological membranes? How do the sugar moieties of membrane components contribute to the functions of membranes?

The Supramolecular Architecture of Membranes

3. Biological membranes are described by the fluid mosaic model. Why are "fluid" and "mosaic" *both* apt terms to describe membrane architecture?

A Lipid Bilayer Is the Basic Structural Element of Membranes

4. Under what conditions are each of the three types of amphipathic lipid aggregates favored?

5. What is the evidence that membranes really are organized in this bilayer fashion?

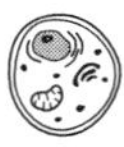

6. *Cell Map:* In the box that represents an enlargement of the lipid bilayer draw in the structures of several different types of membrane lipids. These lipids should be placed in the orientation that would allow them to form a stable bilayer arrangement. Indicate where the different types of noncovalent interactions occur that stabilize the orientation of these lipids in the membrane bilayer.

Membrane Lipids Are in Constant Motion

7. How would an increase in the percentage of saturated fatty acids affect the solid-to-fluid transition temperature of a membrane?

8. Why are sterols capable of both increasing and decreasing the fluidity of membranes according to the surrounding temperature?

9. What forces and/or interactions restrict lipid transbilayer diffusion ("flip-flop")?

Membrane Proteins Diffuse Laterally in the Bilayer

10. Speculate on the functional reasons why some membrane proteins are anchored rather than free to diffuse laterally.

Some Membrane Proteins Span the Lipid Bilayer

11. Why is it important for determining the function of a membrane protein to know if it spans the bilayer or appears on only one face of the membrane?

12. What structural characteristics of glycophorin help maintain its asymmetric orientation? What prevents it from reorienting from one face of the bilayer to another by flip-flop diffusion?

Looking at Membranes

13. What sort of structural information could be gained from the **freeze-fracture** method for electron microscopic viewing of the inside of the erythrocyte membrane?

Peripheral Membrane Proteins Are Easily Solubilized

14. Which forces, interactions, or bonds are most important in binding peripheral proteins to membranes?

15. What is the role of Ca^{2+} ions in the annexin-membrane interaction?

Covalently Attached Lipids Anchor Some Peripheral Membrane Proteins

16. Besides attachment, what other functions may lipid anchors serve?

Integral Proteins Are Held in the Membrane Bilayer by Hydrophobic Interactions with Lipids

17. List and briefly describe the six categories of integral protein spatial arrangements.

18. What forces stabilize α helices formed from hydrophobic amino acid residues in transmembrane proteins? (Hint: Review the section on α-helical structures in Chapter 6).

19. Why would the photosynthetic reaction center of *Rhodopseudomonas viridis* be insoluble in water? How, then, could it be solubilized for study?

The Topology of an Integral Membrane Protein Can Sometimes Be Predicted from Its Sequence

20. Note that hydropathy indices can be determined only if the amino acid sequence of the protein in question is already known.

21. What are some of the similarities and differencs between α-helical and β-barrel transmembrane segments?

22. *Cell Map:* In the schematic diagram of the membrane bilayer at the top of the map, label the different components of cell membranes.

Integral Proteins Mediate Cell-Cell Interactions and Adhesion

23. How is it possible for integrins to have such a wide variety of specificities?

Membrane Fusion Is Central to Many Biological Processes

24. What forces or interactions allow membranes to reorganize without loss of continuity; i.e., what are the interactions that organize amphipathic lipids into the bilayer?

25. What are the similarities between the membrane fusion processes in the entry of the influenza virus and HIV into host cells? What are the differences?

Solute Transport across Membranes

Passive Transport Is Facilitated by Membrane Proteins

26. What are the two components of the electrochemical gradient?

27. In what ways are membrane transporters and enzymes similar? In what ways are they different?

Aquaporins Form Hydrophilic Transmembrane Channels for the Passage of Water

28. Review which amino acid side chains are classified as hydrophilic and which as hydrophobic (Ch. 5, p. 119).

The Glucose Transporter of Erythrocytes Mediates Passive Transport

29. What does the V_0 vs. $[S]_{out}$ plot for glucose entry into a cell tell you about the transport process itself?

30. What sets the upper limit on the concentration of glucose that can be transported into erythrocytes by GluT1?

31. Does GluT2 have a higher or lower affinity for glucose than GluT1? What could be a physiological reason for this?

Defective Glucose and Water Transport in Two Forms of Diabetes

32. These examples of disease states describe the effects of transporters (of either glucose or water). Why can't the transporters already present in the plasma membrane be induced to increase their *rate* of transport? (Hint: What are the hallmarks of passive transport?)

Chloride and Bicarbonate Are Cotransported across the Erythrocyte Membrane

33. The passive transport process mediated by the chloride-bicarbonate exchanger is reversible. Why is this physiologically necessary?

Active Transport Results in Solute Movement against a Concentration or Electrochemical Gradient

34. Keep in mind the definitions of primary and secondary active transport for the rest of the chapter.

35. What information does $\Delta\psi$ supply?

There Are at Least Four General Types of Transport ATPases

36. What activity is common to all types of ATPases?

37. Categorize the different ATPases in Table 12–4 as transporters capable of uniport, symport, or antiport.

A Defective Ion Channel Causes Cystic Fibrosis

38. Phe^{508} is located in the NBF1 domain, which extends from the cytosolic surface. How, then, can its deletion affect the insertion of the transmembrane portion of the protein, and thus its function as a Cl^- channel?

A P-Type ATPase Catalyzes Active Cotransport of Na^+ and K^+

39. If the intracellular concentration of K^+ is 140 mM, as indicated in Figure 12–33, why doesn't intracellular K^+ bind to the empty K^+ sites of the Na^+K^+ATPase shown in part (a) of Figure 12–34?

40. *Cell Map:* Use other portions of the plasma membrane to draw in different transporter proteins and the solutes being transported. The molecular model for the Na^+K^+ ATPase has been provided. You should draw in the details (i.e., solutes and substrates bound and phosphorylation states) of the different conformations of this transporter. Be sure to include various means by which glucose crosses cell membranes.

ATP-Driven Ca^{2+} Pumps Maintain a Low Concentration of Calcium in the Cytosol

41. Why must cytosolic concentrations of calcium be kept low?

Ion Gradients Provide the Energy for Secondary Active Transport

42. How is the Na^+K^+ ATPase involved in transport of glucose and amino acids in intestinal epithelial cells?

43. How does the ionophore valinomycin facilitate the movement of potassium ions across membranes? What is the biological and medical significance of their function?

Ion-Selective Channels Allow Rapid Movement of Ions Across Membranes

44. How are ion channels different from ion transporters?

The Structure of a K^+ Channel Shows the Basis for Its Ion Specificity

45. What types of noncovalent interactions are critical to the ion specificity of the K^+ channel?

The Acetylcholine Receptor Is a Ligand-gated Ion Channel

46. What is the "ligand" for this ligand-gated ion channel?

The Neuronal Na^+ Channel Is a Voltage-gated Ion Channel

47. How are ligand-gated ion channels and voltage-gated ion channels similar? How are they different? How does the nervous system rely on voltage-gated channels?

Ion Channel Function Is Measured Electrically

48. Why is it functionally important that an ion channel remains open for only a very short time?

Defective Ion Channels Can Have Striking Physiological Consequences

49. How can some toxins be so deadly at very low concentrations?

Porins Are Transmembrane Channels for Small Molecules

50. How is the action of the FhuA protein similar to that of a gated channel?

SELF-TEST

Do You Know the Terms?

ACROSS

1. Simultaneous transport of two solutes across a membrane, in either the same or opposite directions.
4. The Na^+K^+ ________ is an example of a cotransporter that is critical to the function of all cells.
6. ________ proteins are very firmly associated with the membrane via hydrophobic interactions with the fatty acid chains of membrane lipids.
7. Proteins rarely exhibit this type of movement in membranes.
8. ________-________ ATPases pump protons, regulating the pH of intracellular compartments.
11. The ________ potential takes into account the effects of both the chemical concentration gradient and the electrical gradient.
13. The structural organization of lipids in biological membranes.
17. Class of lipids containing covalently attached carbohydrates.
18. Face of the lipid bilayer where 2 K^+ ions are released by the Na^+K^+ ATPase.
19. The evocative name of the model describing the structure of biological membranes is the ________ mosaic model.
21. ________ diffusion is mediated by an integral membrane protein that lowers the activation energy for transport; this process exhibits saturation kinetics.
22. A membrane protein in an intact erythrocyte that reacts with IEA must have at least one domain exposed on the ________ face of the lipid bilayer.
24. Facilitated diffusion is also called ________ transport.
26. SNAREs are proteins required for membrane ________ in the process of exocytosis.
27. ________ interactions among lipid molecules in water drive the formation of micelles, bilayers, and liposomes.
28. Type of rapid diffusion exhibited by both lipids and proteins in membranes.

DOWN

1. Membrane sterol responsible for modulating membrane fluidity, depending on the ambient temperature.
2. "Flip-flop" of lipids in membrane bilayers is also known as ________ diffusion; facilitated by flippases.
3. A member of the family of peripheral membrane proteins that associates with the polar head groups of membrane lipids in a Ca^{2+}-dependent manner; may be involved in membrane fusion events.
4. The transport of solutes against a concentration or electrochemical gradient that requires the input of energy is known as ________ transport.
5. The major class of membrane lipids, in terms of weight percent.
9. Ion-selective ________ provide a route for the rapid movement of ions across membranes.
10. Type of membrane protein that is water soluble and can be isolated from membranes by changing the pH or ionic strength of the surrounding solution.
12. A plot of ________ index vs. amino acid residue number in a protein predicts potential membrane-spanning α-helical regions of integral membrane proteins; such a plot is *not* useful for predicting β-barrel transmembrane segments.
14. ________ port refers to the simultaneous transport of two solutes across a membrane in opposite directions.

15. Describes the polar head groups of membrane lipids and peripheral membrane proteins.
16. ________port; transport of a single solute across a membrane.
17. An ion ________ is a source of potential energy that drives secondary transport processes in cells.
19. One category of ________-________ ATPases that are responsible for the production of ATP in mitochondria and chloroplasts; they are also known as ATP synthases.
20. ________-________ ATPases are reversibly phosphorylated by ATP as part of the transport process.
23. Type of diffusion that occurs down a concentration gradient.
25. ________port; transport of two solutes across a membrane in the same direction.
28. An example of a(n) ________-gated ion channel is the acetylcholine receptor.
29. The family of integral proteins that provide channels for rapid movement of water across plasma membranes is known as the ________porins.

Do You Know the Facts?

1. Would the following changes in the structure and/or composition of membrane lipids increase, decrease, or have no effect on the solid-to-fluid transition temperature? (Hint: also see Chapter 11, p. 364–365.)

(a) Unsaturated hydrocarbon chains are replaced by saturated hydrocarbon chains.

(b) The hydrocarbon chains are shortened.

In questions 2–7, decide whether the statement is true or false, then explain your answer.

2. Proteins and lipids account for almost all of the mass of biological membranes.

3. Proteins and lipids are the only components of biological membranes.

4. The relative proportions of protein and lipid are the same in all biological membranes.

5. All the cellular membranes in a particular organism contain the same lipids and the same percentage of lipid to protein.

6. Membranes with different functions have different proteins.

7. Mitochondrial membranes commonly include covalently bound carbohydrate moeities.

8. Which of the following is *not* a membrane lipid? Why?

A. cholesterol
B. triacylglycerol
C. phosphatidylglycerol
D. cerebrosides
E. sphingomyelin

9. At low temperatures, cultured cells increase the sterol content of their membranes to:

A. increase membrane fluidity.
B. decrease membrane fluidity.
C. increase the production of lipid intracellular signaling molecules.
D. stabilize membrane proteins.
E. increase permeability to water.

10. Which of the following is *not* a reason why the term "fluid" is appropriate to describe the fluid mosaic model of the biological membrane?

A. Many lipid and protein components are not stationary, but undergo constant lateral motion.
B. The fatty acid chains of the membrane lipids are able to move by rotation about their carbon–carbon single bonds.
C. Individual lipid molecules can diffuse laterally in the plane of the bilayer.
D. Individual lipid molecules often flip-flop from one side of the bilayer to the other.
E. All of the above are reasonable.

11. Proteins A, B, and C are membrane proteins associated (either as peripheral or integral proteins) with the intact biological membranes of a culture of cells. After the cells containing these proteins are exposed to a severe drop in ionic strength of the cell culture medium, only protein A remains associated with the cells. One can conclude that:

A. B and C are peripheral proteins.
B. B and C are integral proteins.
C. A is a peripheral protein.
D. A is either an integral protein or a peripheral protein covalently attached to a membrane lipid.
E. Both answers A and D apply.

12. Which of the following statements is true concerning integral membrane proteins?

A. Hydrophobic interactions anchor them within the membrane.
B. Ionic interactions and hydrogen bonds occur between the protein and the fatty acyl chains of the membrane lipids.
C. These proteins can be solubilized by a solution of high ionic strength.
D. Hydropathy plots can be used to determine the amino acid sequence of the protein.
E. All of the above.

13. Ruptured biological membranes are "self-sealing" due to all of the following *except*:

A. the amphipathic character of the lipids.
B. hydrophobic interactions between lipids.
C. hydrogen bonding between the head groups of the lipids and H_2O.
D. an increase in entropy of the system upon sealing.
E. covalent interactions among lipids.

14. The hydropathy plot below provides which of the following information?

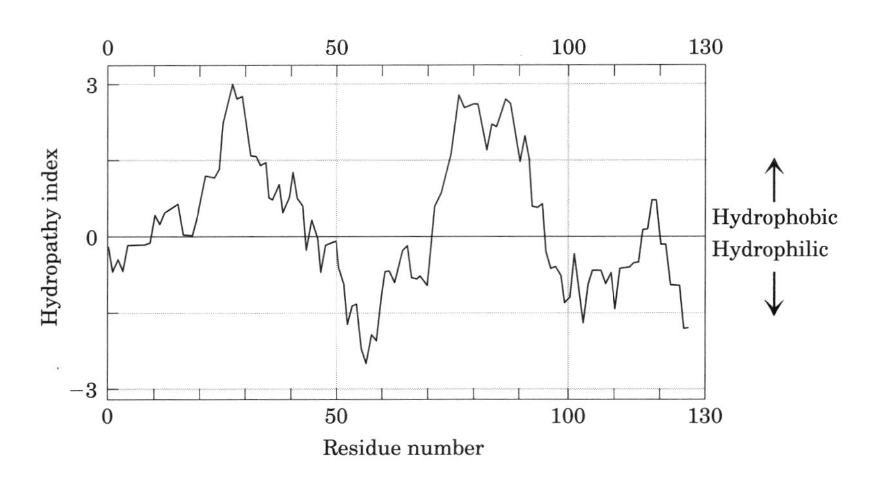

A. The protein being analyzed is a globular, peripheral membrane protein.
B. The protein being analyzed may be an integral membrane protein with two membrane-spanning domains.
C. The membrane being analyzed has a high degree of fluidity.
D. The membrane being analyzed has a high proportion of long-chain (as opposed to short-chain) fatty acids.
E. The protein being analyzed has extensive β-barrel domains.

15. The Na^+K^+ ATPase:

A. mediates active transport.
B. mediates cotransport of Na^+ and K^+.
C. is an integral membrane protein.
D. creates a transmembrane potential.
E. has all of the characteristics above.

16. Describe the lipid and protein content of the following membranes that are described in the text. How do these differences relate to their corresponding functions?

(a) The myelin sheath

(b) Vertebrate retina rod-cell

(c) Plasma membrane of an erythrocyte

17. What experimental evidence supports the lipid bilayer as the basic structure of biological membranes?

18. Explain how sterols act to moderate the extremes of solidity and fluidity of the membranes that contain them.

19. Describe the effect of each of the treatments below on the various types of membrane proteins. Would the treatment release a peripheral protein from the membrane? An integral protein? A peripheral protein covalently attached to a lipid anchor?

(a) Changes in pH

(b) Changes in ionic strength

(c) Detergent

(d) Urea

(e) Phospholipase C

20. Describe the similarities and differences between transporter-mediated facilitated diffusion and transport via an ion-selective channel.

21. For each of the inhibitors listed below, describe first their *direct effect*, and then indicate the type of the transport mechanism(s) that they block.

(a) Vanadate

(b) Cyanide

(c) Valinomycin

Applying What You Know

1. Two different species of bacteria have been isolated from very disparate environments: one, a hot spring with an average water temperature of ~40°C, the other a glacial lake with an average water temperature of ~4°C.

(a) Which of the two bacterial species would be expected to have more unsaturated fatty acids in its membrane lipids?

(b) Which would have longer-chain fatty acids?

(c) At 27°C, which species would have a more fluid membrane?

2. You want to purify an integral membrane protein from the mitochondrial membranes of beef heart. (Because it is large and pumps constantly, it contains many energy-generating mitochondria per cell.) Compared with the precautions taken in a cytosolic protein purification, what extra precautions must you take to purify the integral membrane protein?

3. In the experiment described in Figure 12–7, suppose the chloride-bicarbonate exchange protein were one of the human cell proteins that was followed for location on the fused mouse-human cell. Would these proteins mix evenly on the surface of the fused cell? Why or why not?

4. Ouabain is a specific inhibitor of the Na^+K^+ ATPase. You wish to determine how ouabain accomplishes this inhibition and perform a series of experiments; the results are presented below. What does each result tell you about the mechanism of inhibition of ouabain on the Na^+K^+ ATPase?

(a) Ouabain added to a solution containing intact cells inhibits movement of the ions across the plasma membrane (K^+ in and Na^+ out). Ouabain injected *inside* individual cells does *not* inhibit the movement of K^+ and Na^+ in those cells.

(b) Ouabain is added to a solution containing intact cells and allowed to effect full inhibition of the transporter. The cells are then collected, their membranes solubilized, and the membrane proteins are studied. All the molecules of Na^+K^+ ATPase from the ouabain-treated cells are found to be in the phosphorylated form (P-Enz_{II}).

(c) Based on the above results, which of the forms of the Na^+K^+ ATPase shown in Fig. 12–34 represents the conformation that ouabain brings about?

5. Glucose uptake into adipocytes is mediated by the glucose transporter GluT4 (see Box 12–2). The majority of GluT4 transporters are normally located in the membranes of intracellular vesicles but are translocated to the plasma membrane via membrane fusion in response to insulin; the resulting increase in the number of glucose transporters in the plasma membrane allows increase in uptake of glucose into the cells. Recently, a role for specific SNARE proteins has been described in the insulin-induced translocation of GluT4 to the plasma membrane.

Before this mechanism was determined, other hypotheses existed to explain the behavior of GluT4 uptake of glucose. Briefly describe an experiment that would investigate each of the following possible alternative explanations for the increase in glucose uptake by GluT4 in response to insulin.

(a) Insulin increases the GluT4 V_{max}, or decreases its K_m for glucose.

(b) Many molecules of GluT4 are already present, but they are free in the cytoplasm and not located in the internal membranes.

(c) Membrane fusion is not a necessary part of the translocation process.

(d) SNAREs are not a necessary part of this process.

Biochemistry on the Internet

Nerve cells are able to transmit information due to their ability to generate electrical signals. These signals are the result of rapid changes in the electrical potential across the plasma membrane. If the change in the electrical potential (voltage change) is large enough, it creates a domino effect resulting in similar voltage changes that are transmitted along the entire length of the nerve cell. Such voltage changes are due to rapid changes in the distribution of charged molecules, primarily Na^+ and K^+ ions. Ion channels provide a favorable environment for the movement of ions through the uncharged and hydrophobic environment of the lipid bilayer.

The K^+ channel protein is an example of an integral membrane protein that facilitates the movement of ions through cell membranes. This channel is composed of four subunits, each of which has multiple membrane-spanning segments. A portion of this channel's structure has been determined and is depicted in the model with the PDB ID 1BL8. Try to answer the following questions based on this molecular information.

(a) How many subunits of the tetrameric potassium channel are represented in this model? How many amino acid residues are in each subunit shown in the model?

(b) How would you expect this portion of the K^+ channel protein to be oriented in a plasma membrane; that is, which portions would face the aqueous environment and which portions would be associated with the membrane lipids?

(c) This ion channel is relatively selective for potassium ions. What portion of this structure contributes to its ability to allow the passage of K^+ while excluding other similarly charged ions?

(d) What factors are important in preventing this structure from rotating within the plasma membrane and in ensuring that the openings of the channel remain exposed to the aqueous environment?

ANSWERS

Do You Know the Terms?

ACROSS

1. cotransport
4. ATPase
6. integral
7. flip-flop
8. V-type
11. electrochemical
13. bilayer
17. glycolipid
18. inner
19. fluid
21. facilitated
22. outer
24. passive
26. fusion
27. hydrophobic
28. lateral

DOWN

1. cholesterol
2. transbilayer
3. annexin
4. active
5. phospholipid
9. channels
10. peripheral
12. hydropathy
14. anti
15. hydrophilic
16. uni
17. gradient
19. F-type
20. P-type
23. simple
25. sym
28. ligand
29. aqua

Do You Know the Facts?

1. (a) increase; (b) decrease
2. T; see Table 12–1.
3. F; carbohydrates are also present, usually as glycoproteins and glycolipids.
4. F; proportions differ by organism and function; see Table 12–1.
5. F; content and proportion differ according to function, organ, role in cell, and location.
6. T; the proteins determine the function to a large degree.
7. F; plasma membranes contain glycoproteins on their outer face, but intracellular membranes rarely do.
8. B; triacylglycerols are not amphipathic and so do not aggregate into bilayers.
9. A
10. D
11. E
12. A
13. E
14. B
15. E
16. **(a)** The myelin sheath consists primarily of lipids; it functions as a passive insulator wrapped around the axons of myelinated neurons, and lipids are excellent insulators (see Chapter 11, p. 367).
    (b) In vertebrate retina rod-cell outer segments, a very large percentage (>90%) of the membrane proteins consists of just one single protein, the light-absorbing protein rhodopsin. The function of these cells is to detect light.
    (c) The plasma membrane of an erythrocyte has many prominent proteins, not one specific one. This membrane is not specialized, but has many functions.
17. **(a)** In water, amphipathic lipids spontaneously form bilayers that are stabilized by hydrophobic interactions.
    (b) The thickness of biological membranes is about that expected for a lipid bilayer with protruding proteins on either side.
    (c) Distribution of electron density, as shown by x-ray diffraction, is as expected for a bilayer structure.
    (d) Laboratory-made liposomes behave similarly in terms of their impermeability to polar solutes.
    (e) Membranes prepared and stained using a variety of techniques have been viewed by transmission and scanning electron microscopy. These techniques, especially freeze-fracture (see Box 12–1), reveal views of the biological membranes all consistent with the lipid bilayer model.
18. The bulky steroid structure (Fig. 11–14) fluidizes the membrane below the transition temperature by preventing the highly ordered packing of neighboring fatty acyl chains. Above the transition temper-

ature, the steroid nucleus reduces the freedom of rotational movement of the carbon–carbon bonds of its fatty acyl neighbors in the bilayer, contributing to solidity.

19. The treatments described would release the membrane proteins that are: **(a)** peripheral; **(b)** peripheral; **(c)** integral and peripheral with anchor; **(d)** peripheral; **(e)** peripheral with anchor. Peripheral proteins associate with the membrane through electrostatic interactions and hydrogen bonds. They can therefore be released by changes in pH or ionic strength (which disrupt weak, noncovalent interactions), or by addition of urea (which breaks hydrogen bonds). Some peripheral proteins are attached to the membrane by a covalent lipid "anchor" as well; these can be released by the use of an appropriate enzyme. Integral proteins are more firmly attached, through hydrophobic interactions with the lipid bilayer. These can only be released from the membrane with detergents, organic solvents, or denaturants. Detergents form micelles with the individual proteins, keeping them in native conformation.

20. Similarities: Both types of transport require a protein that spans the bilayer (probably several times) and forms a transmembrane channel that provides a path through which a *specific* substrate can cross the membrane. Both only allow flow of substrate *down* its concentration gradient. Differences: Transporters are saturatable; the rate of transport reaches a V_{max} at high substrate concentrations. Ion channels are not saturatable. While transporters speed up the rate at which their specific substrate can cross the membrane, the rate of flux through ion-selective channels can be orders of magnitude greater than for transporters. Transporters are permanently "open" to their substrate, as long as the concentration gradient exists. Ion channels are transient gates; they open or close in response to a specific signal.

21. (a) Vanadate is an analog of phosphate; it interferes with reversible phosphorylation, inhibiting P-type ATPases. All functions listed for P-type ATPases (in Table 12–4) would be blocked.

(b) Cyanide (CN^-) blocks energy-yielding oxidation reactions that produce a H^+ (proton) gradient. It inhibits both F-type ATPases (ATP synthases) and secondary active transport processes powered by proton gradients.

(c) Valinomycin binds K^+ in a hydrophilic central cavity, while presenting a hydrophobic exterior. This allows diffusion of the complex across the lipid bilayer of membranes, dissipating K^+ ion gradients. Valinomycin disrupts secondary active transport systems.

Applying What You Know

1. (a) The glacial lake bacterial species would have more unsaturated fatty acids in its membrane.

(b) The hot spring species would contain more long-chain fatty acids in its membranes.

(c) At 27°C, the glacial lake species would most likely have a more fluid membrane.

2. Detergent must be included in all buffers, chromatographic columns, etc., or the integral protein will precipitate out as an insoluble aggregate.

3. The chloride–bicarbonate exchange protein would *not* mix evenly over the surface of the fused mouse–human cell, because this integral membrane protein is tethered to an internal cytoskeletal protein.

4. (a) Ouabain must bind specifically to a portion of the Na^+K^+ ATPase that is on the *outside* face of the plasma membrane.

(b) Ouabain binds to and "locks" the Na^+K^+ ATPase in its phosphorylated state.

(c) P-Enz_{II} with 2 K^+ bound and no Na^+ bound (the phosphorylated form has a high affinity for K^+ and a low affinity for Na^+), facing outside.

5. (a) Perform a series of [S] vs. V_0 trials (see Chapter 8) using a *specific number* of purified GluT4 molecules in the presence and absence of insulin; determine the K_m, V_{max}, and K_{cat} (turnover number) values.

(b) Break apart adipocytes, separate soluble proteins from membrane-bound proteins, and determine the locale of the GluT4 protein (using an antibody to GluT4, for example). This strategy could also be used to quantify movement of GluT4 transport proteins from intracellular membranes to the plasma membranes by looking at insulin-induced vs. non-insulin-induced cells.

(c) Lower the temperature of the system and determine if the movement of GluT4 to the plasma membrane slows down (glucose transport slows). Membrane fusion, like all processes that are based on the fluidity of the membrane, will slow down or halt at low temperatures. (This would not be the only explanation, though; other cellular processes are affected by temperature as well.)

(d) Generate anti-SNARE antibodies (see Chapter 7); use them as tools to block interaction of SNAREs in insulin-induced adipocytes, and determine if the rate of glucose transport slows down.

Biochemistry on the Internet

(a) A portion of all four subunits is shown. Display the structure in a **Ribbon** or **Backbone** format and then choose **Color → Chain.** You will see four differently colored segments of the protein backbone displayed. You can determine the number of amino acid residues in each chain by clicking on both ends of the polypeptide backbone. The information displayed indicates that each of the four identical chains comprising the K^+ channels from *Streptomyces lividans* includes residues 23 (Ala)–119 (Gln) of the actual protein. Thus, there are 97 amino acid residues in each chain.

(b) You can try to predict which portion of the polypeptide chain would be exposed to the aqueous environment and which would be associated with lipids based on the properties of the individual amino acid residues. If you click on each end of an alpha helix in these subunits, it is clear that they contain enough amino acid residues to span the lipid bilayer (24–26 residues). Select the hydropho-bic amino acids **(Select → Protein → Hydrophobic)** and **Display** them in a **Ribbon** or **Ball and Stick** format. Then select the charged amino acids **(Select → Protein → Charged)** and **Display** them in **Spacefill** format. The hydrophobic amino acid residues are primarily located in the α-helical segments of the structure. The charged (and therefore hydrophilic) residues are clustered in two regions located on each end of the α-helical segments. These charged resi-dues form ring-shaped structures that would be expected to associate with the aqueous environment on both sides of the lipid bilayer. Display the polar amino acid residues in **Spacefill** format. The polar residues (which are also capable of forming favorable interactions with water molecules) show a similar distribution to that of the charged residues.

Knowing the structure's orientation within a membrane, rotate the molecule so that you are viewing it from one of the ends that are in contact with the aqueous environment. It should be clear from this view how the protein complex allows the movement of molecules from one side of the membrane to the other.

(c) Redisplay the entire molecule in a backbone format **(Select → Select All** then **Display → Backbone).** Then **Select** the **Protein → Acidic** residues. **Display** these residues in a **Spacefill** format. Since these residues are negatively charged, they would be expected to selectively interact with positively charged ions such as K^+. A total of four residues per subunit should be highlighted. These residues are located exclusively at each portion of the channel that is in contact with the aqueous environment. The presence of negative charges on the larger opening facilitates the attraction of positively charged K^+. The narrowing of the other opening combined with the presence of additional negative charges ensures that only molecules of a specific size, as well as charge, can pass through the opening.

In order to verify your assumptions, click **Select → Hetero** and **Display** the molecules in a **Spacefill** format. You will see three K^+ and an O_2 atom located in the region of the negatively charged residues.

(d) It should be clear from the above discussions that the presence of a significant number of hydrophilic residues at each end of the cylindrical channel will ensure that the opening of the channel is not buried in the lipid bilayer and remains associated with the aqueous environment. The hydrophobic residues ensure that the structure remains embedded in the lipid bilayer. Because these interactions are thermodynamically favorable, no energy needs to be expended to keep the molecule bound and oriented properly.

chapter

13

Biosignaling

STEP-BY-STEP GUIDE

Major Concepts

In order to regulate and integrate their metabolic processes, cells have signaling mechanisms that relay information about the external environment to the interior of the cell.

Cells and their metabolic pathways do not exist in a vacuum. To function efficiently cells must be able to coordinate their intracellular metabolism with the prevailing extracellular environment. In order to respond to information about their environment, cells must possess some kind of "sensor" or **receptor** that can interact with extracellular molecular signals. The production of a specific receptor in a cell, coupled with a fairly high affinity (low K_d) of that receptor for a given molecular signal, makes it possible for cells to be selective about the signals they respond to.

Some of the extracellular signaling molecules that cells can respond to are lipid soluble, which allows them to pass unhampered through the lipid bilayer to the intracellular compartment. Receptors for lipid-soluble signals are found in the cytoplasmic and nuclear compartments of many cells. Some signaling molecules are hydrophilic; consequently, cells must have receptors for these signaling molecules on their plasma membrane. This latter signaling system requires a transducer element, which is "turned on" by the binding of the external signal, and which relays information by affecting the cellular machinery. In general, all signaling systems must be able to **amplify** the external signal, **integrate** information from a variety of sources, and turn themselves off when appropriate.

Cells use four basic mechanisms to transmit signals: (1) ion channels, (2) receptor enzymes, (3) G protein–linked receptors, and (4) steroid hormone receptors.

Ion channels allow the movement of ions across the plasma membrane in excitable cells.

Ion channels are integral membrane proteins that span the lipid bilayer several times, thus forming pores that allow charged molecules (e.g., Na^+, K^+, Ca^{2+}) to cross the membrane. The energy for the transmembrane movement through channels comes from the ion gradients established by the Na^+/K^+ ATPase. Channels are usually gated, that is, they can be opened and closed in response to specific signals. These signals include the binding of hormones to cell membrane receptors or changes in the electrical potential (voltage) across the cell membrane. The influx or efflux of ions through these pores can have a variety of effects on cellular metabolism. The nicotinic acetylcholine receptor that mediates muscle contraction in response to neural stimulation is an example of a receptor channel.

Overview of Neuronal Function

To understand signaling activity in neurons, it is helpful to categorize neuronal function into four distinct regions:

The **input region** is usually made up of cellular projections called dendrites. Dendrites contain **ligand-gated channels,** which respond to extracellular signals (e.g., neurotransmitter molecules). Binding of signaling molecules causes a conformational change that opens the associated channel. The movement of ions through channels in the plasma membrane results in a change in the **membrane potential (V_m),** which is passively transmitted by diffusion of ions away from the site where they entered the cell.

At the **integrating region**, called the spike-initiating zone, all of the changes in V_m caused by the opening of ligand-gated channels in the input region are added together. If the summed potentials are above a "set point," called the threshold potential, then an **action potential (AP)** is produced.

An AP is the rapid shift in V_m from a resting potential (the inside of the cell is negatively charged with respect to the outside) to a more positive intracellular potential. An AP is generated primarily by the activity of two **voltage-dependent ion channels,** the Na^+ and K^+ voltage-dependent channels. The initial, rapid and large depolarization of an AP is the result of the opening of **voltage-dependent Na^+ channels**. When these channels open, Na^+ ions enter the cell and flow down their concentration gradient. The V_m in the region of the open channels will move toward the equilibrium potential for Na^+ (which can be calculated using the Nernst equation). The depolarization that occurs during an action potential ultimately causes the voltage-dependent Na^+ channels to close. **Voltage-dependent K^+ channels** open more slowly than do the Na^+ channels and therefore are not fully functional until after the initial Na^+-induced depolarization has occurred. Because the intracellular $[K^+]$ is higher than the extracellular $[K^+]$, this ion will exit the cell. Thus, the V_m will move toward the equilibrium potential for K^+, and the cell will be repolarized to the original resting potential.

The **conducting region** of a neuron is known as the axon and contains many voltage-dependent Na^+ and K^+ channels along its length. The sequential opening of these channels results in an AP that is conducted along the entire length of an axon.

The **output region** of a neuron is usually the nerve terminus. Here, the plasma membrane contains its own contingent of voltage-dependent channels; however, these channels are primarily permeable to Ca^{2+}. The Ca^{2+} ions that enter the terminal initiate a series of events that lead to the fusion of intracellular membrane-bound vesicles with the cell membrane. This fusion causes the release of thousands of neurotransmitter molecules into the extracellular space, called the **synapse,** that occurs between adjacent neurons.

Receptor enzymes have an extracellular receptor for the signal and a catalytic domain on the portion facing the cytosol.

In receptor enzymes, the binding of hormone to the extracellular domain often activates the enzymatic activity on the inside of the cell. The specific enzymatic activity that is turned on may vary, but often involves that of protein kinases. **Kinases** affect cellular metabolism by **phosphorylating** different target proteins such as metabolic enzymes. The insulin receptor is an example of a receptor enzyme in which a protein kinase activity is activated in response to insulin binding.

G protein–linked receptors use a GTP-binding protein to relay information to the interior of cells.

In some cases the message carried by extracellular signals is transmitted to the interior of a cell through the generation of a second signaling molecule inside the cell. These molecules are referred to as **second messengers.** This type of signaling involves the binding of a hormone to an extracellular receptor, which activates a G protein. The G protein, in turn, activates an intracellular target protein (e.g., an enzyme or ion channel). It is this target protein that is responsible for generating the second messenger.

The G proteins can activate three major second messenger systems.

Cyclic nucleotides: Activated **G proteins** can affect the activity of a class of membrane-associated enzymes known collectively as cyclases. Two examples are adenylyl cyclase and guanylyl cyclase, which produce the second messengers **adenosine 3′,5′-cyclic nucleotide monophosphate (cAMP)** and **guanosine 3′,5′-cyclic nucleotide monophosphate (cGMP)** respectively. Many of the effects of cAMP and cGMP are exerted through their activation of specific cyclic nucleotide–dependent protein kinases. These activated kinases then affect cell metabolism by phosphorylating target proteins within the cell.

Phosphatidylinositol bisphosphate metabolites: Hormonal activation of a different class of G proteins activates membrane-associated **phospholipase C (PLC).** This enzyme can cleave the membrane lipid **phosphatidylinositol bisphosphate (PIP_2)** producing two distinct intracellular messengers. One is **diacyl glycerol (DAG),** which is lipid soluble and therefore must exert its effects in or near the lipid bilayer. Typically DAG (along with Ca^{2+}) activates the enzyme protein kinase C (PKC), which like other protein kinases exerts its effects on cells by phosphorylating target proteins. The second is **inositol triphosphate (IP_3),** which is water soluble and exerts its effects through binding to its own intracellular receptor on the surface of the endoplasmic reticulum membrane (ER) or sarcoplasmic reticulum membrane (SR). Binding to these intracellular receptors opens channels in the ER or SR membrane that allow Ca^{2+} that is stored in these structures to diffuse into the cytosol.

Calcium: In addition to the IP_3-mediated release of stored Ca^{2+} described above, there are other routes for getting this second messenger into cells. Normally the level of Ca^{2+} inside cells is very low ($< 1 \mu M$). Extracellular hormones as well as changes in V_m open Ca^{2+} channels in the plasma membrane, resulting in an increase in the intracellular [Ca^{2+}] as this ion passively flows down its concentration gradient into the cell. Ca^{2+} affects cell function in a number of ways, e.g., by binding directly to enzymes and activating or inhibiting them, or by binding to structural proteins and affecting their function. Alternatively, Ca^{2+} can also bind to its own intracellular receptor, **calmodulin.** This Ca^{2+}-protein complex then binds to cellular proteins, affecting their function.

Steroid hormones do not require intracellular messengers because they are lipid soluble and can diffuse across the membrane.

Because **steroid hormones** are relatively hydrophobic they usually must associate with a water-soluble molecule in order to be transported in the blood. When they enter the cell cytoplasm they must bind to specific hormone receptors in the nucleus before they can influence cellular function. Binding to these receptors causes conformational changes that allow the complex to associate with specific regions of DNA and influence gene expression.

Phosphorylation of cellular components is a molecular mechanism used by all cells to regulate cellular processes.

Kinases play a ubiquitous role in the regulation of cellular metabolism. Their influence is due to their ability to phosphorylate cellular components, thereby altering the conformation and thus the function of these substrates. The complicated pathways through which kinases are activated and through which they exert their effects provide multiple points where signal amplication as well as **integration** can occur. The specificity of kinase activity results primarily from the **localization** of the kinases and/or their substrates to specific regions in the cell.

What to Review

Answering the following questions and reviewing the relevant concepts, which you have already studied, should make this chapter more understandable.

- Review the topography of integral membrane proteins (Fig. 12–14 and 12–15).
- The Na^+/K^+ ATPase is critical to the function of all cells and especially for neuronal signaling. Review the model for the Na^+/K^+ ATPase and be sure you understand why this pump is electrogenic (Fig. 12–34).
- Be sure you understand what is meant by membrane potential (Fig. 12–33).
- Hormonal signaling is critical to the regulation of metabolic pathways that you will learn about in coming chapters. The role of second messengers in signaling by peptide hormones is crucial. Two second messengers used by cells are derived from membrane lipids. Review the general structure of lipids (Fig. 11–8) and note the sites of cleavage by lipases (Fig. 11–13).
- Many lipid soluble hormones are derived from cholesterol. Review its structure and be sure you understand why it is lipid soluble. (Fig. 11–14).
- Protein phosphorylation plays a critical role in the regulation of enzyme activity by extracellular hormones and their intracellular messengers. Review which of the amino acids have available —OH groups that can be phosphorylated (Fig. 5–5).

Topics for Discussion

Answering each of the following questions, especially in the context of a study group discussion, should help you understand the important points of this chapter.

Molecular Mechanisms of Signal Transduction

1. The human visual system is sensitive enough to be able to detect a single photon of light. What are the various biochemical mechanisms that allow relatively weak extracellular signals to trigger a significant cellular response?

2. In general terms, what are the four mechanisms cells use to relay extracellular signals to the interior of the cell?

3. How do hormones regulate metabolism only in the cells of specific target organs and not in cells of adjacent nontarget organs?

4. *Cell Map:* By the end of this chapter you should be able to illustrate the various components of the three different classes of cellular receptors in the plasma membrane portion of your cell map.

Scatchard Analysis Quantifies the Receptor-Ligand Interaction

4. What Michaelis-Menten constant is analogous to the K_a for a receptor? B_{max}?

Gated Ion Channels

Ion Channels Underlie Electrical Signaling in Excitable Cells

5. How does the movement of ions through ion channels differ from the movement of ions across membranes achieved by the Na^+/K^+ cotransporter?

6. Why do cells use ion channels instead of the Na^+/K^+ cotransporter to generate the membrane voltage changes that are the basis for neuronal signaling?

7. What are the two gradients that produce the "potential" in the electrochemical potential?

The Nicotinic Acetylcholine Receptor Is a Ligand-Gated Ion Channel

8. *Internet Question:* To see a model of a ligand bound to a receptor go to the Protein Data Bank and retrieve the file with PDB ID 1A22. This is a model of growth hormone bound to the extracellular portion of its receptor. What forces allow the ligand to associate with its binding site on the extracellular portion of the receptor? (Hint: Try displaying the receptor chain in a Spacefill format and the growth hormone in Ribbon structure. Then select charged amino acid residues and display them in a Ball and Stick format.)

9. What will happen to the V_m of a neuron when a neurotransmitter opens a channel that is permeable to more than one ionic species, e.g., Na^+ *and* Ca^{2+}? (Hint: Can you use the Nernst equation to predict the V_m when the membrane is permeable to more than one ionic species? What additional information might you need to know about the ionic permeabilities?)

10. How would V_m be affected by the opening of channels that are *equally* permeable to both Na^+ *and* Cl^-? (Hint: Use the Nernst equation to determine the separate equilibrium potentials and then factor in relative permeability.)

11. In what ways is the binding of acetylcholine to the nicotinic acetylcholine receptor similar to the binding of O_2 to hemoglobin?

Voltage-Gated Ion Channels Produce Neuronal Action Potentials

12. How might ion channel proteins "sense" voltage changes across the plasma membrane? What classes of amino acid residues are likely to be involved in the sensing process?

13. How would the membrane potential, V_m, and AP production be affected by opening voltage-dependent Na^+ channels and voltage-dependent K^+ channels *simultaneously?*

Neurons Have Receptor Channels That Respond to a Variety of Neurotransmitters

14. The interior of ion channels often includes a region referred to as a "selectivity filter". How is it that ion channels can "select" the type of ions that pass through them?

Receptor Enzymes

15. How does the structure of receptor enzymes differ from the structure of receptor channels?

The Insulin Receptor Is a Tyrosine-Specific Protein Kinase

16. *Cell Map:* Draw a diagram that illustrates the general mechanism whereby insulin exerts its effects on cell metabolism.

Guanylyl Cyclase Is a Receptor Enzyme That Generates the Second Messenger cGMP

17. What is “cyclic” about cyclic GMP?

18. How is it that cGMP can generate different responses in different tissues?

G Protein–Coupled Receptors and Second Messengers

19. *Cell Map:* Draw a molecular model showing how a G protein becomes activated in response to binding of a hormone to a G protein–coupled receptor.

The β-Adrenergic Receptor System Acts through the Second Messenger cAMP

20. How is the intracellular signaling pathway mediated by the β-adrenergic receptor different from that of ANF? In what ways are these two pathways similar?

21. In what ways can the effects of epinephrine, acting through the β-adrenergic receptor system, be turned off?

22. How does the G_s protein function as a transducer?

23. *Cell Map:* Draw a model illustrating how the activated G protein from Question 21 turns on the production of cAMP; include the effect of cAMP on its cellular target.

The β-Adrenergic Receptor Is Desensitized by Phosphorylation

24. What is the molecular basis for desensitization of the β-adrenergic receptor?

Cyclic AMP Acts as a Second Messenger for a Number of Regulatory Molecules

25. At which points in the molecular cascade (between receptor binding and kinase activation) can the cAMP second messenger system be regulated?

26. A large number of neurotransmitters and hormones exert their effects on cells by increasing cellular levels of cAMP. How is it possible for each signaling compound to produce a distinct cellular effect when they all use the same second messenger? (Hint: Because of their small size we tend to think of cells as homogeneous "bags of enzymes" but at the microscopic levels they have a great deal of structure and organization.)

Two Second Messengers Are Derived from Phosphatidylinositols

27. *Cell Map:* Both cleavage products of the membrane phospholipid phosphotidylinositol 4,5-bisphosphate (PIP_2) are important intracellular signaling compounds. These products are diacylgylcerol (DAG) and phosphatidylinositol 3,4,5-triphosphate (IP_3). Using your cell map, draw a molecule of PIP_2. To the right of this membrane lipid draw the products of its cleavage by PLC. Which portion of the original membrane lipid do each of these signaling compounds correspond to and where would you expect to find them in the cell?

28. What are the molecular consequences to the cell of increasing the [DAG] in a cell? IP_3?

Calcium Is a Second Messenger in Many Signal Transductions

29. What is calmodulin and in what ways is its function similar to that of G proteins in signal transduction?

30. In what ways is the action of Ca^{2+} similar to that of cAMP in mediating signal transduction?

31. In what ways is ATP critical to the function of this second messenger system?

32. How are the effects mediated by each of the described second messenger systems being turned off?

Sensory Transduction in Vision, Olfaction, and Gustation

Light Hyperpolarizes Rod Cells of the Vertebrate Eye

33. What happens to V_m when a cell is hyperpolarized and how does a decrease in cGMP levels produce this effect?

34. In most neurons, it is the depolarization of membrane potential that leads to an influx of Ca^{2+} and the release of neurotransmitter by the excited neuron. Try to devise a cellular strategy to explain how hyperpolarization in rod cells leads to activation of the visual pathway. (Hint: neuronal signaling in higher organisms is mediated by a *series* of neurons, which relay information to higher brain centers. The signals released by these neurons can have both excitatory as well as inhibitory effects on the cells that follow.)

Light Triggers Conformational Changes in the Receptor Rhodopsin

35. Why are both retinol and opsin needed in order to transduce light energy into something that can affect cellular processes?

Excited Rhodopsin Acts through the G Protein Transducin to Reduce the cGMP Concentration

36. Compare the components of the β-adrenergic receptor system to that of rhodopsin; in what ways are they similar, in what ways do they differ?

Signal Amplification Occurs in Rod and Cone Cells

37. At which points in the transduction of light energy to a neuronal signal is the energy of the original light stimulus amplified?

The Visual Signal Is Terminated Quickly

38. What are the different cellular components that contribute to the rapid termination of the visual signal? (Hint: It may be helpful to construct a cellular model when discussing this question.)

Rhodopsin Is Desensitized by Phosphorylation

39. The protein recoverin plays a role that is analogous to which other important intracellular signaling protein discussed in this chapter?

40. Why do Ca^{2+} levels *decrease* during stimulation of the rod cell by light?

Cone Cells Specialize in Color Vision

41. What is meant by the phrase "slightly different environments" in this section?

Vertebrate Olfaction and Gustation Use Mechanisms Similar to the Visual System

42. List all the different forms of G proteins that have been discussed in this chapter.

43. In what ways are the actions of the different G proteins similar? In what ways are they different?

44. What is the cellular location of each of the different forms of G protein and what are the immediate cellular targets of each form?

45. Why does the closing of K^+ channels cause depolarization in neurons?

G Protein–Coupled Serpentine Receptor Systems Share Universal Features

46. The relatively complex G protein–coupled serpentine receptor system appears to have been evolutionarily "chosen" as the basis for so many different transmembrane signaling pathways. What are the likely reasons that cells don't use much simpler pathways for signaling?

Disruption of G-Protein Signaling Causes Disease

47. Predict the effect of a cholera-like toxin on the visual system if it could interact with $G_{T\alpha}$ in the same way that it interacts with $G_{s\alpha}$. What disease symptoms would you expect to see in a patient exposed to this hypothetical toxin?

Phosphorylation as a Regulatory Mechanism

48. Why is protein phosphorylation such a common regulatory mechanism?

49. Would you expect the consensus phosphorylation sites to be a variable or conserved region of a protein (see p. 139)?

Localization of Protein Kinases and Phosphatases Affects the Specificity for Target Proteins

50. Would you expect a single neurotransmitter or hormone (e.g., epinephrine) to simultaneously activate both protein kinases and protein phosphatases in the same type of cell?

51. How does the existence of proteins such as AKAP help to explain why relatively few intracellular messengers (cAMP, cGMP, Ca^{2+}, IP_3, DAG) are needed to mediate the effects of a wide variety of extracellular signaling molecules?

52. Protein dephosphorylation by phosphatases is a potent regulatory mechanism. What characteristics must phosphatases possess if they are to play regulatory roles in cells?

Regulation of Transcription by Steroid Hormones

53. Why do steroid hormones not require transducing elements such as G proteins to exert their effects on cellular metabolism?

54. Why must the drug tamoxifen, used in the treatment of breast cancer, be lipid soluble?

Regulation of the Cell Cycle by Protein Kinases

The Cell Cycle Has Four Stages

55. How would the daughter cells be affected by a genetic mutation that results in the shortening of the G2 stage of the cell cycle?

Levels of Cyclin-Dependent Protein Kinases Oscillate

56. The cyclin-dependent protein kinase, CDK, is regulated in a manner that is exactly the opposite of the cAMP-dependent protein kinase, PKA. Explain.

Regulation of CDKs by Phosphorylation

57. What are the two molecular effects of CDK phosphorylation that result in the inhibition of catalytic activity?

58. Under what cellular circumstances would it be important to be able to inhibit this enzyme's activity?

Controlled Degradation of Cyclin

59. What would happen to the length of the cell cycle in a cell with a mutation in the gene for DBRP phosphatase? (Assume the mutation renders the gene product inactive.)

Regulated Synthesis of CDKs and Cyclins

60. Progression through the cell cycle depends not only upon the increased synthesis of CDKs but also on the increased synthesis of which other key proteins?

CDKs Regulate Cell Division by Phosphorylating Critical Proteins

61. You have already seen how phosphorylation of enzymes can affect cell metabolism by increasing or decreasing enzyme activity. What are some of the other functional consequences of protein phosphorylation that are discussed in this chapter?

Oncogenes, Tumor Suppressor Genes, and Programmed Cell Death

Oncogenes Are Mutant Forms of the Genes Encoding Proteins that Regulate the Cell Cycle

62. Many of the chemotherapy agents used to treat cancer work by destroying all rapidly dividing cells in an individual. Why is this a good strategy for controlling cancer and why don't these drugs destroy all the other cells in an individual with cancer?

63. Suggest a reason why so many of the known oncogenes code for proteins involved in transmembrane signaling? Why would you expect alterations in signaling to result in cancer?

Defects in Tumor Suppressor Genes Remove Normal Restraints on Cell Division

64. In molecular terms, why are pRb, p53, and p21 all tumor suppressor genes?

Apoptosis Is Programmed Cell Suicide

65. Why must cell death be programmed?

SELF-TEST

Do You Know the Terms?

ACROSS

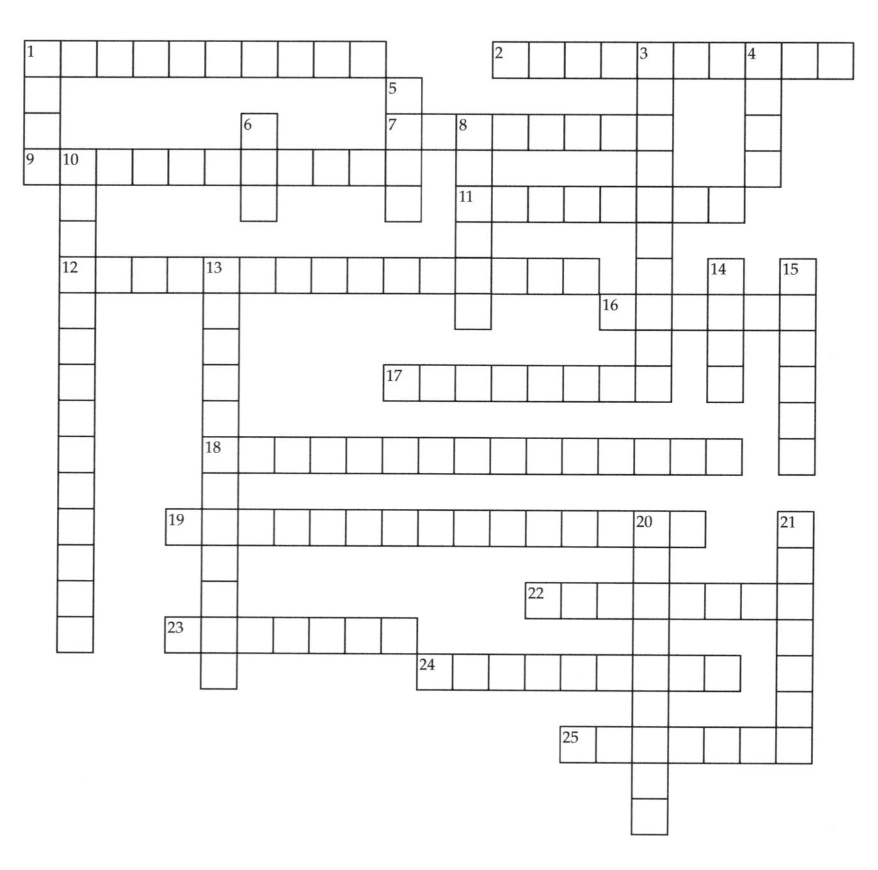

1. This protein mediates many of the actions of the intracellular messenger Ca^{2+}.
2. The G protein that transduces light signals into an electrical signal in the vertebrate rod.
7. This structure is necessary for the cellular reception of all biologically relevant signals.
9. Activation of signaling pathways often results in the phosphorylation of specific cellular proteins; signaling in this type of pathway can be reversed through the actions of this class of enzyme.
11. Ion ________ provide a thermodynamically favorable route for the movements of ions across the lipid bilayer.
12. The direction in which Na^+ ions move through voltage-dependent ion channels is determined by the ________ potential, which is a combination of both the chemical and charge gradients that exists across the cell membrane.
16. The insulin receptor can also be classified as this type of enzyme.
17. "Regulates the Cell Cycle" describes the normal function of this class of mutant gene.
18. One way to reduce a cell's response to an extracellular signal is to reduce the ability of the cell's receptors to respond to the signal; a process referred to as ________.
19. In neurons, a(n) ________ is initiated by a localized redistribution of ions across the cell membrane, resulting in a shift of V_m to more positive values. (2 words)
22. GTP-binding transducer molecule. (2 words)
23. The channels that permit entry of Na^+ and initiates an action potential is an example of a(n) ________-gated channel.
24. When phosphorylated, the β-adrenergic receptor binds this protein, arresting the flow of information through this signaling pathway. (2 words)
25. Hormonal signals in the circulatory system of vertebrates are usually found in relatively low concentrations. These weak extracellular signals are usually amplified inside the cell as a result of an enzyme ________.

DOWN

1. The activation of $G_{s\alpha}$ by epinephrine results in the production of this second messenger.
3. The β-adrenergic receptor is one example of this class of integral membrane protein receptor that contains multiple membrane-spanning regions.
4. The receptor-enzyme complex that synthesizes this second messenger molecule can be found either in the cytosol or associated with the membranes of cells. (abbr.)
5. Extracellular signals that are lipid soluble bind to intracellular receptors. The resulting complex then influences gene expression by binding to specific regulatory sequences in DNA called ________. (abbr.)
6. G protein–mediated activation of phospholipase C results in the production of two intracellular second messengers, IP_3 and ________. (abbr.)
8. The R subunit is to PKA as ________ is to CDK.
10. When V_m changes to a more negative value a cell is said to be ________.
13. The process of converting a light stimulus into an electrical signal in neurons or a hormonal signal into altered cellular metabolism in hepatocytes are examples of signal ________.
14. A protein kinase that regulates the β-adrenergic receptor. (abbr.)
15. IP_3, cAMP, cGMP, and Ca^{2+} are all examples of ________ messengers.
20. Ducks do not form distinct claws during embryogenesis, as do most other birds; this is because the cells present between the forming toe bones do not experience programmed cell death or ________.
21. The cell membranes of most neurons do not actually touch; usually there is a small space between them called a(n) ________ through which neurotransmitters must diffuse.

Do You Know The Facts?

1. All the membrane proteins known as G proteins:
A. hydrolyze GTP.
B. have a subunit that activates adenylate cyclase.
C. are gated ion channels.
D. are homotrimeric proteins.
E. mediate the effects of insulin on cells.

2. GTP plays all of the following roles in the cAMP-mediated second messenger pathway *except:*
A. it associates with the β subunit of the G-protein complex.
B. hydrolysis of GTP drives the adenylate cyclase catalyzed reaction.
C. its binding causes dissociation of the G protein.
D. its effect is terminated by the GTPase activity of the G protein.
E. it transduces information from a hormone–receptor complex to an intracellular system that produces an allosteric effector.

3. Second messengers, such as cAMP, are important to cell function for all of the following reasons *except* they:
A. provide information about the extracellular environment.
B. permit a large intracellular response to relatively weak extracellular signals.
C. provide a site for integrating the metabolic requirements of cells.
D. reversibly regulate enzymatic activity in cells.
E. are a source of energy for cells.

4. Which of the following is *not* a signal transducer?
A. G_i
B. β-adrenergic receptor
C. rhodopsin
D. guanylin
E. PLC

5. cAMP acts as an intracellular second messenger in neurons. The intracellular levels of this molecule can be increased through which of the following routes?
A. by opening voltage-gated channels.
B. ligand binding to G_s-coupled receptors
C. binding of neurotransmitters to ligand-gated channels
D. increasing the activity of the G_i α subunit G proteins
E. by hyperpolarizing the neuron

6. Which statement correctly describes steroid hormones?
A. Their intracellular actions are mediated by integral membrane proteins.
B. Their effects on cells require a water-soluble intracellular signal.
C. Their effects are mediated by binding to a water-soluble receptor protein.
D. They are synthesized directly from amino acid precursors.
E. Their effects usually involve the activation of other intracellular enzymes.

7. Which of the following correctly describes the role of phospholipase C in mediating the cellular effects of hormones?
A. It directly activates a protein kinase.
B. It degrades triacylglycerols to fatty acids and glycerol.
C. It indirectly increases intracellular calcium levels.
D. It directly activates adenylate cyclase.
E. It indirectly increases cAMP levels.

8. Which of the following statements does *not* describe the phosphoinositide signaling pathway?

A. Inositol triphosphate activates protein kinase C.
B. One effect of activation of this pathway is to increase intracellular [Ca^{2+}].
C. The pathway requires GTP.
D. Protein kinase C is inactive in its soluble form.
E. Phospholipase C activation produces two intracellular messengers.

9. G_p-mediated activation of phospholipase C would be turned off by increasing:

A. the concentration of α_p subunits.
B. the concentration of α_s subunits.
C. hydrolysis of GTP.
D. the concentration of β subunits.
E. the concentration of γ subunits.

10. A specific extracellular neurotransmitter or hormone (e.g., epinephrine) may have differing effects on different target tissues because:

A. different types of receptors are activated in different tissues.
B. channels may be opened in one tissue and closed in another in response to the same signaling molecule.
C. different second messenger pathways can be activated in different tissues.
D. the target proteins of the activated intracellular messenger pathway may differ.
E. All of the above are true.

11. Some neurons in the human body have cellular projections (axons) which extend up to a meter in length, for example, the nerves that run from your spinal cord to your toes. These neurons are able to send signals to the muscles in your toes because:

A. release of neurotransmitter in the spinal cord depolarizes toe neurons so strongly that they can act as an electrical cable and can passively transmit the signal for several meters.
B. small depolarizations by input from spinal cord neurons trigger a "domino effect" in the voltage-dependent Na^+ channels, allowing the depolarizing signal to proceed undiminished along long axon lengths.
C. Diffusion of Ca^{2+} stored in the endoplasmic reticulum in the cell body down the axon to the cell terminal causes release of neurotransmitter at the synapse.
D. Second messengers, produced as a result of signaling from spinal cord neurons, can diffuse to effector enzymes in nerve terminals which produce neurotransmitter.
E. Both B and C.

12. Neurons transmit electrical signals due to the presence of voltage-dependent Na^+ channels in the axonal membrane. Opening these channels causes V_m to depolarize to the Na^+ equilibrium potential. The V_m does not remain depolarized but returns very rapidly to its original resting potential so that the neuron is ready to generate another signal. This repolarizing phase of the action potential is most likely due to which of the following?

A. Voltage-dependent K^+ channels open allowing K^+ to exit the neuron flowing passively down their electrochemical gradient.
B. The voltage-dependent Na^+ channels eventually close.
C. Ca^{2+} must be actively pumped out of the cell.
D. Cl^- flows out of the cell down its concentration gradient.
E. Both A and B.

13. Which of the following is *true* for a light stimulated rod cell?

A. Na^+/ Ca^{2+} cation channels are closed.
B. The intracellular levels of cGMP are increased.
C. PDE is tightly bound to an inhibitory subunit.
D. $T\alpha$ has GDP bound.
E. The membrane potential is depolarized.

14. Which of the following is *not* an example of signal amplification?

A. the activation of PKA by cAMP in adipose cells
B. the phosphorylation of IRS-1 by the insulin receptor in liver cells
C. the conversion of cGMP to 5′-GMP by PDE in rod cells
D. the activation of Raf kinase by Ras in hepatocytes
E. both A and D

Applying What You Know

1. Caffeine is an inhibitor of the enzyme phosphodiesterase. How would this inhibitor affect the production of free glucose in cultured hepatocytes?

2. Normally, synaptic input to a neuron occurs at the dendritic or "input" region of the cell, which is distinct from the axon or "output" region. Depolarization of the cell membrane generally occurs unidirectionally, starting with neurotransmitter induced depolarization at the input end, followed by the production of an action potential at the spike initiating zone, which is then transmitted along the full length of the axon to the cell terminus. What would you predict would be the effect of artificially depolarizing an axon in the *middle* of its length?

3. The ability of a squid to avoid becoming a calamari appetizer depends upon the rapid transmission of action potentials by its giant neuron that initiates its escape reflex. The initiation and transmission of an action potential depends upon the opening of voltage-dependent Na^+-selective ion channels in the axon of the neuron.

(a) Given the ion concentrations shown in Table 13–1, what will be the membrane potential (V_m) for a squid axon when these channels are open? (Assume these are the only channels that are open and $T = 18$ °C.)

(b) What was the membrane potential before an action potential was initiated? (Hint: At rest, neuron cell membranes are primarily permeable to K^+.)

4. Olfactory stimuli, acting through serpentine receptors and G proteins, trigger cellular responses either through an increase in the second messenger [cAMP] or an increase in the second messenger [Ca^{2+}].

(a) Outline the steps involved from receptor activation to the production of the second messenger for both cAMP and Ca^{2+}.

(b) In many second messenger pathways, the "goal" is to activate protein kinases, which affect cellular metabolism by phosphorylating cellular proteins (e.g., enzymes). What protein kinase(s) are activated in each of the pathways described above?

(c) If you separated activated olfactory cells into membrane and cytosolic fractions, where would you expect to find the labeled kinases?

5. Peptide hormones and other extracellular signaling molecules are used by cells to regulate growth and development. For example, an immortal cell line cultured from an adrenal tumor, called PC12 cells, can be induced to differentiate into neurons in response to the extracellular application of peptide nerve growth factor (NGF). Bradykinin, a peptide hormone, which can act as a vasodilator, also induces differentiation in this cell line. It has been proposed that the effects of both of these signaling molecules are mediated by the intracellular second messenger, cAMP.

In order to test this hypothesis you decide to measure the levels of cAMP in these cells following the extracellular application of NGF and/or bradykinin. In addition, you examine the effects of these agents on cells that have been injected with βγ subunits of G proteins and obtain the following data:

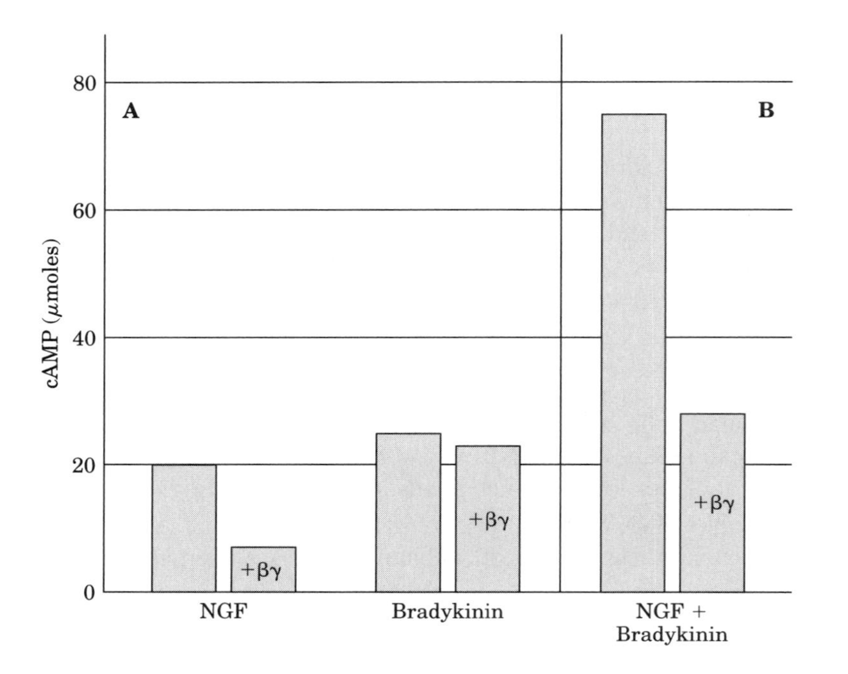

(a) At the molecular level, what would be the effect of adding excess $G_{\beta\gamma}$ on the cAMP pathway?

(b) Based on the results in part A of this figure, would you conclude that both bradykinin and NGF activate adenylate cyclase through a G protein? Explain your reasoning!

(c) What do the combined effects of NGF and bradykinin suggest about this system?

(d) Suggest a molecular model to explain the effects of these two hormones on cAMP levels in these cells.

ANSWERS

Do You Know the Terms

ACROSS

1. calmodulin
2. transducin
7. receptor
9. phosphatase
11. channels
12. electrochemical
16. kinase
17. oncogene
18. desensitization
19. action potential
22. G protein
23. voltage
24. β arrestin
25. cascade

DOWN

1. cAMP
3. serpentine
4. cGMP
5. HREs
6. DAG
8. cyclin
10. hyperpolarized
13. transduction
14. βARK
15. second
20. apoptosis
21. synapse

Do You Know the Facts?

1. A
2. B
3. E
4. D
5. B
6. C
7. C
8. A
9. C
10. E
11. B
12. E
13. A
14. E

Applying What You Know

1. Phosphodiesterase cleaves cAMP to 5′-AMP (inactive). Inhibition of this enzyme indirectly increases [cAMP] in cells. Increased cAMP levels (normally induced by the hormones epinephrine and glucagon) ultimately result in activation of glycogen phosphorylase and inactivation of glycogen synthase. The net effect is an increase in the availability of free glucose for entry into glycolysis and the citric acid cycle; both pathways supply ATP for cellular activities (e.g., muscle contraction).
2. During an AP, the voltage-dependent Na^+ channels open in advance of the voltage-dependent K^+ channels, and they close in response to the depolarization of the membrane potential that they produce. This closing is followed by a "refractory period" during which the channels cannot reopen. Since most AP's proceed from the cell body toward the terminus, this refractory period ensures that the depolarizing signal moves in only one direction. When an axon is artificially depolarized in its middle, the channels on either side of the stimulation site are *not* in a refractory state. Therefore a signal can be generated that is propagated in *both* directions from the point of stimulation.
3. **(a)** Assuming that the neuron membrane is only permeable to Na^+ during the depolarizing phase of an action potential (a valid assumption since the permeability to Na^+ is usually *much* greater than for any other ions), then the membrane potential for the squid axon will approach the equilibrium potential for Na^+. This can be calculated from the Nernst equation using the ionic concentrations given in Table 13–1.

   Intracellular $[Na^+] = 50$ mM , Extracellular $[Na^+] = 440$ mM

   $$\begin{aligned} E_{Na^+} &= (RT/Z\mathcal{F}) \ln (C_{outside}/C_{inside}) \\ &= 0.025 \text{ V} [\ln (440) - \ln (50)] \quad \text{or} \\ &\quad 0.058 \text{ V} [\log (440) - \log (50)] \\ &= 0.058 \times (2.643 - 1.699) \\ &= 0.055 \text{ V} \end{aligned}$$

   Thus, the V_m will approach +55 mV when the voltage-dependent Na^+ channels are open.

   (b) Assuming that the neuron membrane is only permeable to K^+ at the resting membrane potential, the V_m for the squid axon will approach the equilibrium potential for K^+. This can be calculated from the Nernst equation using the ionic concentrations given in Table 13–1.

Intracellular $[K^+]$ = 400 mM, Extracellular $[K^+]$ = 20 mM

$$E_{K^+} = (RT/Z\mathcal{F}) \ln (C_{outside}/C_{inside})$$
$$= 0.025 \text{ V} [\ln (20) - \ln (4000)] \quad \text{or}$$
$$0.058 \text{ V} [\log (20) - \log (400)]$$
$$= 0.058 \times (1.301 - 2.602)$$
$$= -0.075 \text{ V}$$

Thus, the V_m will approach −75 mV when only K^+ channels are open.

4. (a) cAMP: Receptor activation results in the activation of G_s protein, the displacement of GDP for GTP, dissociation of G_s protein into $G_{s\alpha}$ and $G_{s\beta\gamma}$, activation of adenylate cyclase, and the production of cAMP from ATP.

Ca^{2+}: Receptor activation results in the activation of G_p protein, the displacement of GDP for GTP, dissociation of G_p protein into $G_{p\alpha}$ and $G_{p\beta\gamma}$, activation of phospholipase C, and the production of the second messengers diacylglycerol (DAG) and inositol triphosphate (IP_3) from phosphatidylinositol bisphosphate (PIP_2). IP_3 then binds to receptors located on the membrane of the endoplasmic reticulum (sarcoplasmic reticulum in muscle cells), where it opens Ca^{2+} channels and releases stored Ca^{2+}.

(b) cAMP pathway: cAMP-dependent protein kinase

Ca^{2+} pathway: Ca^{2+} plus the other product of PLC activation, DAG, co-operate in the activation of protein kinase C. It is also possible for Ca^{2+} to bind to the ubiquitous cellular protein calmodulin, which can then activate the multifunctional Ca^{2+}/calmodulin-dependent protein kinase, CaM kinase.

(c) cAMP pathway: cAMP (like ATP) is a water soluble molecule. PKA is also soluble so this activated kinase would be expected to co-purify with the cytosolic components of a cell. Ca^{2+} pathway: Ca^{2+}, calmodulin, and CaM kinase are all water soluble compounds, so this activated kinase would also be expected to co-purify with the cytosolic components of a cell. However, DAG is a lipid soluble molecule and cannot leave the lipid bilayer. Consequently the activated form of PKC should be associated with the cell membrane and should co-purify with the membane components of these cells.

5. (a) The βγ subunits normally limit the effects of G_α by binding to the α subunit in its GDP bound form and preventing the exchange of GDP for GTP. This "turns off" the effects of the α subunit.

(b) No; inhibiting the effect of $G_{s\alpha}$ by adding excess βγ subunits has no effect on the ability of bradykinin to increase cAMP levels. This indicates that bradykinin does not affect adenylate cyclase through the direct activation of a G protein. NGF's effects are inhibited by the injection of βγ subunits indicating that the effects of this hormone *are* mediated by α_s.

(c) The combined effects of NGF and bradykinin are more than additive, suggesting a synergistic effect between the two signaling molecules. Physiologically this means that the two extracellular signals, when present at the same time, can convey different information than they can individually.

(d) There are a number of possibilities: Bradykinin could enhance the effects of NGF on adenylate cyclase by inhibiting phosphodiesterase, which hydrolyzes cAMP and terminates the intracellular signal. This might occur if bradykinin could reduce the intracellular levels of free calcium (by closing a Ca^{2+} channel, for example), which is a known activator of phosphodiesterase. Another possibility is that bradykinin could activate adenylate cyclase through a non-G-protein–related pathway. Specifically it could activate the cyclase directly by increasing the levels of calcium in the cell. This would turn on the calcium-sensitive form of adenylate cyclase. Synergistic effects of several extracellular signaling molecules on a calcium-sensitive form of cyclase has been proposed as the molecular basis for cellular "learning" in *Aplysia*. Finally, bradykinin could activate a non-cAMP dependent protein kinase (such as the calcium/calmodulin-dependent protein kinase) that phosphorylates components of the NGF signaling pathway.

Principles of Bioenergetics

STEP-BY-STEP GUIDE

Major Concepts

Basic principles of thermodynamics govern the transfer of energy in biological systems.

The first law of thermodynamics states that all energy is conserved; although the form of the energy may change, the total amount in the universe remains constant. The second law of thermodynamics states that the total disorder (entropy) of the universe increases.

A thermodynamic system can be a single, simple chemical reaction or an entire organism.

The transfer of energy in a single reaction or in an organism can be described by three thermodynamic quantities: (Gibbs) **free energy, G; enthalpy, H; and entropy, S.** These quantities are related by the equation

$$\Delta G = \Delta H - (T\,\Delta S)$$

where T is the absolute temperature (in degrees Kelvin, or K). ΔG describes the change in free energy that occurs in a chemical reaction.

Reactions proceed spontaneously only if ΔG is negative, meaning energy is released by the reaction.

If ΔG is positive, the reaction requires an input of energy. Reactions are at equilibrium if $\Delta G = 0$.

The standard free-energy change, $\Delta G'^\circ$ is a physical constant and can be calculated from the equilibrium constant, K'_{eq}: $\Delta G'^\circ = -RT \ln K'_{eq}$

For a system at equilibrium the rates of the forward and reverse reactions are equal; no further net change occurs in the system. For the reaction

$$A + B \rightleftharpoons C + D$$

at equilibrium

$$\frac{[C][D]}{[A][B]} = K'_{eq}$$

The *actual* free-energy change (ΔG) of a specific reaction depends on the concentrations of products and reactants, temperature, and pH, but is not affected by the reaction pathway or by enzyme catalysis. Under *standard* conditions (25 °C, 1 M, 1 atm, pH = 7), $\Delta G'^\circ$ can be measured. The *actual* ΔG for a reaction can be related to the free-energy change $\Delta G'^\circ$ that occurs under *standard* conditions by substituting the actual concentrations of reactants and products:

$$\Delta G = \Delta G^{\circ\prime} + RT \ln \frac{[C][D]}{[A][B]}$$

Since at equilibrium $\Delta G = 0$, the equation reduces to

$$\boldsymbol{\Delta G'^\circ = -RT \ln K'_{eq}}$$

The $\Delta G'^\circ$ values for sequential reactions are additive so that individual reactions with an unfavorable (positive) $\Delta G'^\circ$ can proceed if they are coupled to reactions with a favorable (negative) $\Delta G'^\circ$. (K'_{eq}s are multiplicative.) Biological systems use the free energy of hydrolysis of "high-energy compounds" to drive energetically unfavorable reactions. Also, in biological systems, a displacement from equilibrium can produce a favorable ΔG, *even if $\Delta G'^\circ$ is positive,* by giving the term RT ln (products/reactants) a larger value than $\Delta G'^\circ$.

ATP is a common source of free energy for biological systems.

The hydrolysis of ATP to ADP and P_i (inorganic phosphate) has a $\Delta G'^{\circ}$ of -30.5 kJ/mol (the actual ΔG may vary depending on the microenvironment). Although the hydrolysis of phosphoryl, pyrophosphoryl, or adenylyl groups from ATP provides energy, without some mechanism for **coupling** this hydrolysis to other reactions the energy would be released as heat and would not be available to drive other reactions in cells. The coupling mechanism involves the transfer of these groups from ATP to an enzyme or to its substrate. This is called a **group transfer reaction.** ATP has a high **phosphoryl group transfer potential.**

Compounds other than ATP have high group transfer potentials and are capable of driving biological reactions.

GTP and some other phosphorylated compounds such as phosphorylated sugars have a large, negative $\Delta G'^{\circ}$ of hydrolysis that can be used to drive other reactions. In addition, the hydrolysis of thioesters such as acetyl-CoA has a large, negative $\Delta G'^{\circ}$. Free energy is released by the hydrolysis of these compounds because the products are much more stable than the initial reactants. There are a number of reasons why this is so: electrostatic repulsion (and resulting bond strain) is relieved by hydrolysis, and the products of hydrolysis may be stabilized by ionization, by formation of a resonance hybrid, by isomerization (tautomerization), and/or by solvation. Despite the availability of other "high-energy" compounds, ATP is the one most widely used by cells. ATP occupies a middle ground in terms of its energy-carrying capacity and thus can serve as a link between energy-producing catabolic and energy-requiring anabolic reactions in cells.

Phosphorylation of certain compounds effectively "primes" them for participation in biological reactions.

Glucose is primed for entry into the glycolytic pathway by the addition of a phosphoryl group to form glucose 6-phosphate. Fatty acids used in the synthesis of lipids and amino acids used in the synthesis of proteins are first primed by hydrolysis of ATP and the covalent addition of AMP to the amino acid or fatty acid molecule. Cleavage of PP_i (pyrophosphate) from activated nucleoside triphosphates is required for assembly of the nucleotide units in RNA and DNA synthesis.

ATP has other roles in cells.

It provides energy for active transport and for muscle contraction.

Electron flow in oxidation reduction reactions can be used to do "biological work."

Oxidation–reduction reactions involve two components: an electron donor (reducing agent) and an electron acceptor (oxidizing agent). Therefore, two simultaneous reactions (half-reactions) occur in any oxidation–reduction process. Electrons can be transferred in several different ways: (1) by direct transfer, as when a metal ion is involved in a reaction (e.g., $Fe^{2+} + Cu^{2+} \rightleftharpoons Fe^{3+} + Cu^{+}$); (2) as a hydrogen atom (as opposed to a hydrogen ion); (3) as a hydride ion, H^-; and (4) when oxygen is covalently attached to an organic compound. **Standard reduction potential,** E'°, describes the affinity of any chemical species for electrons *relative* to the affinity of hydrogen ions (H^+) for electrons (at pH 7). The higher the value of E'°, the greater the affinity for electrons.

Electrons are transferred by a variety of cofactors that act as carriers.

Nucleotide carriers include NAD^+, $NADP^+$, FMN, and FAD. NAD^+ and $NADP^+$ are water-soluble, recyclable cofactors that accept a hydride ion to become reduced (to NADH and NADPH). NAD^+ generally functions to accept hydride ions during catabolic reactions, and NADP (as NADPH) usually functions to transfer hydride ions for anabolic reductions. FAD and FMN can accept either one or two electrons in the form of hydrogen atoms. These two enzymes tend to be tightly bound to their enzymes.

What to Review

Answering the following questions and reviewing the relevant concepts, which you have already studied, should make this chapter more understandable.

- The following terms were introduced in Chapter 1 and again in the Introduction to Part III of the text: **autotroph, heterotroph, catabolism, anabolism, Gibbs free energy, enthalpy, entropy.** A complete understanding of their meaning is critical for the material in this chapter.
- Be sure you understand what an equilibrium constant represents (pp. 96, 249).
- Remind yourself what enzymes can and cannot do with respect to reaction rates, equilibria, and other reaction parameters (pp. 247–249).

Topics for Discussion

Answering each of the following questions, especially in the context of a study group discussion, should help you understand the important points of this chapter.

Bioenergetics and Thermodynamics

Biological Energy Transformations Obey the Laws of Thermodynamics

1. Explain how defining the "system" and "surroundings" allows living organisms to operate within the second law of thermodynamics.

Entropy: The Advantages of Being Disorganized

2. Does the oxidation of glucose represent an increase or decrease in entropy? Does it have a positive or negative ΔG?

Cells Require Sources of Free Energy

3. From where do cells acquire their necessary free energy? Why can't cells use heat as a free-energy source?

The Standard Free-Energy Change Is Directly Related to the Equilibrium Constant

4. What exactly does "at equilibrium" mean in terms of (a) the *rate* of both the forward and reverse reactions and (b) the *concentrations* of the reactants and products?

5. What reaction conditions are used to measure the standard free-energy change, $\Delta G'^{\circ}$?

6. If, at equilibrium, the concentration of products is greater than the concentration of reactants, is $\Delta G'^{\circ}$ positive or negative? What can you say about the value of K'_{eq}?

Actual Free-Energy Changes Depend on Reactant and Product Concentrations

7. What effect does the presence of an enzyme have on the $\Delta G'^{\circ}$ of the reaction it catalyzes?

8. Under what circumstances can ΔG be negative if $\Delta G'^{\circ}$ is positive? Could cells use this strategy to drive thermodynamically unfavorable reactions?

9. Does the value of $\Delta G'^{\circ}$ (or ΔG) tell you anything about (a) the rate at which a reaction occurs or (b) the pathway by which the final product is formed?

Standard Free-Energy Changes Are Additive

10. How can the coupling of a thermodynamically unfavorable reaction to a thermodynamically favorable reaction increase the K'_{eq} of the overall equation?

11. Explain why relatively small changes in $\Delta G'^{\circ}$ correspond to large changes in K'_{eq}.

Phosphoryl Group Transfers and ATP

The Free-Energy Change for ATP Hydrolysis Is Large and Negative

12. What physical and chemical factors contribute to the free-energy change of ATP hydrolysis?

The Free Energy of Hydrolysis of ATP within Cells: The Real Cost of Doing Metabolic Business

13. How do ΔG, $\Delta G'^{\circ}$, and ΔG_p differ?

Other Phosphorylated Compounds and Thioesters Also Have Large Free Energies of Hydrolysis

14. What are the main reasons for the high $\Delta G'^{\circ}$ values for hydrolysis of phosphoenolpyruvate, 1,3-bisphosphoglycerate, phosphocreatine, acetyl-CoA, and other similar compounds?

ATP Provides Energy by Group Transfers, Not by Simple Hydrolysis

15. Why is the "single arrow" representation of the conversion of ATP to ADP and P_i deceiving?

16. The reactions that produce conformational changes linked to ATP (or GTP) hydrolysis are different from other ATP-driven reactions. Explain.

17. What is the relative position of ATP in the hierarchy of compounds with phosphoryl group transfer potentials?

18. What is the difference between *thermodynamic* stability (or instability) and *kinetic* stability?

ATP Donates Phosphoryl, Pyrophosphoryl, and Adenylyl Groups

19. Why are adenylylation reactions so thermodynamically favorable?

Assembly of Informational Macromolecules Requires Energy

20. What are the similarities and differences between the assembly of DNA and RNA from their component nucleotides and the activation of fatty acids and amino acids?

ATP Energizes Active Transport and Muscle Contraction

21. What is the *specific* role of ATP in the transport of solutes across membranes (i.e., what does the transfer of a phosphoryl group from ATP to the pump protein accomplish)?

Firefly Flashes: Glowing Reports of ATP

22. Speculate on what use could be made of a sensitive assay for ATP concentration in a biological sample.

Transphosphorylations Occur between Nucleotides

23. What are the cellular roles of nucleoside diphosphate kinase, adenylate kinase, and creatine kinase? Why are they so important?

Inorganic Polyphosphate Is a Potential Phosphoryl Donor

24. Why is polyP an interesting molecule to those who study cellular evolution?

Biochemical and Chemical Equations Are Not Identical

25. When is it important to use chemical rather than biochemical equations?

Biological Oxidation-Reduction Reactions

26. What are the different sources of electrons (i.e., the different electron-donating compounds) that provide energy for the work done by biological organisms?

The Flow of Electrons Can Do Biological Work

27. Why are cell membranes critical to the generation of a proton-motive force in cells?

Oxidation-Reductions Can Be Described as Half-Reactions

28. In the equation $Fe^{2+} + Cu^{2+} \rightleftharpoons Fe^{3+} + Cu^{+}$, which of the iron species is more oxidized? Which of the copper species is more reduced?

Biological Oxidations Often Involve Dehydrogenation

29. What are the eight oxidation states of carbon?

30. Besides the transfer of electrons in the form of hydrogen atoms, in what other ways does electron transfer occur?

Reduction Potentials Measure Affinity for Electrons

31. How is the hydrogen electrode used as *the* reference half-cell?

32. Which has the higher (more positive) reduction potential, NADH or cytochrome *b* (Fe^{3+})? In which direction will electrons flow in a system that contains these two compounds?

Standard Reduction Potentials Can Be Used to Calculate the Free-Energy Change

33. Why is it important to have a universal standard for measuring reduction potentials?

Cellular Oxidation of Glucose to Carbon Dioxide Requires Specialized Electron Carriers

34. What are the coenzymes that carry electrons in the cells? What types of reactions are they generally associated with?

A Few Types of Coenzymes and Proteins Serve as Universal Electron Carriers

35. What is the significance of the various categories of electron carriers (water-soluble versus lipid-soluble; mobile versus bound; associated with peripheral versus integral membrane proteins) in terms of an electron carrier's functions?

NADH and NADPH Act with Dehydrogenases as Soluble Electron Carriers

36. How do the normal concentration ratios of NAD^+/NADH and $NADP^+$/NADPH in cells reflect the different metabolic roles of these electron carriers?

Flavin Nucleotides Are Tightly Bound in Flavoproteins

37. Why are flavoproteins, as a class, involved in a greater variety of reactions than are NAD-linked enzymes?

SELF-TEST

Do You Know the Terms?

ACROSS

1. Describes the NAD^+/NADH pair. (2 words)
4. ΔG_p; also called the compound's ________ potential.
6. The transfer of phosphoryl, pyrophosphoryl, or adenylyl groups from ________ couples the energy of breakdown to endergonic transformations.
7. A category of the dehydrogenases, this class of enzymes catalyzes reactions involving transfer of a hydride ion to NAD^+ or $NADP^+$.
9. Another way of describing energy-consuming biosynthetic reactions.
11. ________ potential: the ability of ATP to donate a phosphate group, or of acetyl-CoA to donate an acetyl group. (2 words)
12. Term for a single electron participating in an oxidation-reduction reaction, regardless of the form in which the electron is transferred. (2 words)
14. Standard ________: the affinity of a compound for electrons, relative to a hydrogen electrode. (2 words)
16. ________ free energy is amount of free energy released in a cell and available to do work. (Compare to 17 down.)
18. Class of enzymes containing a cofactor derived from the vitamin riboflavin, including FAD and FMN.
20. This is minimal in a completed jigsaw puzzle.
21. ________ trophs use ingested nutrients as carbon source for metabolic processes.
22. ________ trophs use CO_2 as carbon source for metabolic processes.
23. General class of enzymes involved in oxidation-reduction reactions.

DOWN

1. Describes reactions involved in the degradation of ingested nutrients.
2. Source of reducing equivalents used primarily in anabolic pathways.
3. Type of bond in acetyl-CoA that has a large, negative $\Delta G'^\circ$ of hydrolysis.
5. Electron transfer in cells occurs as coupled ________-________ reactions.
6. Cells do not rely solely on de novo synthesis to resupply depleted ATP stores; for example, this enzyme catalyzes the generation of ATP and AMP from 2 ADP. (2 words)
8. NADH is electron ________ in the reaction:

$$\text{Acetaldehyde} + \text{NADH} \longrightarrow \text{ethanol} + \text{NAD}^+.$$

10. Value of this quantity is negative in chemical reactions that release heat.
13. $NAD^+ + H^+ + 2e^- \rightleftharpoons NADH$ and $2O_2 + 2H^+ + 2e^- \rightleftharpoons H_2O$ are examples of ________-________.
15. State of a reaction when $\Delta G'^\circ = 0$.
16. Oxygen is electron ________ in the reaction:

$$\text{Glucose} + 6O_2 \longrightarrow 6CO_2 + 6H_2O.$$

17. Describes the available free energy of a chemical reaction occurring under defined conditions of 1 M reactant and substrate concentrations, pH 7, and 25 °C.
19. Electron carrier used primarily in catabolic pathways.

Do You Know the Facts?

1. Indicate whether each of the following statements about ATP is true or false.

_____ **(a)** It contains a β-*N*-glycosyl linkage.
_____ **(b)** It contains a furanose ring.
_____ **(c)** It is the highest-energy compound in cells.
_____ **(d)** It is used as a long-term storage form of energy in cells.
_____ **(e)** Its free energy of hydrolysis can be used to drive other reactions.
_____ **(f)** It is synthesized from ADP and P_i in an exergonic reaction.
_____ **(g)** It contains three phosphoanhydride bonds.
_____ **(h)** Repulsion between the negatively charged phosphoryl groups is reduced when ATP is hydrolyzed.
_____ **(i)** Its sugar moiety is glucose.
_____ **(j)** Its phosphoryl group transfer potential is higher than that of phosphoenolpyruvate.

2. Which of the following BEST describes the relationship between $\Delta G'^\circ$ and the rate of a reaction?
A. $\Delta G'^\circ$ is linearly proportional to the rate.
B. $\Delta G'^\circ$ is inversely proportional to the rate.
C. If $\Delta G'^\circ$ is positive, the reaction is spontaneous in the forward direction.
D. If $\Delta G'^\circ$ is negative, the reaction is at equilibrium.
E. $\Delta G'^\circ$ provides no information about the rate.

3. ATP is intermediate in the hierarchy of phosphorylated compounds with high standard free energies of hydrolysis (see Table 14–6). Why is it advantageous for cells to use ATP as their primary energy-carrying molecule?
A. ATP can be regenerated by coupling with a reaction that releases more free energy than does ATP hydrolysis.
B. ATP is thermodynamically unstable but kinetically stable.
C. Reactions that release more free energy than ATP hydrolysis do not occur in living cells.
D. The phosphoryl groups of other high-energy compounds cannot be removed.
E. The phosphoryl group transfer potential of ATP is higher than that of any other phosphorylated compound.

4. Which of the following does NOT contribute to the high phosphoryl group transfer potential of ATP? ADP and P_i have ________ than does ATP.
A. more resonance stabilization
B. less electrostatic repulsion
C. a greater degree of ionization
D. a greater ability to isomerize
E. a greater degree of hydration

In questions 5–8, match each electron carrier with the correct statement.

5. NAD^+:
A. accepts 2 electrons and 2 hydrogen ions.
B. accepts 2 electrons and 1 hydrogen ion.
C. accepts 1 electron and 1 hydrogen ion.
D. transfers electrons in reductive biosynthesis.
E. in its oxidized form is NADH.

6. $FADH_2$:
A. accepts 2 electrons and 2 hydrogen ions.
B. accepts 2 electrons and 1 hydrogen ion.
C. accepts 1 electron and 1 hydrogen ion.
D. transfers electrons in reductive biosynthesis.
E. in its oxidized form is FAD.

7. NADH:

A. accepts 2 electrons and 2 hydrogen ions.
B. accepts 2 electrons and 1 hydrogen ion.
C. accepts 1 electron and 1 hydrogen ion.
D. can be used as a source of electrons for ATP synthesis.
E. transfers electrons in reductive biosynthesis.

8. NADPH:

A. accepts 2 electrons and 2 hydrogen ions.
B. accepts 2 electrons and 1 hydrogen ion.
C. accepts 1 electron and 1 hydrogen ion.
D. transfers electrons in reductive biosynthesis.
E. in its oxidized form is NAD^+.

9. The bacterium *Pseudomonas saccharophilia* contains sucrose phosphorylase, an enzyme that catalyzes the phosphorolytic cleavage of sucrose:

$$\text{Sucrose} + P_i \longrightarrow \text{glucose 1-phosphate} + \text{fructose}$$

Given the data below, what is the standard free-energy change for the phosphorolysis of sucrose?

$$H_2O + \text{sucrose} \longrightarrow \text{glucose} + \text{fructose} \qquad \Delta G'^\circ = -29 \text{ kJ/mol}$$
$$H_2O + \text{glucose 1-phosphate} \longrightarrow \text{glucose} + P_i \qquad \Delta G'^\circ = -21 \text{ kJ/mol}$$

A. −8 kJ/mol
B. 8 kJ/mol
C. −50 kJ/mol
D. 50 kJ/mol
E. Cannot be determined from the data given.

10. Given the $\Delta G'^\circ$ values for the reactions below, what is the $\Delta G'^\circ$ for the hydrolysis of acetyl-CoA?

Reaction	$\Delta G'^\circ$ (kJ/mol)
Oxaloacetate + acetyl-CoA + H_2O → citrate + CoASH + H^+	−32.2
Malate → fumarate + H_2O	3.8
Oxaloacetate + acetate → citrate	−0.8
Malate + NAD^+ → oxaloacetate + NADH + H^+	29.7
Oxaloacetate + $2H^+$ + $2e^-$ → malate^{2-}	−32.04

A. −2.5 kJ/mol
B. −31.4 kJ/mol
C. 61.9 kJ/mol
D. 31.4 kJ/mol
E. −61.9 kJ/mol

11. Consider the general reaction A → B, where $\Delta G'^\circ = -60$ kJ/mol. Initially, 10 mM of A and 0 mM of B are present. After 24 hours, analysis reveals the presence of 2 mM of B and 8 mM of A. What can you conclude from this result?

A. A and B have reached equilibrium concentrations.
B. Formation of B is thermodynamically unfavorable.
C. The result described is impossible, given the $\Delta G'^\circ$ of the reaction.
D. Formation of B is kinetically slow; equilibrium has not been reached at 24 hours.
E. An enzyme has shifted the equilibrium toward formation of A.

12. What does "at equilibrium" mean in terms of (a) the rates of forward and reverse reactions and (b) the concentrations of reactants and products?

13. (a) Define oxidation and reduction. Can an oxidation occur without a simultaneous reduction? Why or why not?

(b) What is a conjugate redox pair and how does it relate to standard reduction potential?

Applying What You Know

1. In glycolysis, the enzyme phosphofructokinase I catalyzes the reaction

Fructose 6-phosphate + ATP $\longrightarrow$ fructose 1,6-bisphosphate + ADP

Given the data below, calculate the equilibrium constant for this reaction.

$R = 8.315$ J/ mol · K
$T = 25$ °C
ATP $\longrightarrow$ ADP + P_i $\quad \Delta G'^{\circ} = -30.5$ kJ/mol
Fructose 1,6-bisphosphate $\longrightarrow$ fructose 6-phosphate + P_i $\quad \Delta G'^{\circ} = -16.0$ kJ/mol

2. (a) Does the oxidation of glucose, represented by the equation

$$C_6H_{12}O_6 + 6O_2 \longrightarrow 6CO_2 + 6H_2O,$$

involve an increase or decrease in entropy? Why?

(b) Using the relationship $\Delta G = \Delta H - T\ \Delta S$, determine whether ΔG for the above reaction is positive or negative.

3. Consider the reaction

Glucose 1-phosphate $\rightleftharpoons$ glucose 6-phosphate

If, at equilibrium, the concentration of glucose 6-phosphate is greater than the concentration of glucose 1-phosphate, is the $\Delta G'^{\circ}$ of the reaction positive or negative? Explain.

4. The standard free-energy change for the reaction ADP + P_i $\rightarrow$ ATP is +30.5 kJ/mol. If all of the reactants are *initially* present at equal concentrations of 1 M, what can you say about the relative concentrations of ADP, P_i, and ATP *at equilibrium* (assume standard conditions of pH and temperature)?

ANSWERS

Do You Know the Terms?

ACROSS

1. conjugate redox
4. phosphorylation
6. ATP
7. oxidoreductases
9. anabolic
11. group transfer
12. reducing equivalent
14. reduction potential
16. actual
18. flavoproteins
20. entropy
21. hetero
22. auto
23. dehydrogenases

DOWN

1. catabolic
2. NADPH
3. thioester
5. oxidation-reduction
6. adenylate kinase
8. donor
10. enthalpy
13. half-reactions
15. equilibrium
16. acceptor
17. standard
19. NADH

Do You Know the Facts?

1. **(a)** T; **(b)** T; **(c)** F; **(d)** F; **(e)** T; **(f)** F; **(g)** F; **(h)** T; **(i)** F; **(j)** F
2. E
3. A
4. D
5. B
6. E
7. D
8. D
9. A
10. B
11. D
12. **(a)** At equilibrium, the *rates* of the forward and reverse reactions are exactly equal and no further net change occurs in the system.
    (b) The concentrations of reactants and products *at equilibrium* define the equilibrium constant, K'_{eq}. In the general reaction $aA + bB \rightleftharpoons cC + dD$, where a, b, c, and d are the number of molecules of A, B, C, and D participating, the equilibrium constant is given by

$$K'_{eq} = \frac{[C]^c[D]^d}{[A]^a[B]^b}$$

where [A], [B], [C], and [D] are the molar concentrations of the reactants at the point of equilibrium.

13. **(a)** Oxidation is the loss of electrons; reduction is the gain of electrons. Free electrons are unstable and do not exist in biological systems. Whenever an electron is released in the oxidation of one chemical species, an electron must be accepted by another species, which is thus reduced. (Remember from beginning chemistry: "LEO says GER!" **L**oss of **E**lectrons is **O**xidation; **G**ain of **E**lectrons is **R**eduction!)
    (b) In a redox reaction, the general equation is

$$\text{electron donor} \rightleftharpoons e^- + \text{electron acceptor}$$

The electron donor and the electron acceptor together constitute a conjugate redox pair. Reduction potentials are a measure of affinity for electrons, and standard reduction potentials allow for the calculation of free-energy change. When two conjugate redox pairs are together in solution, electron transfer from the electron donor of one pair to the electron acceptor of the other pair may occur spontaneously. The tendency of such a reaction to occur depends on the relative affinity of the electron acceptor of each redox pair for electrons. The standard reduction potential, $E'°$, is a measure of this affinity and can be determined by experiment.

Applying What You Know

1. The reaction catalyzed by phosphofructokinase is the sum of two reactions, for each of which $\Delta G'°$ is given (for the first reaction below, note that the sign changes for $\Delta G'°$):

$$\text{Fructose 6-phosphate} + P_i \longrightarrow \text{fructose 1,6-bisphosphate} \quad \Delta G'° = 16.0 \text{ kJ/mol}$$

$$\text{ATP} \longrightarrow \text{ADP} + P_i \quad \Delta G'° = -30.5 \text{ kJ/mol}$$

Sum: $\text{Fructose 6-phosphate} + \text{ATP} \longrightarrow \text{fructose 1,6-bisphosphate} + \text{ADP} \quad \Delta G'° = -14.5 \text{ kJ/mol}$

Using $\Delta G'^\circ = -RT \ln K'_{eq}$ and solving for K'_{eq}:

$$\ln K'_{eq} = \Delta G'^\circ/RT$$
$$= -(-14.5\ \text{kJ/mol}) / [(8.315\ \text{J/mol} \cdot \text{K})(298\ \text{K})]$$
$$= 5.8518$$
$$K'_{eq} = 348$$

2. (a) Whenever a reaction proceeds so that there is an increase in the number of molecules (in the case of glucose oxidation, for every 7 reactant molecules there are 12 product molecules), and/or when a solid (such as glucose) is converted into liquid or gaseous products (water and carbon dioxide), which have more freedom than a solid to move and fill space, there is an increase in molecular disorder and thus an *increase in entropy.*

(b) For the glucose oxidation reaction, ΔS is positive (because entropy increases); ΔH is negative (assuming this is a combustion reaction that gives off heat to the surroundings); and T is positive (because the units are K). Given that $\Delta G = \Delta H - T\,\Delta S$, the ΔG for this reaction must be negative.

3. If the concentration of products is greater than concentration of reactants at equilibrium, then $K'_{eq} > 1$. Given that $\Delta G'^\circ = -RT \ln K'_{eq}$, if $K'_{eq} > 1$, then $\ln K'_{eq}$ is positive and $\Delta G'^\circ$ must be negative.

Another way of looking at this is in terms of the final ratio of products and reactants. If the concentration of products exceeds the concentration of reactants, the reaction proceeds in the forward direction and thus $\Delta G'^\circ$ must be negative.

4. A positive value of $\Delta G'^\circ$ means that the product(s) of the reaction contain more free energy than the reactants. The reaction will therefore tend to go in the reverse direction if we start with 1.0 M concentrations of all components. At equilibrium the concentrations of the reactants ADP and P_i will be greater than 1 M and the concentration of the product ATP will be less than 1 M. Keep in mind that if the concentrations of ADP and P_i increase, the concentration of ATP must correspondingly decrease, since ATP $\rightarrow$ ADP + P_i.

chapter

15 Glycolysis and the Catabolism of Hexoses

STEP-BY-STEP GUIDE

Major Concepts

Glycolysis is a series of reactions in which one molecule of glucose is degraded to two molecules of pyruvate, yielding biologically useful energy in the form of ATP and NADH.

Glycolysis can be subdivided into a preparatory phase (five reactions) requiring an input of energy and a payoff phase (five reactions) with a net production of energy in the form of ATP and NADH. In the **preparatory phase** two molecules of ATP are used to phosphorylate and thus "activate" each molecule of glucose. The phosphorylated hexose is then cleaved to form two three-carbon molecules. In the **payoff phase** energy is transferred from the three-carbon intermediates to ADP by substrate-level phosphorylation. Two molecules of ATP are formed for each three-carbon molecule passing through this phase; the three-carbon intermediate is ultimately converted to **pyruvate.** NADH is also formed during this phase. The NADH is either oxidized by the reduction of pyruvate to regenerate NAD^+ or is used to generate additional ATP via the electron transfer chain. The net equation for glycolysis is:

$$\text{Glucose} + 2\text{NAD}^+ + 2\text{ADP} + 2\text{P}_i \longrightarrow 2\text{ pyruvate} + 2\text{NADH} + 2\text{H}^+ + 2\text{ATP} + 2\text{H}_2\text{O}$$

Pyruvate has three possible fates after glycolysis.

Under aerobic conditions, pyruvate is oxidized to acetate, which can enter the citric acid cycle and thus be further oxidized to CO_2 and H_2O. Under anaerobic conditions, pyruvate is reduced by NADH to either **lactate** or **ethanol** and CO_2, depending on the organism. These reactions function to regenerate NAD^+ so that glycolysis can continue.

Carbohydrates other than glucose also feed into the glycolytic pathway.

The energy-storage macromolecules **glycogen** and **starch** can be degraded to a phosphorylated form of glucose, which can enter glycolysis. Other monosaccharides, and disaccharides that have been hydrolyzed to monosaccharides, can enter glycolysis if they are first converted to one of the phosphorylated glycolytic intermediates.

Carbohydrate metabolism, including glycolysis, glycogen breakdown, and glucose synthesis by gluconeogenesis, is tightly regulated.

The mechanisms used in the regulation of glycolysis provide models for understanding the control of many other metabolic pathways. The regulatory mechanisms include: substrate- or enzyme-limited reactions; energetically irreversible reactions; allosteric, covalent, and hormonal regulation; end-product inhibition; and tissue-specific isozymes.

There are alternative pathways for the oxidation of glucose.

Glucose breakdown via the **pentose phosphate pathway** produces NADPH and ribose 5-phosphate; the ratios produced vary with tissue type and metabolic demands.

What to Review

Answering the following questions and reviewing the relevant concepts, which you have already studied, should make this chapter more understandable.

- Most organisms use **glucose** as their primary fuel molecule. Review its structure and be sure you remember the difference between an aldose and a ketose (p. 296).
- The principles of bioenergetics (Chapter 14), especially with respect to ATP use by cells, will be important to your understanding of this chapter as well as of the following metabolism chapters.

Topics for Discussion

Answering each of the following questions, especially in the context of a study group discussion, should help you understand the important points of this chapter.

Glycolysis

1. Why does the glycolytic pathway occupy a central position in cell metabolism?

An Overview: Glycolysis Has Two Phases

2. What is accomplished in the preparatory phase of glycolysis? What is accomplished in the payoff phase of glycolysis?

3. How is the phosphorylation of glyceraldehyde 3-phosphate in the payoff phase of glycolysis different from the phosphorylation of glucose in the preparatory phase?

Fates of Pyruvate

4. Describe three catabolic fates of pyruvate.

ATP Formation Coupled to Glycolysis

5. What is the efficiency of recovery, in the form of ATP, of the energy released by glycolysis under *standard* conditions?

6. Why is the efficiency of energy recovery higher under the conditions that exist in living cells?

Energy Remaining in Pyruvate

7. How much biologically available energy remains in a mole of pyruvate?

Importance of Phosphorylated Intermediates

8. What are the three functions of the phosphoryl groups of phosphorylated intermediates?

The Preparatory Phase of Glycolysis Requires ATP

9. *Cell Map:* You should know the structures and names of the intermediates and the names of the enzymes and cofactors involved in each step of the preparatory phase (and payoff phase). These should be entered in the spaces provided on your Cell Map.

① Phosphorylation of Glucose

10. Why is the $\Delta G'^{\circ}$ for this reaction negative?

11. What is the general role of a *kinase?*

12. The $\Delta G'^{\circ}$ for hydrolysis of ATP is −30.5 kJ/mol. How much of this energy is conserved in the phosphorylation of glucose by hexokinase?

② Conversion of Glucose 6-Phosphate to Fructose 6-Phosphate

13. Why is the conversion of glucose 6-phosphate to fructose 6-phosphate readily reversible?

14. What cellular conditions would bias the reaction in the direction of fructose 6-phosphate formation?

③ Phosphorylation of Fructose 6-Phosphate to Fructose 1,6-Bisphosphate

15. Why is phosphofructokinase-1 (PFK-1) a good candidate for a regulatory enzyme?

16. What conditions activate PFK-1? What conditions inactivate PFK-1?

17. If PFK-1 is inhibited, in which direction will the reaction catalyzed by phosphohexose isomerase proceed?

④ Cleavage of Fructose 1,6-Bisphosphate

18. The aldolase reaction has a $\Delta G'^{\circ}$ of +23.8 kJ/mol. What mechanism do cells use to pull this endergonic reaction in the direction of cleavage?

⑤ Interconversion of the Triose Phosphates

19. If C-3 of glucose is radioactively labeled and the glucose is allowed to go through the preparatory phase of glycolysis, which carbon in glyceraldehyde 3-phosphate will carry the radioactive label? Will *all* glyceraldehyde 3-phosphate molecules be labeled?

The Payoff Phase of Glycolysis Produces ATP and NADH

⑥ Oxidation of Glyceraldehyde 3-Phosphate to 1,3-Bisphosphoglycerate

20. What is being oxidized by NAD^+ in this reaction?

21. How does iodoacetate inhibit this reaction?

⑦ Phosphoryl Transfer from 1,3-Bisphosphoglycerate to ADP

22. How does step ⑦ "pull" step ⑥ forward?

23. Why is this process called *substrate-level* phosphorylation?

24. Where in the cell is this ATP produced? (It is important to keep track of the cellular locations of metabolic processes.)

⑧ Conversion of 3-Phosphoglycerate to 2-Phosphoglycerate

25. What is the essential coenzyme in this reaction, and what is its contribution to the formation of 2-phosphoglycerate?

⑨ Dehydration of 2-Phosphoglycerate to Phosphoenolpyruvate

26. What is the significance of this dehydration reaction to the payoff phase of glycolysis?

⑩ Transfer of the Phosphoryl Group from Phosphoenolpyruvate to ADP

27. Why does this reaction have a large, negative $\Delta G'^{\circ}$?

28. Which of the payoff reactions is/are irreversible in the cell?

The Overall Balance Sheet Shows a Net Gain of ATP

29. Use the overall equation of glycolysis under aerobic conditions to remind yourself of the pathways of carbon, phosphoryl groups, and electrons.

Intermediates Are Channeled between Glycolytic Enzymes

30. What are the advantages to the cell of substrate channeling?

Glycolysis Is under Tight Regulation

31. What is the "Pasteur effect"?

Glucose Catabolism Is Deranged in Cancerous Tissue

32. Why do many cancer cells rely on anaerobic glycolysis?

Fates of Pyruvate under Aerobic and Anaerobic Conditions

33. Why is the cell's ability to regenerate NAD^+ critical to glycolysis? Which glycolytic enzyme requires NAD^+?

Pyruvate Is the Terminal Electron Acceptor in Lactic Acid Fermentation

34. What is the most important consequence to the cell of the reduction of pyruvate to lactate by lactate dehydrogenase?

35. What is the net yield of ATP per glucose molecule by this pathway?

Glycolysis at Limiting Concentrations of Oxygen: Athletes, Alligators, and Coelacanths

36. How does the Cori cycle relate to the deep breaths taken by an athlete as she recovers from a sprint?

Ethanol Is the Reduced Product in Alcohol Fermentation

37. What is the most important consequence to the cell of the reduction of pyruvate to ethanol by pyruvate decarboxylase and alcohol dehydrogenase?

38. Do all organisms perform alcohol fermentation and lactic acid fermentation?

39. What is the net yield of ATP per glucose molecule by this pathway?

Brewing Beer

40. Why are both aerobic and anaerobic stages of yeast growth allowed in beer brewing, even though the aerobic stage produces no ethanol?

Thiamin Pyrophosphate Carries "Active Aldehyde" Groups

41. What structural aspect of TPP allows it to act as an "electron sink"?

42. In general, what is the role of TPP in biological reactions?

Microbial Fermentations Yield Other End Products of Commercial Value

43. Why must industrial fermentations be carried out under conditions that exclude all but the desired microorganisms?

Feeder Pathways for Glycolysis

Glycogen and Starch Are Degraded by Phosphorolysis

44. Does the degradation of glycogen (or starch) to a form of glucose that can enter the glycolytic pathway require an input of energy? Explain.

45. What kind of bond is cleaved by the debranching enzyme but not by the phosphorylases?

Other Monosaccharides Enter the Glycolytic Pathway at Several Points

46. Is it more energetically expensive to feed monosaccharides other than glucose into the glycolytic pathway? Explain.

Dietary Polysaccharides and Disaccharides Are Hydrolyzed to Monosaccharides

47. Can individual cells experience lactose intolerance? Explain.

Regulation of Carbohydrate Catabolism

48. Carbohydrate catabolism is a highly regulated process: the rate must be responsive to cellular demands for the ATP produced by glucose oxidation *and* to cellular demands for biosynthetic precursors. What categories of biomolecules require carbohydrates for their synthesis and/or function?

Regulatory Enzymes Act as Metabolic Valves

49. How are substrate-limited and enzyme-limited reactions different?

50. Which of the following mechanisms for enzyme regulation works the fastest? Which takes the longest to exert its regulatory effect?

a. Allosteric regulation
b. Hormonal action
c. Synthesis/degradation of enzymes

51. What is meant by the term "committed step"? Which step in the glycolytic pathway is the committed step?

52. Why does it make sense to the overall economy of the cell that essentially irreversible (highly exergonic) reactions are sites of metabolic regulation?

Glycolysis and Gluconeogenesis Are Coordinately Regulated

53. What are the three essentially irreversible reactions of glycolysis that must be circumvented in gluconeogenesis?

54. What is the role of fructose 2,6-bisphosphate in the coordinate regulation of these two pathways?

55. Plants lack glucagon. Why is there no role for glucagon in plants?

Phosphofructokinase-1 Is under Complex Allosteric Regulation

56. PFK-1 serves as a model for cellular control of regulatory enzymes and is a critical enzyme in the glycolytic pathway. Take the time to be certain that you completely understand the material in this section.

Hexokinase Is Allosterically Inhibited by Its Reaction Product

57. Which of the two isozymes, muscle hexokinase or liver hexokinase (glucokinase), reaches $\frac{V_{max}}{2}$ at lower substrate concentrations? Why does this make sense physiologically?

58. Why does it make sense for product inhibition of a regulatory enzyme to be reversible?

Isozymes: Different Proteins, Same Reaction

59. List four factors for the differential distribution of isozyme forms in cells and organs.

Pyruvate Kinase Is Inhibited by ATP

60. Which two compounds, in addition to ATP, inhibit pyruvate kinase activity? Why does this make sense?

Glycogen Phosphorylase Is Regulated Allosterically and Hormonally

61. What are the differences between the main functions of glucose metabolism in muscle and in liver?

62. What are the rapid (in milliseconds) versus less rapid (seconds to minutes) regulatory strategies for glycogen phosphorylase in muscle? Which is/are hormonally induced and which is/are allosterically induced?

63. What is the hormone that triggers glycogen breakdown in the muscle? In the liver?

64. How does the hormonal and allosteric regulation of glucose metabolism in muscle differ from that in liver? (This is a preview of some of the connections explored in greater depth in Chapter 23.)

The Pentose Phosphate Pathway of Glucose Oxidation

65. In what tissues is the pentose phosphate pathway most active?

66. What are the products of this pathway and what roles do they play in cellular metabolism?

Glucose 6-Phosphate Dehydrogenase Deficiency: Why Pythagoras Wouldn't Eat Falafel

67. In what specific reaction(s) is NADPH crucial for protection of cellular structures?

SELF-TEST

Do You Know the Terms?

ACROSS

2. The hexokinase reaction is the first step in glycolysis but is *not* the ________ step because the product, glucose 6-phosphate, can enter either the glycolytic or the pentose phosphate pathway.
4. Process in which lactate or pyruvate is used to form new molecules of glucose.
5. Balance achieved in the rate of formation and rate of utilization of, for example, glucose 6-phosphate achieved by the feedback inhibition of hexokinase.
8. The ________ pathway produces NADPH, the source of reducing equivalents for biosynthetic processes, and ribose 5-phosphate, an essential precursor for nucleotide synthesis. (2 words)
9. AMP is a(n) ________ regulator of phosphorylase *b*.
10. Activated by phosphorylation, this enzyme relays information about the binding of epinephrine to the enzymes responsible for glycogen breakdown. (2 words)
12. Hexokinase and glucokinase are examples.
13. The toxic ingredient of fava beans.
14. Its formation in very active muscle regenerates NAD^+ for use in glycolysis.
16. Enzyme regulated by allosteric mechanisms and by covalent phosphorylation/dephosphorylation. Differentially regulated in the liver and muscles.
17. ________-________ reactions are catalyzed by enzymes with high activities restrained only by the availability of substrate.
19. Direct transfer of substrate from the active site of one enzyme to the active site of another without involving a diffusion step.
21. Process whereby NAD^+ needed for glycolysis is regenerated by reduction of acetaldehyde. (2 words)
22. ________-________ reactions are not affected by substrate availability; often act as metabolic valves.

DOWN

1. Enzyme that transfers a phosphoryl group between two compounds.
3. Type of enzyme that transfers a functional group from one position to another in the same molecule; for example, the transfer of a phosphoryl group from C-3 to C-2 of phosphoglycerate.
6. Physiological state indicated by excess lactate in the blood. (2 words)
7. Class of enzymes including that which catalyzes the rearrangement of dihydroxyacetone phosphate to glyceraldehyde 3-phosphate.
8. Key rate-limiting glycolytic enzyme.
11. Hormone that increases the activity of phosphorylase kinase in muscle cells.

15. Its general structure is R—CO—OPO_3^{-2}. Has a high $\Delta G'^{\circ}$ of hydrolysis; for example, 1,3-bisphosphoglycerate. (2 words)
16. Initial product of phosphorolysis of glycogen by phosphorylase.
18. Phosphoenolpyruvate + ADP → pyruvate + ATP is an example of ________-________ phosphorylation.
19. Metabolic cycle that "repays" the oxygen debt created by muscle tissue. (2 words)
20. Increased concentrations of intracellular free ________ cause muscle contraction and breakdown of stored glycogen.

Do You Know the Facts?

For questions 1–10, identify the enzyme that catalyzes each of the ten glycolytic reactions.

GLUCOSE
1. ↓
Glucose 6-phosphate
2. ↓
Fructose 6-phosphate
3. ↓
Fructose 1,6-bisphosphate
4. ↓
Glyceraldehyde 3-phosphate + dihydroxyacetone phosphate
5. ↓
Glyceraldehyde 3-phosphate
6. ↓
1,3-Bisphosphoglycerate
7. ↓
3-Phosphoglycerate
8. ↓
2-Phosphoglycerate
9. ↓
Phosphoenolpyruvate
10. ↓
PYRUVATE

In questions 11–15, refer to the numbered reactions in the previous diagram.

11. Which require(s) an input of energy in the form of ATP?

12. Which involve(s) substrate-level phosphorylation?

13. Which reduce(s) NAD^+?

14. Which is/are irreversible under intracellular conditions?

15. Which is the committed step?

In questions 16–19, indicate the effect (increase, decrease, or no effect) of each situation on the rate of glycolysis. If there is a regulatory effect, identify the enzyme(s) involved and the regulatory mechanism(s) employed.

16. Increased concentration of glucose 6-phosphate.

17. Increased concentration of fructose 1,6-bisphosphate.

18. Increased concentration of fructose 2,6-bisphosphate.

19. Increased concentration of ATP.

20. Which statement is *not* true of phosphofructokinase-1?

A. It is inhibited by fructose 2,6-bisphosphate.
B. It is activated by AMP.
C. It is inhibited by citrate.
D. It is inhibited by ATP.
E. ATP increases its $K_{0.5}$ for fructose 6-phosphate.

21. All of the following statements describe glycolysis *except:*

A. It produces a net of 2 ATP molecules for each molecule of glucose.
B. Its rate is regulated by hexokinase.
C. The glycolytic enzymes are found in the cytosol.
D. Two molecules of glyceraldehyde 3-phosphate are produced for each glucose molecule.
E. Its overall rate is regulated by the energy level of the cell.

22. Indicate whether each of the following statements about the pentose phosphate pathway is true or false.

_____ **(a)** It generates NADH for reductive biosyntheses.
_____ **(b)** The reactions occur in the cytosol.
_____ **(c)** Transketolase and transaldolase link this pathway to gluconeogenesis.
_____ **(d)** It is more active in muscle cells than in fat-storage cells.
_____ **(e)** It interconverts trioses, tetroses, pentoses, hexoses, and heptoses.
_____ **(f)** Through this pathway, excess ribose 5-phosphate can be completely converted into glycolytic intermediates.

23. In human tissues, ALL of the pathways that pyruvate can take:

A. are aerobic.
B. eventually reduce NAD^+ to NADH.
C. eventually reoxidize NADH to NAD^+.
D. provide equal amounts of ATP to the cell.
E. lower the pH of the cell.

Applying What You Know

1. The glycolytic reaction:

$$\text{Glyceraldehyde 3-phosphate} + P_i \longrightarrow \text{1,3-bisphosphoglycerate}$$

is endergonic ($\Delta G'^\circ = 6.3$ kJ/mol). In cells, the reaction is driven in the direction of 1,3-bisphosphoglycerate formation by the next reaction in the glycolytic pathway:

$$\text{1,3-Bisphosphoglycerate} + \text{ADP} \longrightarrow \text{3-phosphoglycerate} + \text{ATP}$$

How are these two reactions connected? Which component in the equation $\Delta G = \Delta G'^{\circ} + RT \ln K'_{eq}$ is being affected?

2. Pyruvate and ATP are the end products of glycolysis. In active muscle cells, pyruvate is converted to lactate. Lactate is transported in the blood to the liver where it is recycled by gluconeogenesis to glucose, which is transported back to muscle for additional ATP production. Why don't active muscle cells export pyruvate, which can also be converted to glucose via gluconeogenesis?

3. If C-1 of glucose is radioactively labeled and then enters glycolysis, which carbon of glyceraldehyde 3-phosphate will be labeled? What fraction of the molecules will carry the radioactive label?

4. The compound 2,3-bisphosphoglycerate (BPG) acts as a coenzyme in the glycolytic reaction catalyzed by phosphoglycerate mutase. Though in most cells BPG is present only in trace amounts—enough to act in its role as coenzyme—it is present in relatively high concentration in erythrocytes, where it acts as a regulator of the affinity of hemoglobin for oxygen (see Chapter 7, pp. 218 and 219). Because erythrocytes synthesize and degrade BPG via a detour from the glycolytic pathway (see Fig. 15–6), the rate of glycolysis and therefore the rate of generation of glycolytic intermediates has an impact on the concentration of BPG. It follows that defects in the glycolytic pathway in erythrocytes can affect the ability of hemoglobin to carry oxygen.

(a) How would the concentration of BPG, and therefore the affinity of hemoglobin for oxygen, be affected in erythrocytes with a deficiency of hexokinase?

(b) How would a pyruvate kinase deficiency affect hemoglobin's affinity for oxygen?

ANSWERS

Do You Know the Terms?

ACROSS

2. committed
4. gluconeogenesis
5. homeostasis
8. pentose phosphate
9. allosteric
10. phosphorylase kinase
12. isozymes
13. divicine
14. lactate
16. glycogen phosphorylase
17. substrate-limited
19. channeling
21. alcohol fermentation
22. enzyme-limited

DOWN

1. kinase
3. mutase
6. oxygen debt
7. isomerases
8. phosphofructokinase-1
11. epinephrine
15. acyl phosphate
16. glucose 1-phosphate
18. substrate-level
19. Cori cycle
20. calcium

Do You Know the Facts?

1. hexokinase
2. phosphohexose isomerase
3. phosphofructokinase-1
4. aldolase
5. triose phosphate isomerase
6. glyceraldehyde 3-phosphate dehydrogenase
7. phosphoglycerate kinase
8. phosphoglycerate mutase
9. enolase
10. pyruvate kinase
11. 1 and 3
12. 7 and 10
13. 6
14. 1, 3, and 10
15. 3
16. Decrease, by product inhibition of hexokinase
17. Increase, due to mass action effects
18. Increase, by allosteric activation of phosphofructokinase-1
19. Decrease, by allosteric inhibition of phosphofructokinase-1 and pyruvate kinase
20. A
21. B
22. **(a)** F; **(b)** T; **(c)** F; **(d)** F; **(e)** T; **(f)** T
23. C

Applying What You Know

1. Because of the rapid removal of 1,3-bisphosphoglycerate by the phosphoglycerate kinase reaction, the preceding glycolytic reaction catalyzed by glyceraldehyde 3-phosphate dehydrogenase

$$\text{Glyceraldehyde 3-phosphate} + P_i \longrightarrow \text{1,3-bisphosphoglycerate} \qquad \Delta G'^{\circ} = 6.3 \text{ kJ/mol}$$

has a *negative* ΔG. Under cellular conditions, the free-energy change can have a negative value, even though $\Delta G'^{\circ}$ is positive, when the value for $RT \ln \frac{[\text{products}]}{[\text{reactants}]}$ is large and negative. Because of the rapid removal of the product 1,3-bisphosphoglycerate, the glyceraldehyde 3-phosphate dehydrogenase reaction has a $\frac{[\text{products}]}{[\text{reactants}]}$ ratio of less than one and the value for $RT \ln \frac{[\text{products}]}{[\text{reactants}]}$ is negative.

2. For glycolysis to proceed, cells must have a constant supply of NAD^+, that is, a way of regenerating NAD^+ from NADH. In active muscle cells, pyruvate is the oxidizing agent used to oxidize NADH. If the cells were to export pyruvate, the continued production of ATP by glycolysis would not be possible.

3. The C-3 of glyceraldehyde 3-phosphate will be labeled. This carbon carries a phosphoryl group, and was *either* C-1 or C-6 in the six-carbon molecule fructose 1,6-bisphosphate. Thus *one-half* of the glyceraldehyde 3-phosphate molecules will be labeled, because half were derived from C-1 of glucose (which was radioactively labeled) and half from C-6 (unlabeled).

4. (a) When hexokinase activity is lowered, the concentrations of glycolytic intermediates decrease, and correspondingly, the concentration of BPG drops. This results in a higher affinity of hemoglobin for oxygen; the oxygen-binding curve is shifted to the left.

(b) If pyruvate kinase activity is decreased, the intermediates of glycolysis build up, increasing the BPG concentration and its binding to hemoglobin, which decreases the affinity for oxygen; the oxygen-binding curve is shifted to the right.

chapter 16

The Citric Acid Cycle

STEP-BY-STEP GUIDE

Major Concepts

Respiration is the complete oxidation of organic fuels to CO_2 and H_2O.

Cellular respiration occurs in three major stages. The first stage oxidizes fuel molecules to the two-carbon molecule **acetyl-CoA.** In the second stage (the citric acid cycle), acetyl-CoA is oxidized to CO_2 and the electron carriers NAD^+ and FAD are reduced. In the third stage (oxidative phosphorylation, discussed in Chapter 19), electrons from the oxidation of fuel molecules are transferred to O_2; ATP is formed as a result of this electron transfer process.

The first stage of cellular respiration includes glycolysis and the production of acetate.

Pyruvate, the product of glycolysis, is converted into acetyl-CoA and CO_2 by the pyruvate dehydrogenase complex. This large enzyme complex consisting of three different enzymes is regulated by allosteric mechanisms and by covalent modification. This is a good example of substrate channeling by an enzyme.

The citric acid cycle is a series of eight chemical transformations.

For each turn of the cycle, two carbons enter as an acetyl group and two carbons leave as molecules of CO_2. **Oxaloacetate** combines with acetyl-CoA to form **citrate,** which "carries" the carbons from acetyl-CoA into the cycle. Oxaloacetate is regenerated at the end of the cycle. The citric acid cycle produces energy for cells in the form of nucleotide triphosphates (ATP and GTP) and reduced electron carriers $FADH_2$ and NADH. Citric acid cycle intermediates also provide precursors for the synthesis of a number of biomolecules. Intermediates "lost" from the cycle in this way are replenished through a variety of processes called anaplerotic reactions.

The citric acid cycle, as befits a reaction series with diverse functions, is under tight regulation.

Regulation occurs at the conversion of pyruvate to acetyl-CoA and at three points in the cycle itself: the entry of acetyl-CoA into the cycle and both of the oxidative decarboxylation steps. The primary enzymes involved in regulation are the pyruvate dehydrogenase complex, citrate synthase, isocitrate dehydrogenase, and α-ketoglutarate dehydrogenase.

The glyoxylate cycle is a variation of the citric acid cycle that occurs in some microorganisms and all plants.

This pathway (which occurs in addition to the citric acid cycle, the two being coordinately regulated) results in the net formation of oxaloacetate from two molecules of acetate. The glyoxylate cycle makes it possible to convert the carbons of acetyl-CoA into glucose molecules, an anaplerotic pathway not present in vertebrate systems.

What to Review

Answering the following questions and reviewing the relevant concepts, which you have already studied, should make this chapter more understandable.

- Review the roles of **FAD** and **NAD^+** (pp. 518–521) as electron carriers. Reduced forms of these nucleotides are generated in the citric acid cycle and play a critical role in the oxidative phosphorylation of ADP (Chapter 19).
- **TPP** is an important cofactor used by both glycolytic and citric acid cycle enzymes. Review its role in glycolysis (pp. 545–546).

- Review what is meant by **reaction equilibria** and the relationships among ΔG, $\Delta G'^{\circ}$, and K'_{eq} (pp. 494–499). These concepts are important to an understanding of the regulation of metabolic pathways such as the citric acid cycle.

Topics for Discussion

Answering each of the following questions, especially in the context of a study group discussion, should help you understand the important points of this chapter.

Production of Acetate

Pyruvate Is Oxidized to Acetyl-CoA and CO_2

1. The oxidative decarboxylation of pyruvate is a highly exergonic reaction. How is the energy released by this reaction conserved?

The Pyruvate Dehydrogenase Complex Requires Five Coenzymes

2. Make sure you can recognize the structure of coenzyme A; pay attention to the components of the molecule and to its general function. Why is the thioester bond important to the function of coenzyme A?

3. What are the two possible roles of lipoate in enzymatic reactions?

The Pyruvate Dehydrogenase Complex Consists of Three Distinct Enzymes

4. Be sure you understand how the different enzyme subunits of pyruvate dehydrogenase and their cofactors act together to catalyze the decarboxylation of pyruvate.

Intermediates Remain Bound to the Enzyme Surface

5. What is the result of the first three reactions (steps ① to ③) of the overall pyruvate dehydrogenase complex reaction? What is the source of energy for this set of reactions?

6. What is the result of steps ④ and ⑤ in this process?

7. Why is it advantageous to have three different enzymatic activities clustered into a single enzyme complex?

8. Why is thiamine deficiency a serious condition?

Reactions of the Citric Acid Cycle

9. What are the roles of the citric acid cycle?

10. Where in eukaryotes do these reactions take place? Where do they occur in prokaryotes?

The Citric Acid Cycle Has Eight Steps

11. *Cell Map:* Draw in the structures and names of the intermediates and the names of the enzymes and cofactors for each step of the citric acid cycle on your cell map at this point.

① Formation of Citrate

12. The concentration of oxaloacetate in the cell is normally quite low. What factor in the citrate synthase reaction prevents this from hindering the operation of the cycle?

② Formation of Isocitrate via *cis*-Aconitate

13. The $\Delta G'^{\circ}$ of the aconitase reaction is 13.3 kJ/mol. What drives the aconitase reaction forward?

14. What is the general function of the iron-sulfur center in aconitase?

③ Oxidation of Isocitrate to α-Ketoglutarate and CO_2

15. Why do eukaryotic cells need two isozymes of isocitrate dehydrogenase?

④ Oxidation of α-Ketoglutarate to Succinyl-CoA and CO_2

16. How is the energy of oxidation of α-ketoglutarate conserved?

17. What are the similarities between the α-ketoglutarate dehydrogenase complex and the pyruvate dehydrogenase complex? What are the differences?

⑤ Conversion of Succinyl-CoA to Succinate

18. What is the source of the energy used to drive the substrate-level phosphorylation of GDP?

19. What are the differences between substrate-level phosphorylation and respiration-linked phosphorylation?

20. Given that the $\Delta G'^{\circ}$ of the nucleoside diphosphate kinase reaction is 0 kJ/mol, what factor would encourage the reaction to proceed in the direction of ATP formation?

Synthases and Synthetases; Ligases and Lyases; Kinases, Phosphatases, and Phosphorylases: Yes, the Names Are Confusing!

21. You can more easily learn and remember the functions of specific enzymes if you learn the names and functions of the general classes to which they belong. In your text, find an enzyme for each of the enzyme classes listed in the title above, and write the reaction it catalyzes.

⑥ Oxidation of Succinate to Fumarate

22. Explain why it is significant that the electron carrier FAD is covalently bound to succinate dehydrogenase whereas NAD^+ is in a soluble, unconjugated form.

⑦ Hydration of Fumarate to Malate

23. This reaction is readily reversible, with a $\Delta G'^\circ$ of -3.8 kJ/mol. Why do you think it proceeds in the direction of malate formation in vivo?

⑧ Oxidation of Malate to Oxaloacetate

24. The $\Delta G'^\circ$ of the malate dehydrogenase reaction is 29.7 kJ/mol. What drives this reaction forward?

The Energy of Oxidations in the Cycle Is Efficiently Conserved

25. Which reaction(s) of the citric acid cycle store(s) the energy derived from oxidations as NADH? Which store(s) it as $FADH_2$?

Citrate: A Symmetrical Molecule That Reacts Asymmetrically

26. What is necessary in an enzyme's active site in order for a symmetric molecule to react asymmetrically with the enzyme?

27. What is the net equation for one turn of the citric acid cycle?

28. What is the net energy yield per molecule of glucose for the combined reactions of glycolysis, the pyruvate dehydrogenase reaction, and the citric acid cycle?

Why Is the Oxidation of Acetate So Complicated?

29. Why is a process as complex as the citric acid cycle actually an economical way for cells to do their metabolic business?

Citric Acid Cycle Components Are Important Biosynthetic Intermediates

30. *Cell Map:* Because the citric acid cycle is the "hub" of cellular metabolism, it is important that you have a firm grasp of all the reactions and intermediates involved in this pathway. Add in arrows from the CAC intermediates out to their various cellular products on your cell map.

Citrate Synthase, Soda Pop, and The World Food Supply

31. How does citrate act to immobilize Al^{3+}?

Anaplerotic Reactions Replenish Citric Acid Cycle Intermediates

32. Under what cellular circumstances is pyruvate carboxylase activity stimulated?

Biotin in Pyruvate Carboxylase Carries CO_2 Groups

33. What is the actual source of CO_2 groups for biotin?

34. What structural component of biotin is directly involved in carboxylation reactions?

Regulation of the Citric Acid Cycle

Production of Acetyl-CoA by the Pyruvate Dehydrogenase Complex Is Regulated by Allosteric and Covalent Mechanisms

35. What do all the allosteric *activators* of the pyruvate dehydrogenase complex signal about the cell's energy state? What do the allosteric *inhibitors* signal?

36. How does covalent modification regulate the pyruvate dehydrogenase complex?

37. Does the covalent regulation provide any information about the cell's energy state?

The Citric Acid Cycle Is Regulated at Its Three Exergonic Steps

38. Why are citrate synthase, isocitrate dehydrogenase, and α-ketoglutarate dehydrogenase good candidates for regulatory enzymes?

39. What is the role of calcium ions in regulation of the citric acid cycle?

The Glyoxylate Cycle

The Glyoxylate Cycle Produces Four-Carbon Compounds from Acetate

40. What is the net equation of the glyoxylate cycle?

41. What is the advantage to plants of using fatty acids as energy storage molecules in seeds, rather than complex carbohydrates such as starch?

42. In what plant tissue types are glyoxysomes abundant?

The Citric Acid and Glyoxylate Cycles Are Coordinately Regulated

43. Why is the coordinated regulation of isocitrate lyase and isocitrate dehydrogenase of advantage to the organism?

44. Do the intermediates of the citric acid cycle and glycolysis exert their regulatory effects directly or indirectly on these enzymes?

SELF-TEST

Do You Know the Terms?

ACROSS

1. Enzymes, including synthetases, that catalyze condensation reactions requiring an input of energy.

3. Disease resulting from a dietary deficiency of the vitamin thiamin.

6. Enzymes that catalyze condensation reactions that do not require a nucleotide triphosphate for an energy source, such as: oxaloacetate + acetyl-CoA + $H_2O \rightarrow$ citrate + CoA-SH.

7. Organelles containing enzymes of fatty acid degradation and the glyoxylate cycle.

9. Describes pathways used in both anabolism and catabolism; for example, the citric acid cycle, in which oxaloacetate is an intermediate in both the degradation and the synthesis of glucose.

10. Cofactor involved in the decarboxylation of pyruvate and the initial binding of the resulting acetyl group to pyruvate dehydrogenase. (abbr.)

11. Describes symmetric molecules that bind to a substrate-binding site in only one of two possible orientations.

14. Process of passing electrons from fuel molecules to O_2.

15. General type of reaction catalyzed by isocitrate dehydrogenase and by α-ketoglutarate dehydrogenase complex. (2 words)

16. Enzymes that catalyze group-transfer reactions involving a phosphoryl group.

17. Acyl groups are linked to coenzyme A through a ________ bond and are thus activated for group transfer.

DOWN

1. Enzymes that catalyze cleavage reactions involving double bonds and electron rearrangements.

2. Enzymes that catalyze reactions involving a nucleoside triphosphate; for example:

$$\text{Succinyl-CoA} + \text{GDP} + P_i \rightleftharpoons \text{succinate} + \text{GTP} + \text{CoA-SH}$$

4. Consuming too much eggnog or too many raw eggs may cause a deficiency in the vitamin ________, which is required as a cofactor for pyruvate carboxylase.

5. Compound formed by the condensation of glyoxylate and acetyl-CoA.

8. Pyruvate dehydrogenase cofactor involved in the transfer of acyl groups and of electrons in the form of hydrogen.

9. Describes reactions that produce citric acid cycle intermediates.

12. End product of glycolysis.

13. Citric acid cycle intermediate that inhibits phosphofructokinase-1.

Do You Know the Facts?

In questions 1–10, fill in the names of the missing intermediates in the diagram below.

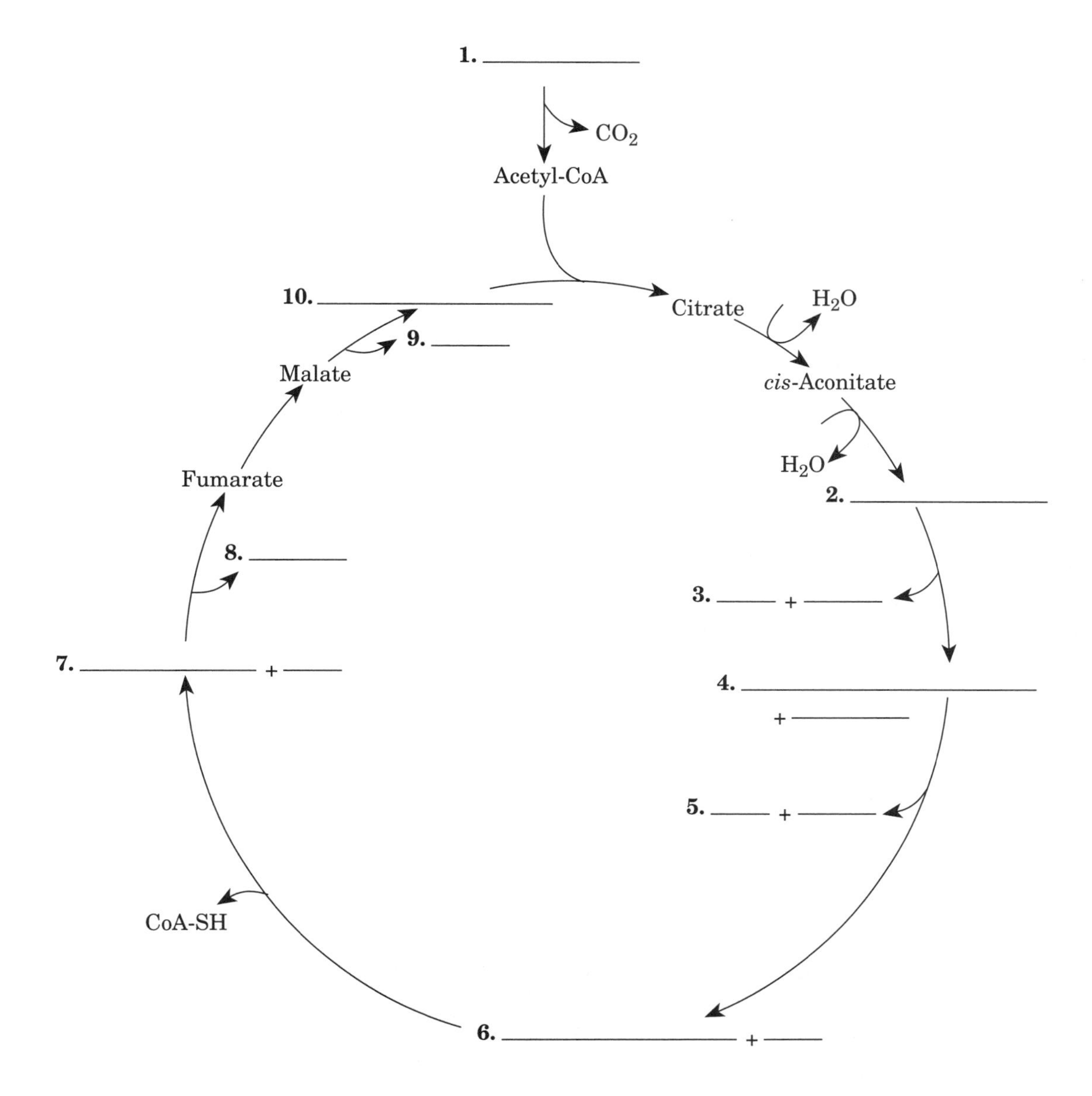

11. Indicate whether each of the following metabolic effects would or would not occur (yes/no) in cells lacking the phosphoprotein phosphatase of the pyruvate dehydrogenase complex.

_____ **(a)** Increased [lactate]

_____ **(b)** Increased pH

_____ **(c)** Increased [citrate]

_____ **(d)** Decreased [NADH] and [ATP]

12. Which of the following correctly describes the citric acid cycle?

A. Oxygen is required to regenerate electron acceptors.
B. Citrate is dehydrated and hydrated by the same enzyme, aconitase.
C. Ten high-energy phosphate bonds are eventually formed as a result of one round of the cycle.
D. Succinate dehydrogenase links the citric acid cycle to oxidative phosphorylation.
E. All of the above are true.

13. In order to examine the citric acid cycle you have obtained a *pure preparation of isolated,* ***intact*** *mitochondria.* You add some succinyl-CoA to the suspension of mitochondria. How many moles of ATP would you expect to be generated in one turn of the citric acid cycle from each mole of succinyl-CoA added to the test tube?

A. 3
B. 4
C. 5
D. 5.5
E. No ATP would form under these conditions.

14. You have discovered a compound that inhibits fumarase. How many moles of ATP would you expect to be generated from one mole of acetyl-CoA in the presence of this inhibitor?

A. 5
B. 6
C. 6.5
D. 7.5
E. No ATP would form under these conditions.

15. All of the following enzymes are linked to the reduction of NADH *except:*

A. isocitrate dehydrogenase.
B. lactate dehydrogenase.
C. succinate dehydrogenase.
D. pyruvate dehydrogenase.
E. α-ketoglutarate dehydrogenase.

In questions 16–22, match the descriptions to the appropriate enzyme(s) from the list below. (There can be more than one answer to a question.)

(a) Citrate synthase
(b) Isocitrate dehydrogenase
(c) Pyruvate dehydrogenase
(d) Succinate dehydrogenase
(e) Succinyl-CoA synthetase
(f) α-Ketoglutarate dehydrogenase

_____ **16.** Allosterically activated by calcium.
_____ **17.** Catalyzes the committed step in the citric acid cycle.
_____ **18.** The only membrane-bound enzyme in the citric acid cycle.
_____ **19.** Catalyzes the substrate-level phosphorylation of ADP or GDP.
_____ **20.** Regulated by reversible phosphorylation.
_____ **21.** Inhibited by NADH.
_____ **22.** Activated by AMP or ADP.

In questions 23–26, indicate the effect (increase, decrease, no change) of each situation on the overall rate of the citric acid cycle. If there is a regulatory effect, identify the enzyme(s) involved and the type of regulatory mechanism(s) employed.

23. Increased concentration of citrate.

24. Increased concentration of ATP.

25. Increased concentration of glucose 6-phosphate.

26. Increased concentration of succinyl-CoA.

27. Indicate whether each of the following statements about the glyoxylate cycle is true or false.

_____ **(a)** It allows the products of fatty acid oxidation to be converted, eventually, to glucose.

_____ **(b)** It provides intermediates for the citric acid cycle during periods when amino acids are being synthesized.

_____ **(c)** It depletes cellular stores of oxaloacetate.

_____ **(d)** It occurs instead of the citric acid cycle in plants.

28. Although molecular oxygen does not participate directly in any of the reactions in the citric acid cycle, the cycle only operates when oxygen is present. This is because O_2:

A. is necessary as an activator for several enzymatic dehydrogenations (oxygenations) in the cycle.
B. is necessary for producing water, which is crucial for all cellular processes.
C. accepts electrons from the electron transfer chain, allowing reoxidation of NADH to NAD^+.
D. removes toxic by-products of the citric acid cycle.
E. has all of the above functions.

Applying What You Know

1. The flow of metabolites into the glycolytic pathway and the citric acid cycle is reduced during periods of DNA replication. How might cells coordinate DNA synthesis and glucose metabolism?

2. You and your fellow students find it difficult to believe that the oxidation of a two-carbon acyl compound requires a pathway as complex as the citric acid cycle. In order to verify the metabolic pathways you have been learning about, you conduct an experiment using a sample of glucose radioactively labeled with ^{14}C at C-1. If the information you have learned is correct, in which glycolytic or citric acid cycle intermediate, and on which carbon, should you find the label:

(a) if O_2 were unavailable?

(b) in the presence of malonate?

(c) after one round of the citric acid cycle?

(d) after three rounds of the citric acid cycle?

3. You wish to determine the metabolic fate of oxaloacetate in cells. In an experiment using oxaloacetate radioactively labeled with ^{14}C at C-1, you find that much of the label ends up in CO_2. Can you conclude from this result that the overall cellular levels of oxaloacetate in the sample are being depleted?

4. Calculate the amount of metabolic energy produced from the following carbon sources under the specified cellular conditions. Which of these yields the most energy and which yields the least?

(a) 1 mol of glyceraldehyde 3-phosphate; O_2 is available.

(b) 1 mol of succinate and 1 mol of pyruvate; O_2 is available.

(c) 8 mol of glucose; anaerobic conditions.

(d) 1 mol of acetyl-CoA; O_2 is available.

ANSWERS

Do You Know the Terms?

ACROSS

1. ligases
3. beriberi
6. synthases
7. glyoxysomes
9. amphibolic
10. TPP
11. prochiral
14. respiration
15. oxidative decarboxylation
16. kinases
17. thioester

DOWN

1. lyases
2. synthetases
4. biotin
5. malate
8. lipoate
9. anaplerotic
12. pyruvate
13. citrate

Do You Know the Facts?

1. pyruvate
2. isocitrate
3. CO_2 + NADH
4. α-ketoglutarate + CoA-SH
5. CO_2 + NADH
6. succinyl-CoA + GDP
7. succinate + GTP
8. $FADH_2$
9. NADH
10. oxaloacetate
11. **(a)** Y; **(b)** N; **(c)** N; **(d)** Y
12. E
13. E; succinyl-CoA could not enter the *intact* mitochondrion
14. D
15. C
16. (b), (c), and (f)
17. (a)
18. (d)
19. (e)
20. (c)
21. (a), (c), and (f)
22. (a), (b), and (c)
23. Decrease. Citrate synthase is inhibited by a negative feedback mechanism through product accumulation. The amount of acetyl-CoA available for entry into the cycle will decrease due to allosteric inhibition of PFK-1.
24. Decrease. ATP inhibits citrate synthase and isocitrate dehydrogenase through product accumulation. ATP also allosterically inhibits PFK-1, pyruvate kinase, and pyruvate dehydrogenase.
25. No effect.
26. Decrease. α-Ketoglutarate dehydrogenase is inhibited by product accumulation.
27. **(a)** T; **(b)** T; **(c)** F; **(d)** F
28. C

Applying What You Know

1. DNA synthesis syphons off oxaloacetate for the production of the pyrimidine bases thymine and cytosine. As levels of oxaloacetate decrease, acetyl-CoA accumulates, decreasing the activity of pyruvate dehydrogenase. The net result is a slowing of the overall rate of the citric acid cycle. The pentose phosphate pathway, an alternative route of glucose catabolism, reroutes glucose 6-phosphate from glycolysis to generate ribose 5-phosphate, slowing glycolysis. This pentose is necessary for the biosynthesis of nucleotides, which are needed for DNA synthesis.
2. **(a)** Lactate at C-3, or ethanol in the methyl group, depending on the organism.
   (b) Succinate at either C-2 or C-3.
   (c) Oxaloacetate in the carbonyl carbon and at C-3.
   (d) CO_2 and oxaloacetate, at all carbons. Remember that fumarate is a symmetric molecule; in hydration of the double bond, a hydroxyl group is added to either C-2 or C-3 with equal probability.
3. You cannot conclude that oxaloacetate levels are being depleted. Although the oxaloacetate carbons (including the ^{14}C label) are lost as CO_2 in the citric acid cycle, new carbons are introduced when acetyl-CoA condenses with oxaloacetate. A molecule of oxaloacetate is regenerated after each turn of the citric acid cycle, but this contains two new carbon atoms obtained from the incoming acetyl-CoA.
4. **(a)** 1 mol of glyceraldehyde 3-phosphate produces 16–17 mol of ATP under aerobic conditions (depending on the shuttle used to transfer electrons into the mitochondrion from the NADH formed during glycolysis).
   (b) 1 mol of succinate and 1 mol of pyruvate produce 16.5 mol of ATP under aerobic conditions.
   (c) 8 mol of glucose produce 16 mol of ATP under anaerobic conditions.
   (d) 1 mol of acetyl-CoA produces 10 mol of ATP under aerobic conditions.

   Thus under the conditions specified, (a) and (b) yield the most energy, (d) the least. However, on a *per mole* basis, (c) produces the least energy.

chapter

17 Oxidation of Fatty Acids

STEP-BY-STEP GUIDE

Major Concepts

The insolubility of triacylglycerols poses a problem for the absorption of dietary lipids and for the mobilization of triacylglycerols stored in adipocytes.

Absorption of dietary fats is dependent on the presence of amphipathic bile salts, which disrupt fat globules into small micelles. The triacylglycerols in the micelles are then cleaved by **lipases** into monoacyl- and diacylglycerols, free fatty acids, and glycerol, which can be absorbed by the intestinal epithelial cells. These components are then reassembled into triacylglycerols and transported in the blood by complexing to lipid-binding proteins, **apolipoproteins,** to form **chylomicrons,** which can form favorable associations with the aqueous environment.

This process of triacylglycerol cleavage and passage of the components into cells is repeated in the capillaries of muscle and adipose tissue. In muscle cells, fatty acids are oxidized for energy; in adipocytes they are reesterified to triacylglycerols for storage. Stored triacylglycerols are subsequently mobilized when hormonal cues signal a need for metabolic energy. The fatty acids liberated from adipocytes are carried in the bloodstream by serum albumin.

Fatty acids are oxidized in mitochondria.

Mitochondria have two distinct membrane layers: an outer membrane that faces the cytosol and an inner membrane that is thrown into numerous folds called **cristae** (see Fig. 19–1, p. 660). The space between the outer and inner membranes is called the **intermembrane space,** and the interior of the mitochondrion is called the **matrix.** Fatty acids are oxidized in the matrix.

Fatty acids must be transported across the inner mitochondrial membrane.

The fatty acids are first activated by enzymes in the outer mitochondrial membrane, the **acyl–CoA synthetases.** These enzymes catalyze the formation of a thioester linkage between the fatty acid and coenzyme A (CoA-SH) to form a **fatty acyl–CoA,** a reaction coupled to the cleavage of ATP to AMP and PP_i. Passage of the fatty acyl-CoA across the inner membrane requires the assistance of a carrier compound, **carnitine.** The fatty acyl–carnitine ester, formed by the action of **carnitine acyltransferase I,** is carried into the matrix by facilitated diffusion via the **acyl-carnitine/carnitine transporter.** This entry process is rate limiting and commits the fatty acid to oxidation in the mitochondrial matrix.

Oxidation of fatty acids in the mitochondria has three stages.

The first stage is **β oxidation.** This involves oxidation of the fatty acid at the β carbon (C-3), followed by a hydration, an oxidation, and a thiolysis, a series of reactions resulting in the removal of C-1 and C-2 as a molecule of acetyl-CoA. One round of β oxidation (removal of one two-carbon unit) produces one molecule each of $FADH_2$ and NADH. In the second stage of fatty acid oxidation, the acetyl-CoA is further oxidized in the citric acid cycle, producing three NADH, one $FADH_2$, and one GTP. In the third stage, the NADH and $FADH_2$ formed during stages 1 and 2 pass their electrons to O_2 via the electron transfer chain. Counting each $FADH_2$ as equivalent to 1.5 ATP, and each NADH as equivalent to 2.5 ATP; the yield of ATP produced from all three stages of oxidation for an even-numbered, saturated fatty acid containing n carbons is $14(n/2) - 4$. To account for the two ATP equivalents used to activate the fatty acid to fatty acyl-CoA, two more ATPs must be subtracted; the net equation is therefore $14(n/2) - 6$.

Oxidation of unsaturated and odd-chain fatty acids requires additional reactions.

For monounsaturated fatty acids, an isomerization reaction changes the cis double bond to a trans; for polyunsaturated fatty acids, an additional reduction reaction is required. These reactions produce a saturated fatty acid that can enter β oxidation. Odd-chain fatty acids proceed through the same pathway as do the even-numbered chains until the final round of β oxidation, which produces acetyl-CoA and a three-carbon compound, propionyl-CoA. Propionyl-CoA is converted to succinyl-CoA in three steps: a carboxylation, a group-transfer reaction, and an isomerization requiring coenzyme B_{12}. Succinyl-CoA is then oxidized in the citric acid cycle.

Fatty acid oxidation is highly regulated.

The carnitine acyltransferase I reaction is inhibited by malonyl-CoA, the first intermediate in the *synthesis* of fatty acids. In addition, two of the β-oxidative enzymes are inhibited when the overall energy levels of the system are high.

Peroxisomes and glyoxysomes are organelles that carry out some oxidative breakdown of fatty acids.

The oxidation of fatty acids in peroxisomes differs from that in mitochondria. The initial reaction produces a double bond between the α and β carbons, with electrons transferred directly to O_2, producing H_2O_2. No ATP is produced. The H_2O_2 is cleaved by catalase to H_2O and O_2. In glyoxysomes, β-oxidation allows conversion of stored lipids into carbohydrates.

Omega oxidation is another pathway for fatty acid breakdown.

This pathway, which occurs in the endoplasmic reticulum, produces dicarboxylic acids such as succinate. Genetic defects in the enzyme medium-chain acyl-CoA dehydrogenase (MCAD) can produce a serious disease characterized by the inability to oxidize medium-chain fatty acids via β oxidation. It is indicated by high urinary concentrations of medium-chain dicarboxylic acids produced via omega oxidation.

In mammals, an alternative pathway for acetyl-CoA produces ketone bodies.

The key to the fate of acetyl-CoA is the concentration of oxaloacetate, the intermediate that is required for entry of acetyl-CoA into the citric acid cycle and is used in the formation of glucose (gluconeogenesis). When the demand for glucose is high, oxaloacetate levels are reduced. Under these conditions, acetyl-CoA does not enter the citric acid cycle; it is converted to the ketone bodies **acetoacetate, D-β-hydroxybutyrate,** and **acetone** in the liver. These molecules can be transported in the blood to tissues such as muscle and brain, where they are converted back to acetyl-CoA and oxidized via the citric acid cycle.

What to Review

Answering the following questions and reviewing the relevant concepts, which you have already studied, should make this chapter more understandable.

- Review the basic structure of a triacylglycerol (Fig. 11–2).
- Where do lipases cleave triacylglycerols (Fig. 11–13)? Review the properties of fatty acids in aqueous solution (Fig. 4–7).
- The oxidation of fatty acids produces **acetyl-CoA,** which can be further metabolized to produce additional energy for cells. Review how acetyl-CoA enters the citric acid cycle (Fig. 16–7). Review the **glyoxylate cycle.** What is the role of the glyoxylate cycle in plant seeds (Fig. 16–16, p. 589)?
- A number of steps in the oxidation of fatty acids result in the production of NADH and $FADH_2$, as well as compounds that can be further metabolized in the citric acid cycle. Be sure you remember how much ATP is generated from reoxidation of NADH and $FADH_2$ (p. 582, Table 16–1). And be sure you remember which steps in the citric acid cycle produce reduced electron carriers and high-energy compounds (Fig. 16–7).
- Fatty acids must be transported across mitochondrial membranes in order to be metabolized. Make sure you understand the characteristics of the type of membrane transport called **facilitated diffusion** (pp. 409–413).

Topics for Discussion

Answering each of the following questions, especially in the context of a study group discussion, should help you understand the important points of this chapter.

Digestion, Mobilization, and Transport of Fats

Dietary Fats Are Absorbed in the Small Intestine

1. How do the chemical properties of lipids affect the way they must be mobilized in biological systems?

2. How are the different types of lipids transported in the blood *to* adipose tissue?

Hormones Trigger Mobilization of Stored Triacylglycerols

3. How are lipids transported in the blood *from* adipose tissue?

4. Why does it make sense that low blood glucose triggers mobilization of fat stores?

Fatty Acids Are Activated and Transported into Mitochondria

5. *Cell Map:* Find on your cell map the acyl-carnitine/carnitine transporter. Fill in the enzymes that activate fatty acids for transport, that attach fatty acyl groups to carnitine, and that re-attach acyl groups to CoA in the matrix.

6. What is the energetic "cost" of the formation of a fatty acyl-CoA?

7. What are the roles of the two pools of coenzyme A? How does the acyl-carnitine/carnitine transporter maintain the separation of these pools?

β Oxidation

The β Oxidation of Saturated Fatty Acids Has Four Basic Steps

8. *Cell Map:* To the right of the citric acid cycle, enter onto your cell map the four steps of β oxidation. Where do the acetyl-CoA, NADH, and $FADH_2$ go from here?

9. What *types* of reactions make up the four basic steps of β oxidation?

10. What is the overall effect of the first three reactions of β oxidation on the bond between the α and β carbons?

11. At what steps are reduced electron carriers generated in the β oxidation of fatty acids?

12. To which enzyme complexes do the reduced electron carriers pass their electrons?

The Four Steps Are Repeated to Yield Acetyl-CoA and ATP

13. How many ATPs are generated by the reoxidation of each type of electron carrier?

14. How many ATPs are generated from this first stage (i.e., β oxidation) of the overall process of fatty acid oxidation?

Fat Bears Carry on β Oxidation in Their Sleep

15. By what reaction is water generated during hibernation? How is blood glucose produced?

Acetyl-CoA Can Be Further Oxidized in the Citric Acid Cycle

16. Why are *two* ATP equivalents subtracted in the calculation of the net yield of fatty acid oxidation?

Oxidation of Unsaturated Fatty Acids Requires Two Additional Reactions

17. What structural property of mono- and polyunsaturated fatty acids prevents oxidation of these compounds by the β-oxidation pathway? What additional *types* of reactions are needed?

18. Is there any difference in the amount of ATP formed by saturated and unsaturated fatty acid oxidation?

Complete Oxidation of Odd-Number Fatty Acids Requires Three Extra Reactions

19. Is there any difference in the amount of ATP formed by even-chain and odd-chain fatty acid oxidation?

20. What two cofactors are necessary in the extra reactions required for odd-chain fatty acid oxidation?

Coenzyme B_{12}: A Radical Solution to a Perplexing Problem

21. How does coenzyme B_{12} illustrate the necessity for trace elements in our diet?

Fatty Acid Oxidation Is Tightly Regulated

22. The processes of fatty acid oxidation and synthesis take place in separate cellular compartments. What molecule acts as the regulatory signal to prevent the catabolic and anabolic processes from occurring simultaneously?

Peroxisomes Also Carry Out β Oxidation

23. Is there any difference in the amount of ATP formed by fatty acid oxidation in a mitochondrion and in a peroxisome?

Plant Peroxisomes and Glyoxysomes Use Acetyl-CoA from β Oxidation as a Biosynthetic Precursor

24. What is the overall role of plant glyoxysomes? Which metabolite connecting β oxidation to the glyoxylate cycle makes this role possible?

The β-Oxidation Enzymes of Different Organelles Have Diverged during Evolution

25. What is a possible advantage of a bifunctional protein with two enzymatic activities?

Omega Oxidation Occurs in the Endoplasmic Reticulum

26. What are the oxidation steps that occur in omega oxidation?

Genetic Defects in Fatty Acyl-CoA Dehydrogenases Cause Serious Disease

27. Why do individuals with a mutation in the MCAD gene have high urinary levels of medium-chain dicarboxylic acids?

Ketone Bodies

Ketone Bodies Formed in the Liver Are Exported to Other Organs

28. Where in the cell, and from what metabolite, are ketone bodies made?

Extrahepatic Tissues Use Ketone Bodies as Fuels

29. How and where are ketone bodies used?

Ketone Bodies Are Overproduced in Diabetes and during Starvation

30. How does formation of ketone bodies allow continuation of fatty acid oxidation in the liver?

31. What do starvation and untreated diabetes have in common as a trigger to induce increased production of ketone bodies?

SELF-TEST

Do You Know the Terms?

ACROSS

1. Association of hydrophobic lipids with these molecules permits lipid transport in the blood.

4. Organelles in which hydrogen peroxide (H_2O_2) is produced in the first step of β oxidation.

6. Enzyme catalyzing the second step in β oxidation: removal of a double bond by the addition of water. (2 words)

8. Enzyme catalyzing activation of the fatty acid in which ATP is converted to AMP + PP_i. (2 words)

10. Hormone-sensitive ________ lipase links hormonal signaling to mobilization of stored energy in adipose tissue.

13. Enzyme that transfers the fatty acyl group from Co-A to a carrier molecule. (2 words)

15. Alternative name for acyl-CoA acyltransferase, enzyme catalyzing the fourth reaction in β oxidation, removing a two-carbon unit and transferring remainder of the fatty acid to new CoA-SH.

16. In animal cells, fatty acid synthesis is carried out in the ________.

19. Site of β oxidation of fatty acids, citric acid cycle, and electron transfer chain is the mitochondrial ________.

20. Intermediate in synthesis of long-chain fatty acids that inhibits carnitine acyltransferase I, thus regulating fatty acid oxidation at the fatty acid transport step.

21. In the third step of β oxidation, electrons are transferred to NAD^+; NADH then donates its electrons to ________ in the electron-transfer chain in a reaction catalyzed by β-hydroxyacyl-CoA dehydrogenase.

22. First step in β oxidation transfers electrons from fatty acyl-CoA to the FAD prosthetic group of acyl-CoA ________.

23. When oxaloacetate concentration is low, these compounds are formed from excess acetyl-CoA produced by β oxidation of fatty acids. (2 words)

24. The final three-carbon compound generated by oxidation of odd-chain fatty acids is converted from methylmalonyl-CoA to succinyl-CoA by methylmalonyl-CoA ________.

25. ________ oxidation of fatty acids in the endoplasmic reticulum produces dicarboxylic acids such as succinate.

DOWN

2. Before it can be oxidized, the final three-carbon compound generated by oxidation of odd-chain fatty acids is carboxylated by this biotin-containing enzyme. (2 words)

3. ________ lipase: extracellular enzyme in muscle and fat tissues that hydrolyzes triacylglycerols in blood.

5. Oxidation of unsaturated fatty acid produces a fatty acid with a cis double bond that can undergo further β oxidation only after a cis to trans conversion by enoyl-CoA ________.

7. General name for enzyme that hydrolyzes triacylglycerols.

9. Carrier compound required for transport of fatty acids across the inner mitochondrial membrane.

11. Fatty acids, in the form of acyl-carnitine esters, enter the mitochondrial matrix and are transferred to intramitochondrial ________ by carnitine acyltransferase II.
12. Enzyme that catalyzes the production of H_2O and O_2 from the H_2O_2 produced by extramitochondrial β oxidation of fatty acids.
14. The rate-limiting step in fatty acid oxidation is the transport of fatty acids from the ________ space into the mitochondrial matrix.
17. Glycerol ________ converts the glycerol backbone of triacylglycerols to glyceraldehyde 3-phosphate, which can enter the glycolytic pathway.
18. Pathway for conversion of fatty acids to acetyl-CoA.
21. Structures formed by folding of the inner mitochondrial membrane.

Do You Know the Facts?

1. Which of the following describes fatty acid transport into the mitochondrial matrix?
A. It is the rate-limiting step in fatty acid oxidation.
B. It is regulated by [malonyl-CoA].
C. The cytosolic and matrix pools of CoA are distinct and separate.
D. Once fatty acyl groups have entered the matrix, they are committed to oxidation to acetyl-CoA.
E. All of the above are true.

2. Which of the following is true of β oxidation of fatty acids?
A. In a single round, one molecule of $FADH_2$ and one molecule of NADPH are produced.
B. It is the same for both saturated and unsaturated fatty acids.
C. Lipoprotein lipase catalyzes the first step.
D. Fatty acids are oxidized at C-3 to remove a two-carbon unit.
E. It occurs in the intermembrane space of mitochondria.

3. Place the following steps in lipid oxidation in their proper sequence. (Note: this is *not* a complete list of steps.)
(a) Thiolysis
(b) Reaction of fatty acyl-CoA with carnitine
(c) Oxidation requiring NAD^+
(d) Hydrolysis of triacylglycerol by lipase
(e) Activation of fatty acid by joining to CoA
(f) Hydration

In questions 4–8, match the role in fatty acid oxidation and/or mobilization to the appropriate component from the list below.
(a) Bile salt
(b) Serum albumin
(c) ApoC-II
(d) Apolipoprotein
(e) Carnitine

______ **4.** Acts as a "carrier" of fatty acids across the inner mitochondrial membrane.
______ **5.** Acts as a biological detergent, disrupting fat globules into small mixed micelles.
______ **6.** Binds and transports triacylglycerols, phospholipids, and cholesterol between organs.
______ **7.** Activates lipoprotein lipase, which cleaves triacylglycerols into their components.
______ **8.** Binds some fatty acids mobilized from adipocytes and transports them in the blood to heart and skeletal muscle.

9. (a) At which steps in the β oxidation of saturated fatty acids are reduced electron carriers generated?

(b) For each two-carbon increase in the length of a saturated fatty acid chain, how many additional moles of ATP can be formed upon complete oxidation of one mole of the fatty acid to CO_2 and H_2O?

10. What is the net result of the additional reactions required for the β oxidation of unsaturated fatty acids?

11. How do the functions for which plants use the products of triacylglycerol oxidation differ from those in mammals?

12. How does the production of ketone bodies allow continued oxidation of fatty acids? How does this affect the cytosolic coenzyme A concentration?

Applying What You Know

1. Compare the net ATP production in the following situations:

(a) Complete oxidation of palmitate (16-C, saturated fatty acid).

(b) Complete oxidation of palmitoleic acid (16-C, monounsaturated fatty acid with a cis double bond between C-9 and C-10).

(c) Partial oxidation of palmitate to the ketone body acetoacetate.

(d) Oxidation of palmitate in a liver peroxisome.

ANSWERS

Do You Know the Terms?

ACROSS

1. apolipoproteins
4. peroxisomes
6. enoyl-CoA hydratase
8. acyl-CoA synthetase
10. triacylglycerol
13. carnitine acyltransferase I
15. thiolase
16. cytoplasm
19. matrix
20. malonyl-CoA
21. Complex I
22. dehydrogenase
23. ketone bodies
24. mutase
25. omega

DOWN

2. propionyl-CoA carboxylase
3. lipoprotein
5. isomerase
7. lipase
9. carnitine
11. coenzyme A
12. catalase
14. intermembrane
17. kinase
18. β oxidation
21. cristae

Do You Know the Facts?

1. E
2. D
3. (d), (e), (b), (f), (c), (a)
4. (e)
5. (a)
6. (d)
7. (c)
8. (b)
9. **(a)** $FADH_2$ is formed in the first step of β oxidation, the oxidation reaction catalyzed by acetyl-CoA dehydrogenase. NADH is formed in the oxidation reaction catalyzed by β-hydroxyacyl-CoA dehydrogenase. Passage of electrons from the $FADH_2$ and NADH through the electron transfer chain produces ATP.

(b) For each extra two-carbon unit oxidized, 14 more ATP molecules are formed. The two oxidations of the β-oxidation pathway produce 1 $FADH_2$, and 1 NADH, which yield 1.5 and 2.5 ATP, respectively, by oxidative phosphorylation. The extra acetyl-CoA, when oxidized via the citric acid cycle, yields another 10 ATP equivalents: 3 NADH, 1 $FADH_2$ and 1 ATP (or GTP).

10. The two additional reactions, an isomerization and a reduction reaction, produce a saturated fatty acid, which can pass through the β-oxidation pathway.

11. *Plants* do not use the oxidation of triacylglycerols to generate metabolic energy in any significant amount; plant mitochondria lack the appropriate enzymes for this process. Triacylglycerols can be degraded to fatty acids, which are activated to their CoA derivatives, then oxidized in peroxisomes and glyoxysomes (found in germinating seeds). This oxidation produces acetyl-CoA, which is routed through the glyoxylate cycle and into gluconeogenesis. The glucose so formed is the precursor of a variety of metabolic intermediates. *Mammals* use triacylglycerol breakdown and fatty acid oxidation to generate large amounts of ATP; this takes place in the mitochondria. Mammals cannot use acetyl-CoA as a precursor of glucose; they lack crucial enzymes of the glyoxylate cycle.

12. The generation of ketone bodies (which contain no CoA moiety) from acetyl-CoA frees up mitochondrial coenzyme A for continued fatty acid oxidation. The separation of the coenzyme A pools of the cytosol and the mitochondrial matrix is maintained by the acyl-carnitine/carnitine transport system. Ketone body production therefore does not affect the cell's cytosolic CoA concentration.

Applying What You Know

1. **(a)** Complete oxidation of palmitate involves 7 rounds of β oxidation producing:

- 7 $FADH_2$ and 7 NADH
- 8 acetyl-CoA, which are oxidized in the citric acid cycle to produce 24 NADH, 8 $FADH_2$, and *8 ATP* (from GTP).

Thus 15 $FADH_2$ and 31 NADH are produced. Electron transfer and oxidative phosphorylation produce:

- 1.5 ATP per $FADH_2$ = *22.5 ATP*

- 2.5 ATP per NADH = *77.5 ATP*

  Thus a total of 22.5 + 77.5 + 8 = *108 ATP* is produced. From this we must *subtract 2 ATP*, because the activation of palmitate to palmitoyl-CoA required 2 ATP equivalents.
- Net energy produced = **106 ATP.**

(b) Complete oxidation of palmitoleic acid involves:

- 3 rounds of β oxidation producing 3 $FADH_2$ and 3 NADH.
- 1 round with a "miss" of formation of $FADH_2$ during the acyl-CoA dehydrogenase step; the cis double bond must be converted to trans in order for it to be acted on by enoyl-CoA hydratase. Enoyl-CoA isomerase catalyzes this conversion and allows the rest of this round of β oxidation, producing 1 NADH.
- 3 more rounds of β oxidation producing 3 $FADH_2$ and 3 NADH.
- 8 acetyl-CoA formed, which (as above) produce 24 NADH, 8 $FADH_2$, and *8 ATP*.

  Thus 14 $FADH_2$ and 31 NADH are produced, which give rise to:
- *21 ATP* and *77.5 ATP*

  Thus a total of 21 + 77.5 + 8 = *106.5 ATP* is produced. Again, we must subtract *2 ATP*:
- Net energy produced = **104.5 ATP.**

(c) Partial oxidation of palmitate to acetoacetate involves 7 rounds of β oxidation producing:

- 7 $FADH_2$ and 7 NADH
- 8 acetyl-CoA, which combine to form 4 acetoacetate.
- Thus 7 $FADH_2$ and 7 NADH are produced; these give rise to:
- *10.5 ATP* and *17.5 ATP*
- Thus the total is 10.5 + 17.5 = *28 ATP*. Again, *subtracting 2 ATP*:
- Net energy produced = **26 ATP.**

(d) Oxidation of palmitate in a liver peroxisome involves 7 rounds of β oxidation producing:

- $FADH_2$, which does not pass its electrons through the respiratory chain to O_2, so *no ATP* is derived from this step; instead, the electrons are passed directly to O_2 to form H_2O_2.
- 7 NADH, which cannot be reoxidized within the peroxisome; NADH is exported to the cytosol, from where it eventually passes its reducing equivalents into mitochondria.
- 8 acetyl-CoA, which are exported to the cytosol as acetate because liver peroxisomes do not contain the enzymes of the citric acid cycle.

  Thus *within the peroxisome itself*, **no ATP** is produced. Assuming no energetic cost for transporting NADH and acetyl-CoA out of the peroxisome and into the mitochondrial matrix, the 7 NADH and 8 acetyl-CoA could eventually produce **95.5 ATP.**

Amino Acid Oxidation and the Production of Urea

chapter

18

STEP-BY-STEP GUIDE

Major Concepts

Dietary proteins are the primary source of biologically useful nitrogen in animals.

Ingested proteins are hydrolyzed to individual amino acids by a series of enzymatic reactions beginning in the stomach and continuing in the small intestine. The enzymes involved in proteolysis are usually secreted into the digestive tract as inactive precursors called **zymogens,** which are then activated by other enzymes; this protects the secretory cells from the action of the proteolytic enzymes. Free amino acids are transported across the intestinal epithelium and carried in the circulatory system to the liver.

The general scheme for the further metabolism of these amino acids involves the transfer of the amino group to α-ketoglutarate, forming glutamate plus an α-keto acid.

These reactions are catalyzed by **aminotransferases,** or **transaminases.** All aminotransferases have a common prosthetic group, **pyridoxal phosphate (PLP),** which acts as a carrier of the amino group and as an electron sink stabilizing the transamination reaction.

The glutamate produced is transported to liver mitochondria and deaminated by glutamate dehydrogenase.

Glutamate dehydrogenase, present only in the mitochondrial matrix, is a complex allosteric enzyme using NAD^+ or $NADP^+$ as the electron acceptor. The enzyme is positively modulated by high [ADP] in the cell, negatively modulated by high [GTP], and catalyzes the deamination of glutamate to α-ketoglutarate and NH_4^+.

Glutamine and alanine transport ammonia formed in other tissues to the liver.

Ammonia produced in extrahepatic tissues can be transported in the blood to the liver in the form of glutamine. This amino acid converted to glutamate and NH_4^+ in liver mitochondria. In active muscle cells the excess pyruvate formed from glycolysis is combined with NH_4^+ to form alanine. This nontoxic carrier of ammonia is then transported in the circulatory system to the liver, where it is reconverted to pyruvate and ammonia. Through gluconeogenesis, pyruvate is a source of additional glucose molecules that can be transported back to and used as a fuel source by muscles.

Nitrogen is excreted as ammonia, urea, or uric acid.

Amino groups resulting from the oxidative degradation of amino acids cannot be released as ammonia because this is highly toxic to cells. Therefore amino groups are either reused or excreted. Aquatic organisms, including bacteria, that have access to unlimited quantities of water can excrete nitrogen as ammonia because it is diluted in the aqueous surroundings. Terrestrial organisms for which conserving water is especially important excrete nitrogen in the form of **uric acid,** a semisolid. Other terrestrial organisms excrete nitrogen in the form of **urea,** which is less toxic than ammonia but nevertheless can be tolerated only at low concentrations in most biological systems.

Urea is formed from ammonia in a series of reactions called the urea cycle.

The urea cycle begins in the mitochondrial matrix. NH_4^+ is first combined with HCO_3^- to yield **car-**

bamoyl phosphate. This reaction requires hydrolysis of two molecules of ATP and is essentially irreversible. In the urea cycle itself, the carbamoyl moiety of carbamoyl phosphate is passed to **ornithine,** forming **citrulline,** which passes to the cytosol. Here, citrulline acquires an amino group from **aspartate** to yield **argininosuccinate,** in a reaction requiring ATP. The argininosuccinate is cleaved to **arginine** and **fumarate.** The fumarate can be shunted to the citric acid cycle (thus linking the two cycles) where it is converted to oxaloacetate; this can be aminated to regenerate aspartate. The arginine is cleaved to form **urea** and ornithine, which reenters the mitochondrial matrix. **(Note that aspartate and ornithine are not consumed in the urea cycle but act as carriers of ammonia in the form of amino groups. The final product is one molecule of urea, which contains two nitrogen atoms and a single oxidized carbon atom.)**

Urea cycle activity is regulated.

On a long-term basis, urea cycle activity can be altered by regulating the synthesis of the urea cycle enzymes. Short-term regulation involves allosteric activation of the first enzyme in the pathway (carbamoyl phosphate synthetase I) by *N*-acetylglutamate.

Deaminated amino acids produce carbon skeletons that can be funneled into the citric acid cycle.

Each of the 20 standard amino acids produces a different carbon skeleton upon deamination. However, these compounds ultimately enter the citric acid cycle in one of only five different forms: as **acetyl-CoA, α-ketoglutarate, succinyl-CoA, fumarate,** or **oxaloacetate.** Although the specific reaction pathways for the degradation of each amino acid differ, they have two types of reactions in common: **transaminations** by enzymes requiring PLP and **one-carbon transfer reactions.** The one-carbon transfers involve one of several cofactors: **biotin,** which transfers CO_2; **tetrahydrofolate,** which usually transfers —HC=O, —HCOH, or sometimes $—CH_3$ groups (covalently linked to N-5 or N-10 of tetrahydrofolate); and ***S*-adenosylmethionine,** which transfers methyl groups, $—CH_3$.

Some amino acids are ketogenic, some are glucogenic, and some are both.

Ketogenic amino acids are degraded to acetoacetyl-CoA and/or acetyl-CoA, which can be converted to ketone bodies. Glucogenic amino acids are degraded to compounds that can be converted into glucose via gluconeogenesis.

What to Review

Answering the following questions and reviewing the relevant concepts, which you have already studied, should make this chapter more understandable.

- Be sure to know the structures of amino acids (p. 120). Be able to see which carbons of their skeletons will be part of citric acid cycle intermediates.
- The carbon skeletons generated by deamination of amino acids can be funneled into the citric acid cycle for energy production. A review of the structures of the intermediates in the citric acid cycle (Fig. 16–7) will help you learn the specific points in the cycle where these carbon skeletons enter.
- What are **ketone bodies** and what is their function (pp. 615–618). Recall that ketone bodies can be used as an alternative fuel source by muscle, kidney, and brain tissue. Ketone bodies are also produced by the catabolism of amino acids.
- Once again, review the structure of mitochondria (see Fig. 19–1 for the best view). Some of the urea cycle reactions occur in the cytosol and some in the mitochondrial matrix.

Topics for Discussion

Answering each of the following questions, especially in the context of a study group discussion, should help you understand the important points of this chapter.

Metabolic Fates of Amino Groups

1. What are the fates of proteins degraded in plants? In carnivorous animals? In herbivorous animals?

Dietary Protein Is Enzymatically Degraded to Amino Acids

2. What is the role of the low pH of gastric juice in degrading dietary proteins?

3. Why is it important that proteases, which cleave proteins to their individual amino acid components, are produced and secreted as zymogens?

4. Why is a *series* of enzymes required for degradation of dietary proteins to free amino acids?

Pyridoxal Phosphate Participates in the Transfer of α-Amino Groups to α-Ketoglutarate

5. What are the common characteristics of many aminotransferases?

6. What are the two substrates, and in what order do they react, in the Ping-Pong reaction catalyzed by an aminotransferase?

Assays for Tissue Damage

7. Why would you *not* expect to observe measurable quantities of aminotransferases in the blood in the absence of tissue damage?

Glutamate Releases Ammonia in the Liver

8. How is the glutamate dehydrogenase reaction an “intersection” of carbon and nitrogen metabolism?

Glutamine Transports Ammonia in the Bloodstream

9. What are the different roles of glutamine and glutamate in amino group metabolism and transport?

Alanine Transports Ammonia from Muscles to the Liver

10. Why is the use of alanine as a transporter of ammonia from muscle to liver a “kill-two-birds-with-one-stone” solution?

Ammonia Is Toxic to Animals

11. Why would a depletion in cellular ATP be so dangerous in brain cells?

Nitrogen Excretion and the Urea Cycle

Urea Is Produced from Ammonia in Five Enzymatic Steps

12. Why is the reaction catalyzed by carbamoyl phosphate synthetase I also a "kill-two-birds-with-one-stone" situation?

13. The role of ornithine in the urea cycle is analogous to the role of oxaloacetate in the citric acid cycle. To what citric acid cycle intermediate is citrulline analogous?

14. Which four metabolites involved in the urea cycle must traverse the mitochondrial membrane?

15. *Cell Map:* Add to your cell map the structures and enzymes of the urea cycle below and to the right of the citric acid cycle, near the mitochondrial inner membrane.

The Citric Acid and Urea Cycles Can Be Linked

16. Aspartate is a participant in the urea cycle. From where is its amino group derived?

17. *Cell Map:* Add to your cell map the interconnections between the citric acid and urea cycles as indicated in Figure 18–11.

The Activity of the Urea Cycle Is Regulated at Two Levels

18. What are the short- and long-term strategies for coping with changes in the demand for urea cycle activity?

Pathway Interconnections Reduce the Energetic Cost of Urea Synthesis

19. The urea cycle produces one molecule of urea at the expense at least 1.5 ATPs overall. Why is this a worthwhile, and necessary, investment?

Genetic Defects in the Urea Cycle Can Be Life-Threatening

20. How does dietary administration of aromatic acids remove ammonia from the bloodstream?

Natural Habitat Determines the Pathway for Nitrogen Excretion

21. What types of organisms in what types of habitats excrete nitrogen as ammonia? As uric acid? As urea?

22. Why is it energetically more favorable (if possible) for an organism to excrete nitrogen as ammonia than as urea?

Pathways of Amino Acid Degradation

23. How significant is the contribution of amino acid catabolism to overall energy production in humans?

24. How does the degradation of amino acids feed into the central catabolic processes of the cell?

Several Enzyme Cofactors Play Important Roles in Amino Acid Catabolism

25. Which cofactors involved in amino acid degradation carry which one-carbon groups?

Ten Amino Acids Are Degraded to Acetyl-CoA

26. List the amino acids degraded to acetyl-CoA. Note that several degradative pathways share certain reactions, enzymes, and coenzymes.

Phenylalanine Catabolism Is Genetically Defective in Some People

27. Persons with phenylketonuria sometimes exhibit very light coloration of the skin, eyes, and some internal organs where melanin is normally accumulated. Suggest a reason for this.

Five Amino Acids Are Converted to α-Ketoglutarate

28. Which five amino acids are degraded to the citric acid cycle intermediate α-ketoglutarate? Note that they do so through the initial production of glutamate.

Four Amino Acids Are Converted to Succinyl-CoA

29. What are these four amino acids? Note once again the use of certain reactions and enzymes for more than one pathway. Some reactions occurring in the oxidation of odd-chain fatty acids are also used in the degradation of some amino acids to succinyl-CoA.

Branched-Chain Amino Acids Are Not Degraded in the Liver

30. Where in the body are branched-chain aminotransferases found?

Asparagine and Aspartate Are Degraded to Oxaloacetate

31. What are the overall final products of amino acid degradation?

Scientific Sleuths Solve a Murder Mystery

32. Why are very limited amounts of isoleucine and valine *not* fatal to individuals with methylmalonic acidemia (MMA)?

Some Amino Acids Can Be Converted to Glucose, Others to Ketone Bodies

33. Why are ketogenic amino acids especially important under conditions of starvation or untreated diabetes?

34. Which reaction pathway converts the carbon skeleton of alanine to a ketone body?

SELF-TEST

Do You Know the Terms?

ACROSS

3. Compound regulating activity of the urea cycle in the short term.
5. Prosthetic group of all aminotransferases; acts as an electron sink. (2 words)
8. Cofactor used in methyl group transfers; reduced form of the vitamin folate.
9. Enzymes that remove α-amino groups from amino acids.
12. Describes amino acids that can be converted to intermediates in gluconeogenesis.
13. Synthesis of a molecule of urea requires four high-energy phosphate groups; two of these come from a single molecule of ATP in the form of _______, which is hydrolyzed to yield two P_i.
15. Cofactor used in methyl group transfers; its synthesis from ATP and methionine releases a *triphosphate.*
17. In the liver glutamate undergoes _______ deamination, catalyzed by glutamate dehydrogenase.
18. The formation of citrulline, the first intermediate in the urea cycle, is catalyzed by _______ transcarbamoylase.
19. Intermediate directly linking the urea and citric acid cycles.
20. The enzyme _______ dehydrogenase, present only in the mitochondrial matrix, can use either NAD^+ or $NADP^+$ as acceptor of reducing equivalents.
22. Transaminases are also referred to as _______ transferases.
24. Keto sugar that accepts an amino group to form aspartate, which eventually donates the amino group to urea.
26. Argininosuccinate is cleaved to arginine and fumarate by argininosuccinate _______.
28. Pepsinogen is to pepsin as a _______ is to an active enzyme.
29. Describes organisms with access to (relatively) unlimited supplies of water that use an energy-saving method of nitrogen excretion.
30. In most animals, excess nitrogen produced by amino acid breakdown is transported in this form; contains *two* amino groups.
31. Describes organisms that use $CO(NH_2)_2$ as nitrogenous waste product.

DOWN

1. Cellular location of carbamoyl phosphate synthetase II.
2. Electron acceptor not generally used by enzymes in catabolism but *can* be used by an enzyme involved in the urea cycle.
4. Intermediate that carries NH_4^+ generated in liver mitochondria into the urea cycle. (2 words)

6. Produced, directly or indirectly, by breakdown of the carbon skeletons of ten amino acids.
7. Carbamoyl phosphate synthetase I is located in the _______ of mitochondria.
10. Urea cycle intermediate containing four nitrogen-containing groups, one contributed by aspartate and three by citrulline.
11. Ornithine is to oxaloacetate as citrulline is to _______.
14. Mitochondrial enzyme that "releases" transported amino group from glutamine in the liver.
16. Cofactor used in one-carbon transfers that carries carbon groups in their most oxidized form, CO_2.
21. Describes organisms that use uric acid as nitrogenous waste product, relying on its insolubility to minimize its toxicity.
22. Urea contains two amino groups: one derived from deamination of glutamate in mitochondria; the other from deamination of _______, also generated in mitochondria.
23. Describes amino acids that are degraded to acetyl-CoA or acetoacetyl-CoA.
25. Enzyme that catalyzes regeneration of ornithine and formation of urea.
27. In mitochondria, this molecule condenses with bicarbonate to form an "activated" compound that can enter the urea cycle.

Do You Know the Facts?

1. Indicate whether each of the following is or is not (yes/no) common to the degradation of *all* amino acids.

_______ **(a)** Separation of the amino group(s) from the carbon skeleton.
_______ **(b)** Use of the amino groups for synthesis of new amino acids or other nitrogenous products.
_______ **(c)** Excretion of excess amino groups in a form appropriate to the organism and its environment.
_______ **(d)** Passage of the carbon skeletons to the gluconeogenic pathway.
_______ **(e)** Conversion to α-ketoglutarate.
_______ **(f)** No associated net gain of energy for the cell.
_______ **(g)** In mammals, a process occurring mainly in the liver.
_______ **(h)** Usually, transfer of amino groups to pyruvate or glutamate for transport to the liver.

2. Which of the following is a characteristic of many aminotransferase reactions?
A. They have a large, negative $\Delta G'^{\circ}$.
B. The amino group is transferred to an α-keto acid (such as α-ketoglutarate) to form the corresponding amino acid.
C. The amino group is transferred from an ammonia molecule.
D. They are catalyzed by the same enzyme.
E. They require the cofactor *S*-adenosylmethionine.

3. In some respects the urea and citric acid cycles are analogous processes. Ornithine and citrulline have roles that are similar to those of oxaloacetate and what other citric acid cycle intermediate?
A. acetyl-CoA
B. citrate
C. CO_2
D. malate
E. ammonia

4. Which compound can serve as a direct acceptor of an additional amino group derived from amino acid catabolism?
A. glutamine
B. asparagine
C. α-ketoglutarate
D. fumarate
E. glycerol

5. The carbon skeleton produced by deamination of alanine enters the citric acid cycle as:
A. malate.
B. oxaloacetate.
C. fumarate.
D. succinate.
E. acetyl-CoA.

6. *Not* taking into account the NADH generated in the malate dehydrogenase reaction, how many high-energy phosphate bonds are used to form a molecule of urea, starting from ammonia and HCO_3^-?
A. 1
B. 2
C. 3
D. 4
E. 0

7. The citric acid cycle and the urea cycle overlap to form what has sometimes been called the "Krebs bicycle." Which of the following statements is relevant to the interactions between these two metabolic cycles?

A. Oxaloacetate is converted to aspartate.
B. Aspartate combines with citrulline to produce argininosuccinate in the cytosol.
C. Argininosuccinate is cleaved to fumarate and arginine.
D. Fumarate is a citric acid cycle intermediate.
E. All of the above are true.

8. Early in its life, a tadpole lives in an aqueous environment and excretes much of its excess nitrogen as ammonia. Once it matures into an adult, the frog spends more time on dry land and becomes ureotelic. Which of the following enzyme activities would be most likely to increase drastically in the adult frog?

A. carbamoyl phosphate synthetase I
B. glutamine synthetase
C. glutaminase
D. α-ketoglutarate dehydrogenase
E. carboxypeptidase

9. The reactions of the urea cycle occur in two different cellular compartments. Which urea cycle intermediate(s) must be transported across the inner mitochondrial membrane?

A. argininosuccinate
B. citrulline
C. ornithine
D. A and C
E. B and C

10. Which cofactor involved in amino acid degradation is correctly matched to the one-carbon group it transfers?

A. biotin: CHO
B. tetrahydrofolate: CHOH
C. *S*-adenosylmethionine: CH_2OH
D. pyridoxal phosphate: CH_3
E. pepsinogen: CO_2

11. Degradation of amino acids yields compounds that are common intermediates in the major metabolic pathways. Explain the distinction between ketogenic and glucogenic amino acids in terms of their metabolic fates. Give three examples each of ketogenic and glucogenic amino acids.

12. Describe the role and reactions of the glucose-alanine cycle. What cofactor is required for this and all other transamination reactions?

Applying What You Know

1. Suppose you are stranded on a desert island and have nothing but protein to eat. Outline *briefly* how your body will use this source of food for its energy needs.

2. Define "zymogen" and describe the role of one zymogen in protein digestion. Why is it important that proteases, which cleave proteins to their individual amino acid components, are produced as zymogens?

3. Why does a mammal go to the energetic expense of making urea from ammonia rather than simply excreting ammonia, as do bacteria?

4. Many Europeans enjoy eating "sweetbreads," which are organ meats, including pancreas tissue. The pancreas is the source of the polypeptide hormone insulin, which is important in regulating carbohydrate metabolism. Why don't such gourmets have problems with their carbohydrate metabolism?

ANSWERS

Do You Know the Terms?

ACROSS

3. *N*-acetylglutamate
5. pyridoxal phosphate
8. tetrahydrofolate
9. transaminases
12. glucogenic
13. pyrophosphate
15. *S*-adenosylmethionine
17. oxidative
18. ornithine
19. fumarate
20. glutamate
22. amino
24. oxaloacetate
26. lyase
28. zymogen
29. ammonotelic
30. glutamine
31. ureotelic

DOWN

1. cytosol
2. NADP
4. carbamoyl phosphate
6. acetyl-CoA
7. matrix
10. argininosuccinate
11. citrate
14. glutaminase
16. biotin
21. uricotelic
22. aspartate
23. ketogenic
25. arginase
27. ammonia

Do You Know the Facts?

1. **(a)** Y; **(b)** Y; **(c)** Y; **(d)** N; **(e)** N; **(f)** N; **(g)** Y; **(h)** Y
2. B
3. B
4. C
5. E
6. D
7. E
8. A
9. E
10. B
11. *Ketogenic* amino acids are catabolized to yield acetyl-CoA or acetoacetyl-CoA, the precursors of ketone bodies. Examples are Trp, Phe, Tyr, Ile, Leu, and Lys. *Glucogenic* amino acids are catabolized to intermediates that can serve as substrates for gluconeogenesis: that is, pyruvate and/or any of the four- or five-carbon intermediates of the citric acid cycle. Examples are Ala, Asp, Asn, Cys, Arg, Gly, Ser, Trp, Phe, Tyr, Ile, Met, Thr, Val, Gln, His, and Pro.
12. Alanine is the nontoxic form in which ammonia from amino acid catabolism in muscle is transported to the liver. Alanine is formed in muscle by transamination of pyruvate, produced from the breakdown of glucose in glycolysis; glutamate is the amino group donor. In the liver, alanine is reconverted to pyruvate by transamination; its amino group is eventually converted to urea. The pyruvate is converted to glucose by gluconeogenesis then exported back to the muscle. The necessary cofactor for the transaminations is PLP.

Applying What You Know

1. On the desert island your body will: (a) Break down ingested proteins to their constituent amino acids. (b) Remove or transaminate amino groups. (c) Synthesize needed proteins and other compounds that incorporate nitrogen. (d) Use excess amino acid carbon skeletons for energy (they enter the citric acid cycle at various points), *or* for gluconeogenesis (for export of glucose to specific tissues that require glucose as fuel), *or* for production of ketone bodies (also for export to specific tissues.)
2. A zymogen is an inactive form of an enzyme, which can be activated by proteolytic cleavage. The pancreatic enzymes chymotrypsinogen, trypsinogen, and procarboxypeptidases A and B are inactive (zymogen) forms of proteases, which are activated by proteolytic cleavage after their release into the small intestine. Formation and secretion of proteases as zymogens protects the secretory cells from being digested by the enzymes they produce and secrete.
3. Because bacteria are unicellular organisms they release their waste nitrogen, in the form of ammonia, directly into the surrounding medium. The excreted ammonia is immediately diluted to nontoxic levels (1 g of nitrogen, in the form of ammonia, requires 300–500 mL of H_2O to dilute it to nontoxic levels). The ammonia produced by amino acid catabolism in mammals cannot be sufficiently diluted in the tissues and the blood to avoid accumulation to toxic levels. Dissolved urea is much less toxic than ammonia (about 50 mL of H_2O is required to dilute 1 g of nitrogen, in the form of urea, to nontoxic levels).
4. The biological activity of insulin (a polypeptide) is destroyed by the low pH in the stomach and by proteases in the small intestine. Even if insulin were not degraded to its constituent amino acids in the intestine, it could not enter the bloodstream; the epithelial cells that line the intestine transport (absorb) free amino acids, not whole polypeptides or proteins.

chapter

19 Oxidative Phosphorylation and Photophosphorylation

STEP-BY-STEP GUIDE

Major Concepts

Oxidative phosphorylation is the enzymatic synthesis of ATP coupled to the transfer of electrons to molecular oxygen.

This process, occurring in the inner mitochondrial membrane of eukaryotes and in the plasma membrane of prokaryotes, involves the transfer of electrons from NADH and $FADH_2$ to O_2 via a series of protein complexes and coenzymes. The energy for ATP production is ultimately driven by this electron flow.

The mitochondrial respiratory chain is an ordered array of electron carriers arranged in complexes.

Electrons flow through this chain in a series of steps, from carriers of relatively low, to relatively high, $E'°$. The flow is spontaneous, but the serial transfer from one carrier to another ensures that the electron transfer occurs in a controlled and energetically useful fashion.

Complex I is a transmembrane protein complex of the inner membrane.

The complex is positioned so that it interacts with NADH produced in reactions catalyzed by enzymes located in the matrix. Reducing equivalents are transferred from NADH, first to the **FMN** prosthetic group and then to the **Fe-S** centers of Complex I. The complex then passes its electrons to **ubiquinone (Q).** Complex I functions as a **proton pump,** moving protons from the matrix to the intermembrane space.

Complex II, or succinate dehydrogenase, transfers electrons from succinate to the complex's covalently bound FAD and then to its Fe-S centers.

Electrons that first enter the electron transfer chain at Complex II are passed to Q. However, Complex II does not span the bilayer, and no protons are pumped into the intermembrane space. Other proteins that transfer electrons to Q include **ETFP,** which transfers electrons from the $FADH_2$ generated by fatty acid oxidation in the matrix to Q. The production of dihydroxyacetone phosphate from glycerol 3-phosphate in the cytosol catalyzed by **glycerol 3-phosphate dehydrogenase** also results in the transfer of electrons to Q. Because NADH cannot cross the inner mitochondrial membrane, this latter reaction pathway shuttles reducing equivalents generated in the cytosol to the inner mitochondrial membrane, where they can be used to generate ATP. (The malate-aspartate shuttle can also be used to transport reducing equivalents from cytosolic NADH into the matrix.)

Complex III transfers electrons from QH_2 to cytochrome c.

The protons generated when QH_2 is reoxidized to Q via the Q-cycle are released into the intermembrane space, further elevating the $[H^+]$ of the intermembrane space relative to the mitochondrial matrix.

Cytochrome c is a mobile protein that "shuttles" electrons, in this case carrying them from Complex III to Complex IV.

Complex IV contains **cytochromes *a* and a_3,** which accept electrons from cytochrome *c* and transfer them to oxygen, forming H_2O. In the process, additional protons are pumped from the matrix to the intermembrane space.

The chemiosmotic model explains in molecular terms how the proton gradient generated by the flow of electrons through the respiratory chain drives the synthesis of ATP.

Both a chemical (ΔpH) and an electrical (Δψ) component make up the **proton-motive force.** ATP is synthesized from ADP and P_i by **ATP synthase** in the inner mitochondrial membrane. This enzyme is made up of two complexes: $\mathbf{F_o}$ is a membrane-spanning protein complex, and $\mathbf{F_1}$ is located on the matrix side of the inner membrane. Because the inner membrane is impermeable to H^+, the only path for protons to reenter the matrix is through the pore formed by the F_o subunit of ATP synthase. ADP is phosphorylated by the enzymatic activity of the F_1 complex, driven by the proton-motive force.

The proton-motive force also drives other cell activities.

The transfer of electrons to O_2 and concomitant pumping of H^+ into the intermembrane space accounts for most of the ATP produced in aerobic organisms. The proton gradient that is established also provides the energy for the active transport of some molecules and ions across membranes, for the generation of heat in certain tissues, and for the rotary motion of bacterial flagella.

In photosynthesis, light energy produces electron flow that is coupled to the phosphorylation of ADP to ATP.

The critical difference between oxidative phosphorylation and photophosphorylation is that in the former process, the energy needed to drive the phosphorylation of ADP is provided by the input of a good electron donor (NADH or $FADH_2$), whereas in the latter process, the energy provided by light *creates* a good electron donor (in higher plants, P680* and P700*).

The photosystems of plants contain pigment molecules that absorb photons, then release the absorbed energy in a controlled fashion.

Plant photosystems are located in the inner membranes of chloroplasts. These inner membranes, like the inner membrane of mitochondria, are relatively impermeable and form saclike structures called **thylakoids.** Chloroplasts also have an outer membrane that is permeable to small molecules and ions. In **photosystem II, chlorophyll** and **accessory pigments** absorb light energy and transfer this energy to the **reaction center, P680.** The excited reaction center (P680*) loses an electron, which is transferred through a series of electron carriers. The electrons lost by P680* are replaced by electrons from H_2O, resulting in the production of O_2. In **photosystem I,** the excited **P700*** reaction center also donates an electron which is passed through a series of carriers, ultimately reducing $NADP^+$ to NADPH. The electron lost by P700* is replaced by an electron from the soluble protein **plastocyanin.**

Electrons from photosystem II are transferred to photosystem I by the cytochrome b_6f complex.

The pathway of electron flow from photosystem II to photosystem I is diagramed by the **Z scheme.** In this cooperative process, H^+ ions are pumped across the **thylakoid** membrane into the lumen of the thylakoid. The proton gradient generated is analogous to that generated across the inner mitochondrial membrane in oxidative phosphorylation but is oriented in the "opposite" direction. The ATP synthase of thylakoids is very similar to the complex in mitochondria.

ATP can be generated by photosystem I without production of NADPH or O_2 in a process called cyclic photophosphorylation.

In a "detour" from the noncyclic route from ferredoxin to NADPH, electrons loop back to the cytochrome b_6f complex. Protons are pumped across the thykaloid membrane by the cytochrome b_6f complex, generating a proton gradient that can be used to synthesize ATP.

Photosynthesis can occur with alternative hydrogen donors.

Hydrogen sulfide (H_2S) or reduced organic compounds (depending on the organism) can be substituted for H_2O in the general equation for photosynthesis:

$$2H_2D + CO_2 \xrightarrow{\text{light}} (CH_2O) + H_2O + 2D$$

What to Review

Answering the following questions and reviewing the relevant concepts, which you have already studied, should make this chapter more understandable.

- What is indicated by a low or high standard reduction potential, E'° (pp. 515–517)? If you had a mixture consisting only of ferredoxin, cytochrome *c*, and oxygen, which would be the final electron acceptor?
- This chapter deals with the fate of the reducing equivalents carried by NADH and $FADH_2$. In which reaction(s) of the citric acid cycle (Chapter 16) and the β oxidation of fatty acids (Chapter 17) are NADH and $FADH_2$ produced?
- ATP synthases are members of a group of proteins known as **F-type ATPases.** What are the characteristics of these enzymes (Fig. 12–31 and Table 12–4)?
- The glycolytic reaction catalyzed by glyceraldehyde 3-phosphate dehydrogenase provides a

model of an oxidation coupled to a phosphorylation. The product of this reaction led researchers to look for a similar chemical intermediate in mitochondrial oxidation coupled to phosphorylation. What are the reactant(s) and product(s) of the glyceraldehyde 3-phosphate dehydrogenase reaction (pp. 535–536)?

- **NADH** is a carrier of reducing equivalents. Review its structure in order to understand why the NADH molecule cannot be transported across the inner mitochondrial membrane (p. 519).

Topics for Discussion

Answering each of the following questions, especially in the context of a study group discussion, should help you understand the important points of this chapter.

1. What are the three fundamental similarities between oxidative phosphorylation and photophosphorylation? Keeping these similarities in mind (and noting further ones) as you read this chapter will help make overall sense of many of the details.

Electron-Transfer Reactions in Mitochondria

2. Why does it make physiological sense that heart mitochondria have more sets of electron transfer components than liver mitochondria?

3. How does the structure of the mitochondrion provide for physical separation of metabolic processes in the cell?

Electrons Are Funneled to Universal Electron Acceptors

4. Categorize the electron carriers NAD^+, $NADP^+$, FMN, and FAD according to the number of hydrogen atoms and/or electrons they carry and the type of association with dehydrogenases.

5. What is the cellular role of NADH? Of NADPH?

Electrons Pass Through a Series of Membrane-Bound Carriers

6. As you proceed through this section, categorize each electron carrier in the respiratory chain according to which of the four types of electron transfer it accomplishes.

7. Which chemical characteristics of ubiquinone make it uniquely useful in the electron transfer chain?

8. What are the similarities among cytochromes, iron-sulfur proteins, and flavoproteins as electron carriers? What are the differences?

9. What proteins have you already encountered that contain heme prosthetic groups?

10. How was the order of electron carriers in the mitochondrial inner membrane determined?

Electron Carriers Function in Multienzyme Complexes

11. Categorize each of the mitochondrial electron-carrier complexes in terms of the overall reaction it catalyzes and whether or not it acts as a proton pump. Use the subscripts "P" and "N" to keep track of the location of the protons.

12. By which route do electrons from the β oxidation of fatty acids enter the respiratory chain? What is the route from glycerol 3-phosphate oxidation?

13. What is the net equation (and the net effect) of the Q cycle?

The Energy of Electron Transfer Is Efficiently Conserved in a Proton Gradient

14. Why is the *actual* free-energy change of the $NADH + H^+ \rightarrow H_2O + NAD^+$ reaction not the same as the standard free-energy change?

15. What are the two components of the proton-motive force in mitochondria?

Plant Mitochondria Have Alternative Mechanisms for Oxidizing NADH

16. What happens to the energy in NADH in plant mitochondria when NAD^+ must be regenerated?

17. *Cell Map:* Find in the mitochondrion in the map the components of the electron transfer chain. Show where electrons from NADH and $FADH_2$ enter; indicate where protons are pumped.

Alternative Respiratory Pathways and Hot, Stinking Plants

18. What are the three routes for entry of electrons into the respiratory pathway that result in the generation of heat rather than ATP?

ATP Synthesis

19. What is the mechanism of uncoupler action? What would occur if DNP were added to a cyanide-treated system as described in Figure 19–17(a)?

20. How are both components of the proton-motive force induced in the experimental set-up shown in Figure 19–19?

ATP Synthase Has Two Functional Domains, F_o and F_1

21. Why are both F_1 and F_o necessary for ATP synthesis?

22. How would the protein purification protocols for F_1 and F_o differ?

ATP Is Stabilized Relative to ADP on the Surface of F_1

23. The free energy change for ATP synthesis with purified F_1 acting as the catalyst is close to zero. From where does the energy come to drive the equilibrium of the reaction toward the formation of product?

The Proton Gradient Drives the Release of ATP from the Enzyme Surface

24. Where is the “hill” of activation energy in the ATP synthase reaction?

Each β subunit of ATP Synthase Can Assume Three Different Conformations

25. Which subunit(s) of the ATP synthase complex contain the catalytic site(s) for ATP synthesis?

Rotational Catalysis Is Key to the Binding-Change Mechanism for ATP Synthesis

26. What are three conformations that the β subunits can assume?

27. Which subunit of F_o is postulated to act like a drive shaft?

28. Which other allosteric protein that you have encountered shows alterations in one ligand-binding site depending on the ligand-binding state of other protein subunits?

Chemiosmotic Coupling Allows Nonintegral Stoichiometries of O_2 Consumption and ATP Synthesis

29. Why is it so technically difficult to measure P/O? To measure proton fluxes?

30. How many protons are pumped out per NADH? Per succinate? How many protons must flow in to produce one ATP? Finally, after all these metabolism chapters, the math becomes clear for the P/O values of 2.5 with NADH as the electron donor and 1.5 for succinate!

31. *Cell Map:* Find in the mitochondrion an ATP synthase; indicate the direction of H^+ movement.

The Proton-Motive Force Energizes Active Transport

32. Which component of the proton-motive force is dissipated by the adenine nucleotide translocase? Which component is dissipated by the phosphate translocase?

33. *Cell Map:* Find the two translocases to your map; indicate what enters and exits.

Shuttle Systems Are Required for Mitochondrial Oxidation of Cytosolic NADH

34. What characteristics of NADH itself and of the NADH dehydrogenase make it necessary to have shuttle systems for transporting cytosolic reducing equivalents into the matrix?

35. What are the similarities and differences between the two shuttle systems in terms of tissue location, number of ATPs made, and involvement of membrane transport proteins?

36. *Cell Map:* Indicate on the two shuttle systems provided on the map how reducing equivalents from cytosolic NADH are transported into the respiratory chain.

Regulation of Oxidative Phosphorylation

37. In the brewing of beer, as described in Chapter 15, an aerobic and then an anaerobic stage of yeast growth are encouraged and allowed. Why is the aerobic phase important? (Hint: think in terms of *numbers* of cells and efficiency of catabolism.)

Oxidative Phosphorylation Is Regulated by Cellular Energy Needs

38. Why is the cellular concentration of ATP maintained at such a steady level? Why doesn't the cell "store" ATP just as it stores polysaccharides and lipids?

Uncoupled Mitochondria in Brown Fat Produce Heat

39. Do you think brown fat mitochondria have sufficient molecules of thermogenin so that no ATP is formed by oxidative phosphorylation in these specialized adipose cells? Why or why not?

ATP-Producing Pathways Are Coordinately Regulated

40. Figure 19–29 may be one of the most useful in the textbook. Make sure all the accelerations and inhibitions of the various metabolic steps by ADP, ATP, NAD^+, and NADH make sense to you.

Mutations in Mitochondrial Genes Cause Human Disease

41. Why are neurons especially vulnerable to defects in their ATP-generating systems?

Mitochondria Probably Evolved from Endosymbiotic Bacteria

42. During operation of bacterial electron transfer chains, into what space are protons pumped? Does this create a problem in terms of "retrieval" of these protons for energy generation?

Photosynthesis: Harvesting Light Energy

43. What is the overall equation for photosynthesis in higher plants?

General Features of Photophosphorylation

44. Which processes of photosynthesis occur only in the light? Which processes do not require light?

Photosynthesis in Plants Takes Place in Chloroplasts

45. Do plants have mitochondria? Explain.

Light Drives Electron Flow in Chloroplasts

46. What is the biological electron acceptor in chloroplasts that corresponds to the acceptor "A" in the Hill equation?

47. What is the ultimate electron donor and the ultimate electron acceptor in chloroplasts? In mitochondria? In each organelle, is energy necessary as an input to the process, or is it produced as a consequence?

Light Absorption

48. Why do you think that infrared and ultraviolet light are not useful for photosynthesis?

Chlorophylls Absorb Light Energy for Photosynthesis

49. What structural property of the photopigments accounts for their ability to absorb light?

50. Why are plants green?

Accessory Pigments Extend the Range of Light Absorption

51. What advantage would a plant gain by having several different types of light-absorbing pigments?

Chlorophyll Funnels Absorbed Energy to Reaction Centers by Exciton Transfer

52. What is the series of reactions that occurs to bring about electric charge separation in the reaction-center chlorophyll? It is complicated, but very important to understand.

The Central Photochemical Event: Light-Driven Electron Flow

Bacteria Have One of Two Types of Single Photochemical Reaction Centers

53. What is the end result of the electron transfer process through the Type II reaction center?

54. What additional product (over the Type II reaction center) does the Type I reaction center produce?

Kinetic and Thermodynamic Factors Prevent Energy Dissipation by Internal Conversion

55. How are diffusion and random collisions minimized in reaction centers?

In Higher Plnats, Two Reaction Centers Act in Tandem

56. What is the connecting protein between photosystems II and I?

57. What does the "Z" shape of the "Z scheme" actually represent?

58. How many photons are required to oxidize a molecule of water?

59. Which component of photosystem II finally receives the four electrons abstracted from water?

Spatial Separation of Photosystems I and II Prevents Exciton Larceny

60. Where in the thylakoid membrane is PSII located? Where are PSI and ATP synthase? Where is the cytochrome b_6f complex?

The Cytochrome b_6f Complex Links Photosystems II and I

61. What are the similarities and differences between the cytochrome b_6f complex of chloroplasts and Complex III of mitochondria?

Cyanobacteria Use the Cytochrome b_6f Complex and Cytochrome c_6 in Both Oxidative Phosphorylation and Photophosphorylation

62. What is the "double role" played by cytochrome *c*?

Water Is Split by the Oxygen-Evolving Complex

63. What is the stoichiometry of number of photons absorbed to number of electrons "boosted" in each photosystem?

ATP Synthesis by Photophosphorylation

A Proton Gradient Couples Electron Flow and Phosphorylation

64. In order for this coupling to occur, why must the thylakoid membrane, like the inner mitochondrial membrane, be impermeable to protons?

The Approximate Stoichiometry of Photophosphorylation Has Been Established

65. What component of the proton motive force is most important in chloroplasts? Why?

66. What is the overall equation for noncyclic photophosphorylation?

Cyclic Electron Flow Produces ATP but Not NADPH or O_2

67. Under what cellular circumstances does cyclic electron flow rather than noncyclic electron flow occur?

The ATP Synthase of Chloroplasts Is Like That of Mitochondria

68. On which side of the thylakoid membrane is the CF_1 protein complex? How is this different from and similar to the placement of the mitochondrial F_1?

69. *Cell Map:* In the chloroplast, place the components of the proton and electron circuits of thylakoids. (The most useful Figure for this is 19–52.)

Chloroplasts Probably Evolved from Endosymbiotic Bacteria

70. What characteristics of chloroplasts link them to their probable photosynthetic prokaryotic ancestors?

Diverse Photosynthetic Organisms Use Hydrogen Donors Other Than Water

71. Can photosynthetic bacteria produce ATP? NADPH? O_2? Explain.

In Halophilic Bacteria, a Single Protein Absorbs Light and Pumps Protons to Drive ATP Synthesis

72. How is the phototransducing machinery of *H. salinarum* dissimilar to that of cyanobacteria and plants?

SELF-TEST

Do You Know the Terms?

ACROSS

2. An electrochemical ______-______ force is used to drive the synthesis of ATP in both oxidative phosphorylation and photosynthesis.

5. Passes an electron to an Fe-S protein in PSI.

6. Compound that supplies electrons to replace those "lost" by photosystem II in noncyclic photophosphorylation.

8. Electrons are transferred from ferredoxin to $NADP^+$ by the enzyme ferredoxin-$NADP^+$ ______.

10. The catalytic portion of Complex I of the respiratory chain. (2 words)

13. ______ molecules: pigments that absorb light of wavelengths other than those absorbed by chlorophylls *a* and *b,* thus "grabbing" stray light energy that would otherwise be missed.

15. Photosynthetic pigments are located in the ______ membrane of chloroplasts; analogous to the cristae of mitochondria.

17. Involved in mitochondrial electron transfers, this compound relies, in part, on its mobility within membranes for its function.

18. Only enzyme of the citric acid cycle that is membrane-bound; directly links the cycle to the electron transfer chain. (2 words)

19. The ______-______ shuttle translocates reducing equivalents from cytosolic NADH into the mitochondrion, at no cost in terms of ATP.

24. Iron-sulfur protein that can transfer its electron *either* to the cytochrome b_6f complex and then P700 *or* to a flavoprotein and then $NADP^+$.

25. Mobile electron transfer protein that links photosystem II and photosystem I.

26. The ______ model explains how a proton gradient is used to drive ATP synthesis.

27. Diagram of the interaction between photosystem I and photosystem II.

28. The ______ complex interacts with cytochrome *c*; a site of proton pumping.

29. Respiratory complex that transfers electrons to molecular oxygen. (2 words)

DOWN

1. The ______ complex interacts with the quinone PQ_B; a site of proton pumping.

2. Electron transfer compound receiving its electrons directly from $Pheo^-$.

3. Photosystem ______ is responsible for the production of NADPH. (word)

4. Adenine nucleotide ______ is required for cells to use the ATP generated inside mitochondria by oxidative phosphorylation.

5. Without this *transmembrane* transport system, ATP synthase would quickly run out of substrate. (2 words)

7. P680* and P700* are excellent ______ in photosystems II and I, respectively. (2 words)

9. Causes the transport of ions across a membrane, down their concentration gradient; can increase the intracellular levels of calcium, for example.

11. Net production of carbohydrate occurs during the carbon-fixation, or _____ reactions of photosynthesis.
12. ATP and NADPH are produced in the _____-dependent reactions of photosynthesis.
14. Photosystem _____ is involved only in noncyclic photophosphorylation. (word)
16. Enzyme complex containing F_1 and F_o.
20. Location in chloroplast where carbon fixation takes place.
21. Substance involved in electron transfer in photophosphorylation; receives an electron from P680*.
22. Proteins present in mitochondria, chloroplasts, and bacterial membranes; contain iron-containing prosthetic groups.
23. Prosthetic group involved in electron transfers; bound to proteins through Cys residues. (2 words)

Do You Know the Facts?

1. List (or construct a diagram showing) the sequence of electron transfer complexes and mobile electron carriers that are associated with the inner mitochondrial membrane. (Be sure to include the original electron donor and the ultimate electron acceptor.) Which has the highest, and which the lowest, E'°?

2. Indicate whether each of the following statements about the mitochondrial electron transfer chain and oxidative phosphorylation is true or false.

_____ **(a)** NADH dehydrogenase complex, cytochrome bc_1 complex, and cytochrome oxidase all are transmembrane proteins.
_____ **(b)** Synthesized ATP must be transported into the intermembrane space before it can enter the cytosol.
_____ **(c)** Cytochrome *c* and the F_1 subunit of ATPase are peripheral membrane proteins.
_____ **(d)** Complexes I, II, III, and IV all are proton pumps.
_____ **(e)** Ubiquinone is a hydrophilic molecule.
_____ **(f)** Ubiquinone and the F_o subunit of ATP synthase are both peripheral membrane proteins.
_____ **(g)** The final electron acceptor is H_2O.

3. Which of the following experimental observations would *not* support the chemiosmotic model of oxidative phosphorylation?

A. If mitochondrial membranes are ruptured, oxidative phosphorylation cannot occur.
B. Raising the pH of the fluid in the intermembrane space results in ATP synthesis in the matrix.
C. Transfer of electrons through the respiratory chain results in formation of a proton gradient across the inner mitochondrial membrane.
D. The orientation of the enzyme complexes of the electron transfer chain results in a unidirectional flow of H^+.
E. Radioactively labeled inorganic phosphate is incorporated into cytosolic ATP only in the presence of an H^+ gradient across the inner mitochondrial membrane.

4. The proton-motive force generated by the electron transfer chain:

A. includes a pH-gradient component.
B. includes an electrical-potential-gradient component.
C. is used for active transport processes.
D. is used to synthesize ATP.
E. has all of the above characteristics.

5. Where do the electrons from NADH and $FADH_2$, respectively, enter the electron transfer chain? Why must they enter at different sites? What effect does this have on the number of ATPs produced from reoxidation of each carrier?

6. Why do various cell types differ in the maximum number of ATPs they can produce per molecule of glucose aerobically oxidized?

7. Of all the components of the "Z scheme," which has the lowest $E'°$ (i.e., is the best reducing agent)?

A. P700
B. P700*
C. O_2
D. H_2O
E. NADPH

8. Which of the following statements about photosystem II is correct?

A. It is located in the inner mitochondrial membrane.
B. It contains the electron carrier with the most negative $E'°$ in the entire photosynthetic system.
C. It contains an Mn-containing complex that splits water.
D. Its final electron acceptor is NADPH.
E. H_2O is the only electron donor capable of regenerating P680 from P680*.

9. Which of the following statements about cyclic photophosphorylation and noncyclic photophosphorylation is correct?

A. Cyclic photophosphorylation involves only photosystem II and produces only ATP; noncyclic photophosphorylation involves both photosystems I and II and produces only ATP.
B. Both pathways liberate oxygen.
C. Both pathways involve photosystems I and II.
D. Cyclic photophosphorylation reduces $NADP^+$ and liberates oxygen; noncyclic photophosphorylation reduces $NADP^+$ but does not liberate oxygen.
E. Noncyclic photophosphorylation reduces $NADP^+$ and liberates oxygen; cyclic photophosphorylation produces ATP but does not liberate oxygen.

10. For each of the following statements, indicate whether it is true of only chloroplasts, only mitochondria, or both.

	Chloroplasts	Mitochondria	Both
(a) One source of electrons is NADH.	____	____	____
(b) Electron transfer leads to establishment of a proton gradient.	____	____	____
(c) The organelle requires a system of intact membranes to generate ATP.	____	____	____
(d) The organelle contains cytochromes and flavins in its electron transfer chain.	____	____	____
(e) The final electron acceptor is $NADP^+$.	____	____	____

11. Some photosynthetic prokaryotes use H_2S, hydrogen sulfide, instead of water as their photosynthetic hydrogen donor. How does this *change* the ultimate products of photosynthesis?

A. Carbohydrate (CH_2O) is not produced.
B. H_2O is not produced.
C. Oxygen is not produced.
D. ATP is not produced.
E. The products do not change.

Applying What You Know

1. Oxidative phosphorylation and photophosphorylation resemble each other in certain respects. Describe the ways in which the two processes are similar, then list the significant differences.

2. Artificial, but functional, electron transfer systems can be made in the lab by building artificial membrane-bound vesicles. This is done by combining detergent-solubilized, purified respiratory complexes and membrane lipids. When the mixture is dialyzed to remove the detergent, liposomes spontaneously form that contain the protein complexes integrated into the "membrane." The central cavity of the liposome can be made to contain certain molecules in aqueous solution; the surrounding medium can also be manipulated.

Using this protocol, you create the following electron transfer systems in liposomes, containing the listed set of components (not necessarily in their functioning order) along with the specified initial electron donors. Place the components in their correct functional sequence and indicate the final electron acceptor in each case.

(a) NADH as initial electron donor; Q and Complexes I, III, and IV in the liposomes; oxygen is present.

(b) NADH as initial electron donor; Complexes I, II, and IV in the liposomes; oxygen is present.

(c) Succinate as initial electron donor; Q, cytochrome *c*, and Complexes II, III, and IV in the liposomes; oxygen is present.

3. (a) If the orientation of the mitochondrial ATP synthase were reversed so that the F_1 unit were on the opposite side of the inner mitochondrial membrane, and assuming that nothing else is changed in the cell, what would be the consequences to the cell? Explain.

(b) What would be the consequences if just the F_o complex were flipped within its membranous surroundings, so that its normally matrix-facing surface faced the intermembrane space? Assume the F_1 complex is still on the matrix side.

4. You have been given three different species of bacteria by your research advisor. You are told that one species lives by fermentation and is a facultative anaerobe; one lives by oxidation of glucose via the citric acid cycle and oxidative phosphorylation; and one lives by photosynthesis. All are capable of consuming glucose. You manage, somehow, to mix up the three bacterial cultures. Without resorting to any biochemical tests (which would require help from a grouchy graduate student), how can you use *different laboratory growth conditions* to determine which culture is which?

ANSWERS

Do You Know the Terms?

ACROSS

2. proton-motive
5. phylloquinone
6. water
8. oxidoreductase
10. NADH dehydrogenase
13. antenna
15. thylakoid
17. ubiquinone
18. succinate dehydrogenase
19. malate-aspartate
24. ferredoxin
25. plastocyanin
26. chemiosmotic
27. Z scheme
28. cytochrome bc_1
29. cytochrome oxidase

DOWN

1. cytochrome b_6f
2. plastoquinone
3. one
4. translocase
5. phosphate translocase
7. electron donors
9. ionophore
11. dark
12. light
14. two
16. ATP synthase
20. stroma
21. pheophytin
22. cytochromes
23. Fe-S center

Do You Know the Facts?

1. The sequence of complexes and mobile carriers is as follows:

(a) Complex I, NADH dehydrogenase: transfers electrons from NADH (lowest $E'°$) to Q.

(b) Ubiquinone (Q)

or

(a) Complex II, succinate dehydrogenase: transfers electrons from $FADH_2$ to Q.

(b) Ubiquinone

then

(c) Complex III, cytochrome bc_1 complex: transfers electrons from Q to cytochrome *c*.

(d) Complex IV, cytochrome oxidase: transfers electrons from cytochrome *c* to O_2 (highest $E'°$).

2. **(a)** T; **(b)** T; **(c)** T; **(d)** F; **(e)** F; **(f)** F; **(g)** F

3. B

4. E

5. Electrons from NADH enter at the NADH dehydrogenase complex, Complex I. Electrons from the $FADH_2$ of FAD-linked enzymes, which can be derived from several different reactions, enter through Complex II (succinate dehydrogenase), glycerol 3-phosphate dehydrogenase, or ETFP. All of these enzymes transfer electrons to Q. The $E'°$ of $FADH_2$ in each of these enzymes is higher than that of NADH, so electrons must enter at a "later" point in the respiratory chain, missing the first proton pumping site, Complex I. Thus whereas electrons derived from NADH yield 2.5 ATP, those from $FADH_2$ yield only 1.5 ATP.

6. The maximum number of ATPs depends on the shuttle system. NADH from glycolysis must be reoxidized in the mitochondrial matrix, but it cannot traverse the inner mitochondrial membrane as NADH. Some types of cells use the malate-aspartate shuttle, which results in the production of 2.5 ATP per NADH. Other cells use the glycerol 3-phosphate shuttle, which yields only 1.5 ATP per NADH. This results in two fewer ATPs produced per glucose molecule.

7. B

8. C

9. E

10. **(a)** mitochondria; **(b)** both; **(c)** both; **(d)** both; **(e)** chloroplasts

11. C

Applying What You Know

1. Similarities between oxidative phosphorylation and photophosphorylation include:

- A sequential chain of membrane-bound electron carriers.
- Cytochromes and flavins in the electron transfer chains.
- Electron transfer leading to establishment of a proton gradient.
- A system of intact membranes to separate protons "inside" and "outside."
- An ATP synthase as a coupling factor.

- The F_1 portion of the ATP synthase located on the more alkaline side of the membrane, so the direction of flow of protons is down their concentration gradient.

Differences include:

- The organelle in which the process occurs: mitochondrion vs. chloroplast.
- The initial source of electrons: NADH and $FADH_2$ vs. H_2O (or H_2S, or organic hydrogen donors such as lactate).
- Source of energy for the electron donors: fuel molecules vs. photons.
- Final electron acceptor: O_2 vs. $NADP^+$.
- Sidedness of the F_1 component of the ATP synthase: "inside" facing the matrix vs. "outside" facing the stroma.

2. (a) Electrons will pass from NADH to Complex I to Q to Complex III, the final electron acceptor, since there is no cytochrome *c* to transfer the electrons to Complex IV.

(b) Complex I is the first and final electron acceptor, since no Q is present.

(c) Succcinate donates electrons to the FAD of Complex II. Electrons are then transferred to Q, then to Complex III, then to cytochrome *c*, then to Complex IV, which finally passes the electrons to O_2.

3. (a) Protons could not flow down their concentration gradient into the matrix because they can only pass through the inner mitochondrial membrane via the *normally* oriented ATP synthase, first through the F_o channel then through F_1 (which catalyzes the formation of ATP from ADP and P_i). Here, the F_1 units would be on the opposite side of the inner membrane. Furthermore, ADP + P_i are at a high concentration in the matrix, not in the intermembrane space, and would not be readily available to the F_1 subunit to make ATP. Thus the cell would be unable to generate ATP by oxidative phosphorylation. Furthermore, as electron transfer continued, the protons would continue to increase in concentration in the intermembrane space; a limit would be reached, and electron transfer, proton pumping, and oxidation of electron carriers such as NADH would cease. Substrate-level phosphorylation in glycolysis would also eventually stop because the cell would be unable to regenerate NAD^+. Having lost its ability to form ATP, the cell would die.

(b) Although F_o acts as a proton channel or pore through the membrane, it is not symmetrical, or equal, on each side of the membrane. F_1 is held to the matrix side of the F_o unit by specific interactions. Because most biomolecules have sidedness or direction, the same interactions could not occur on the surface that normally faces the intermembrane space. In addition, the ability of protons to enter the proton channel is likely to be influenced by the nature of the amino acid residues present in the portion of the channel that faces the intermembrane space. Therefore, flipping the F_o complex would, most likely, inhibit the ability of protons to flow through the pore.

4. You should transfer three samples of each culture into separate test tubes. For each culture, put one test tube in each of the following growth conditions: dark with no oxygen; dark with oxygen present; and light with no oxygen. Wait a few days, and see what grows and what dies. A chart, like the one below, will help organize the information:

	Dark, no O_2	Dark, O_2	Light, no O_2
Culture A			
Culture B			
Culture C			

The culture that grows under all three conditions is the fermenter, the facultative anaerobe. The culture that grows only in light is the photosynthesizer. The culture that grows only in the presence of oxygen is the third species, which depends on aerobic glucose oxidation.

chapter

20 Carbohydrate Biosynthesis

STEP-BY-STEP GUIDE

Major Concepts

Gluconeogenesis is the ubiquitous pathway for synthesis of glucose from noncarbohydrate precursors.

Biosynthesis of glucose is essential in mammals because some tissues, including the brain, use glucose as their sole or primary fuel source. Gluconeogenesis is not simply the reversal of glycolysis, although the two pathways do share seven enzymes. Three irreversible reactions of glycolysis must be bypassed, or circumvented, in gluconeogenesis: these three steps of gluconeogenesis are conversion of pyruvate to phosphoenolpyruvate; conversion of fructose 1,6-bisphosphate to fructose 6-phosphate by hydrolysis of the C-1 phosphate; and conversion of glucose 6-phosphate to free glucose. The enzyme catalyzing this third reaction is found in the liver, where most gluconeogenesis occurs, but not in the muscle or brain.

The formation of glucose is an energy-requiring process, and gluconeogenesis and glycolysis are reciprocally regulated to avoid futile cycling in cells.

In the liver of animals this regulation is largely mediated by **fructose 2,6-bisphosphate,** an allosteric effector of both the glycolytic enzyme **phosphofructokinase-1** and the gluconeogenic enzyme **fructose 1,6 bisphosphatase.** The concentration of this effector is controlled by the hormone **glucagon.**

The metabolism of both fats and proteins can provide carbon atoms for the synthesis of glucose.

Plants, but not vertebrate animals, can use even-carbon fatty acids as precursors of glucose. Both glyoxysomes and mitochondria are integral to this process, which provides the plant with carbohydrates during germination.

Glycogen synthesis takes place under conditions of excess glucose availability in animals.

The process uses **UDP-glucose** (a **sugar nucleotide**); the enzyme **glycogen synthase;** a branching enzyme, **glycosyl-(4 → 6)-transferase;** and an unusual protein called **glycogenin,** which serves as both primer and catalyst. The synthesis and breakdown of glycogen is regulated by hormones. The regulatory mechanism involves the reciprocal activation and inactivation of both glycogen synthase and glycogen phosphorylase by phosphorylation and dephosphorylation. This is an important example of metabolic control strategies, and illustrates a recurring theme in metabolic regulation. The biosynthetic pathways of the polysaccharide starch and the disaccharides sucrose and lactose also use sugar nucleotides as intermediates; these sugar derivatives have properties that make them especially suitable for biosynthetic reactions.

Photosynthetic organisms synthesize carbohydrates from carbon dioxide and water.

Carbon assimilation occurs in three stages. In the first stage, CO_2 condenses with ribulose 1,5-bisphosphate, a reaction catalyzed by **ribulose 1,5-bisphosphate carboxylase/oxygenase,** or **rubisco.** This enzyme, the key enzyme in the formation of organic biomolecules from inorganic carbon (CO_2), is regulated by substrate availability, cofactor concentration, covalent modification, and pH. The six-carbon intermediate formed in the fixation reaction is broken down to two molecules of 3-phosphoglycerate. The second stage of carbon assimilation, the conversion of 3-phosphoglycerate to glyceraldehyde 3-phosphate, uses the ATP and NADPH synthesized in the light reactions. Glyceraldehyde 3-phosphate has several fates in the plant cell, including entry into glycolysis or conversion into hexoses. The third stage regenerates ribulose 1,5-bisphosphate via a pathway that is essentially the reversal of the oxidative pentose phosphate pathway. **The net reaction of the carbon fixation pathway (the Calvin cycle) is synthesis of one molecule of glyceraldehyde 3-phosphate from three CO_2 molecules, at a cost of six NADPH and nine ATP.**

Carbon assimilation requires a P_i–triose phosphate antiport system in the chloroplast inner membrane.
This system exchanges cytosolic P_i for stromal dihydroxyacetone phosphate, allowing continued carbon fixation; it effectively shuttles ATP and reducing equivalents as well.

Regulation of carbohydrate metabolism in plants is complicated.
The major enzyme of carbon fixation, rubisco, is under tight control as noted above. Specific enzymes of the Calvin cycle are also subject to regulation by light (mediated by changes in pH), alterations in cofactor concentrations, thioredoxin, and covalent modification. Gluconeogenesis and glycolysis are also regulated. Light plays a regulatory role in that the triose phosphates produced by photosynthesis inhibit phosphofructokinase-2 activity; this lowers the levels of fructose 2,6-bisphosphate, resulting in a decrease in glycolysis and an increase in gluconeogenesis.

Besides its carboxylase activity, rubisco also has an oxygenase function that catalyzes the condensation of O_2 with ribulose-1,5-bisphosphate.
This reaction is followed by a series of carbon "salvage" reactions, which, along with the oxygenase activity, consume O_2 and produce CO_2 in an energy-expending process known as **photorespiration. C_4 plants,** so called because they initially fix CO_2 into a four-carbon compound, have decreased levels of photorespiration. By increasing the local concentration of CO_2 in certain specialized cells, C_4 plants are able to outcompete **C_3 plants** under hot and bright environmental conditions.

What to Review

Answering the following questions and reviewing the relevant concepts, which you have already studied, should make this chapter more understandable.

- Because this chapter is the first to deal with biosynthetic pathways, it would be useful to review the section opener of Part III (pp. 485–492), which introduces the basic concepts and discusses the coordination of metabolism.
- Rubisco, the key enzyme of the carbon fixation reactions in photosynthetic organisms, has two substrates, oxygen and carbon dioxide, each with its own K_m value. What does the K_m of an enzyme mean? If an enzyme has two substrates, does just knowing the K_m of each substrate tell you at what rate the enzyme catalyzes each reaction (pp. 259–264)?
- In plant cells, the P_i–triose phosphate antiporter in the chloroplast inner membrane provides a conduit for the products of photosynthesis to enter the cytosol. What are the characteristics of antiport systems (p. 413)?
- **Sucrose** is the principal form in which sugar is transported from leaves to the rest of the plant (p. 303); **glycogen** and **starch** are the storage polysaccharides of animals and plants, respectively (pp. 304–306). What are the monosaccharide units of these compounds? What type(s) of bonds connect the monomers?
- The regulation of glycogen phosphorylase has been introduced already (p. 557); it will be treated in more detail in this chapter. What levels of control are active in this regulation?
- The **glyoxylate cycle** allows plants and some bacteria to synthesize carbohydrates from acetyl-CoA, an ability that vertebrate animals lack. What is the net reaction of the glyoxylate cycle? What enzymes are present in this pathway that are not present in the citric acid cycle (pp. 588–589)?
- The end product of **fatty acid oxidation** is acetyl-CoA, which is not a starting point for carbohydrate biosynthesis. However, **degradation of triacylglycerols,** the storage form of lipids, does provide a molecule usable in glycolysis *and* in gluconeogenesis. What is this molecule? How does it enter the carbohydrate metabolic pathways (pp. 601–602)?
- Certain **amino acids** are considered glucogenic, and others ketogenic. Recall which amino acids are converted to which citric acid cycle intermediates; note that only leucine and lysine are degraded only to acetyl-CoA and are therefore exclusively ketogenic (p. 639).

Topics for Discussion

Answering each of the following questions, especially in the context of a study group discussion, should help you understand the important points of this chapter.

1. Make sure you understand the "organizing principles of biosynthesis" presented in the introduction. You will find these very useful to keep in mind for this and the following chapters.

Gluconeogenesis

2. Why is the biosynthesis of glucose so important in mammals?

3. How can both glycolysis and gluconeogenesis be irreversible processes in cells?

Conversion of Pyruvate to Phosphoenolpyruvate Requires Two Exergonic Reactions

4. What is the energetic cost of converting pyruvate to PEP?

5. Some of the reactions for conversion of pyruvate to PEP occur in the mitochondrial matrix, but the end product is made in the cytosol. Besides the transfer of oxaloacetate to the cytosol, what other important molecule is made available for gluconeogenesis by this path?

6. What cellular factor allows the shorter pathway from lactate to PEP to occur?

Conversion of Fructose 1,6-Bisphosphate to Fructose 6-Phosphate Is the Second Bypass

7. What drives this reaction forward?

Conversion of Glucose 6-Phosphate to Free Glucose Is the Third Bypass

8. Why can't gluconeogenesis occur in muscle and brain cells?

Gluconeogenesis Is Expensive

9. Compare the net equations of glycolysis and gluconeogenesis. Why is this high cost of synthesis of glucose necessary?

Citric Acid Cycle Intermediates and Many Amino Acids Are Glucogenic

10. Compare the structures of the amino acids alanine and glutamine and their entry points for glucose synthesis, pyruvate and α-ketoglutarate. Why are some amino acids, and fatty acids, *not* glucogenic?

11. How does fatty acid oxidation contribute to gluconeogenesis?

Futile Cycles in Carbohydrate Metabolism Consume ATP

12. How do bees use futile cycling?

Gluconeogenesis and Glycolysis Are Reciprocally Regulated

13. What are the control points for reciprocal regulation of gluconeogenesis and glycolysis?

14. What is the mechanism of glucagon's action on blood glucose?

15. The discussion of the control of phosphofructokinase-1 is a continuation from Chapter 15. You will find it extremely helpful to review this earlier material (pp. 554–555). How is the level of fructose 2,6-bisphosphate controlled?

16. *Cell Map:* On your map, draw in the gluconeogenic bypasses next to the reactions of glycolysis. We suggest using a different color to highlight these reactions.

Gluconeogenesis Converts Fats and Proteins to Glucose in Germinating Seeds

17. Why are seeds so high in protein and fat?

18. How does the glycerol from triacylglycerols enter gluconeogenesis?

19. How does the compartmentation provided by glyoxysomes and mitochondria aid in gluconeogenesis in plant seeds?

Biosynthesis of Glycogen, Starch, Sucrose, and Other Carbohydrates

20. What properties of sugar nucleotides make them particularly suitable for biosynthetic reactions?

UDP-Glucose Is the Substrate for Glycogen Synthesis

21. What drives the UDP-glucose pyrophosphorylase reaction forward?

22. What are the necessary components (substrates, enzymes, and cofactors) for building glycogen, beginning with the primer glycogenin?

23. Glycogenin remains buried in the glycogen particle it helped to build. How is this fate different from the fate of most enzymes?

Glycogen Synthase and Glycogen Phosphorylase Are Reciprocally Regulated

24. Compare the mechanisms of regulation of glycogen synthase and glycogen phosphorylase. What are the roles of glucagon, epinephrine, and insulin in these mechanisms? Which are allosteric controls and which are covalent? This is a complex regulatory scheme and serves as a model for regulatory mechanisms used in other pathways. (Hint: You will find pp. 445–455 helpful here.)

ADP-Glucose Is the Substrate for Starch Synthesis in Plants and Glycogen Synthesis in Bacteria

25. What are the similarities and differences between glycogen and starch synthesis?

UDP-Glucose Is the Substrate for Sucrose Synthesis in Plants

26. What makes sucrose a good "transport" form of carbon?

Lactose Synthesis Is Regulated in a Unique Way

27. Describe the mechanism by which lactose synthesis is turned on in mammary tissue.

UDP-Glucose Is an Intermediate in the Formation of Glucuronate and Vitamin C.

28. What are the (very important) roles of UDP-glucuronate?

Sugar Nucleotides Are Precursors in Bacterial Cell Wall Synthesis

29. Note that the construction of bacterial cell walls involves the participation of nucleotides, amino acids, heteropolysaccharides, and isoprenoids: all of biochemistry in one macromolecular construct!

Penicillin and β-Lactamase: The Magic Bullet versus the Bulletproof Vest

30. Clavulanate doesn't kill bacteria on its own, but when given in combination with penicillin or other β-lactam-containing antibiotics, it makes the antibiotic effective even against bacterial strains containing β-lactamases. Clavulanate is a competitive inhibitor of β-lactamases. From this information, what can you guess about the structure of clavulanate?

Photosynthetic Carbohydrate Synthesis

31. Does net CO_2 "fixation" occur in the pyruvate carboxylase, acetyl-CoA carboxylase, or carbamoyl phosphate synthetase I reactions?

Carbon Dioxide Assimilation Occurs in Three Stage

32. What is the net reaction, and the actual stoichiometry, for each stage of the Calvin cycle?

Stage 1: Fixation of CO_2 into 3-Phosphoglycerate

33. Where in the ribulose 1,5-bisphosphate molecule is the CO_2 that is being "fixed" added?

34. Where is rubisco located in plants? (Although "rubisco" is a nice short name, the full name gives all the information about the two activities of this enzyme.)

Stage 2: Conversion of 3-Phosphoglycerate to Glyceraldehyde 3-Phosphate

35. Could glycolysis occur in the chloroplast stroma?

36. What are the various roles of glyceraldehyde 3-phosphate in plant cells?

Stage 3: Regeneration of Ribulose 1,5-Bisphosphate from Triose Phosphates

37. What is the general role of the prosthetic group thiamin pyrophosphate? (See Fig. 15–9a.)

38. Compare the final products of the "reductive" pentose phosphate pathway with those of the oxidative pentose phosphate pathway.

Each Triose Phosphate Synthesized from CO_2 Costs Six NADPH and Nine ATP

39. *Cell Map:* What is the net reaction of the Calvin cycle? What is its stoichiometry? Add the (abbreviated) version of the Calvin cycle in Fig. 20–30 to the chloroplast on your cell map; connect the inputs of ATP and NADPH from photophosphorylation.

40. What is the source of energy for this process?

41. Why is the term "carbon assimilation reactions" much more accurate than "dark reactions"?

42. What prevents animals from being able to assimilate CO_2?

A Transport System Exports Triose Phosphates from the Chloroplast and Imports Phosphate

43. Why is the chloroplast membrane (or any membrane) impermeable to phosphorylated compounds?

44. *Cell Map:* What are the functions of the P_i–triose phosphate antiport system in the inner chloroplast membrane? Add this transporter to the chloroplast membrane on your map.

Regulation of Carbohydrate Metabolism in Plants

Rubisco Is Subject to Both Positive and Negative Regulation

45. Why does it make biological sense that high CO_2 concentrations activate rubisco?

46. What is the advantage of rubisco activity being decreased in the dark by the "nocturnal inhibitor"?

Certain Enzymes of the Calvin Cycle Are Indirectly Activated by Light

47. How are stromal environmental conditions influenced by the presence of light? What enzymes are affected by these conditions?

48. Describe the mechanism of action of thioredoxin.

The Use of Triose Phosphates for Sucrose and Starch Synthesis Is Tightly Regulated in Plants

49. What proportion of the triose phosphates produced during carbon fixation is consumed in starch synthesis? In sucrose synthesis? How does this proportion relate to the P_i–triose phosphate antiport system?

50. How does the regulation of fructose 2,6-bisphosphate concentration differ in plants and animals (see also pp. 722–723)?

Photorespiration Results from Rubisco's Oxygenase Activity

51. What is the net result of photorespiration?

52. Suppose you were conducting an experiment on rubisco in intact plant cells, with environmental conditions of 9 μm CO_2 and 350 μm O_2 (and a temperature under 28 °C). What more would you need to know in order to determine the proportion of rubisco activity involved in carbon assimilation and the proportion involved in oxygen "fixation"?

Some Plants Have a Mechanism to Minimize Photorespiration

53. Which of the following mechanisms is used by C_4 plants to increase the efficiency of the carbon-fixing activity of rubisco in bundle-sheath cells: increasing the local concentration of CO_2, decreasing the local concentration of O_2, decreasing the K_m for CO_2, or increasing the K_m for O_2?

54. Which uses more energy, the fixation of CO_2 in C_4 plants or in C_3 plants?

SELF-TEST

Do You Know the Terms?

ACROSS

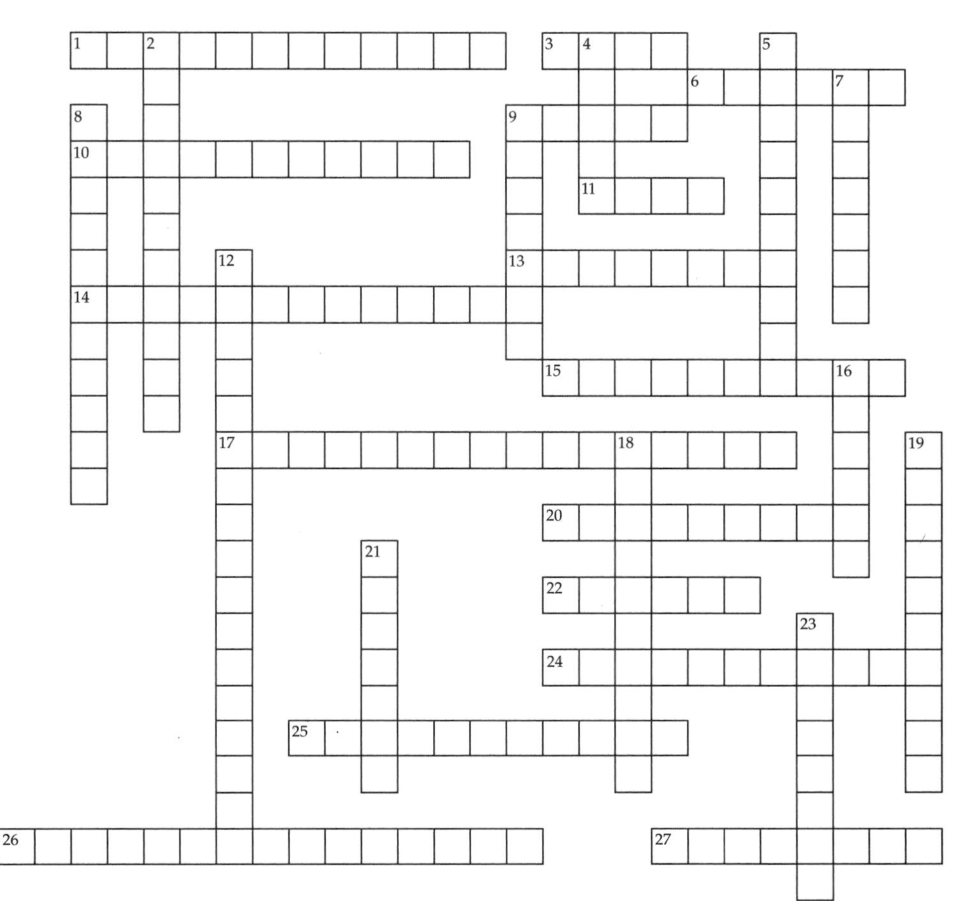

1. Sometimes used by organisms to generate heat; for the most part these pathways result only in the net hydrolysis of ATP. (2 words)
3. Enzyme catalyzing an irreversible step in glycolysis that is circumvented by the activity of fructose 1,6-bisphosphatase. (abbr.)
6. Carbon assimilation is also known as the _______ cycle.
9. Product of the committed step in glycolysis; must be dephosphorylated in gluconeogenesis. (abbr.)
10. General name for an enzyme that reverses the effects of a kinase.
11. Activation of this enzyme indirectly activates phosphofructokinase-1. (abbr.)
13. Glycogen _______ catalyzes the formation of $(\alpha 1 \rightarrow 4)$ glycosidic bonds.
14. Glycogen _______ catalyzes the cleavage of $(\alpha 1 \rightarrow 4)$ glycosidic bonds.
15. Substrate that binds to glycogen synthase.
17. Carbon fixation and _______ are competitive pathways in plants.
20. Describes amino acids that can be degraded to acetyl-CoA.
22. Cells "trap" glucose molecules by phosphorylating them in an irreversible reaction catalyzed by hexokinase. Liver cells release free glucose into the blood by reversing this process in a _______ reaction catalyzed by glucose 6-phosphatase.
24. In the synthesis of hexoses from nonhexose precursors, the first step toward rephosphorylation of pyruvate to phosphoenolpyruvate is catalyzed by pyruvate _______.
25. Sugar _______ : activated forms of sugars.
26. Pathway by which the glycerol backbone of triacylglycerols can be made into fuel for brain cells; occurs primarily in liver cells.
27. Describes a class of cellular reactions; an example is 26 ACROSS.

DOWN

2. Molecule that couples light energy to enzyme activation.
4. Allosteric activator of 3 ACROSS. (abbr.)
5. Describes amino acids that can be degraded to one of the citric acid cycle intermediates.
7. Peptide hormone that regulates the metabolism of both glucose and triacylglycerols.
8. Hormone that stimulates phosphorylation of phosphorylase *b* kinase, glycogen synthase, and glycogen phosphorylase; its receptors are found primarily on muscle cells.
9. Activation of this enzyme results in decreased [fructose 2,6-bisphosphate]. (abbr.)
12. First compound to become labeled by carbon from $^{14}CO_2$ in carbon assimilation.
16. Glycogen is the glucose storage molecule in animals; _______ is the glucose storage molecule in plants.
18. Type of transporter; examples include Na^+K^+ ATPase and adenosine nucleotide translocase.
19. Dual function enzyme; acts as both substrate and catalyst.
21. Ribulose 1,5-bisphosphate carboxylase/oxygenase.
23. Hormone that stimulates phosphorylation of phosphorylase *b* kinase, glycogen synthase, and glycogen phosphorylase; its receptors are located mainly on liver cells.

Do You Know the Facts?

1. Indicate whether each of the following statements about gluconeogenesis is true or false.

_____ **(a)** It occurs only in the inner mitochondrial membrane.

_____ **(b)** Pyruvate carboxylase, catalyzing an anaplerotic reaction, is the first regulatory enzyme in the pathway.

_____ **(c)** Precursors of hexoses include lactate, pyruvate, glycerol, and alanine.

_____ **(d)** The three bypass reactions have ΔG values near zero, whereas other gluconeogenic reactions have large, negative ΔG values.

_____ **(e)** Some reactions occur in the mitochondrial matrix and some in the cytoplasm.

_____ **(f)** Fructose 2,6-bisphosphate is an intermediate.

_____ **(g)** Oxidation of fatty acids provides energy for this process during starvation.

In questions 2–4, indicate the (immediate) effect (increase, decrease, no change) of each situation on the rate of gluconeogenesis.

2. Decreased concentration of acetyl-CoA.

3. Phosphorylation of phosphofructokinase-2.

4. Increase in [AMP]/[ATP] ratio.

5. Which of the following is true of the control of gluconeogenesis?

A. Glucagon stimulates adenylyl cyclase, causing the formation of cAMP.
B. Cyclic AMP stimulates the phosphorylation and thus increases the activity of FBPase-2.
C. FBPase-2 activity lowers the level of fructose 2,6-bisphosphate, thus increasing the rate of gluconeogenesis.
D. It is reciprocally linked to the control of glycolysis.
E. All of the above are true.

6. Which of the following correctly matches the *glycolytic reaction* with the *gluconeogenic enzyme* used in the corresponding bypass reaction?

A. Glucose → glucose 6-phosphate; glucose 6-phosphatase
B. Fructose 6-phosphate → fructose 1,6-bisphosphate; phosphofructokinase-2
C. Fructose 1,6-bisphosphate → dihydroxyacetone phosphate + glyceraldehyde 3-phosphate; glyceraldehyde 3-phosphate dehydrogenase
D. 2-Phosphoglycerate → phosphoenolpyruvate; phosphoglycerate kinase
E. Phosphoenolpyruvate → pyruvate; pyruvate kinase

7. Indicate whether each of the following descriptions of glycogen and its synthesis and breakdown is true or false.

_____ **(a)** Glycogen synthase is regulated by hormone-dependent phosphorylation and is active in its phosphorylated form.

_____ **(b)** Branching increases the solubility of glycogen.

_____ **(c)** Glycogen phosphorylase and glycogen synthase are usually active simultaneously.

_____ **(d)** Glycogen synthase and glycogen phosphorylase are reciprocally regulated.

_____ **(e)** Phosphoprotein phosphatase removes phosphate groups from Ser residues on glycogen synthase *b*, converting it to the active form.

_____ **(f)** Glycogen synthesis requires conversion of glucose 1-phosphate to UDP-glucose.

_____ **(g)** The primer protein glycogenin is required for initiation of glycogen synthesis.

8. Which of the following has the (immediate) effect of increasing the rate of glycogen breakdown? (Hint: Information from p. 557 is also necessary here.)

A. increased concentration of cAMP
B. increase in the [AMP]/[ATP] ratio
C. increased secretion of glucagon
D. A and C
E. A, B, and C

9. Which of the following has the (immediate) effect of increasing the rate of glycogen synthesis? (Hint: Information from p. 557 is also necessary here.)

A. increased concentration of cAMP
B. increase in the [AMP]/[ATP] ratio
C. increased secretion of insulin
D. A and C
E. A, B, and C

10. Given that plants can produce ATP and NADPH from photophosphorylation, for what purpose(s) do they reduce CO_2 to glucose?

A. At night, plants need the ATP produced by glycolysis and the citric acid cycle in the dark.
B. Plants need glucose to produce starch and cellulose.
C. Plants need glucose as a precursor of components of nucleic acids, lipids, and proteins.
D. B and C
E. A, B, and C

11. Carbon fixation involves a condensation reaction between CO_2 and:

A. 3-phosphoglycerate.
B. phosphoglycolate.
C. ribulose 1,5-bisphosphate.
D. fructose 6-phosphate.
E. ribose 5-phosphate.

12. Which equation correctly summarizes the carbon fixation reactions of photosynthesis?

A. $6H_2O + 6CO_2 \rightarrow 6O_2 + C_6H_{12}O_6$
B. $6CO_2 + 12NADPH + 18ATP + 12H^+ \rightarrow 12NADP^+ + 18ADP + 18P_i + C_6H_{12}O_6$
C. $6H_2O + 6ADP + 6P_i + 6NADP \rightarrow 6O_2 + 6ATP + NADPH + 6H^+$
D. $2H_2O \rightarrow O_2 + 4H+ + 4e^-$
E. $2C_3O_3H_5 + CO_2 \rightarrow 2C_3O_3H_3 + H_2O + CH_2O$

13. Ribulose 1,5-bisphosphate carboxylase/oxygenase is arguably the most important enzyme on earth because nearly all life is dependent, ultimately, on its action. The reactions catalyzed by this enzyme are influenced by:

A. pH.
B. substrate concentration.
C. Mg^{2+}concentration.
D. temperature.
E. all of the above.

14. Under hot, dry conditions, temperate grass species (C_3 grasses) are outcompeted by other grasses, notably crabgrass (C_4 plants). Why?

A. C_4 plants do not use as much ATP to fix CO_2 to hexose as do C_3 plants.
B. C_4 plants use CO_2 more efficiently under hot conditions because they increase $[CO_2]$ in bundle-sheath cells.
C. The K_m of the oxygenase reaction of rubisco is higher in C_3 plants.
D. C_4 plants are able to bypass the Calvin cycle and so save energy.
E. All of the above are true.

15. Sugar cane is a C_4 plant. Which of the following is true for these (often) tropical plants?

A. The affinity of rubisco for oxygen decreases with increasing temperature.
B. "C_4" refers to the four-carbon compounds by which CO_2 is shuttled from mesophyll cells to bundle-sheath cells.
C. The net result of the C_4 strategy is to lower the K_m of rubisco for CO_2.
D. C_4 plants can outcompete C_3 plants in hot environments because they have a different enzyme for carbon fixation.
E. All of the above are true.

16. Sugar nucleotides are the substrates for polymerization of monosaccharides into di- and polysaccharides. What properties of sugar nucleotides make them especially suitable for biosynthetic reactions?

17. Explain how both glycolysis and gluconeogenesis can be irreversible processes in the cell.

18. What is the net reaction of the Calvin cycle? Include energy inputs and electron carriers.

19. Describe the reciprocal regulation of glycogen phosphorylase *a* and *b* and glycogen synthase *a* and *b*.

Applying What You Know

1. Many bacterial species can carry out the reactions of gluconeogenesis. Would you expect to find the gluconeogenic enzyme glucose 6-phosphatase in such bacteria? Why or why not?

2. How do levels of acetyl-CoA in the cell affect gluconeogenesis? At what step of the pathway does this molecule play an important role?

3. Explain fully the regulatory role of fructose 2,6-bisphosphate in gluconeogenesis and glycolysis, and the mechanism by which it affects fructose 1,6-bisphosphatase-1 and phosphofructokinase-1. How is the concentration of fructose 2,6-bisphosphate regulated?

4. Why is it important, and logical, that some enzymes of the Calvin cycle are regulated by the presence or absence of light?

ANSWERS

Do You Know the Terms?

ACROSS

1. futile cycles
3. PFK-1
6. Calvin
9. F1,6-BP
10. phosphatase
11. PFK-2
13. synthase
14. phosphorylase
15. UDP-glucose
17. photorespiration
20. ketogenic
22. bypass
24. carboxylase
25. nucleotides
26. gluconeogenesis
27. anabolic

DOWN

2. thioredoxin
4. F2,6-BP
5. glucogenic
7. insulin
8. epinephrine
9. FBPase-2
12. 3-phosphoglycerate
16. starch
18. antiporter
19. glycogenin
21. rubisco
23. glucagon

Do You Know the Facts?

1. **(a)** F; **(b)** T; **(c)** T; **(d)** F; **(e)** T; **(f)** F; **(g)** T
2. Decrease
3. Increase
4. Decrease
5. E
6. A
7. **(a)** F; **(b)** T; **(c)** F; **(d)** T; **(e)** T; **(f)** T; **(g)** T
8. E
9. C
10. E
11. C
12. B
13. E
14. B
15. B
16. The properties of sugar nucleotides include the following: (a) Their formation splits one high-energy bond, releasing PP_i, which is then hydrolyzed by inorganic pyrophosphatase. The large, negative free-energy change makes the synthetic reaction thermodynamically favorable. (b) They offer many potential groups for noncovalent interactions with enzymes. (c) The nucleotidyl group is an excellent leaving group. (d) "Tagging" some hexoses with nucleotidyl groups specifies the use of those hexoses for a specific cellular purpose.

17. Seven of the ten enzymatic steps of gluconeogenesis are the reverse of glycolytic reactions. Three glycolytic reactions are essentially irreversible in cells and cannot be used in gluconeogenesis: the conversion of glucose to glucose 6-phosphate by hexokinase; the phosphorylation of fructose 6-phosphate to fructose 1,6-bisphosphate by phosphofructokinase; and the conversion of phosphoenolpyruvate to pyruvate by pyruvate kinase. These three reactions are characterized by a large, negative free-energy change, whereas the other reactions of glycolysis (also used in gluconeogenesis) have ΔG values near zero. The three irreversible glycolytic steps are bypassed by a separate set of enzymes in gluconeogenesis; these steps also have large, negative free-energy changes, but in the direction of glucose synthesis rather than glucose breakdown. Gluconeogenesis and glycolysis are independently regulated through controls exerted at the enzymatic steps that are not common to both pathways.

18. The net reaction is

$$9\text{ATP} + 3\text{CO}_2 + 6\text{NADPH} + \text{H}^+ \longrightarrow 1\text{G 3-P} + 9\text{ADP} + 8\text{P}_\text{i} + 6\text{NADP}^+$$

where G 3-P represents glyceraldehyde 3-phosphate. This is not given in the text as such. Make sure you understand that the "lost" P_i is part of glyceraldehyde 3-phosphate, and that 3 CO_2 are necessary to make one triose molecule.

19. Glycogen phosphorylase is the enzyme that cleaves glycogen into individual glucose residues (glucose 1-phosphate). It exists in two forms: glycogen phosphorylase *b* (a relatively inactive form) and glycogen phosphorylase *a* (a phosphorylated and fully active, glycogen-degrading form). Only the *b* form of this enzyme can be *allosterically* regulated; the allosteric *activator* is AMP and the *inhibitor* is ATP, making this enzyme responsive to cellular energy levels. Glycogen phosphorylase is also covalently modified in response to hormonal stimulation. Binding of epinephrine to receptors on muscle cells or glucagon binding to receptors on hepatocytes results in the production of cAMP, which activates an enzyme cascade. This enzyme cascade ultimately results in the activation of phosphorylase *b kinase* and the phosphorylation of glycogen phosphorylase *b*. This is the covalent modification that converts glycogen phosphorylase to its fully active *a* form. In addition both hormones inactivate phosphorylase *a phosphatase*, which reverses the effects of covalent modification by removing the phosphate group.

Glycogen synthase is the enzyme that synthesizes glycogen. This enzyme also exists in two forms: glycogen synthase *b* (the inactive form) and glycogen synthase *a* (the active, glycogen-synthesizing form). In contrast to glycogen phosphorylase, the active form of glycogen synthase is *not* phosphorylated whereas the inactive, *b* form *is* phosphorylated. The inactivation of glycogen synthase *a* is the result of phosphorylation by a cAMP-dependent protein kinase. Thus, the same hormones that activate glycogen phosphorylase by increasing cAMP levels *also* inactivate glycogen synthase. Due to the opposing effects of phosphorylation on glycogen phosphorylase and glycogen synthase, the anabolic and catabolic pathways can be reciprocally regulated by the same hormonal input.

Applying What You Know

1. Bacteria, as single-celled organisms, have no need to export free glucose to other organs or cells. Formation of free glucose by hydrolysis of glucose 6-phosphate would essentially waste the energy of a phosphoanhydride bond. It is more likely that glucose 6-phosphate is converted to glucose 1-phosphate for incorporation into a storage polysaccharide or some other compound.

2. Acetyl-CoA is a positive allosteric modulator of pyruvate carboxylase, stimulating the first step of gluconeogenesis. Biosynthesis of glucose is promoted only when there is an excess of acetyl-CoA in mitochondria, indicating that the cell does not need energy from the citric acid cycle. Acetyl-CoA also inhibits the pyruvate dehydrogenase reaction.

3. Fructose 2,6-bisphosphate (F 2,6-BP; not to be confused with fructose 1,6-bisphosphate) is *not* an intermediate in gluconeogenesis or glycolysis, but an allosteric effector that acts as a mediator for the hormonal regulation of these processes. High [F 2,6-BP] stimulates glycolysis by activating PFK-1 and inhibiting FBPase-1, which slows gluconeogenesis. Low [F 2,6-BP] stimulates gluconeogenesis.

The concentration of F 2,6-BP is controlled by the relative rates of its formation and breakdown: by the activity of *phosphofructokinase-2* (PFK-2), which catalyzes the formation of F 2,6-BP by phosphorylation of fructose 6-phosphate; and by *fructose 2,6-bisphosphatase* (FBPase-2), which catalyzes the dephosphorylation of F 2,6-BP. These two enzymes are part of a single protein. The balance of these two activities in the liver, and therefore the level of F 2,6-BP, is regulated by glucagon. Glucagon stimulates adenylyl cyclase, causing the formation of cAMP. Cyclic AMP stimulates a kinase, which transfers a phosphate group from ATP to the PFK-2/FBPase-2 protein. Phosphorylation leads to increased FBPase-2 activity, lowering [F 2,6-BP] and thus inhibiting glycolysis and stimulating gluconeogenesis. Glucagon therefore inhibits glycolysis, increasing gluconeogenesis and enabling the liver to replenish blood glucose levels.

4. Plants get energy (ATP) and reducing power (NADPH) from the light reactions of photosynthesis, both of which are necessary for CO_2 fixation. If the plant lacks sufficient cellular energy (ATP) and reducing power (NADPH), it is of no advantage for the Calvin cycle to be running. Plants have developed mechanisms to signal when light reactions are occurring, such as making the stromal compartment alkaline so that enzymes can work at their optimal pH.

chapter

21 Lipid Biosynthesis

STEP-BY-STEP GUIDE

Major Concepts

Biosynthesis of fatty acids does not proceed as simple reversal of fatty acid oxidation.

Malonyl-CoA, a three-carbon intermediate, is formed from acetyl-CoA and bicarbonate by the biotin-containing enzyme **acetyl-CoA carboxylase.** The assembly of fatty acids from these malonyl-CoA units and an initial acetyl group is catalyzed by a multienzyme complex, **fatty acid synthase.** A four-step sequence is repeated seven times to build palmitate, the 16-carbon saturated fatty acid that is the principal product of fatty acid synthesis in animals. The sequence includes: **condensation** of an activated acetyl group and malonyl group; **reduction** of the carbonyl group at C-3; **dehydration** at C-2 and C-3, yielding a double bond; and finally, **reduction** of the double bond. Further reactions after the fatty acid synthase series can include elongation and/or desaturation of the palmitate product. Some fatty acids are not made in mammals; these essential fatty acids must be obtained in the diet. The signaling lipids known as **eicosanoids** are formed from the essential fatty acid linoleate in response to hormonal or other stimuli.

These reactions are under tight control because the process is energetically expensive.

The first enzyme, acetyl-CoA carboxylase, is inhibited by palmitoyl-CoA, the final product of biosynthesis, and activated by citrate, which carries acetyl-CoA from the mitochondrial matrix into the cytosol where the synthetic reactions take place. This enzyme is also regulated by covalent alteration, which is under hormonal control.

Fatty acid synthesis and oxidation are coordinately regulated so that they do not occur simultaneously.

They take place in separate cellular locations (synthesis in the cytosol and breakdown in the mitochondrial matrix), and the breakdown process is inhibited by malonyl-CoA, the first intermediate in the synthetic process.

Synthesis of storage and membrane lipids from fatty acids is determined by the metabolic needs of the organism.

Triacylglycerols and glycerophospholipids are synthesized from common precursors, fatty acyl–CoA and glycerol 3-phosphate. Triacylglycerol synthesis is regulated by hormones; synthesis occurs when the fuel intake exceeds the energy needs of the organism. Membrane lipids are synthesized when the cell is growing rapidly. Glycerophospholipid and sphingolipid biosynthesis share some precursors and enzyme mechanisms.

Cholesterol is synthesized from acetyl-CoA.

This process occurs in four stages: the first stage is the condensation of three acetyl-CoA molecules to form **HMG-CoA.** This compound is reduced to mevalonate by **HMG-CoA reductase,** a reaction that is the committed step of the pathway. The second stage is the conversion of mevalonate to two **activated isoprene** compounds; this requires the input of phosphate groups from ATP. In the third stage, six isoprene units condense to form **squalene,** a 30-carbon compound. Finally, squalene is rearranged into the four-ring steroid nucleus. Cholesterol synthesis is controlled by feedback inhibition and hormonal signals, largely at the rate-limiting step catalyzed by HMG-CoA reductase.

Once synthesized, cholesterol has several fates.

It is a component of cell membranes; it is converted into bile salts for use in digestive processes; and it is the precursor of steroid hormones, which have diverse and crucial functions. These hormones are produced from cholesterol by alteration of its side chain and introduction of oxygen into the steroid ring system. In addition, intermediates in cholesterol biosynthesis have a wide variety of critical biological roles; these include vitamins, hormones, and electron carriers.

Cholesterol and other lipids are transported in the blood as lipoproteins, which have a variety of densities, compositions, and specific functions.

Chylomicrons carry exogenous (dietary) lipids from the small intestine to peripheral tissues and are the largest, but least dense, of the lipoproteins. **VLDLs** carry endogenous lipids (synthesized in the liver) to muscle and adipose tissue. **LDLs** are derived from VLDLs and are very rich in cholesterol and cholesteryl esters; high levels of LDLs have been correlated with **atherosclerosis. HDLs** scavenge cholesterol from chylomicrons and VLDLs, returning it to the liver, which is the only route for its ultimate excretion. There is a negative correlation between HDL levels and arterial disease.

Uptake of cholesterol by cells is accomplished by receptor-mediated endocytosis.

This uptake, as well as the biosynthesis and dietary intake of cholesterol, influences the concentration of this lipid in human blood.

What to Review

Answering the following questions and reviewing the relevant concepts, which you have already studied, should make this chapter more understandable.

- Aspirin and ibuprofen work to alleviate pain and/or fever by interfering in the synthesis of **prostaglandins** and **thromboxanes.** Review the functions of these eicosanoids (pp. 378–379).
- **Membrane fusion** processes are important both in the intracellular transport of newly synthesized membrane lipids from the site of synthesis to the site of function, and in receptor-mediated endocytosis of LDLs (Fig. 12–20).
- **Biotin** functions as a prosthetic group in the carboxylation reaction that produces malonyl-CoA from acetyl-CoA and bicarbonate. What other metabolic roles have already been discussed for this vitamin (Fig. 16–14; Fig. 17–11)?
- *Some* of the reactions of fatty acid synthesis are simple reversals of those in fatty acid oxidation. The control of fatty acid synthesis and oxidation must be coordinated to prevent futile cycling. Outline the sequence of reaction steps and the regulation of **fatty acid oxidation;** be sure to note *where* in the cell these reactions take place (Fig. 17–8, p. 612).
- The acyl carrier protein that is part of the fatty acid synthase complex carries acyl groups attached through a thioester linkage. Why are **thioesters** important in these carrying functions (pp. 503, 569)?
- The fatty acid synthase complex represents the ultimate in substrate channeling. Review what is meant by **substrate channeling** and be sure you understand its advantages for cell function (pp. 540–541).
- When a cell or organism has excess fuel, the excess is converted to triacylglycerols for storage. Why are triacylglycerols such good stores of energy (pp. 366–367)?

Topics for Discussion

Answering each of the following questions, especially in the context of a study group discussion, should help you understand the important points of this chapter.

Biosynthesis of Fatty Acids and Eicosanoids

Malonyl-CoA Is Formed from Acetyl-CoA and Bicarbonate

1. What are the three functional regions of the acetyl-CoA carboxylase enzyme?

Fatty Acids Are Synthesized by a Repeating Reaction Sequence

2. From this preview, how are fatty acid oxidation and biosynthesis similar and different?

The Fatty Acid Synthase Complex Has Seven Different Active Sites

3. What is the role of the acyl carrier protein?

Fatty Acid Synthase Receives the Acetyl and Malonyl Groups

4. What reactions must take place before the condensation reaction occurs?

5. *Cell Map:* Fatty acid synthesis takes place in the cytosol and starts with acetyl-CoA. Indicate the subsequent steps in the pathway by drawing the appropriate structures on the schematic diagram of the enzyme complex. Next to each reaction, indicate the reaction type (e.g., condensation, reduction, etc.).

Step ① Condensation

6. What is the source of energy that drives the condensation reaction? To which reaction in gluconeogenesis is this step similar?

7. Is any CO_2 actually "fixed" in this process?

8. What is the product of the condensation reaction?

Step ② Reduction of the Carbonyl Group

9. What is the electron donor in the reduction of both the carbonyl group and the double bond?

Step ③ Dehydration

10. From where are the elements of water removed in the dehydration step?

Step ④ Reduction of the Double Bond

11. What is the final product after one pass through the fatty acid synthase complex?

The Fatty Acid Synthase Reactions Are Repeated to Form Palmitate

12. How many carbons are in the acyl chain after one pass through the fatty acid synthase complex? After two passes? After seven passes?

13. For which processes are ATP and NADPH required?

The Fatty Acid Synthase of Some Organisms Is Composed of Multifunctional Proteins

14. Note the importance of channeling of substrate/product in this extremely complex reaction.

Fatty Acid Synthesis Occurs in the Cytosol of Many Organisms but in the Chloroplasts of Plants

15. Why is the location of fatty acid synthesis (and other anabolic processes) in the cytosol "logical" for animal cells but not for plant cells?

Acetate Is Shuttled out of Mitochondria as Citrate

16. What are the sources of acetyl-CoA for fatty acid synthesis?

17. In addition to transporting acetyl-CoA out of mitochondria, the shuttle systems described in this section "feed" a number of metabolic pathways in both the cytosol and the mitochondrial matrix. How many alternative pathways can you list for the various intermediates shown in Figure 21–11?

18. *Cell Map:* Fatty acid synthesis is fueled by intermediates from the citric acid cycle. Show how the citrate, malate–α-ketoglutarate, and pyruvate shuttle systems are involved in providing necessary components.

Fatty Acid Biosynthesis Is Tightly Regulated

19. What are the allosteric and covalent mechanisms controlling the rate-limiting step of fatty acid synthesis in vertebrates? Which hormones are involved?

20. What factors activate acetyl-CoA carboxylase in plants?

21. Why don't bacteria store triacylglycerols as energy reserves?

22. How is the simultaneous functioning of fatty acid oxidation and synthesis avoided?

Long-Chain Saturated Fatty Acids Are Synthesized from Palmitate

23. Where in the cell does elongation of fatty acids occur?

24. What is the acyl carrier in the elongation of palmitate?

Some Fatty Acids Are Desaturated

25. What type of reaction is the fatty acyl–CoA desaturase reaction?

26. Linoleate and α-linolenate are considered "essential fatty acids." Why?

Mixed-Function Oxidases, Oxygenases, and Cytochrome P-450

27. What is accomplished by the hydroxylation of substances such as drugs and xenobiotics by cytochrome P-450?

Eicosanoids Are Formed from 20-Carbon Polyunsaturated Fatty Acids

28. How do aspirin and ibuprofen alleviate pain and/or fever, and possibly prevent stroke? (Hint: see also pp. 378–379 in the textbook.)

Cyclooxygenase Isozymes and the Search for Better Aspirin: Relief Is in (the Active) Site

29. Why does aspirin use cause stomach irritation? How would a new NSAID avoid this side effect?

Biosynthesis of Triacylglycerols

Triacylglycerols and Glycerophospholipids Are Synthesized from the Same Precursors

30. What are the precursors of triacylglycerols?

Triacylglycerol Biosynthesis in Animals Is Regulated by Hormones

31. Why do people with severe diabetes nellitus lose weight?

Biosynthesis of Membrane Phospholipids

32. What steps are required for assembly of phospholipids? Where in the cell does this assembly occur?

There Are Two Strategies for Attaching Head Groups

33. What does CDP contribute to the glycerophospholipid product in both strategies of head-group attachment?

Phospholipid Synthesis in *E. coli* Employs CDP-Diacylglycerol

34. When in the synthetic process does head-group modification occur?

Eukaryotes Synthesize Anionic Phospholipids from CDP-Diacylglycerol

35. Remind yourself of the general function of kinases.

Eukaryotic Pathways to Phosphatidylserine, Phosphatidylethanolamine, and Phosphatidylcholine Are Interrelated

36. Phosphatidylcholine is a major membrane lipid; note that in mammals there are two routes to its production.

Plasmalogen Synthesis Requires Formation of an Ether-Linked Fatty Alcohol

37. What type of enzyme is responsible for the formation of the double bond in plasmalogens?

Sphingolipid and Glycerophospholipid Synthesis Share Precursors and Some Mechanisms

38. How do the sugar moieties of cerebrosides and gangliosides enter these molecules during their biosynthesis?

Polar Lipids Are Targeted to Specific Cell Membranes

39. Why must membrane lipids be specially transported from their site of synthesis?

Biosynthesis of Cholesterol, Steroids, and Isoprenoids

40. What are the functions of cholesterol in humans?

Cholesterol Is Made from Acetyl-CoA in Four Stages

Stage ① Synthesis of Mevalonate from Acetate

41. Which is the committed step of cholesterol synthesis? What enzyme catalyzes this step?

Stage ② Conversion of Mevalonate to Two Activated Isoprenes

42. How are the two activated isoprene species structurally related?

Stage ③ Condensation of Six Activated Isoprene Units to Form Squalene

43. What drives these reactions? Are they irreversible?

Stage ④ Conversion of Squalene to the Four-Ring Steroid Nucleus

44. Note the participation once again of a mixed-function oxidase.

Cholesterol Has Several Fates

45. For what reason(s) is cholesterol converted into cholesteryl esters?

Cholesterol and Other Lipids Are Carried on Plasma Lipoproteins

46. Why is a *mono*layer of phospholipids found on the surface of lipoproteins, rather than the *bi*layer occurring in membranes?

47. Which are the most dense lipoproteins? The least dense? How does this relate to their lipid:protein ratios? (It may help to remember that in chicken soup, the fat floats on the surface but the meat (mostly protein) sinks.)

48. Categorize the various lipoproteins according to: site of origin; lipids transported within them; apolipoproteins contained; lipid:protein ratio; function; and site of degradation or uptake.

Apolipoprotein E Alleles Predict Incidence of Alzheimer's Disease

49. What is the known function of apol?

Cholesteryl Esters Enter Cells by Receptor-Mediated Endocytosis

50. What part of the LDL particle is recognized by the LDL receptor?

Cholesterol Biosynthesis Is Regulated by Several Factors

51. What are the allosteric, hormonal, and covalent strategies for regulation of cholesterol biosynthesis? At what stage in the process of cholesterol synthesis do lovastatin and compactin exert their inhibition?

52. Based on the function of HDLs, why would you expect a negative correlation between HDL levels and arterial disease?

Steroid Hormones Are Formed by Side Chain Cleavage and Oxidation of Cholesterol

53. What are the general functions of mineralocorticoids, glucocorticoids, and the sex hormones?

Intermediates in Cholesterol Biosynthesis Have Many Alternative Fates

54. Be aware of how the use of a single starting material to make many different end products is an energy-saving strategy in cells.

55. What does prenylation of a protein accomplish?

SELF-TEST

Do You Know the Terms?

ACROSS

1. Class of hormones derived from cholesterol that affect the ability of the kidney to reabsorb Na^+ and Cl^- from blood.
4. Cytochrome in membranes of the endoplasmic reticulum (*not* in mitochondrial membranes); involved in hydroxylations that render substances more water soluble and excretable.
5. Acetylsalicylate acts through inhibition of enzymes that synthesize this class of compounds.
7. General name for enzyme that oxidizes compounds by direct incorporation of oxygen atoms.
10. The rate-limiting step of fatty acid biosynthesis is catalyzed by acetyl-CoA ______.
11. Protein containing phosphopantetheine, similar to coenzyme A; carries fatty acyl groups attached by a thioester bond. (abbr.)
13. General name for enzyme that oxidizes substrates by transferring electrons to oxygen, producing H_2O or H_2O_2.
15. Enzyme responsible for the cholesterol and lipid scavenging ability of high-density lipoproteins. (abbr.)
17. Regulated by phosphorylation triggered by glucagon and ephinephrine. (2 words)
18. Starting molecule for triacylglycerol synthesis; formed from either dihydroxyacetone phosphate or glycerol.
22. Competitive inhibitor of key enzyme in cholesterol synthesis; shows promise for treatment of hypercholesterolemia.
24. Receptor-mediated ______: mechanism of uptake of lipoproteins by peripheral (nonhepatic) tissues.
26. Class of hormones derived from cholesterol that increase the synthesis of PEP carboxykinase, fructose 1,6-bisphosphate, glucose 6-phosphate, and glycogen synthase.
27. Lipoproteins synthesized in the liver that transport primarily triacylglycerols. (abbr.)
28. Complex of multifunctional proteins involved in lipid synthesis; channels intermediates between its different active sites. (3 words)
29. ______ reductase catalyzes the committed step in cholesterol biosynthesis. (abbr.)

DOWN

2. Compound important in synthesis of regulatory lipids; formed only from essential fatty acids in the diet.
3. Precursor of cholesterol; formed from acetyl-CoA.
4. Isoprenyl derivatives, which appear to serve as signals targeting proteins to cell membranes, are attached to proteins through a ______ reaction.
5. Another name for diacylglycerol 3-phosphate.
6. Process that increases efficiency of multiple enzyme-catalyzed reactions. (2 words)
8. Estrogen, testosterone, and cholesterol are examples of this class of enzymes.
10. Attachment of this compound in final step of glycerophospholipid synthesis forms an activated compound to which the polar head group is attached. (abbr.)
11. Enzyme that converts cholesterol to cholesteryl esters in the liver. (abbr.)

12. Primary carrier of dietary triacylglycerols in the blood; synthesized in epithelial cells of the small intestine.
14. Compounds derived from arachidonic acid; include several classes of "local" hormones released by most body cells that cause smooth muscle contraction, vasoconstriction, localized tissue swelling, and platelet aggregation.
16. Lipoproteins rich in cholesterol and cholesteryl esters; synthesized in the liver.
19. Macromolecular complexes that associate with lipids, making them water soluble for transport in the blood.
20. Lipoproteins that transport lipids and cholesterol to the liver, where they can be excreted.
21. Product of the committed step in cholesterol biosynthesis.
23. Compound formed by condensation of six isoprene units.
25. Cofactor for key enzymes of both fatty acid synthesis and gluconeogenesis; involved in one-carbon transfers.

Do You Know the Facts?

1. A sample of malonyl-CoA synthesized from radioactive (^{14}C-labeled) HCO_3^- and unlabeled acetyl-CoA is used in fatty acid synthesis. In which carbon(s) will the final fatty acid be labeled? (Recall that the carboxyl carbon is C-1.)

A. every carbon
B. every odd-numbered carbon
C. every even-numbered carbon
D. only the carbon farthest from C-1
E. No part of the molecule will be labeled.

2. Fatty acid synthase of mammalian cells:

A. uses a four-step sequence to lengthen a growing fatty acyl chain by one carbon.
B. contains ACP, which carries acyl groups attached through thioester linkages.
C. requires chemical energy in just one form: the reducing power of NADPH.
D. is activated by glucagon and epinephrine.
E. is activated by palmitoyl-CoA, the principal product of fatty acid synthesis.

3. List the steps of fatty acid synthesis in their correct order.

(a) The double bond is reduced to form butyryl-ACP.
(b) Condensation of the acetyl and malonyl groups produces an acetoacetyl group bound to ACP.
(c) The elements of water are removed to yield a double bond.
(d) The carbonyl group of acetoacetyl-ACP is reduced.
(e) The fatty acid synthase complex is charged with the correct acyl groups.

4. The synthesis of palmitate requires:

A. 8 acetyl-CoA.
B. 14 NADH.
C. 7 ATP.
D. A and C.
E. A, B, and C.

5. In mammalian calls, fatty acid synthesis occurs in the cytosol. The beginning substrate for this series of reactions is acetyl-CoA, which is formed in the mitochondrial matrix. How does the acetyl-CoA move from matrix to cytosol?

A. Acetyl-CoA reacts with oxaloacetate and leaves the matrix as citrate via the citrate transporter.
B. Acetyl-CoA combines with bicarbonate and is transported out of the matrix as pyruvate.
C. There is a specific transport protein for acetyl-CoA.
D. Acetyl-CoA is a nonpolar molecule and can diffuse across all membranes.
E. Both inner and outer mitochondrial membranes are freely permeable to acetyl-CoA due to the presence of transmembrane pores or channels.

6. How are fatty acid oxidation and synthesis controlled so that futile cycling does not occur?

A. They occur in different cellular compartments.
B. They employ different electron carriers.
C. The product of the first oxidation reaction inhibits the rate-limiting step of biosynthesis.
D. A and B
E. A, B, and C

7. Which of the following reactions is catalyzed by a mixed-function oxidase?

A. $AH + BH_2 + O_2 \longrightarrow A\text{–}OH + B + H_2O$
B. $RH + O_2 + 2NADPH \longrightarrow R\text{–}OH + H_2O + 2NADP^+$
C. $R^1\text{–}O\text{–}CH_2\text{–}CH_2\text{–}R^2 + NADH + H^+ + O_2 \longrightarrow R^1\text{–}O\text{–}CH{=}CH\text{–}R^2 + 2H_2O + NAD^+$
D. A and B
E. A, B, and C

8. Which of the following is *not* true of cholesterol synthesis?

A. Reduction of HMG-CoA to mevalonate catalyzed by HMG-CoA reductase is the committed step.
B. Acetyl-CoA is the ultimate source of all 27 carbon atoms of cholesterol.
C. It is inhibited by elevated levels of intracellular cholesterol.
D. It is hormonally inactivated by glucagon and activated by insulin.
E. It occurs in the mitochondrial matrix.

9. Cholesterol and other lipids are transported in the human body by lipoproteins. Indicate whether each of the following statements about these molecules is true or false.

____ **(a)** The more lipid in a particular lipoprotein, the denser is the lipoprotein.
____ **(b)** Lipoproteins "solubilize" hydrophobic lipids by surrounding them with a monolayer of amphipathic lipids.
____ **(c)** Endogenous (internally synthesized) triacylglycerols follow the same route in the body as exogenous triacylglycerols (obtained from the diet).
____ **(d)** Both cholesterol and cholesteryl esters are transported in lipoproteins.
____ **(e)** Chylomicrons are the largest but least dense lipoproteins, containing a high proportion of triacylglycerols.
____ **(f)** Chylomicrons carry fatty acids obtained in the diet to tissues for use or storage.
____ **(g)** Very low-density lipoproteins perform the same role as chylomicrons, but with endogenous lipids rather than dietary lipids.
____ **(h)** As triacylglycerols are removed from VLDLs, the VLDLs become chylomicrons.
____ **(i)** VLDLs, LDLs, and HDLs are progressively denser versions of the same initial molecule.
____ **(j)** Low-density lipoproteins are very high in cholesterol and cholesteryl esters.
____ **(k)** High-density lipoproteins carry the enzyme LCAT, which removes cholesterol from chylomicrons and VLDLs, and transport cholesterol back to the liver.
____ **(l)** Some of the apolipoproteins of lipoproteins act as enzymes for degradation of the lipids once they reach their target tissue.
____ **(m)** The liver is the only route through which cholesterol leaves the body (through bile salts lost in digestive processes).

10. Plasmalogens are:

A. membrane lipids.
B. synthesized from fatty acyl–CoA.
C. synthesized from the glycolytic intermediate dihydroxyacetone phosphate.
D. A and B.
E. A, B, and C.

Applying What You Know

1. Which two hormones affect the activity of acetyl-CoA carboxylase in fatty acid biosynthesis? Why does this make physiological sense?

2. What are the two possible fates of newly synthesized fatty acids in vertebrates? Under what conditions is one fate more likely than the other?

3. Why do untreated diabetics lose weight?

4. Glucagon has been shown to reduce the activity of HMG-CoA reductase. Why does this make physiological sense?

5. High levels of cholesterol in the blood have been positively correlated with the incidence of atherosclerosis. Recently, the LDL:HDL ratio has been shown to be a better indicator. Explain why this ratio might be a predictor of coronary artery obstruction.

6. List the ways in which fatty acid oxidation and fatty acid synthesis differ. (Suggestions: What are the electron carriers? What are the "activating groups"? Where in the cell do the reactions take place? What group is added or taken away?)

ANSWERS

Do You Know the Terms?

ACROSS

1. mineralocorticoids
4. P-450
5. prostaglandins
7. oxygenase
10. carboxylase
11. ACP
13. oxidase
15. LCAT
17. acetyl-CoA carboxylase
18. glycerol 3-phosphate
22. lovastatin
24. endocytosis
26. glucocorticoids
27. VLDLs
28. fatty acid synthase
29. HMG-CoA

DOWN

2. arachidonate
3. isoprene
4. prenylation
5. phosphatidate
6. substrate channeling
8. sterols
10. CDP
11. ACAT
12. chylomicrons
14. eicosanoids
16. LDLs
19. lipoproteins
20. HDLs
21. mevalonate
23. squalene
25. biotin

Do You Know the Facts?

1. E
2. B
3. (e), (b), (d), (c), (a)
4. D
5. A
6. D
7. E
8. E
9. **(a)** F; **(b)** T; **(c)** F; **(d)** T; **(e)** T; **(f)** T; **(g)** T; **(h)** F; **(i)** F; **(j)** T; **(k)** T; **(l)** T; **(m)** T
10. E

Applying What You Know

1. Acetyl-CoA carboxylase, which forms malonyl-CoA from acetyl-CoA and HCO_3^-, is inactivated by glucagon and epinephrine (through enzyme phosphorylation, which causes dissociation of the enzyme into its inactive monomeric subunits). Glucagon and epinephrine both indicate the body's need for blood glucose. It makes sense that fatty acid synthesis does not proceed when an immediately available source of energy (i.e., glucose) is needed, rather than a storage form of energy (i.e., fatty acids).

2. The two possible fates of fatty acids are storage as metabolic energy or incorporation into phospholipids destined for membranes. During rapid cell growth, membrane phospholipid synthesis is the more likely fate. Cells not undergoing growth, and having a surplus of energy, will store the fatty acids (as triacylglycerols in humans, for example).

3. The rate of triacylglycerol biosynthesis is affected by the action of several hormones, one of which is insulin. Insulin promotes the conversion of carbohydrates to triacylglycerols. People with severe diabetes, due to failure of insulin secretion or action, not only are unable to use glucose properly but also fail to synthesize fatty acids from carbohydrates or amino acids. Untreated diabetics have increased rates of fat oxidation and ketone body formation and, as a consequence, lose weight.

4. The presence of glucagon is a signal that blood glucose is low and must be diverted to those organs that require it to generate energy. The need for catabolically generated energy in general suppresses biosynthetic reactions; for example, the inhibition of HMG-CoA reductase suppresses cholesterol biosynthesis.

5. LDLs shuttle cholesterol from the liver to other sites in the body (including the coronary arteries), and can be viewed as "bad" cholesterol. They contribute to the high blood cholesterol levels associated with atherosclerotic plaque formation. HDLs, on the other hand, scavenge cholesterol from the various lipoproteins in the bloodstream, removing it from LDLs and VLDLs and transporting it back to the liver. The liver is the only route for excretion of cholesterol (via bile salts lost in digestive processes). A high LDL:HDL ratio indicates a high level of circulating cholesterol; a low LDL:HDL ratio, even if overall cholesterol in the body is high, indicates that HDL levels are sufficient to keep the blood cholesterol concentration low.

6.

	Fatty acid synthesis	**Fatty acid oxidation**
Electron carrier	NADPH	NADH
Acyl groups bound to:	ACP	CoA
Cellular location	Cytosol (in animal cells)	Mitochondrial matrix
Process involves:	Addition of malonyl groups	Removal of acetyl groups

Biosynthesis of Amino Acids, Nucleotides, and Related Molecules

STEP-BY-STEP GUIDE

Major Concepts

Prokaryotes cycle nitrogen into and out of biologically available forms.

Nitrogen fixation is carried out only by prokaryotes, both symbiotic and free-living, which convert atmospheric nitrogen (N_2) to ammonia through activity of the enzymes of the **nitrogenase complex.** The NH_4^+ produced can be oxidized to nitrite and nitrate by the bacterial process known as **nitrification.** Plants and microorganisms incorporate ammonia into amino acids, and these amino acids are used by animals as a nitrogen source. When organisms die, it is again prokaryotes that degrade the proteins to ammonia and, ultimately, to N_2.

α-Ketoglutarate and glutamate act as the entry point for nitrogen in biosynthetic reactions.

The incorporation of ammonia into these biomolecules is catalyzed by glutamine synthetase and glutamate synthase. As in amino acid degradation, glutamate and glutamine are the primary transport forms of NH_4^+ for biosynthetic reactions. **Glutamine synthetase** is a key regulatory enzyme in nitrogen metabolism. This enzyme is regulated through allosteric inhibition by the end products of glutamine metabolism, and through covalent modification: Removal of an AMP (deadenylylation) activates and addition of AMP (adenylylation) decreases its activity. Glutamine sythetase is also under additional, indirect control imposed by regulation of the enzyme **adenylyl transferase,** which catalyzes the adenylylation and deadenylylation. The activity of adenylyl transferase is modulated by a regulatory protein that is, itself, covalently modified in response to various metabolites and end products. In bacteria, transfer of an amino group from glutamine to α-ketoglutarate to form two molecules of glutamate is catalyzed by **glutamate synthase.** The net effect of these two enzymatic reactions is the synthesis of one glutamate molecule from ammonia and α-ketoglutarate.

Amino acids are synthesized from intermediates in glycolysis, the citric acid cycle, and the pentose phosphate pathway.

Ten amino acids are derived from the citric acid cycle intermediates **α-ketoglutarate** and **oxaloacetate.** Nine amino acids are derived from the glycolytic intermediates **3-phosphoglycerate, phosphoenolpyruvate,** and **pyruvate.** Glucose 6-phosphate can be diverted from glycolysis into the pentose phosphate pathway to produce **ribose 5-phosphate,** a precursor of histidine. PLP is an important cofactor in many of these pathways, and NADPH is the primary source of reducing power. Several of the amino acid synthetic pathways have a common intermediate, **phosphoribosyl pyrophosphate (PRPP),** which is also of importance in nucleotide biosynthesis.

Amino acid biosynthesis is regulated by a variety of feedback inhibition strategies.

In **concerted inhibition,** many modulators act on the enzyme catalyzing the first, usually irreversible, reaction in a sequence, and the effect of these modulators is more than additive. **Enzyme multiplicity** often occurs when there are several possible fates for the metabolic intermediates in a pathway. In this case, different isozymes are independently controlled by differing sets of modulators; consequently, the entire reaction pathway is not shut down by one biosynthetic product when other products of the same pathway are needed. In **sequential feedback inhibition** a metabolite inhibits its own formation at several different points in a pathway; these inhibitory effects *overlap*, since they usually inhibit several of the isozymes and affect different enzymatic steps in the pathway.

Amino acids are the precursors of many extremely important biomolecules, in addition to proteins.

Porphyrins, such as heme, and **bilirubin,** the degradation product of heme, are derived from glycine and succinyl-CoA. **Phosphocreatine** and **glutathione** have roles in energy generation and redox reactions, respectively. Plant products such as **lignin, tannins,** and **auxin** are derived from aromatic amino acids. Neurotransmitters are often primary or secondary amines derived from amino acids; these include **dopamine, norepinephrine, epinephrine,** and **GABA**. The vasodilator **histamine** is formed from histidine, and the polyamines **spermine** and **spermidine** used in DNA packaging are also derived from amino acids.

Nucleotides are synthesized de novo and through salvage pathways.

In **de novo purine synthesis,** AMP and GMP are synthesized from PRPP, glutamine, glycine, N^{10}-formyl H_4 folate, aspartate, and fumarate. The purine ring structure is built upon 5-phospho-β-D-ribosylamine one or a few moieties at a time. **Inosinate** (IMP) is the intermediate common to both AMP and GMP formation. IMP, AMP, and GMP all regulate the activity of key enzymes in the pathway by feedback inhibition. In **de novo pyrimidine synthesis,** CMP and UMP are synthesized from aspartate, PRPP, and carbamoyl phosphate. The pyrimidine ring structure, orotate, is constructed first and then attached to PRPP to form **orotidylate.** The pyrimidines are synthesized from this intermediate. The carbamoyl phosphate component of the pyrimidine ring is produced in the cytosol by a different enzyme from that which synthesizes carbamoyl phosphate in the mitochondrion for the urea cycle. As for purine biosynthesis, this pathway is regulated by feedback inhibition.

Nucleotide monophosphates are phosphorylated by phosphate transfer from ATP.

AMP is converted to ADP by **adenylate kinase.** ADP is then converted to ATP by oxidative phosphorylation or by glycolytic enzymes. ATP is used to convert the other nucleotide monophosphates to triphosphates by the action of **nucleoside monophosphate kinases** and **nucleoside diphosphate kinase.**

Ribonucleotides are reduced to deoxyribonucleotides by ribonucleotide reductase.

The reaction catalyzed by this enzyme involves the generation of a free radical. Both the activity and substrate specificity of the reductase are regulated. Activity is regulated by the binding of either ATP (activator) or dATP (inactivator). Substrate specificity is regulated by the binding of ATP, dATP, dTTP, or dGTP (acting as effector molecules), which induce different conformations in the enzyme. These controls all ensure a balanced pool of precursors for DNA synthesis.

Purines are degraded to uric acid and pyrimidines to urea.

Uric acid is excreted by a few organisms (notably birds), but in most organisms it is further degraded to urea or ammonia. Metabolic degradation of nucleotides produces free bases that can be recycled in **salvage pathways.** AMP is synthesized from adenine and PRPP by the action of **adenosine phosphoribosyltransferase.** GMP is synthesized from guanine and hypoxanthine by **hypoxanthine-guanine phosphoribosyltransferase.** Pyrimidines are not salvaged in significant amounts in mammals.

Because cancer cells grow more rapidly than normal cells and therefore have a greater need for nucleotides, they are vulnerable to disruptions in nucleotide synthetic processes.

Pharmaceuticals that take advantage of this fact include **azaserine, fluorouracil, methotrexate,** and **aminopterin.**

What to Review

Answering the following questions and reviewing the relevant concepts, which you have already studied, should make this chapter more understandable.

- **Aminotransferases** are important in the transfer of α-amino groups in amino acid biosynthesis; review their role in amino acid oxidation, including the importance of their cofactor **pyridoxal phosphate** (PLP) in the transfer of amino groups (pp. 628–630).
- The enzyme glutamine synthetase is central to ammonia assimilation in bacteria. Glutamine also plays an important role in mammals, that of ammonia transporter. Why must NH_4^+ be converted into glutamine for transport in the bloodstream (pp. 632)?
- Other cofactors involved in the reactions discussed in this chapter include **tetrahydrofolate** and ***S*-adenosylmethionine.** What other reactions require these cofactors (pp. 640–643)?
- Amino acids are built from precursors that are derived from the central carbohydrate metabolic pathways. Review the structures of these precur-

sors from glycolysis (Chapter 15), the citric acid cycle (Chapter 16), and the pentose phosphate pathway (Chapter 15).

- **Mechanism-based inactivators,** or **suicide inhibitors,** are coming into use as pharmaceutical agents for attacking parasites or processes that use certain pathways and intermediates at a greater rate than do the host's normal processes. How do they work? Why are they so powerful and specific in their inhibition (p. 269)?

Topics for Discussion

Answering each of the following questions, especially in the context of a study group discussion, should help you understand the important points of this chapter.

Overview of Nitrogen Metabolism

The Nitrogen Cycle Maintains a Pool of Biologically Available Nitrogen

1. What are the forms of nitrogen in the nitrogen cycle, from the most oxidized to the most reduced form?

Nitrogen Is Fixed by Enzymes of the Nitrogenase Complex

2. What organisms are involved in nitrogen fixation? From where do they get the energy to fix nitrogen? From where do they get the reducing power?

3. How is the role of ATP in the process of nitrogen fixation unusual?

4. How do the various nitrogen-fixing organisms cope with the "oxygen problem" during nitrogen fixation?

Ammonia Is Incorporated into Biomolecules through Glutamate and Glutamine

5. Why is the nitrogen required for amino acid and nucleotide synthesis usually supplied by glutamate or glutamine?

Glutamine Synthetase Is a Primary Regulatory Point in Nitrogen Metabolism

6. Why is the activity of glutamine synthetase so tightly regulated?

7. What are the various levels of regulation of this enzyme?

Several Classes of Reactions Play Special Roles in the Biosynthesis of Amino Acids and Nucleotides

8. What is the main difference between glutamine amidotransferases and glutaminase?

Biosynthesis of Amino Acids

9. Note that the terms "essential" and "nonessential" as applied to amino acids refer to the ability or inability of *mammals* to synthesize these compounds.

10. What are the six primary precursor molecules from which all 20 standard amino acids are derived?

11. From what is PRPP synthesized?

α-Ketoglutarate Gives Rise to Glutamate, Glutamine, Proline, and Arginine

12. What is the other product of the argininosuccinase reaction that produces arginine?

Serine, Glycine, and Cysteine Are Derived from 3-Phosphoglycerate

13. What is the role of PLP in the biosynthesis of these amino acids?

Three Nonessential and Six Essential Amino Acids Are Synthesized from Oxaloacetate and Pyruvate

14. Which amino acids are synthesized by simple transamination of carbohydrate metabolites?

15. What contributes the reducing power for these biosynthetic reactions?

Chorismate Is a Key Intermediate in the Synthesis of Tryptophan, Phenylalanine, and Tyrosine

16. Why is tyrosine not considered an essential amino acid?

Histidine Biosynthesis Uses Precursors of Purine Biosynthesis

17. How is the role of ATP in this pathway unusual?

Amino Acid Biosynthesis Is under Allosteric Regulation

18. What three types of regulation are used in coordinating amino acid biosynthesis? Under what conditions is each type of regulation more useful than the other two?

Molecules Derived from Amino Acids

Glycine Is a Precursor of Porphyrins

19. What structural characteristic of porphyrin-type molecules causes them to react strongly to ultraviolet light?

Biochemistry of Kings and Vampires

20. Why do you think that α amino β ketoadipate does not accumulate in individuals with acute intermittent porphyria?

Degradation of Heme Yields Bile Pigments

21. Why is bilirubin sufficiently hydrophobic to require transport by serum albumin, yet sufficiently hydrophilic to be water soluble?

Amino Acids Are Required for the Biosynthesis of Creatine and Glutathione

22. What are the functions of phosphocreatine and glutathione?

D-Amino Acids Are Found Primarily in Bacteria

23. What is the enzyme target of the antibacterial agent cycloserine?

Aromatic Amino Acids Are Precursors of Many Plant Substances

24. What are the functions of the plant compounds lignin and auxin?

Amino Acids Are Converted to Biological Amines by Decarboxylation

25. What do all the amino acid decarboxylases have in common?

Curing African Sleeping Sickness with a Biochemical Trojan Horse

26. Why does DFMO cause no harm to human patients, yet kill the trypanosome parasites?

Arginine Is the Precursor for Biological Synthesis of Nitric Oxide

27. What category of hormones does nitric oxide represent?

28. What intracellular signal turns on the production of nitric oxide?

Biosynthesis and Degradation of Nucleotides

29. What are the roles of nucleotides? Do these functions provide clues as to why their synthesis includes both de novo and salvage pathways?

De Novo Purine Nucleotide Synthesis Begins with PRPP

30. Which molecules contribute the nitrogen atoms in purine synthesis? What is the source of energy? From where are the carbons derived?

Purine Nucleotide Biosynthesis Is Regulated by Feedback Inhibition

31. Both sequential and concerted feedback inhibition regulate purine biosynthesis. Is enzyme multiplicity a control mechanism here as well?

Pyrimidine Nucleotides Are Made from Aspartate, PRPP, and Carbamoyl Phosphate

32. What are the general similarities and the differences in the de novo synthetic schemes of purines and pyrimidines?

Pyrimidine Nucleotide Biosynthesis Is Regulated by Feedback Inhibition

33. Why is it logical that ATP prevents the inhibition of aspartate transcarbamoylase by CTP?

Nucleoside Monophosphates Are Converted to Nucleoside Triphosphates

34. Why is the relative nonspecificity of nucleoside monophosphate kinases and of nucleoside diphosphate kinase so important in cells?

Ribonucleotides Are the Precursors of Deoxyribonucleotides

35. What are the sources of reducing power for the reduction of the D-ribose moiety of ribonucleotides?

36. How many active sites are there in a functional dimer of ribonucleotide reductase? How many regulatory sites?

37. How do ATP, low [dATP], and high [dATP] each affect the activity of ribonucleotide reductase?

38. Why is it crucial that deoxyribonucleotide concentrations are balanced within the cell?

Thymidylate Is Derived from dCDP and dUMP

39. Why must the concentration of dUTP be kept low?

Degradation of Purines and Pyrimidines Produces Uric Acid and Urea, Respectively

40. In humans, is the excretion of uric acid a large fraction of the total excreted nitrogen? (Hint: See also p. 634.)

Purine and Pyrimidine Bases Are Recycled by Salvage Pathways

41. Calculate the input of energy necessary to synthesize nucleotide bases; this will give you an idea about the need for salvage pathways for these bases.

Overproduction of Uric Acid Causes Gout

42. How are gout and Lesch-Nyhan syndrome similar, and how are they different?

Many Chemotherapeutic Agents Target Enzymes in the Nucleotide Biosynthetic Pathways

43. What is the mechanism for azaserine's inhibition of nucleotide biosynthetic pathways?

44. Which enzymes are inactivated by fluorouracil and methotrexate?

45. *Cell Map:* We recommend that you don't include all the amino acid and nucleotide synthesis and degradation pathways on your map since it would become incomprehensible. Instead, indicate the glycolytic pentose phosphate pathway, amino acid breakdown, and/or citric acid cycle intermediates that are used to generate the various amino acids and nucleotides.

SELF-TEST

Do You Know the Terms?

ACROSS

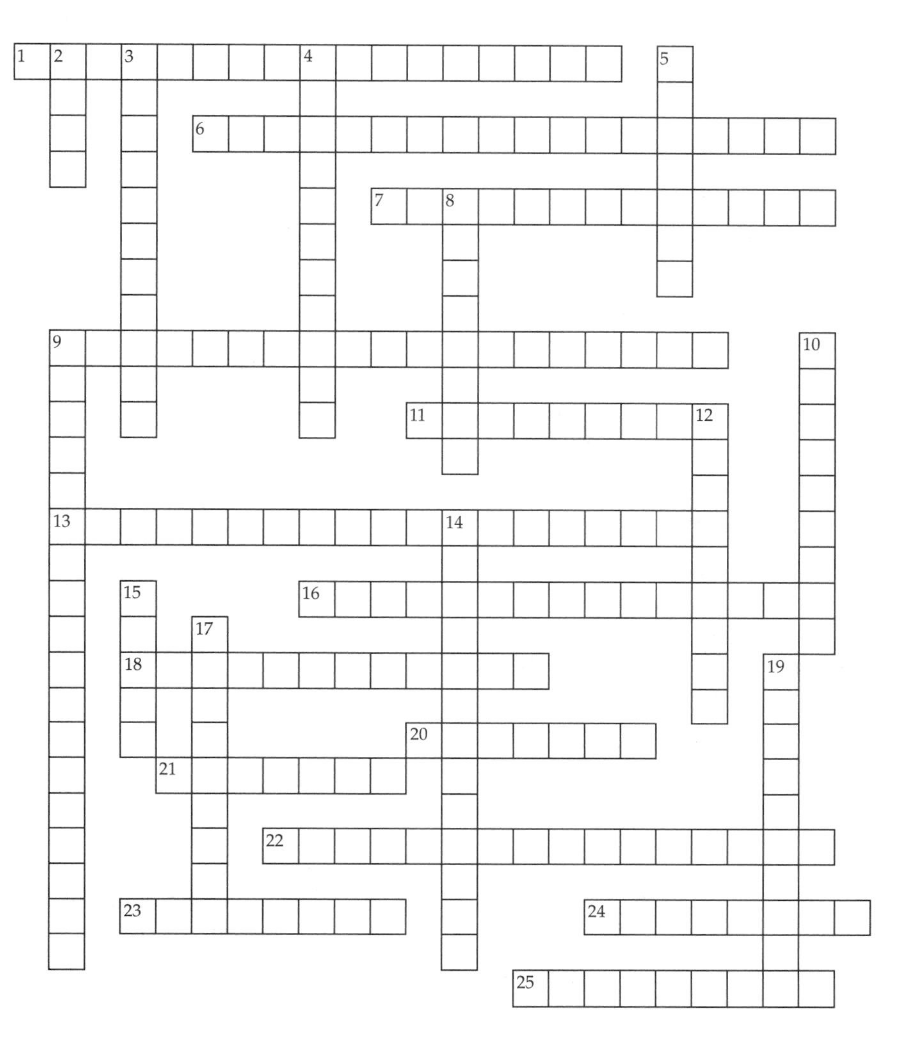

1. Cys, Ser, and Gly are all derivatives of this common intermediate.
6. Type of inhibition in which each enzyme of an enzymatic series is inhibited by the product of its reaction. (2 words)
7. Series of reactions by which nitrogen in the atmosphere and in the proteins of organisms is circulated, salvaged, and reused. (2 words)
9. Product of the committed step in pyrimidine biosynthesis.
11. Common metabolic intermediate in the synthesis of the purines, AMP, GMP, and IMP.
13. Enzyme with 12 subunits that are regulated allosterically and covalently; produces an intermediate for at least six separate metabolic pathways. (2 words)
16. Process in which electrons from oxidative phosphorylation are transferred to NO_3^-.
18. Regulatory strategy known as enzyme ______ uses different isozymes, which can be regulated independently, to catalyze the same reaction.
20. Class of enzymes catalyzing synthesis of di- and triphosphates from nucleoside monophosphates.
21. Recycling is something biological systems have been doing for millennia; an example is the ______ pathways for purine and pyrimidine bases.
22. Type of enzyme involved in biosynthesis of most amino acids.
23. ______ transferase "senses" [ATP] in cells and activates glutamine synthetase by deadenylylation when [ATP] is high.
24. Common precursor for the biosynthesis of Ile, Val, Leu, and Ala.
25. ______ inhibition: synergistic (more than additive) effects of feedback, covalent, and allosteric inhibition of an enzyme.

DOWN

2. Important intermediate in biosynthesis of all nucleotides; generated from glucose via the pentose phosphate pathway. (abbr.)
3. In pyrimidine biosynthesis, compound formed from attachment of the ribose 5-phosphate moiety to orotate.
4. Important reducing agent when present as a tripeptide; when oxidized, consists of two tripeptides linked by a disulfide bond.
5. *Fixation* of atmospheric nitrogen is a complex process that ______ N_2 to NH_3 or NH_4^+.
8. A number of important signaling molecules are derived from the decarboxylation of amino acids; for example, this amino acid is decarboxylated to yield dopamine, norepinephrine, and epinephrine.
9. Supramolecular structure containing dinitrogenase reductase and dinitrogenase. (2 words)
10. More efficient nitrogen carrier than glutamate.
12. In mammals, most amino acids are derived from a few common precursor molecules using relatively simple enzymatic pathways; the remainder are ______ amino acids that cannot be synthesized endogenously.
14. Process of converting "fixed" nitrogen to NO_2^- and NO_3^-.
15. Aspartate can be used to generate these five amino acids. (one-letter abbrs.)
17. The bacterial enzyme ______ synthase incorporates reduced nitrogen into proteins and nucleotides.
19. Phe, Trp, and Tyr are all derived from this common precursor.

Do You Know the Facts?

1. Which of the following correctly describes the nitrogen cycle?

A. Fixation of atmospheric nitrogen by nitrogen-fixing bacteria yields bioavailable nitrate.
B. Nitrate is reduced to ammonia in a process known as denitrification.
C. Biological nitrogen fixation is carried out by a complex of proteins called the nitrogenase complex.
D. A and B
E. A, B, and C

2. Which of the following is *not* true of nitrogen fixation by the nitrogenase complex?

A. This enzyme complex is inactivated by exposure to oxygen.
B. ATP binding causes a change in the conformation of the reductase moiety.
C. The primary role of ATP is to drive nitrogen fixation through the hydrolysis of PP_i.
D. Nitrogen fixation occurs only in prokaryotes.
E. The final electron acceptor in this process is N_2.

3. Which of the following is the immediate product of nitrogen fixation (i.e., the product of the reaction catalyzed by the nitrogenase complex)?

A. N_2
B. NO_3^-
C. NH_4^+
D. urea
E. amino acids

4. Which cofactor is essential to all transamination reactions?

A. PRPP
B. CoA
C. ATP
D. PLP
E. adoMet (*S*-adenosylmethionine)

5. Synthesis of the amino acid glutamine, catalyzed by glutamine synthetase, has been studied extensively in bacteria. Indicate whether each of the following statements about glutamine synthetase and/or the reaction it catalyzes is true or false (in bacteria, unless otherwise noted).

____ **(a)** The enzyme catalyzes the reaction: glutamate + NH_4^+ + ATP → glutamine + ADP + P_i + H^+.
____ **(b)** In mammals, this is the main pathway for converting toxic free ammonia into nontoxic glutamine for transport in the blood.
____ **(c)** Each of the enzyme's 12 subunits is subject to allosteric regulation.
____ **(d)** The enzyme is covalently inactivated by the addition of a phosphate group.
____ **(e)** The enzyme is activated by the covalent attachment of AMP.
____ **(f)** The complex of adenylyl transferase and P_{II}-UMP catalyzes deadenylylation of the enzyme.
____ **(g)** ATP is a reactant in the reaction.
____ **(h)** ATP is a regulator in the reaction.
____ **(i)** Uridylylation of the enzyme is stimulated by α-ketoglutarate.
____ **(j)** High concentrations of glutamine itself, and of end products of glutamine metabolism, decrease the enzyme's activity.
____ **(k)** The enzyme is covalently activated by uridylylation.

6. Which of the following does *not* provide a carbon skeleton for the synthesis of amino acids?

A. succinate
B. α-ketoglutarate
C. pyruvate
D. oxaloacetate
E. ribose 5-phosphate

7. Shown below are the structures of three metabolic intermediates that can be used in amino acid synthesis. Which is correctly paired with its amino acid end product(s)?

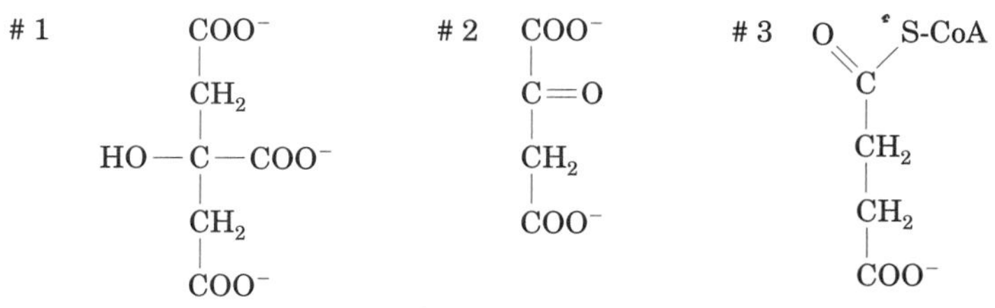

1

$$\begin{array}{c} COO^- \\ | \\ CH_2 \\ | \\ HO-C-COO^- \\ | \\ CH_2 \\ | \\ COO^- \end{array}$$

2

$$\begin{array}{c} COO^- \\ | \\ C=O \\ | \\ CH_2 \\ | \\ COO^- \end{array}$$

3

$$\begin{array}{c} O \quad S\text{-}CoA \\ \backslash\backslash \;\; / \\ C \\ | \\ CH_2 \\ | \\ CH_2 \\ | \\ COO^- \end{array}$$

A. #1; serine, glycine, cysteine
B. #2; alanine, valine, leucine
C. #3; glutamate, glutamine, proline
D. #1; histidine
E. #2; methionine, threonine, lysine

8. What does the term "essential" mean in terms of amino acids in the human diet?
A. Necessary for all protein synthesis.
B. Only available in animal protein.
C. Cannot be synthesized by humans.
D. Cannot be coded for by DNA.
E. Cannot be degraded in the liver.

9. Which of the following is *not* a physiological role of nucleotides?
A. allosteric regulators
B. intermediates for biosynthetic processes
C. components of many proteins
D. components of the coenzymes NAD, FAD, and CoA
E. intracellular signaling molecules

10. How is the type of feedback inhibition known as enzyme multiplicity correctly described?
A. Different isozymes in a pathway are independently controlled by different modulators; this prevents one biosynthetic product from shutting down key steps in a pathway when other products of the same pathway are needed.
B. A metabolite inhibits its own formation at several different points with multiple, overlapping inhibition in the synthetic pathway.
C. Metabolite synthesis is inhibited at one enzymatic point by more than one end product of its own metabolism; effects of the different inhibitors on the enzyme are more than additive, and all the inhibitors together more or less shut down the enzyme.
D. Enzyme multiplicity refers to an enzyme cascade that results in the covalent modification of a key regulatory enzyme.
E. All of the above are true.

11. Consider the following series of enzymatic reactions involved in the synthesis of two newly discovered bacterial amino acids:

Amino acid 1, vivekine, is formed by oxygenation of a compound called previvekine in a reaction catalyzed by previvekine oxygenase; this enzyme is inhibited by high concentrations of the product vivekine.

Amino acid 2, avanic acid, is formed by carboxylation of vivekine, catalyzed by vivekine carboxylase; this enzyme is inhibited by high concentrations of avanic acid, which also inhibits previvekine oxygenase.

Previvekine is synthesized from a compound called proctorol in a reaction catalyzed by proctorol reductase; this enzyme is inhibited by vivekine and by avanic acid; when both of these products are present, the inhibitory effect is more than additive.

Regulation of this reaction sequence is an example of:

A. sequential feedback inhibition.
B. enzyme multiplicity.
C. concerted feedback inhibition.
D. A and C
E. A, B, and C

12. Indicate (yes/no) whether each of the following describes a way in which orotate is used in pyrimidine biosynthesis.

——— **(a)** It condenses with PRPP to form orotidylate and then UMP.
——— **(b)** It contributes a portion of its carbon atoms to the formation of PRPP.
——— **(c)** It is constructed on PRPP through 10 separate reactions.
——— **(d)** It contributes a portion of its carbon atoms to the formation of the pyrimidine base via tetrahydrofolate.
——— **(e)** An amino group is transferred to the C-1 of orotate to form CTP.
——— **(f)** It is formed from cytosolic carbamoyl phosphate.
——— **(g)** It is formed from mitochondrial carbamoyl phosphate.

13. Indicate (yes/no) whether each of the following is a way in which inosinate is used in purine biosynthesis.

——— **(a)** It condenses with PRPP to form IMP.
——— **(b)** An amino group is transferred to inosinate to form AMP.
——— **(c)** It is deaminated to form GMP.
——— **(d)** It contributes a portion of its carbon atoms to the formation of the purine base via tetrahydrofolate.
——— **(e)** It is a precursor of PRPP.
——— **(f)** It is constructed on PRPP through 10 separate reactions.
——— **(g)** It is formed from cytosolic carbamoyl phosphate.

14. In nucleotide metabolism, all of the following are true *except:*

A. The committed step in purine biosynthesis is the transfer of an amino group to PRPP.
B. Both purine and pyrimidine biosynthesis are regulated by end-product inhibition.
C. Nucleotides can be synthesized in a single reaction via salvage pathways.
D. De novo pyrimidine synthesis begins with a molecule of PRPP.
E. Orotidylate is the common precursor in the biosynthesis of pyrimidines, and inosinate is the common precursor in the biosynthesis of the purines ATP and GTP.

15. Which of the following describes the activity and regulation of ribonucleotide reductase and/or its importance to the cell?

A. Both its activity and its substrate specificity are regulated by the binding of effector molecules.
B. ATP increases the overall activity of the enzyme.
C. Control of the enzyme's activity ensures a balanced pool of precursors for DNA synthesis.
D. Balanced pools of deoxyribonucleotides are necessary in DNA synthesis, given the complementary base-pairing of nucleotides in double-stranded DNA.
E. All of the above are true.

Applying What You Know

1. The hereditary condition called orotic aciduria has symptoms of retarded growth and severe anemia, and high levels of orotate (orotic acid) are excreted. In this condition, either or both of the enzymes orotate phosphoribosyltransferase and orotidylate carboxylase are absent. When patients with this disease are fed uridine or cytidine, there is a significant improvement in the anemia and a decrease in the production of orotate. How can these effects be explained?

ANSWERS

Do You Know the Terms?

ACROSS

1. 3-phosphoglycerate
6. sequential feedback
7. nitrogen cycle
9. *N*-carbamoylaspartate
11. inosinate
13. glutamine synthetase
16. denitrification
18. multiplicity
20. kinases
21. salvage
22. aminotransferase
23. adenylyl
24. pyruvate
25. concerted

DOWN

2. PRPP
3. orotidylate
4. glutathione
5. reduces
8. tyrosine
9. nitrogenase complex
10. glutamine
12. essential
14. nitrification
15. KIMNT (or a variation of this)
17. glutamate
19. chorismate

Do You Know the Facts?

1. C
2. C
3. C
4. D
5. **(a)** T; **(b)** T; **(c)** T; **(d)** F; **(e)** F; **(f)** T; **(g)** T; **(h)** T; **(i)** F; **(j)** T; **(k)** F
6. A
7. E
8. C
9. C
10. A
11. D
12. **(a)** Y; **(b)** N; **(c)** N; **(d)** N; **(e)** Y; **(f)** Y; **(g)** N
13. **(a)** N; **(b)** Y; **(c)** N; **(d)** N; **(e)** N; **(f)** Y; **(g)** N
14. D
15. E

Applying What You Know

1. Orotate phosphoribosyltransferase catalyzes the conversion of orotate to orotidylate and orotidylate carboxylase converts orotidylate to uridylate (UMP), a precursor in the biosynthesis of both UTP and CTP. Uridine (UMP) and cytidine (CMP) circumvent these enzymatic steps and can be phosphorylated to provide the necessary nucleotides, UTP and CTP, for DNA synthesis in rapidly growing red blood cells. This would improve the anemia. A logical explanation for the decrease in orotate excreted might seem to be that the CTP produced could act as a feedback inhibitor of the enzyme aspartate transcarbamoylase, so that orotate would not be synthesized in high concentrations. In *animals*, however, ATCase is not the regulatory enzyme, as it is in bacteria. Pyrimidine synthesis is controlled at the level of carbamoyl phosphate synthetase II, which is inhibited by UTP, and this is the reason for the decrease in orotic acid excretion in the treated patients.

chapter

23 Integration and Hormonal Regulation of Mammalian Metabolism

STEP-BY-STEP GUIDE

Major Concepts

Mammalian tissues are specialized for different metabolic functions.

Although the anabolic and catabolic pathways that are critical to the function of organisms occur within individual cells, many cells are specialized for particular functions. Cells may associate into tissues that specialize in different aspects of cellular metabolism.

The liver monitors and maintains the blood levels of key metabolites, specifically sugars, amino acids, and fatty acids.

Glucose 6-phosphate has a central role in five major metabolic pathways: (1) Only liver cells contain the enzyme that can dephosphorylate glucose 6-phosphate to glucose, which can be released into the blood for use by other tissues; alternatively, (2) glucose 6-phosphate is converted into glycogen for storage. (3) In a pinch, glucose 6-phosphate can be oxidized to produce ATP, although liver cells prefer to use fatty acids. (4) Excess glucose 6-phosphate is used to make acetyl-CoA for the production of triacylglycerols and lipids. (5) Glucose 6-phosphate can be shunted into the pentose phosphate pathway, producing NADPH and/or ribose 5-phosphate. **Amino acids** can be (1) used as synthetic precursors for proteins, nucleotides, and hormones in the liver itself, or (2) exported in the blood to other tissues; alternatively, (3) they are degraded to acetyl-CoA and citric acid cycle intermediates. (4) Alanine can also be deaminated to pyruvate and converted to glucose for export in the blood. **Fatty acids** from lipids provide (1) a major source of acetyl-CoA for energy production in liver, (2) a source of ketone bodies for energy production in extrahepatic tissues, (3) precursors for liver lipid synthesis, and (4) precursors for cholesterol biosynthesis. They can also be exported (5) to adipose tissue for storage or (6) for use as an energy source by muscle.

Adipose tissue stores and releases fatty acids.

This tissue is specialized for the uptake of lipids from the blood and for lipid storage in large fat droplets within cells. Under the appropriate conditions, stored triacylglycerols can be hydrolyzed by lipases and their components released into the blood.

Muscle tissue contains contractile proteins that use ATP to do mechanical work.

Glucose, fatty acids, and ketone bodies can all be converted to acetyl-CoA, which is completely oxidized to CO_2 via the citric acid cycle. Glucose from stored glycogen can be rapidly converted to lactic acid, producing ATP by anaerobic fermentation. The high phosphate transfer potential of stored phosphocreatine is also used to generate ATP from ADP.

Brain tissue is a specialized signal transduction organ.

It receives information from body systems as well as from external sources, integrates this information, and then releases hormones that regulate the metabolic demands of different tissues. Brain cells rely primarily upon glucose metabolism to meet the high energy demands of nerve cells.

Blood cells connect the various body tissues by relaying information from the brain (via hormones) and by transporting metabolites and oxygen.

Signals from the brain cause cells of the neuroendocrine system to release hormones into the circulatory system. The circulatory system carries hormones to peripheral tissues where they regulate metabolic pathways.

The neuroendocrine system has a hierarchical structure.

The **hypothalamus** is the interface between the informational input from the brain and the hormonal output of the pituitary. The hypothalamus secretes releasing factors into a local network of blood vessels; these factors bind to receptors on the **anterior pituitary** causing them to release hormones into the general circulation. The **posterior pituitary** contains the endings of hypothalamic neurons, which release hormones directly into the circulatory system.

Hormone molecules are often grouped into the following major categories: peptide hormones, amine hormones (also called biogenic amines), steroid hormones, thyroid hormones, and eicosanoids.

Peptide hormones have 3 to 200 amino acid residues and interact with specific target cells via membrane-bound receptors. **Amine hormones** that affect metabolism are primarily the catecholamines epinephrine, norepinephrine, and dopamine. They also interact with specific target cells via membrane receptors. **Steroid hormones** are lipid soluble and therefore do not require membrane-bound receptors to influence metabolic processes. Nevertheless, they do bind to receptors inside their target cells. **Eicosanoids** are derivatives of arachidonic acid and, like steroids, are lipid soluble. These "local" hormones are not transported by the blood but are released into the interstitial fluid, affecting cells near the site of release.

Metabolic processes are tightly regulated by the concerted actions of hormones.

For example, the amine hormone **epinephrine** causes a number of physiological and metabolic effects that result in increased mobilization of glucose (i.e., it promotes glycogen breakdown) and a decrease in enzyme activity in fuel storage pathways. **Glucagon,** a peptide hormone, also increases blood glucose levels. Like epinephrine, it promotes the breakdown of glycogen but it also increases gluconeogenesis in the liver. The net effect is that more glucose is exported to the blood. The peptide hormone insulin opposes the effects of both epinephrine and glucagon. **Insulin** stimulates the uptake of glucose from the blood and stimulates glycogen and fatty acid synthesis. **Diabetes mellitus,** a disease caused by a deficiency in the secretion or action of insulin, causes numerous metabolic changes that can be life-threatening, illustrating the importance of balanced control of carbohydrate metabolism.

Regulation of body mass involves multiple signaling pathways.

Control of food consumption represents regulation at the behavioral level. Leptin, a peptide hormone, is a feedback signal produced by adipose cells that alters feeding behavior by acting on the brain to reduce food intake. Insulin, in addition to its effects on enzyme activity, also acts on the brain to decrease food consumption. Conversely, neuropeptide Y (NPY) is a signal released during starvation that stimulates feeding behavior.

Regulation of food intake involves a complex cascade of regulatory events.

For example, binding of leptin to receptors in the hypothalamus causes the phosphorylation of proteins that regulate transcription of genes involved in metabolism. An increase in the production of the mitochondrial uncoupling protein in adipocytes leads to thermogenesis and the expenditure of energy. In addition, many of leptin's effects can be attributed to an inhibition of the release of NPY. Leptin also stimulates activity in the sympathetic nervous system, increasing blood pressure and heart rate, which results in the use of metabolic energy. Finally leptin appears to play a role in regulating the release of sex hormones, thereby coordinating reproductive behaviors with the levels of stored fat.

What to Review

Answering the following questions and reviewing the relevant concepts, which you have already studied, should make this chapter more understandable.

- All of the following metabolic pathways must be integrated and regulated in cells in order to keep the supplies of metabolites balanced with cellular demand. You should now be familiar with all of these pathways; an overview or summary of each can be found in the figure indicated.
  - Glycolysis (Fig. 15–2)
  - Fates of pyruvate (Fig. 15–3)
  - Pentose phosphate pathway, an alternative to glycolysis (Fig. 15–21)
  - Citric acid cycle (Fig. 16–7)
  - Glyoxylate cycle, a variation of the citric acid cycle in some organisms (Fig. 16–16)
  - Fatty acid oxidation (Fig. 17–8)
  - Amino acid oxidation (Fig. 18–2)
  - Urea cycle (Fig. 18–9)
  - Oxidative phosphorylation:
    - Electron transfer chain (Fig. 19–14)
    - ATP synthesis and the chemiosmotic model (Fig. 19–16; Fig. 19–23)
  - Shuttle systems for cytosolic NADH generated by glycolysis:

- Malate-aspartate shuttle (Fig. 19–26)
- Glycerol 3-phosphate shuttle (Fig. 19–27)
- Photophosphorylation (Fig. 19–46)
- Gluconeogenesis, compared with the glycolytic pathway (Fig. 20–2)
- Carbon fixation, overview (Fig. 20–22)
- Fatty acid biosynthesis (Fig. 21–2)
- Triacylglycerol synthesis (Fig. 21–18; Fig. 21–19)
- Nitrogen fixation (Fig. 22–2)
- Amino acid synthesis (Fig. 22–9)
- Nucleotide synthesis:
  - Purines (Fig. 22–33)
  - Pyrimidines (Fig. 22–34)
- Nucleotide degradation:
  - Purines (Fig. 22–43)
  - Pyrimidines (Fig. 22–44)

- You should now have a greater appreciation of Figures 16–1 and 16–13. These figures show how a number of key metabolic pathways are interconnected.
- Several **molecular shuttles** are used to transport metabolic intermediates between tissues. Review the glucose-alanine cycle (Fig. 18–8) and the Cori cycle (Box 15–1). Note how these shuttles contribute to the specialization of tissue functions.
- Information carried by hormones is critical to the regulation of metabolic pathways. This information is often relayed to the interior of cells by intracellular messengers; two intracellular messengers are derived from membrane lipids. Review the structure of **membrane lipids** (p. 370) and note the sites of cleavage by lipases (p. 374).
- Protein phosphorylation plays a critical role in the regulation of enzyme activity by hormones and their intracellular messengers. Review which of the amino acids have an available —OH group that can be phosphorylated (p. 120).

Topics for Discussion

Answering each of the following questions, especially in the context of a study group discussion, should help you understand the important points of this chapter.

Tissue-Specific Metabolism: The Division of Labor

The Liver Processes and Distributes Nutrients

1. Which pathways for energy production in the liver use glucose 6-phosphate?

2. What are the possible fates of acetyl-CoA produced by the liver?

3. What tissue specializations make the liver such a critical organ in the regulation of blood glucose levels?

Adipose Tissue Stores and Supplies Fatty Acids

4. What types of lipids are stored in adipocytes?

5. Under what conditions are these lipids "mobilized" (i.e., hydrolyzed to release fatty acids that can be oxidized, producing acetyl-CoA and ultimately ATP)? How are the enzymes (lipases) that hydrolyze lipids activated?

Muscle Uses ATP for Mechanical Work

6. Which compounds are used as energy sources in muscle tissue?

7. How do muscle cells use the energy in ATP to do mechanical work?

8. What adaptations do muscles employ to ensure adequate supplies of energy during periods of intense activity?

The Brain Uses Energy for Transmission of Electrical Impulses

9. Which compounds are used as energy sources by the brain?

10. Why does the brain consume as much as 20% of the O_2 required by a resting human (i.e., what is the major energy-consuming metabolic process in brain cells)?

Blood Carries Oxygen, Metabolites, and Hormones

11. Name some of the metabolic intermediates you would expect to find in the blood.

12. Why is the ability of blood cells to transport the metabolite H^+ so critical to organisms? How does blood transport H^+?

Hormonal Regulation of Fuel Metabolism

Epinephrine Signals Impending Activity

13. How is it possible for a single hormone, such as epinephrine, to affect enzyme activity in three separate tissue types (liver, muscle, and adipose tissue)?

14. What is the *molecular mechanism*, initiated by epinephrine, that results in a stimulation of glycogen breakdown in liver, a stimulation of glycolysis in muscle, and an increase in fat mobilization in adipose tissue?

15. How does epinephrine *coordinate* enzyme activity in both the gluconeogenic and glycogen breakdown pathways?

Glucagon Signals Low Blood Glucose

16. What is the effect of glucagon, at the molecular level, on cell metabolism?

17. At which point(s) is it possible for the actions of epinephrine and glucagon to be integrated (i.e., where do these two systems interact)?

During Fasting and Starvation, Metabolism Shifts to Provide Fuel for the Brain

18. Why does the citric acid cycle in hepatocytes slow and eventually stop during fasting, even though acetyl-CoA (from lipid oxidation) is available?

19. What happens to the acetyl-CoA from stored fats that becomes available during starvation?

Insulin Signals High Blood Glucose

20. Insulin affects cells by binding to a tyrosine kinase receptor on the cell surface. How does insulin affects a cell's metabolism?

21. Based on what you have learned about the effects of epinephrine on glycogen phosphorylase, predict how insulin regulates this enzyme's activity.

Cortisol Signals Stress, Including Low Blood Glucose

22. Cortisol is derived from cholesterol; how does this hormone exert its effects on cell metabolism? What enzyme activities does it influence?

Diabetes Is a Defect in Insulin Production or Action

23. Diabetics have increased levels of acetyl-CoA due to the increased oxidation of lipids, yet this acetyl-CoA is used to produce acetoacetyl-CoA and β-hydroxybutyrate (ketone bodies) rather than energy via the citric acid cycle. Explain why this is so.

Hormones: Diverse Structures for Diverse Functions

24. The only difference between the effects of a neurotransmitter and the effects of a hormone lies in the timing of the response. Do you agree with this statement? Why or why not?

Hormone Discovery and Purification Requires a Bioassay

25. List bioassays for hormones such as tumor necrosis factor or nerve growth factor.

26. What features of the radioimmunoassay technique make it so sensitive?

Hormones Act through Specific High-Affinity Cellular Receptors

27. How does the timing of hormone effects correlate to their molecular mode of action?

Hormones Are Chemically Diverse

28. Why can water-soluble hormones, such as adrenalin, be stored in and released from membrane-bound vesicles whereas hormones such as testosterone and estrogen are not?

29. To which secretory class of proteins do eicosanoids belong? Nitric oxide (NO)?

What Regulates the Regulators?

30. What are the advantages of having a neuroendocrine system that is organized into such a complex hierarchy?

31. In what ways can the hypothalamus be viewed as the coordinating center of the endocrine system?

Long-Term Regulation of Body Mass

Leptin Was Predicted by the Lipostat Theory

32. What is the role of leptin in weight regulation?

Many Factors Regulate Feeding Behavior and Energy Expenditure

33. Based on the information in this section, what are the likely effects of leptin on hormone secretion by the hypothalamus?

Leptin Triggers a Regulatory Cascade

34. Does your response to question 33 fit with the model of leptin action shown in Fig. 23–29 in the text?

The Leptin System May Have Evolved to Regulate the Starvation Responses

35. Why is it thought that leptin plays a role in starvation responses and not primarily in preventing obesity? (Hint: Think in nonhuman terms.)

SELF-TEST

Do You Know the Term?

ACROSS

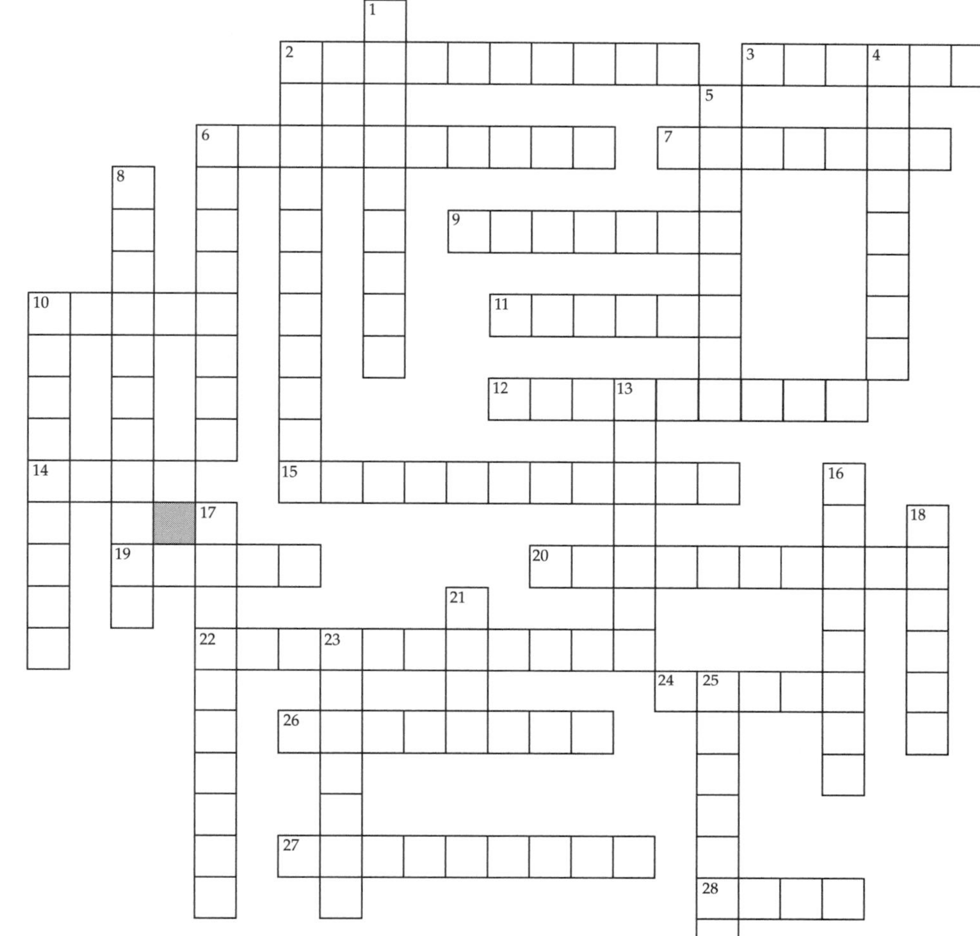

2. The inability to regulate blood glucose levels causes ______, which is the production of copious amounts of urine containing large amounts of glucose.

3. General name for an enzyme that catalyzes protein phosphorylation.

6. Carbon backbones from this class of amino acids can be used by the liver for gluconeogenesis.

7. A peptide hormone that increases the rate of gluconeogenesis and glycogen breakdown in hepatocytes.

9. Oxytocin and vasopressin belong to this class of hormone.

10. Home pregnancy tests are based on this technique, which uses an enzyme-linked immunosorbent to detect small quantities of the hormone HCG present in urine. (abbr.)

11. Overproduction of ketone bodies from metabolized fatty acids occurs in fasting individuals and is characterized by the odor of acetone on their breath; this condition is called ______.

12. Interleukins are examples of local regulators which act in a ______ fashion; they are secreted by immune system cells and diffuse to neighboring target cells where they exert their effects.

14. This second messenger is produced by a cyclase that is located in the cytosol and not in the plasma membrane. (abbr.)

15. Derived from membrane phospholipids, one member of this class of hormones is secreted by cells of the placenta causing the nearby muscle of the uterus to contract, inducing labor.

19. It is believed that chronic overeating leads to a loss of responsiveness to insulin, producing a form of diabetes known as ______. (abbr.)

20. Multiple hormonal "messages" can be carved out of this.

22. One way to reverse the regulatory effects of a protein kinase is through the activation of a protein ______.

24. The leptin signal is mediated by signal transducers and activators of transcription or ______ that move to the nucleus, bind to specific DNA sequences, and stimulate the expression of specific genes.

26. An essential component in the signaling pathway for all hormones and neurotransmitters.

27. A structure in the brain that releases hormones from both neurons as well as endocrine cells.

28. A common metabolic disorder that can be treated by the exogenous replacement of a missing hormone.

DOWN

1. For this class of hormones, the secretory cell is also the target.

2. Low K_m isozyme of hexokinase that serves as a glucose "sensor" for the liver.

4. Uncontrolled diabetes can overwhelm the capacity of the bicarbonate buffering system of blood causing a condition called ______.

5. Endocrine cells in this portion of the pituitary secrete hormones in response to blood-borne signals from the hypothalamus.

6. Although this hormone is secreted when blood glucose levels (and system energy levels) are low, one of its primary effects is to inhibit glycolysis in hepatocytes.

8. A sensitive assay technique that relies on competition between unlabeled and radioactively labeled compounds for binding sites on antibody molecules is the ______ assay.

10. Neurons secrete neurotransmitters; ______ cells secrete hormones.

13. One of the main targets for hormones regulating metabolism and the storage site for triacylglycerols is ______ tissue.

16. Measurements of urine output from kidneys provide a ______ for the identification of hormones that can affect water balance such as the antidiuretic hormone.

17. ______ secrete . . .

18. . . . ______, a peptide hormone that acts on the hypothalamus to curb feeding in mice.

21. This second messenger mediates the effects of epinephrine and/or glucagon on cells.

23. ______ hormones are derived from cholesterol and are lipid soluble; they do not require intracellular, second messengers.

25. Derived from the precursor protein thyroglobulin, ______ hormones stimulate energy-yielding metabolism by activating the expression of genes encoding key catabolic enzymes.

Do You Know the Facts?

1. Which compound links glycolysis, nucleotide synthesis, and glycogen synthesis?

A. acetyl-CoA
B. oxaloacetate
C. citrate
D. glucose 6-phosphate
E. glycerol 3-phosphate

2. Anabolic and catabolic pathways for a given biomolecule almost never involve identical sequences of reactions because:

A. enzyme-catalyzed reactions are usually at equilibrium.
B. catabolic and anabolic pathways do not take place in the same cell.
C. they would be difficult to regulate and futile cycling would most likely occur.
D. the $\Delta G'^{\circ}$ for the reactions always favors the anabolic pathway.
E. it would be too easy for students of biochemistry to learn them.

3. Ingested fatty acids are degraded to acetyl-CoA, which is an allosteric activator of pyruvate carboxylase. What is the product of this enzyme reaction, and what is its effect on metabolism?

A. oxaloacetate; increased rate of gluconeogenesis and the citric acid cycle
B. oxaloacetate; decreased rate of fatty acid synthesis
C. lactate; increased rate of glycolysis and gluconeogenesis
D. acetyl-CoA; increased rate of the citric acid cycle
E. acetoacetyl-CoA; decreased rate of triacylglycerol synthesis

4. Acetyl-CoA is a common metabolic intermediate in a number of metabolic pathways; this intermediate enables mammals to convert:

A. proteins to lipids.
B. sugars to triacylglycerol.
C. lipids to glycogen.
D. A and B only.
E. A, B, and C.

5. Rank the following energy sources according to the order in which they would be used by heart muscle tissue under fasting conditions.

(a) cellular proteins
(b) muscle glycogen
(c) liver glycogen
(d) triacylglycerol in adipose tissue
(e) blood glucose
(f) ATP from other body tissues
(g) blood acetoacetate

6. Which pair correctly matches the enzyme with its allosteric activator?
A. hexokinase; ATP
B. phosphofructokinase-1; AMP
C. pyruvate kinase; ATP
D. pyruvate dehydrogenase; NADH
E. pyruvate carboxylase; ADP

7. How can *both* glycolysis and gluconeogenesis be irreversible processes in cells?
A. They occur in separate cellular compartments.
B. Key reactions in each pathway are characterized by large and negative free-energy changes.
C. All the enzymes in the pathways are different, so the reaction pathways don't overlap.
D. A and B
E. A, B, and C

8. A cell that is deficient in pyruvate dehydrogenase phosphate–phosphatase would exhibit all of the following metabolic effects except:
A. an increase in lactate production.
B. a decrease in pH.
C. an increase in the levels of citrate.
D. a decrease in the levels of NADH and ATP.
E. an increase in glycolytic activity.

9. Mature erythrocytes (red blood cells) are *full* of hemoglobin and have *no* organelles. They metabolize glucose at a high rate, generating lactate, which is transported to the liver for use in gluconeogenesis. Why do erythrocytes metabolize glucose to lactate in order to generate energy?
A. They have no citric acid cycle.
B. Oxygen is not available for the aerobic oxidation of glucose.
C. Anaerobic oxidation of pyruvate regenerates the NAD^+ needed for glycolysis to continue.
D. A and C
E. B and C

10. Predict the effect of a duplication of the gene for phosphorylase kinase in mutant liver cells relative to normal liver cells that contain only a single copy of this gene.
A. Glycogen synthase activity would be higher in mutant cells.
B. The rate of glycolysis would be increased in mutant liver cells.
C. The ratio of unphosphorylated/phosphorylated glycogen phosphorylase will be higher in mutant cells.
D. The activity of glycogen phosphorylate will be increased in mutant cells.
E. All of the above would be seen in mutant cells.

11. Which correctly describes the receptor-mediated effects of epinephrine in glucose metabolism? The hormone:
A. activates phosphorylase kinase.
B. stimulates the activity of glycogen synthase.
C. activates phosphodiesterase.
D. activates phospholipase C.
E. does all of the above.

12. Indicate whether each of the following is true or false.
_____ **(a)** Insulin increases the intracellular concentration of glucose in hepatocytes.
_____ **(b)** Glucagon is secreted in response to low blood glucose levels.
_____ **(c)** Insulin increases the capacity of the liver to synthesize glycogen.
_____ **(d)** Insulin has effects opposite to those of epinephrine and glucagon.
_____ **(e)** Insulin, epinephrine, and glucagon all activate adenylate cyclase.
_____ **(f)** Insulin activates phospholipase C.
_____ **(g)** Glucagon increases triacylglycerol synthesis.
_____ **(h)** Glycogen phosphorylase catalyzes glycogen breakdown in the absence of glycogen.

13. Which statement correctly describes steroid hormones?
A. Their intracellular actions are mediated by integral membrane proteins.
B. Their effects on cells require a water-soluble intracellular signal.
C. Their effects are mediated by binding to a water-soluble receptor protein.
D. They are synthesized directly from amino acid precursors.
E. Their effects usually involve the activation of other intracellular enzymes.

14. The most *efficient* way to turn off glycogen degradation is to:
A. decrease the activity of phosphorylase kinase.
B. increase the activity of phosphodiesterases.
C. increase the activity of phosphatases.
D. increase the activity of glycogen synthase.
E. decrease the intracellular levels of cAMP.

15. Untreated diabetics experience hyperglycemia, which is an excess of glucose in their blood. Any glucose present in the blood normally gets reabsorbed by transporters present in the nephrons of the kidney. In diabetics not all blood glucose is reabsorbed and excreted in the urine (hence the name *diabetes mellitus,* meaning "sweet urine"). What is the most likely molecular explanation for this effect?
A. The transporters are inhibited by the presence of high levels of glucose.
B. The binding affinity of these transporters is increased in diabetics.
C. The V_{max} of the transporter is reached at lower blood glucose levels.
D. The binding capacity of the transporters has been exceeded.
E. Both C and D

16. Chromium is often taken as a dietary supplement and possible appetite suppressant. Data suggests that chromium functions as part of a complex that facilitates the binding of insulin to its receptor, thus enhancing the effect of insulin on cellular metabolism. Which of the following would be expected to occur in someone taking chromium supplements who had previously been deficient in this compound?
A. Enhanced activity of glucokinase, enhanced uptake of dietary fatty acids, and subsequent glycogen and triacylglycerol synthesis.
B. Increased rates of glycogen phosphorylase leading to an increase in the breakdown of stored glycogen.
C. Increased activity of lipases in adipocytes results in the breakdown of stored triacylglycerols.
D. Increased storage of glucose as glycogen in muscle cells.
E. Increased activity of fructose 1,6-bisphosphatase resulting in an increase in the rate of glycolysis in hepatocytes.

17. The *porcine stress syndrome,* where pigs die suddenly on their way to the slaughter house, is due to a genetic defect in metabolism. The symptoms are a dramatic rise in body temperature and both metabolic and respiratory acidosis, which occur in affected pigs that are exposed to high-stress situations. Uncontrolled futile cycling, in which ATP hydrolysis is greatly accelerated, is thought to be the basis for the rise in body temperature. Which of the following pairs of enzymes, assuming both are fully activated, would produce a futile cycle?
A. glucose 6-phosphatase: fructose 1,6-bisphosphatase
B. phosphofructokinase-1: phosphofructokinase-2
C. glycogen phosphorylase kinase: glycogen synthase
D. pyruvate carboxylase: phosphoenolpyruvate carboxykinase
E. glycogen phosphorylase: hexokinase

Applying What You Know

1. Explain, in terms of enzyme kinetics, why glucokinase is responsible for channeling glucose into the glycogen synthesis pathway in the liver, and why hexokinase (an isozyme of glucokinase) regulates the entry of glucose into the glycolytic pathway in muscle.

2. Caffeine is an inhibitor of the enzyme phosphodiesterase. How does this information help to explain the stimulatory effects of caffeinated drinks such as coffee?

3. Two patients exhibiting problems with their glucose metabolism had their blood tested for oxygen-binding ability. The oxygen-binding curves for the hemoglobin of each patient are shown below, along with the data for hemoglobin from a "normal" individual. Both patients were later tested for deficiencies in several glycolytic enzymes. Patient A was shown to be deficient in hexokinase, and patient B deficient in pyruvate kinase.

   Why did the attending physician order the blood test? Why was the oxygen-binding test the first test to be done? Explain why the binding curves for the two patients are shifted relative to that for the normal individual.

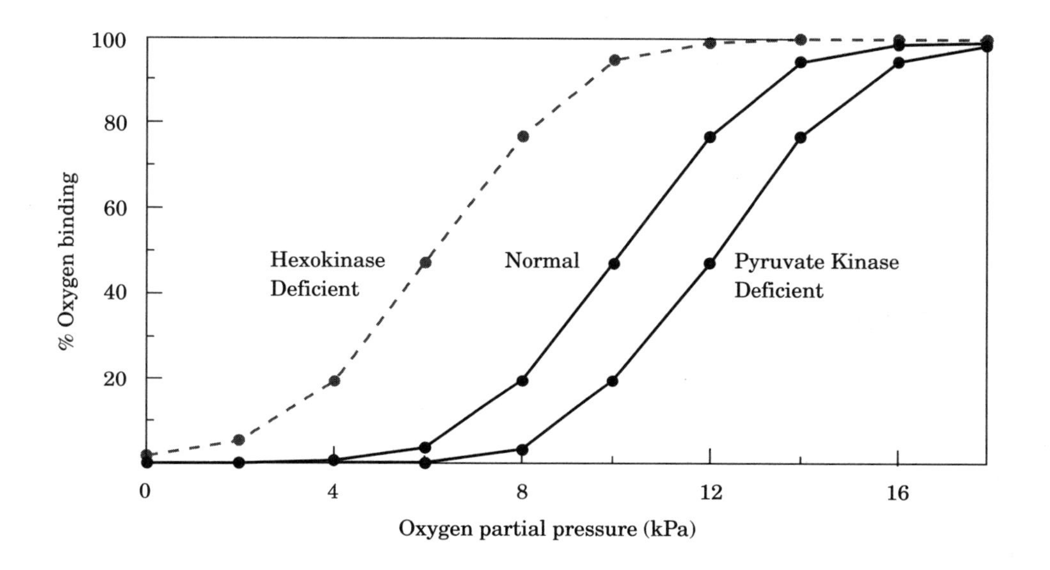

4. One symptom of diabetes is the excessive production of ketone bodies resulting from excessive oxidation of fatty acids. The normal *inhibitory* effects of insulin on the oxidation of fatty acids are absent in diabetics. However, *positive* regulation of fatty acid oxidation also contributes to the increased levels of acetoacetate in diabetics. Which hormone(s) is responsible for the positive regulation of fatty acid oxidation and how is this regulation achieved?

ANSWERS

Do You Know the Terms?

ACROSS

2. glucosuria
3. kinase
6. glucogenic
7. insulin
9. peptide
10. ELISA
11. ketosis
12. paracrine
14. cGMP
15. eicosanoids
19. NIDDM
20. prohormone
22. phosphatase
24. STATS
26. receptor
27. pituitary
28. IDDM

DOWN

1. autocrine
2. glucokinase
4. acidosis
5. anterior
6. glucagon
8. radioimmuno
10. endocrine
13. adipose
16. bioassay
17. adipocytes
18. leptin
21. cAMP
23. steroid
25. thyroid

Do You Know the Facts?

1. D
2. C
3. A
4. D
5. (e), (b), (c), (d), (g), (a). Note that (f) will *not* contribute as an energy source to heart muscle; ATP does not leave individual cells and so is not a "transportable" energy source.
6. B
7. B
8. C
9. D
10. D
11. A
12. **(a)** T; **(b)** T; **(c)** T; **(d)** T; **(e)** F; **(f)** F; **(g)** F; **(h)** T
13. C
14. E
15. D
16. A
17. C

Applying What You Know

1. *Hexokinase* phosphorylates glucose to glucose 6-phosphate. This enzyme binds glucose with a relatively high affinity: it is half-saturated with substrate at [glucose] = 0.1 mM (about 50-fold less than the normal [glucose] in blood). At normal blood glucose levels, hexokinase binds glucose and "traps" it inside cells. In the liver, *glucokinase* (or hexokinase D) is half-saturated with substrate at [glucose] = 10 mM, 100-fold greater than the half-saturation level for hexokinase, but only twice the [glucose] in normal blood. Consequently, glucokinase is only slightly active at normal blood glucose levels, and makes a significant contribution to glucose level in hepatocytes only when blood [glucose] is high.

Insulin is also released when blood [glucose] is elevated, resulting in the activation of glycogen synthase and inactivation of glycogen phosphorylase. The "trapped" glucose is therefore funneled into the glycogen synthesis pathway. This does not happen in muscle cells because they lack glucokinase and their energy demands are quite high, so that most of the available glucose is funneled into the glycolytic pathway.

2. Phosphodiesterase cleaves cAMP to 5′-AMP (inactive). Inhibition of this enzyme indirectly increases [cAMP] in cells. Increased cAMP levels (normally induced by the hormones epinephrine and glucagon) ultimately result in activation of glycogen phosphorylase and inactivation of glycogen synthase. The net effect is an increase in the availability of free glucose for entry into glycolysis and the citric acid cycle; both pathways supply ATP for cellular activities (e.g., muscle contraction).

3. As you now know, carbohydrate metabolism involves many interconnected anabolic and catabolic pathways. A defect in just one of the enzymes in any of these pathways can have profound effects on the way an individual metabolizes carbohydrates. It would be cumbersome (and expensive) to test *each* enzyme in all of these pathways for possible defects. However, there are ways to narrow the field to a small subset of enzymes. For example, defects in any of the glycolytic enzymes would be expected to affect the levels of 1,3-bisphosphoglycerate. This glycolytic intermediate is converted into 2,3-bisphosphoglycerate (BPG) by phosphoglycerate mutase. Because BPG has a profound effect on O_2 binding by hemoglobin, examining the O_2 binding of hemoglobin is a quick way to screen for defects in glycolytic enzymes. BPG binds in the center of the hemoglobin tetramer and stabilizes the deoxy form.

The patient with a leftward shift in the O_2-binding curve (patient A) has hemoglobin that is half-saturated at lower O_2 partial pressures, indicating a higher affinity for O_2. Reduced levels of BPG (reflecting reduced levels of 1,3-bisphosphoglycerate) could account for this shift, suggesting a defect in one of the glycolytic enzymes that catalyzes a reaction *before* the production of 1,3-bisphosphoglycerate. Conversely, increased levels of BPG (reflecting increased levels of 1,3-bisphosphoglycerate) would shift the binding curve to the right (patient B) because more hemoglobin would be stabilized in the deoxy form. A rightward shift in the curve suggests a defect in a glycolytic enzyme that catalyzes a reaction *after* the production of 1,3-bisphosphoglycerate.

4. Epinephrine-induced increases in [cAMP] result in the phosphorylation (and activation) of adipocyte lipases, which hydrolyze lipids and release free fatty acids. In addition, glucagon- and epinephrine-induced increases in [cAMP] result in the inhibition of acetyl-CoA carboxylase. The consequent decrease in [malonyl-CoA] results in concomitant activation of carnitine acyl-transferase I, which increases the rate of fatty acid oxidation.

chapter

24 Genes and Chromosomes

STEP-BY-STEP GUIDE

Major Concepts

All living organisms contain genetic material that directs the processes of life.

Prokaryotic and eukaryotic cells contain DNA as their genetic material; viral genomes are either DNA or RNA. Viral DNA molecules are small because the virus itself does not have to direct functions that the host cell provides. Bacteria carry much more DNA, in a single, circular **chromosome.** They also can contain small extra-chromosomal circular molecules of DNA called **plasmids,** which often code for specialized enzymes such as those involved in the degradation of antibiotics and unusual food sources. Eukaryotic cells have more DNA than prokaryotic cells. The DNA is arranged into chromosomes, with the amount of DNA per chromosome and the number of chromosomes varying with species. Mitochondria and chloroplasts of eukaryotic cells also contain small amounts of functioning, nonchromosomal DNA.

A gene is defined as a portion of the chromosome that codes for a single cellular component.

Structural genes code for polypeptides or for RNAs. Genes coding for proteins include all the DNA sequences required for synthesis of a functional protein, including the **introns,** which are nontranslated interruptions of the coding sequence for a polypeptide, and transcriptional control regions. There are a number of other identifiable sequences on chromosomes. These include **regulatory sequences,** which function as signals to turn on and off DNA-dependent processes; **highly repetitive sequences,** noncoding DNA that is associated with **centromeres** and **telomeres;** and **moderately repetitive DNA,** which is scattered throughout the chromosome and has no known function.

Supercoiling of the DNA molecule aids in compaction but, paradoxically, makes the DNA accessible for replication and transcription processes.

Most DNA in cells is **underwound,** which promotes strand separation but "strains" the DNA molecule. DNA supercoiling both reduces this strain and leaves the DNA in its more accessible, underwound state; this is because supercoiling is energetically more favorable than actual strand separation, which requires the breaking of hydrogen bonds between base pairs. Because the process of supercoiling is critical to both DNA replication and transcription, it is actively controlled by cells, primarily by topoisomerases. The extent of underwinding (and therefore supercoiling) in closed circular DNA is described by a set of parameters including: **linking number,** which specifies the number of helical turns in the DNA molecule; **specific linking difference,** a measure of the degree of underwinding; **twist,** which describes the spatial relationship of neighboring base pairs; and **writhe,** a measure of the coiling of the helical axis. Identical DNA molecules that differ only in their topology (e.g., linking number or linking difference) are **topoisomers.**

Eukaryotic chromosomes are made of chromatin, which includes not only DNA but also significant amounts of protein and small amounts of RNA.

Important chromatin proteins are **histones,** which are small, basic proteins. The strict conservation of the amino acid sequences of histones across species is an indication of their critical functions. Histones act as central cores for the wrapping of the DNA double helix. Such a wrapped kernel makes up a **nucleosome,** which is considered the first level of DNA packaging. Further compression of the DNA is accomplished by formation of **30 nm fibers,** wrapping of the nucleosomes into a structure that resembles the stalk of a brussel sprout plant. Still tighter packaging is accomplished with loops of 30 nm fibers radiating out from a nuclear scaffold. This elaborate compacting strategy is necessary in order to fit the enormous length of DNA into the cell, while still maintaining sufficient organization to allow unwinding, and accessibility, of specific portions of the DNA.

What to Review

Answering the following questions and reviewing the relevant concepts, which you have already studied, should make this chapter more understandable.

- Much of the discussion in this chapter assumes a very clear knowledge of the **structure of DNA** as described in Chapter 10. Review the components that make up DNA, the covalent bonds and noncovalent interactions that stabilize its structure, and the chemical behavior of the molecule as a whole.
- There are many references to the differences between **prokaryotes** and **eukaryotes.** Make sure you understand the critical differences between the two. Where do **viruses** fit into this categorization (pp. 21, 29, 46, 47)?
- **Mutations** are alterations in the DNA sequence that result in a difference in the amino acid sequence of a protein, often leading to decreased or no function. Review the effect of a *single* amino acid alteration on the functioning of hemoglobin in sickle-cell anemia (pp. 219–221).
- Review **cruciform structures** and **Z-DNA** (pp. 338–339), conformations of DNA that are affected by DNA supercoiling.

Topics for Discussion

Answering each of the following questions, especially in the context of a study group discussion, should help you understand the important points of this chapter.

Chromosomal Elements

Genes Are Segments of DNA That Code for Polypeptide Chains and RNAs

1. How is the "classical" definition of a gene different from the most modern, "one gene–one polypeptide" definition? (There will be more on the concept of the "whole" gene in Chapters 26 and 27.)

2. Do regulatory sequences encode a polypeptide product?

3. Why can only the "minimum" size of a gene be estimated from the size of its polypeptide product?

4. How large a polypeptide could be produced from an "average" gene, 1,050 base pairs long?

Eukaryotic Chromosomes Are Very Complex

5. Where in the eukaryotic chromosome are highly repetitive segments located? Where are moderately repetitive segments located? What are their functions?

6. If the function of *Alu* repeats is unknown, why is it important to mention them at all? These *Alu* repeats account for approximately how many nucleotide base pairs in the human genome?

Many Eukaryotic Genes Contain Intervening Nontranscribed Sequences (Introns)

7. Why is it that only the "minimum" size of a gene can be estimated from the size of its polypeptide product?

8. Are introns essential components of eukaryotic genes?

The Size and Sequence Structure of DNA Molecules

Viral DNA Molecules Are Relatively Small

9. Why can the genomes of viruses be much smaller than those of cells?

10. How many times larger is the human genome compared to that of a medium-sized DNA virus such as bacteriophage λ?

Bacteria Contain Chromosomes and Extrachromosomal DNA

11. What characteristics of plasmids make them similar to bacterial chromosomes? How are they different?

Eukaryotic Cells Contain More DNA Than Do Prokaryotes

12. Is there any relationship between amount of DNA per cell, or number of chromosomes per cell, and relative complexity of organisms? Why or why not?

Organelles of Eukaryotic Cells Also Contain DNA

13. Based on your knowledge of mitochondrial structure, specifically the inner membrane, indicate why some proteins need to be encoded by mitochondrial DNA.

DNA Supercoiling

14. Is DNA supercoiling a random process?

15. What are two cellular functions of DNA supercoiling?

Most Cellular DNA Is Underwound

16. How does underwinding of DNA lead to greater ease in strand separation?

DNA Underwinding Is Defined by Topological Linking Number

17. Why is linking number a topological property?

18. What would be the specific linking difference for a DNA molecule of 420 base pairs from which eight turns were removed?

19. What is the difference between positive and negative supercoiling?

20. Why are twist and writhe *not* topological properties?

Topoisomerases Catalyze Changes in the Linking Number of DNA

21. What are the similarities and differences between the type 1 and type 2 topoisomerases?

DNA Compaction Requires a Special Form of Supercoiling

22. Twisting a long, *double* strand of pearls or beads is another useful illustration of the effects of supercoiling. What do the beads represent?

Chromatin and Nucleoid Structure

Histones Are Small, Basic Proteins

23. DNA is a negatively charged molecule. Does this offer a clue as to why histones have such high percentages of basic amino acids?

24. Are histones the only types of protein associated with DNA in chromatin?

Nucleosomes Are the Fundamental Organizational Units of Chromatin

25. Which histone proteins are part of the nucleosome proper?

26. Is the location of nucleosomes on DNA always random?

27. To visualize how nucleosomes aid in compaction of DNA, try winding a thick, closed circular rubber band around a "nucleosome." Use a thick rubber band, folded so that it is flat, to represent the DNA double helix and a pencil to represent the nucleosome core. Hold one end of the rubber band against the pencil with your thumb and use the other hand to wind the doubled rubber band around the pencil. What happens to the unwound portion of the rubber band? (A compensatory supercoil should form.)

Nucleosomes Are Packed into Successively Higher-Order Structures

28. Why would DNA regions being transcribed need to be less ordered?

29. Once again, the use of a hand-held model may help in understanding the organization of eukaryotic DNA. A single strand of beads, twisted evenly and gently, will form something that looks like the 30 nm structure shown in Fig. 24–27. Twist it harder, and loops will form. Twist it much more, and—unlike DNA—it will break.

30. What is the nucleosome scaffold made of?

Bacterial DNA Is Also Highly Organized

31. Why is bacterial DNA less condensed, in general, than eukaryotic DNA?

SELF-TEST

Do You Know the Terms?

ACROSS

1. State of DNA double helices that is thought to promote transcription by facilitating strand separation.
5. DNA plus two each of histone 2A, 2B, 3, and 4.
7. Supercoiled DNA structure assumed in most cells (due to the association of proteins), known as ________ supercoiling.
10. Higher-order structure of DNA, referred to as a 30 nm fiber, requires the association of the protein ________.
11. Structural units made up of DNA and associated proteins; contain heritable genetic information.
13. Nontranslated regions of DNA in the genes of most eukaryotes.
15. ________ genes are regions of DNA that actually code for proteins or RNA.
17. Describes sequences of DNA usually associated with centromeres and telomeres. (2 words)
18. The linking number for underwound DNA is ________ than that for relaxed DNA.
20. Describes plasmid and bacterial DNA, which lack telomeres.
21. Again, describes plasmid and bacterial DNA.
22. In vivo state of most double-stranded DNA.
24. Circular, extrachromosomal DNA that can replicate independently of genomic DNA.
26. Regions of DNA that influence both *where* gene transcription starts and ends and the *rate* of transcription; called ________ sequences.
27. Proteins that contain a relative abundance of the amino acids H, R, and K.
28. Coding regions of DNA that usually specify amino acid sequences for only a portion of the final protein.

DOWN

2. Organisms that package most of their genetic material in a membrane-bounded nucleus.
3. Region of a bacterial cell that contains its genetic information.
4. All of the information needed to make an entire organism.
6. Property of an organism that is the product of the information contained in the genome.
8. Treating chromatin with DNase degrades ________ DNA.
9. Its genetic information is encoded in either DNA or RNA.
11. Structure essential for the separation of chromosomes during cell division.
12. Extrachromosomal DNA in eukaryotes includes ________ DNA.
14. Enzymes that alter the linking number for a given molecule of DNA; especially important in the process of DNA replication.
16. Formed by repetitive DNA sequences; help to stabilize linear DNA.
19. Bacteria; organisms that lack cellular organelles.
20. The stuff (in nondividing cells) of which chromosomes are made.
23. DNA that is not supercoiled and not underwound is said to be in a ________ state.
25. Most DNA exists as a ________ of polynucleotide strands.

Do You Know the Facts?

1. DNA supercoiling:

A. results in compaction of the DNA structure.
B. makes the DNA molecule relaxed.
C. occurs only in viruses.
D. occurs in only one direction (i.e., is only positive).
E. usually results from overwinding.

2. Human chromosomes are extremely large, complex structures. Which of the following statements correctly describes the organization of human chromosomes?

A. All the genetic information of the cell is encoded in the nuclear, chromosomal DNA.
B. Genes for histones and ribosomal RNAs are structural genes.
C. Most of the chromosomal DNA codes for proteins.
D. A and B.
E. A, B, and C.

3. Plasmids are:

A. part of the bacterial chromosome.
B. found only in mitochondria.
C. composed of RNA and protein.
D. closed circular DNA molecules.
E. found only in viruses.

4. There are several levels of protein architecture, and the same is true for chromosomal DNA. Indicate whether each of the following statements about the structure and packaging of chromosomal DNA is true or false.

_____ **(a)** Histones are basic proteins that make up about half the mass of chromatin.
_____ **(b)** Histones grip the DNA double helix like a fist, with the DNA in the center.
_____ **(c)** The nucleosome structure acts to condense DNA by decreasing its length.
_____ **(d)** Formation of the nucleosome structure is the only condensing step in chromosomal DNA.
_____ **(e)** DNA comprises approximately 90% of the total mass of chromatin; the rest is protein and a small amount of RNA.
_____ **(f)** The amino acid sequences of histones vary widely from species to species.
_____ **(g)** Between the nucleosomes are protease-sensitive regions of DNA.

5. In a certain organism, the gene for the glycolytic enzyme hexokinase has 21,000 bases. The molecular weight of hexokinase is approximately 110,000. Is this organism a prokaryote or eukaryote? How do you know? (Average molecular weight of an amino acid $\approx$ 110; of a base $\approx$ 450.)

Applying What You Know

1. Prokaryotic topoisomerase II, also called DNA gyrase, is specifically inhibited by an antibiotic called novobiocin, which binds to the β subunit of the enzyme. This antibiotic inhibits bacterial DNA replication. Suggest why this is so.

Biochemistry on the Internet

Approximately 2 meters of DNA must be packaged into the nucleus of most eukaryotic cells. In order to achieve this level of compaction, it is necessary to organize the DNA strands into higher-order structures. The structure known as the nucleosome represents the first level of DNA compaction and includes a protein-based core around which double-helical DNA is wound. The nucleosome structure achieves about a seven-fold compaction, reducing the 2 meters of DNA to about 30 cm in length. The protein core of the nucleosome is made up of two copies of each of the four histone proteins: H2A, H2B, H3, and H4. A model of a nucleosome can be obtained from the Protein Data Bank (http://www.rcsb.org/pdb) using the PDB ID code 1AOI.

(a) Proteins in the nucleosome core are arranged as histone dimers. According to this model, which of the histones listed above can form dimers with each other?

(b) The DNA-protein interactions that stabilize the nucleosome structure are noncovalent. Looking at the molecular model of the nucleosome, can you discern any pattern in the way the DNA duplex interacts with the histone proteins in this complex? How might these interactions contribute to the overall structure and function of the nucleosome?

ANSWERS

Do You Know the Terms?

ACROSS

1. underwound
5. nucleosome
7. solenoidal
10. H1
11. chromosomes
13. introns
15. structural
17. highly repetitive
18. less
20. circular
21. closed
22. supercoiled
24. plasmid
26. regulatory
27. histones
28. exons

DOWN

2. eukaryotes
3. nucleoid
4. genome
6. phenotype
8. linker
9. virus
11. centromere
12. mitochondrial
14. topoisomerases
16. telomeres
19. prokaryotes
20. chromatin
23. relaxed
25. duplex

Do You Know the Facts?

1. A
2. B
3. D
4. **(a)** T; **(b)** F; **(c)** T; **(d)** F; **(e)** F; **(f)** F; **(g)** F
5. The gene is ~7 times as long as it needs to be to code for the ~1000 amino acids that make up this protein. The presence of untranslated regions of DNA within the coding sequence indicates that this gene must contain introns and the organism, therefore, is a eukaryote.

Applying What You Know

1. Inhibition of topoisomerase II prevents the necessary addition of negative supercoils during replication. Look at Figure 24–11 and imagine the separation of strands necessary for replication; clearly, inhibition of DNA gyrase would inhibit DNA replication.

Biochemistry on the Internet

(a) According to the PDB Summary Information page, chains A and E represent histone H3, chains B and F represent histone H4, chains C and G represent histone H2A, and chains D and H represent histone H2B. The easiest way to visualize the different chains in this model is to display them in **Ribbon** format and then change the color setting to **Chain.** With these settings, each chain of the nucleosome is given a different color. By clicking on the various chains it is possible to identify each one. Chains A-B and E-F form two of the four dimers in the nucleosome core, indicating that H3 and H4 can dimerize with each other. Chains C-D and G-H form the remaining two dimers, indicating that H2A and H2B can also form dimers.

(b) You can visualize the points of contact between the DNA and protein core by selecting the protein (**Select → Protein → Protein**) and displaying it in **Spacefill** format. Visualized in this format, the points of contact with the DNA double helix become more obvious. Careful examination of the points of interaction shows that almost all involve the minor groove of the DNA double helix. In addition most of the sites of contact occur on the inner face of the helix, suggesting that it is the DNA backbone that is involved. In fact approximately 50% of the noncovalent interactions between DNA and the histone core in a nucleosome are hydrogen bond interactions between the protein main chain and the phosphodiester backbone atoms in the DNA. This means that the formation of nucleosomes on a DNA strand is not completely dictated by the specific nucleotide sequence in that stretch of DNA.

There are also several sites at which the histone tails appear to penetrate the DNA supercoil. These tail regions thus represent additional sites at which the nucleosome core can bind DNA. In fact the H3 tail has been shown to interact not only with DNA but also with the H3 tail from adjacent nucelosomes. In addition, the histone H4 tail has been shown to interact with the H2A-H2B dimer in adjacent nucleosomes. These observations suggest a role for histone tails in both the stabilization of the nucleosome structure as well as the formation of higher-order structures in chromatin.

chapter

25

DNA Metabolism

STEP-BY-STEP GUIDE

Major Concepts

DNA replication adheres to certain precepts in every cellular organism.

This process is semiconservative. It begins at a specific origin, or many specific origins, and proceeds bidirectionally in a $5' \rightarrow 3'$ direction only. It is continuous on one strand and discontinuous on the other strand. The polymerization reaction requires a primer to begin synthesis and a template to direct the addition of each monomer.

Replication is an extremely complicated process and requires more than 20 enzymes and proteins.

The fidelity of transfer of the genetic message is the most important consideration, and, unlike most other biochemical reactions, the energetic expense is *not* a significant factor. Replication proceeds in stages: **initiation, elongation,** and **termination.** Each stage has many specific enzymes, though some enzymes are necessary for more than one stage.

Central to all stages of replication are DNA polymerases.

E. coli has at least three of these enzymes, and eukaryotes also have at least three. The primary property of polymerases is that they catalyze extremely accurate template-driven polymerization of deoxyribonucleotides into polynucleotides in the $5' \rightarrow 3'$ direction. Additional, and extremely important, functions of some DNA polymerases (or polymerase subunits) include **exonuclease activity,** in either the $3' \rightarrow 5'$ or $5' \rightarrow 3'$ direction, and **proofreading.** Polymerases are also categorized by a property called **processivity,** which refers to the number of nucleotides that the polymerase can add to the polymer before it dissociates from the DNA.

Because DNA synthesis can occur only in the $5' \rightarrow 3'$ direction, it proceeds continuously in one direction in the leading strand and in pieces, or discontinuously, in the lagging strand.

These small pieces are called **Okazaki fragments.** Synthesis of the lagging strand requires even more different enzymes and proteins than does leading-strand synthesis. Not surprisingly, replication in eukaryotes is more complex than in prokaryotes, though not currently as well understood.

Mutations are permanent changes in the base sequence of DNA; all cells have mechanisms to detect and repair these before they can be replicated and transmitted to the next generation.

DNA damage on one strand is detectable and repairable because the complementary strand contains the correct directions for restoration. Cells have multiple repair systems. **Mismatch repair** finds and replaces incorrectly matched (i.e., wrongly base-paired) bases after replication. Discrimination between parental and daughter strands is achieved through methylation of the parental strand. **Base-excision repair** recognizes lesions such as those produced by deamination of purines and pyrimidines; it first removes the base itself, then the rest of the nucleotide, and finally refills the gap using DNA polymerase I. Large distortions in the helical structure of DNA require the mechanism of **nucleotide-excision repair,** which uses a special enzyme, **ABC excinuclease.** This enzyme is unusual in that it makes two cuts, not one, in the DNA in order to remove the damaged portion. **Direct repair** does not involve extraction of a base or nucleotide and is thus limited to those types of damage that can be fixed by removing or reversing the damage—usually a methyl group or extra bonds between pyrimidines. In extreme cases of extensive damage to the DNA, an **error-prone repair** pathway is induced. This system allows replication even without a complete template, which increases mutation rates. Though risky, such a strategy is preferable to an inability to replicate at all.

DNA recombination processes maintain genetic diversity, repair DNA damage, and regulate gene expression.

There are different types of recombination, varying in their need for homology between DNA sequences that are recombined and in when the recombination occurs. **Homologous genetic recombination** occurs between pieces of the chromosome that are similar, but not identical, leading to small but often significant differences in the recombinants. This type of recombination is important during the crossing over of chromosomes in meiosis. It is also critical to DNA repair when there is a double-strand break or a lesion in a single strand; this is called **recombinational repair.** A model to explain the physical positioning of the recombination structures is the **Holliday intermediate,** a crossover structure that illustrates how different strands interact during the process. **Site-specific recombination** occurs only at unique DNA sequences and is responsible for a variety of functions, the most notable being the regulation of gene expression. An example of this type of recombination is the process that generates a virtually unlimited number of immunoglobulin proteins (needed to recognize all of the potential antigens that might invade a system) from a relatively limited number of immunoglobulin genes. Recombination by **transposons** is unusual in that neither homology nor specific sites are required; transposition, which must be tightly regulated, allows segments of DNA to hop from one position in the chromosome to another, in mostly random fashion. Each type of recombination requires specialized enzymes.

What to Review

Answering the following questions and reviewing the relevant concepts, which you have already studied, should make this chapter more understandable.

- Why are areas of DNA that have high concentrations of A═T base pairs easier to denature (p. 346)?
- Review what the **5′ end** and **3′ end** of a DNA strand are; this information is essential in understanding replication (and transcription) processes (pp. 329–330).
- How can noncovalent bonds contribute to the energetics (favorable or unfavorable) of a reaction (pp. 250–255)?
- Review the unusual tautomeric forms of nucleotide bases that can cause incorrect base pairing; this incorrect pairing can cause mutations (p. 331).
- NAD^+ can donate an AMP moiety to DNA ligase during catalysis. Review the structure of **NAD** to reassure yourself that this is, in fact, possible (p. 357).
- How do **topoisomerases** relieve the topological stress created by strand separation (pp. 921–922)?
- Repair of several different types of damage that can occur to DNA are discussed in this chapter; review the causes and effects of DNA damage (pp. 348–350).

Topics for Discussion

Answering each of the following questions, especially in the context of a study group discussion, should help you understand the important points of this chapter.

DNA Replication

DNA Replication Is Governed by a Set of Fundamental Rules

DNA Replication Is Semiconservative

1. If replication were *conservative,* how many different DNA bands would appear in a cesium chloride density gradient after 100 rounds of replication?

Replication Begins at an Origin and Usually Proceeds Bidirectionally

2. Why does addition of radioactive thymidine to the growth medium generate the labeled DNA molecules shown in the autoradiograms in Figure 25–3? Make sure that you really understand these experiments.

DNA Synthesis Proceeds in a $5' \rightarrow 3'$ Direction and Is Semidiscontinuous

3. In what direction are leading strands synthesized relative to the direction of movement of the replication fork? Is the synthesis continuous? In what direction are lagging strands synthesized relative to the direction of movement of the replication fork? Is the synthesis continuous?

DNA Is Degraded by Nucleases

4. What is the primary difference between an endonuclease and an exonuclease?

5. As you read through this chapter, it may be useful to make a list of the different roles played by endo- and exonucleases in DNA metabolism.

DNA Is Synthesized by DNA Polymerases

6. What is the general reaction equation of DNA polymerases?

7. What other polymerization reaction have you already learned about that requires a primer for synthesis?

8. Could it be said that a DNA polymerase with high processivity is using substrate channeling?

9. Why are relatively weak, noncovalent interactions of more importance than strong covalent bonds to the energetics of polymerization?

10. Which noncovalent interactions play the most significant roles in the process of DNA polymerization, and what are those roles?

Replication Is Very Accurate

11. List the two strategies for maintaining the very high degree of fidelity of DNA replication during polymerization itself.

E. coli Has at Least Three DNA Polymerases

12. Following the discovery of DNA polymerase I, what experimental evidence suggested that another enzyme is the primary enzyme of replication?

13. What properties of DNA polymerase III make it a more appropriate enzyme than DNA polymerase I for the primary role in *E. coli* DNA replication?

14. Why is the action of DNA polymerase I called "nick translation"?

15. How is $5' \rightarrow 3'$ exonuclease activity different from $3' \rightarrow 5'$ proofreading activity?

DNA Replication Requires Many Enzymes and Protein Factors

16. Luckily for students of biochemistry, most of the major enzymes involved in replication have names that simply describe their function. List the enzymes involved and their functions.

Replication of the *E. coli* Chromosome Proceeds in Stages

Initiation

17. A consensus is a "general agreement or common consent." What does "consensus sequence" mean?

18. Why is it significant that the replication origin contains many A═T base pairs?

19. What category of enzyme is the DnaB protein? What is its role in the initiation of DNA replication?

20. Why is it important that initiation of DNA replication is carefully regulated? What is the proposed role of DnaA in this process?

Elongation

21. What extra proteins are necessary for lagging-strand synthesis that are not needed in leading-strand synthesis?

22. In the synthesis of Okazaki fragments, in which direction does the primosome travel relative to movement of the replication fork? In which direction does DNA polymerase III travel?

23. What is the nature of the bond formed by DNA ligase? What entities does the bond link?

Termination

24. What terminates the bidirectional replication process?

Replication in Eukaryotic Cells Is More Complex

25. How do eukaryotic cells (probably) make up for the relatively slow movement of the replication fork?

26. PCNA of eukaryotic cells is analogous to which protein in prokaryotes? RPA of eukaryotic cells is analogous to which protein in prokaryotes?

DNA Repair

Mutations Are Linked to Cancer

27. Define substitution, insertion, deletion, and silent mutations.

28. Does the Ames test directly measure the carcinogenicity of compounds?

29. Why don't all cells eventually accumulate enough mutations to become cancerous?

All Cells Have Multiple DNA Repair Systems

30. Whenever possible, cells appear to use the fewest possible enzymes and enzyme pathways to accomplish their goals (e.g., consider the overlap in enzymes catalyzing reactions in both glycolysis and gluconeogenesis). Why, then, do cells go to the energetic expense of using multiple enzyme systems to accomplish the single task of DNA repair?

31. What property of DNA molecules makes it possible for DNA damage to be detected and correctly repaired?

Mismatch Repair

32. Why is there a time limit during which mismatch repair must occur?

33. How does the MutH protein mark the correct strand for repair?

Base-Excision Repair

34. Why are four types of enzymes necessary just to clear damaged DNA consisting of altered or incorrect bases?

Nucleotide-Excision Repair

35. In what way is the ABC excinuclease unusual?

Direct Repair

36. How is direct repair different from the other methods of DNA damage repair?

37. Why must the methyltransferase repair occur *before* replication?

The Interaction of Replication Forks with DNA Damage Leads to Recombination or Error-Prone Repair

38. When is error-prone repair the preferable option? What are the other options?

DNA Recombination

39. Why is DNA recombination desirable?

Homologous Genetic Recombination Has Multiple Functions

40. During which stage of meiosis does crossing over at chiasmata occur?

41. Name the three functions of homologous genetic recombination.

42. What types of damage require recombinational repair?

43. Which enzymes does branch migration require?

44. Homologous means "comparable" or "much the same," *not* identical. Homologous recombination does alter the base sequence of the DNA.

Recombination during Meiosis Is Initiated with Double-Strand Breaks

45. Why are the 3′ ends of nicked DNA used to initiate genetic exchange in homologous recombination?

Recombination Requires Specific Enzymes

46. What are the functions of the *chi* sequences?

47. How does the RecA protein interact with DNA?

48. What causes Holliday structures to form, and how are the chromosomes involved in this structure "untangled"?

All Aspects of DNA Metabolism Come Together to Repair Stalled Replication Forks

49. How does homologous recombination provide a new complementary strand from which the damaged strand can be repaired?

Site-Specific Recombination Results in Precise DNA Rearrangements

50. What are the components of a site-specific recombination system?

51. How are the various outcomes of inversion, deletion, or insertion determined?

52. Does the integration of λ phage into a bacterial chromosome disrupt the bacterial genome's functioning? Explain.

Complete Chromosome Replication Can Require Site-Specific Recombination

53. Why may DNA repair by homologous recombination also require site-specific recombination?

Transposable Genetic Elements Move from One Location to Another

54. How is transposition different from the other classes of recombination discussed?

55. What would be the consequences of unregulated transposition?

Immunoglobulin Genes Are Assembled by Recombination

56. Note that assembly of diverse antibodies is made possible by recombination processes (at the DNA level); posttranscriptional processes (at the mRNA level), and protein folding and assembly (at the protein level).

57. In what cells does the antibody-generating recombination process take place?

58. Why is genetic recombination essential to the function of the immune system?

SELF-TEST

Do You Know the Terms?

ACROSS

1. Enzymes used in base-excision repair to create abasic sites.
6. DNA lesions repaired directly or by ABC excinuclease in nucleotide-excision repair. (2 words)
7. Small pieces of DNA synthesized in the direction opposite to the direction of movement of the replication fork. (2 words)
9. Type of mutation involving infiltration of an extra base pair into the DNA sequence.
12. It would take twice as long for an *E. coli* chromosome to be duplicated if replication were not ________.
13. Traveling protein machine that helps in lagging-strand synthesis.
15. Crossover structures in homologous genetic recombination. (2 words)
17. *E. coli* has (at least) three types: I, II and II; eukaryotes have three types: α, δ, ϵ; all types catalyze $5' \rightarrow 3'$ synthesis of DNA.
18. Like glycogen synthesis, DNA polymerization requires this to begin.
19. Similar but not exactly alike; also describes a type of genetic recombination.
20. DNA-binding protein necessary in initiation and elongation steps of replication. (abbr.)
23. The replication ________ is a moving opening that leads the replication process.
24. Parental strand that provides guidance for synthesis of new DNA, using the Watson-Crick base-pairing rules.
25. Describes the DNA strand made in a discontinuous fashion.
26. The Ames test is a simple test for these.
27. Type of repair that discriminates between the parental strand and the newly synthesized strand by recognizing methylation of the template strand.

DOWN

2. In bacteria, when DNA damage is extensive, the ________ response kicks in and initiates error-prone repair.
3. Describes the DNA strand that is continuously synthesized.
4. ________-________ recombination is important in regulation of expression of certain genes; uses a recombinase.
5. Nucleoprotein filament that assembles cooperatively on single-stranded DNA.
6. Number of nucleotides added before dissociation of a polymerase from DNA; a measure of "hold."
8. Enzymes that act after formation of abasic sites.
10. Enzyme that catalyzes $3' \rightarrow 5'$ error correction or $5' \rightarrow 3'$ removal of RNA primers.
11. Describes DNA replication in which the newly synthesized DNA duplex has one newly made strand and one strand from the parental duplex.
12. Process occurring during homologous genetic recombination in which at least one strand is partially paired with each of two complementary strands. (2 words)
14. All the necessary DNA replication enzymes and proteins, in one neat package; has not yet been isolated as such.
16. "Jumping genes."
17. Mechanism that helps to ensure the integrity of DNA; occurs during polymerization.
21. Enzyme that unwinds DNA.
22. Enzyme that catalyzes formation of a phosphodiester bond between a 3′-hydroxyl at the end of one DNA strand and a 5′-phosphate at the end of another.

Do You Know the Facts?

1. One of the most demanding and remarkable of all biological processes is the replication of DNA. More than 20 proteins are involved in this complex process. Some of the more important proteins involved in DNA replication in *E. coli* are listed in the table below. Identify the stage of replication each is involved in and match it with its function from the accompanying list.

Protein	Stage	Function
Primase		
DNA pol I		
SSB		
DnaA protein		
DnaB protein		
DNA gyrase		
DNA ligase		
DNA pol III		
Tus		
Dam methylase		

(a) Erases primer and fills gaps.
(b) Synthesizes DNA.
(c) Stabilizes single-stranded regions.
(d) Helicase that begins unwinding the double helix.
(e) Opens duplex at specific sites.
(f) Introduces negative supercoils.
(g) Synthesizes RNA primers (also known as Dna G).
(h) Protein that binds to a 20 base pair *Ter* sequence.
(i) Selectively methylates DNA at the *ori*C region.
(j) Joins the ends of DNA.

2. In the laboratory, some *E. coli* cells have been grown with transcription inhibitors so that only replication could occur. You break open the cells; gently purify the DNA, separating it from the rest of the cell contents; then treat the intact DNA with radioactively-labeled antibodies to DNA ligase and DNA polymerase I. Where do you expect to find the highest concentration of these enzyme antibodies?

A. on the leading strand, in one direction from the point of origin
B. on the leading strands, in both directions from the point of origin
C. on the lagging strand, in one direction from the point of origin
D. on the lagging strands, in both directions from the point of origin
E. on all parts of the DNA molecule equally

3. Although DNA replication has very high fidelity, mutations do occur. Which of the following types of single base-pair mutations would be most likely to be a lethal mutation?

A. substitution
B. insertion
C. deletion
D. silent
E. B and C

4. You have been sitting in the sun a lot lately, though you know full well that the UV in sunlight can cause the formation of pyrimidine dimers (covalent links between adjacent pyrimidine residues) in the DNA of your skin cells. Luckily, skin cells have repair mechanisms that will most likely fix the lesions formed by UV light. Which of the following describes enzymes or mechanisms used in such repair processes?

A. ABC excinuclease recognizes the lesion and cuts out the damaged DNA.
B. RNA polymerase III fills in the damaged area.
C. DNA glycosylases remove the affected bases.
D. The MutH protein marks the strand for repair.
E. Methyltransferase transfers a methyl group from DNA to its Cys residue.

5. *Both* general recombination *and* transposon-type recombination:

A. occur between homologous regions on chromosomes.
B. are important in the repair of damaged DNA.
C. involve RecA protein.
D. generate genetic diversity.
E. The two types of recombination have nothing in common.

6. Explain how each of the following facets of DNA structure or metabolism has evolved to aid in maintaining the integrity of the genetic message—that is, to prevent alterations in the DNA sequence through repair systems or "quality control."

(a) Thymine instead of uracil in the DNA strand.

(b) Watson-Crick base pairing.

(c) Methylation of the parental strand.

7. The immune system is capable of generating antibody molecules that can bind to (and thereby inactivate) an unlimited number of foreign substances. Discuss the mechanisms used by the immune system to generate diversity in antibody binding sites.

Applying What You Know

1. What kind of results would Meselson and Stahl have seen after one generation on the new, ^{14}N medium if DNA ligase were suddenly inactived in the *E. coli* cells?

2. In eukaryotic DNA replication, DNA polymerase α has a primase activity and relatively low processivity, and DNA polymerase δ has a $3' \rightarrow 5'$ proofreading activity, lacks a primase activity, and has a very high processivity when complexed with PCNA. Both enzymes replicate DNA by extending a primer in the $5' \rightarrow 3'$ direction under the direction of a single-stranded DNA template. Why is it likely that DNA polymerase δ acts as the leading-strand synthesizing enzyme and DNA polymerase α as the lagging-strand synthesizing enzyme?

3. The kind of recombination displayed by transposons (sometimes called RecA-independent recombination) has been regarded as having far greater evolutionary significance than does general (RecA-dependent) recombination. Why would this be so?

4. Why should mutagenesis of *E. coli* using UV radiation as the mutagen be done in a laboratory without windows?

ANSWERS

Do You Know the Terms?

ACROSS

1. glycosylases
6. pyrimidine dimers
7. Okazaki fragments
9. insertion
12. bidirectional
13. primosome
15. Holliday intermediates
17. polymerases
18. primer
19. homologous
20. SSB
23. fork
24. template
25. lagging
26. mutagens
27. mismatch

DOWN

2. SOS
3. leading
4. site-specific
5. RecA
6. processivity
8. AP endonucleases
10. exonuclease
11. semiconservative
12. branch migration
14. replisome
16. transposons
17. proofreading
21. helicase
22. ligase

Do You Know the Facts?

1.

Protein	Stage	Function
Primase	Elongation	(g)
DNA pol I	Elongation	(a)
SSB	Initiation	(c)
DnaA protein	Initiation	(e)
DnaB protein	Initiation	(d)
DNA gyrase	Initiation, elongation	(f)
DNA ligase	Elongation, termination(?)	(j)
DNA pol III	Elongation	(b)
Tus	Termination	(h)
Dam methylase	Initiation	(i)

2. D
3. E
4. A
5. D
6. (a) Under typical cellular conditions, deamination of cytosine in DNA will occur in ~1 of every 10^7 cytosines in 24 hours. The product of cytosine deamination—uracil—is recognized as foreign in DNA and is removed by the base-excision repair system. If DNA normally contained uracil rather than thymine, distinguishing between the "correct" uracils and those formed from cytosine deamination would be difficult.

(b) In synthesis of a daughter strand of DNA, the template DNA strand provides the instruction through formation of the proper hydrogen bonds between A and T and between C and G. These base-pairing rules are the basis for the fidelity of replication.

(c) Methylation of the parental (template) strand distinguishes between the old and the new. Because it is the newly made strand that may have mistakes made during replication, the mismatch repair system has to be able to recognize which strand is new. Methylation also serves to differentiate "self" from "foreign" DNA. For example, restriction enzymes are produced by bacteria to degrade invading DNA; methylation of the bacterial DNA protects it from these enzymes (see p. 351).

7. Recombination mechanisms at the DNA level bring together different combinations of V and J segments in the light-chain genes of different B lymphocytes. Imperfect DNA splicing during the rearrangement of antibody genes increases the already numerous versions of each coding region. The final joining of the V-J combinations to the C region is carried out by an RNA-splicing reaction after transcription; because there are a number of different C genes, this further increases the number of possible combinations. Finally, mutation of the selected V sequences occurs at an unusually high rate during B-cell differentiation, once again increasing the number of different mRNAs (and ultimately polypeptides) that can be produced from a limited set of immunoglobulin genes.

Applying What You Know

1. DNA ligase is required to join together the Okazaki fragments of the lagging strand and to link the ends of the leading strand at the end of replication. Without this enzyme activity, there would be no neat duplexes, each with one ^{15}N strand and one ^{14}N strand. Instead, at least briefly, there would be full-length, circular ^{15}N strands (the parental strands);

a number of long fragments representing the noncircularized ^{14}N leading strands; and many small ^{14}N Okazaki fragments. The single-stranded pieces would be unlikely to exist for very long; nucleases present in many cells recognize single-stranded DNA as foreign and degrade it.

2. The leading-strand replicase (replication enzyme) needs high processivity but only occasionally needs a primer. The lagging-strand replicase requires frequent priming, but processivity is not as important because the lagging strand is synthesized in short pieces. The two eukaryotic DNA polymerases have the appropriate properties for their distinct functions: the δ form constructing the leading strand and the α form the lagging strand.

3. Because general recombination can occur only between homologous segments of chromosomes, the resulting recombinant genome will not contain drastic alterations in the encoded information and the alterations that do occur are unlikely to be lethal. Transposon recombination allows for insertions, deletions, and duplications of genes. These rearrangements, though often lethal, occasionally produce a new combination that provides some evolutionary advantage.

4. UV light induces mutations by causing the formation of pyrimidine dimers, especially between two adjacent (on the same strand) thymines. This distorts the DNA helix and interferes with proper replication, leading to an increased number of mutations in the daughter cells that survive. Many cells are so badly damaged that they are unable to replicate at all. The usual repair mechanism is excision of the damaged area of DNA followed by insertion of new bases to replace the excised portion; this is the nucleotide-excision repair system. Visible light (in this case, sunshine coming in through windows) activates another repair enzyme, a photolyase. The photolyase of *E. coli* binds to the thymine dimers and is activated on exposure to visible light, splitting the dimer and thus repairing it before replication. This decreases the number of mutations induced by the UV light. If the goal is to produce *E. coli* mutants, sunshine should be excluded from the lab.

RNA Metabolism

chapter 26

STEP-BY-STEP GUIDE

Major Concepts

RNA synthesis, or transcription, is a template-dependent process.

The enzymes of transcription, the **DNA-dependent RNA polymerases,** add ribonucleotide units to the 3′ end of the growing RNA chain by following the instructions contained in (i.e., by base pairing with) one strand of a DNA duplex, the **template strand.** The added ribonucleotides adhere to the base-pairing rules except for the addition of U instead of T. The RNA therefore has a sequence identical to that of the **nontemplate strand** of the DNA, except for the substitution of a U for every T in the DNA. Polymerization occurs only in the 5′ → 3′ direction, as does DNA synthesis. No primer is required for transcription.

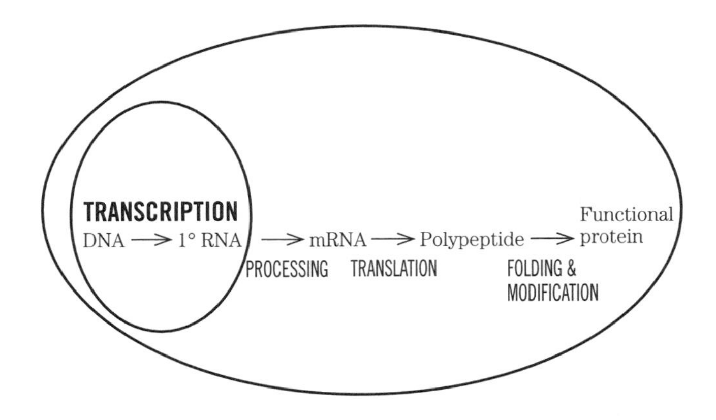

RNA synthesis is initiated at specific DNA sequences called promoters.

The RNA polymerase of *E. coli* has a specific subunit, the σ **subunit,** that directs the enzyme to the correct promoter. Many promoters from different bacterial species have been found to contain similarities in their sequences at positions −35 and −10 base pairs "upstream" from the start site of transcription. It is presumably these **consensus sequences** that are recognized by the σ subunit. Once properly positioned, the rest of the RNA polymerase enzyme unwinds a portion of the DNA, forming the **transcription bubble,** and catalyzes polymerization of ribonucleotides. The unwinding of the DNA duplex generates supercoils in front of the transcribing enzyme, which can be relaxed by topoisomerase. The regulation of gene transcription is probably the most critical regulatory point in the process of gene expression; some of the mechanisms employed by cells to regulate transcription are discussed in Chapter 28.

Termination of transcription is signaled by specific DNA sequences.

One class of termination signal relies on the aid of a protein called **rho** (ρ) and the other class does not. In both types of termination, the formation of an RNA hairpin structure helps in dissociation of the RNA transcript from the DNA template.

Eukaryotic transcription has three types of RNA polymerases.

RNA polymerase I synthesizes preribosomal RNA, which contains the precursors for the 5.8S, 18S, and 28S **ribosomal RNAs (rRNAs).** The major role of RNA polymerase II is to synthesize **messenger RNAs (mRNAs),** which carry the genetic message from the chromosome to the ribosomes. RNA polymerase III synthesizes **transfer RNAs (tRNAs),** which act as adapters between mRNA and amino acids. Eukaryotic promoters are more variable than are those of prokaryotes; eukaryotes also need **transcription factors** to modulate the binding of RNA polymerases to the promoter sequences.

RNA polymerases are subject to inhibition by a variety of substances.

Some inhibitors **intercalate** between base pairs in the DNA, disrupting transcription. Others bind specifically to parts of certain RNA polymerases; these specific inhibitors not only are poisons, but can be used as experimental tools in studying the mechanics of transcription.

Once transcribed, the RNA in eukaryotes is extensively processed.

A variety of alterations to the primary transcripts occur in the nucleus. One process involves removal of noncoding regions of the RNA called **introns** and the careful **splicing** of the remaining **exons** to form a contiguous sequence. There are four classes of introns, and their cutting and splicing reactions vary in their requirement for high-energy cofactors and for specialized RNA-protein complexes. In some cases, the same primary transcript can be spliced in different ways (alternative splicing) to produce a variety of different mRNAs and thus different protein products. One of the biggest surprises in biochemistry came from the study of splicing in introns: in some cases, protein enzymes are not used in catalysis and the catalytic component is the RNA itself. Studies of these **ribozymes** have shown them to have many of the properties of protein enzymes. Other modifications of the primary transcript include addition of a **7-methylguanosine cap** on the 5′ end and a **poly(A) tail** to the 3′ end. Both of these steps are thought to protect the mRNA from enzymatic attack by nucleases in the cytoplasm, to which they are exported after the processing reactions are completed. This is not long-term protection, however; mRNAs must be degraded after their protein products are no longer needed. Additional modifications of RNA molecules include base methylation, deamination, and reduction.

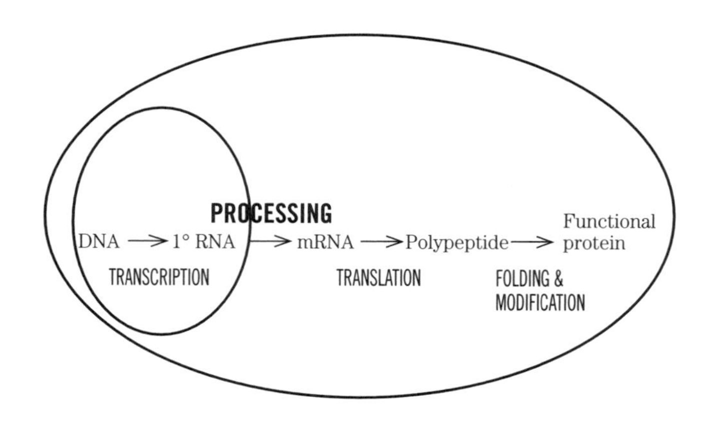

DNA and RNA can be synthesized from an RNA template.

Retroviruses use RNA as their genetic material and can "reverse transcribe" RNA into DNA. They contain the enzyme **reverse transcriptase.** This enzyme has some properties similar to DNA and RNA polymerases, but its job is to catalyze the synthesis of a DNA duplex that is complementary to the viral RNA. The DNA so formed is then integrated into the host chromosome, where it can hide for many cell generations, eventually reemerging to generate new virus particles. All reverse transcriptases, but in particular that of **HIV,** lack any proofreading capability and so have a higher error rate. This means a faster rate of evolution of new viral strains, making treatment of the host's virus-related disease particularly difficult. Some retroviruses contain cancer-causing genes called **oncogenes.** These sequences are derived from normal cellular genes that often encode proteins involved in cell growth and development.

The enzyme **telomerase,** which prevents the progressive shortening of linear eukaryotic chromosomes, acts as a specialized reverse transcriptase in that it uses an internal RNA template for the synthesis of a DNA segment. Some viruses have RNA-dependent RNA polymerases, or **replicases,** which act only on their own RNA. The structural and functional complexity of RNA has led to speculation that the "RNA world" was important in the transition from prebiotic chemistry to life.

What to Review

Answering the following questions and reviewing the relevant concepts, which you have already studied, should make this chapter more understandable.

- The term **consensus sequence** is important in this chapter in relation to promoters. It means "a DNA or amino acid sequence consisting of residues that occur most commonly at each position within a set of similar sequences." These sequences were first mentioned in reference to proteins regulated by cAMP. Why do these proteins have such a sequence (p. 452)?
- The technique of **DNA–RNA hybridization** was used in the discovery of introns. What characteristics of nucleic acids does this procedure take advantage of? How is the hybridization done (p. 347)?
- Transposon insertion in bacteria is similar to integration of retroviral DNA into its host cell DNA. Review the properties of these "hopping genes" (pp. 970–973).
- Cell-growth and cell-division factors such as tyrosine kinases, growth factors and their receptors,

and G proteins all can be altered so that their activity is uncontrolled; cell growth is then unfettered and cancer can result. Oncogenes are cancer-causing genes, some of which are derived from the normal proto-oncogenes for such growth factors. Review the mechanisms by which these factors wield such power (Chapters 13 and 23).

- The discussion about the "RNA world" is extended in this chapter. Review the possible "time-line" for the transition from the prebiotic RNA world to the biotic DNA world presented in Chapter 3 (pp. 75–76).

Topics for Discussion

Answering each of the following questions, especially in the context of a study group discussion, should help you understand the important points of this chapter.

DNA-Dependent Synthesis of RNA

RNA Is Synthesized by RNA Polymerases

1. How is transcription similar to and different from replication?

2. How large is the transcription bubble?

3. What type of supercoil is generated in front of the moving RNA polymerase? Behind the polymerase? How is the strain from this supercoiling relieved?

4. Is the template strand always the same DNA strand for different genes along the chromosome? Explain.

5. Note that the nontemplate (coding) strand of DNA is identical to the RNA transcript except that there is a U in the RNA for every T in the DNA.

6. Why is a mistake in an RNA molecule less of a catastrophe than a mistake in DNA?

RNA Synthesis Is Initiated at Promoters

7. Note that the sequences of the promoters shown in Figure 26–5 are those of the *coding* strand. What would be the DNA sequence of the *template* strand for each of these promoters?

8. What is the function of the σ subunit of RNA polymerase? How can different σ subunits coordinate the expression of different sets of genes?

RNA Polymerase Leaves Its Footprint on a Promoter

9. Does the "footprint" on the gel appear because of *more* cuts in the DNA, or *fewer*?

Transcription Is Regulated

10. Why does it make energetic sense for the cell to regulate transcription at its early stages?

Specific Sequences Signal Termination of RNA Synthesis

11. In ρ-independent termination, in which nucleic acid (DNA or RNA) is the termination sequence? Which nucleic acid forms the hairpin structure?

12. Why might a poly A═U hybrid be relatively unstable?

Eukaryotic Cells Have Three Kinds of Nuclear RNA Polymerases

13. List the functions of the three different eukaryotic RNA polymerases.

RNA Polymerase II Requires Many Other Protein Factors for Its Activity

14. Which of the transcription factors act to unwind the DNA?

15. Which transcription factors use ATP? For what purpose(s)?

16. Why could elongation factors also be named "suppression-of-arrest" factors?

17. What are the different roles of TFIIH?

DNA-Dependent RNA Polymerase Can Be Selectively Inhibited

18. How do intercalators inhibit transcription? Compare this with the mechanisms of inhibition by rifampicin and α-amanitin.

RNA Processing

19. Are transcribed introns removed from primary transcripts?

20. Where in the eukaryotic cell do RNA processing reactions take place?

The Introns Transcribed into RNA Are Removed by Splicing

21. Why did the intron sequences of the chicken ovalbumin DNA not hybridize with any part of the ovalbumin mRNA? Did the mRNA hybridize with the template or nontemplate strand?

RNA Catalyzes Splicing

22. What is unusual about the use of a guanosine nucleotide as a cofactor in group I intron splicing reactions?

23. Which class(es) of introns use catalytic RNAs?

24. Which class(es) of introns use snRNPs?

25. The spliceosome complex represents a major investment by the cell, indicating that the splicing process is very important. Why must the intron splicing reactions be so exquisitely accurate?

Eukaryotic mRNAs Undergo Additional Processing

26. Attaching the poly(A) tail uses a lot of ATP. What is the (probable) function of the poly(A) tail that would make this energy outlay worthwhile to the cell?

27. What percentage of the length of the ovalbumin gene is actually reflected in the mature transcript?

Multiple Products Are Derived from One Gene by Differential RNA Processing

28. What are the two mechanisms for differential processing of complex transcripts?

Ribosomal RNAs and tRNAs Also Undergo Processing

29. What further processing reactions of precursor transcripts are required to produce mature rRNAs and tRNAs?

Some Events in RNA Metabolism Are Catalyzed by RNA Enzymes

30. What are the similarities between ribozymes and protein enzymes? What are the differences?

31. What can L-19 IVS do? Does it do this *in* the cell?

Cellular mRNAs Are Degraded at Different Rates

32. Why is it essential to cell function that mRNAs are eventually degraded?

Polynucleotide Phosphorylase Makes Random RNA-like Polymers

33. How does polynucleotide phosphorylase differ from other enzymes that synthesize nucleic acids?

RNA-Dependent Synthesis of RNA and DNA

34. What is the "central dogma" concerning the flow of genetic information? What processes had to be added when the role of RNA templates was discovered?

Reverse Transcriptase Produces DNA from Viral RNA

35. What are the three typical retroviral genes? What protein product(s) does each code for?

36. How are reverse transcriptases similar to DNA and RNA polymerases? How are they different?

37. Why is the lack of a proofreading function in reverse transcriptases medically important?

Retroviruses Cause Cancer and AIDS

38. What is the *mechanism* by which overexpression of the *src* gene in Rous sarcoma virus causes unregulated cell division? (Hint: See Chapter 13.)

39. Are proto-oncogenes "normal" genes?

40. How is the HIV reverse transcriptase different from other known reverse transcriptases? Why is this of concern?

Fighting AIDS with Inhibitors of HIV Reverse Transcriptase

41. Why must AZT be given as the *unphosphorylated* nucleoside?

Many Transposons, Retroviruses, and Introns May Have a Common Evolutionary Origin

42. Why are retrotransposons trapped within a single cell?

43. What is the evidence that introns originated as molecular "parasites"?

Telomerase Is a Specialized Reverse Transcriptase

44. What problem does telomerase solve?

45. What is the function of a T loop?

Some Viral RNAs Are Replicated by RNA-Dependent RNA Polymerase

46. Why aren't RNA replicases generally useful in recombinant DNA technology?

RNA Synthesis Offers Important Clues to Biochemical Evolution

47. What is the evidence that implies that adenine was the first nucleotide constituent?

48. Evolution is defined as the genetic change in a population over generations. It occurs when natural selection produces changes in the relative frequencies of alleles in a population's gene pool. What is the (*unnatural*) selection present in the SELEX technique?

49. What is DNA a better molecule than RNA for the purpose of long-term storage of genetic information? (Hint: See Chapter 10.)

SELF-TEST

Do You Know the Terms?

ACROSS

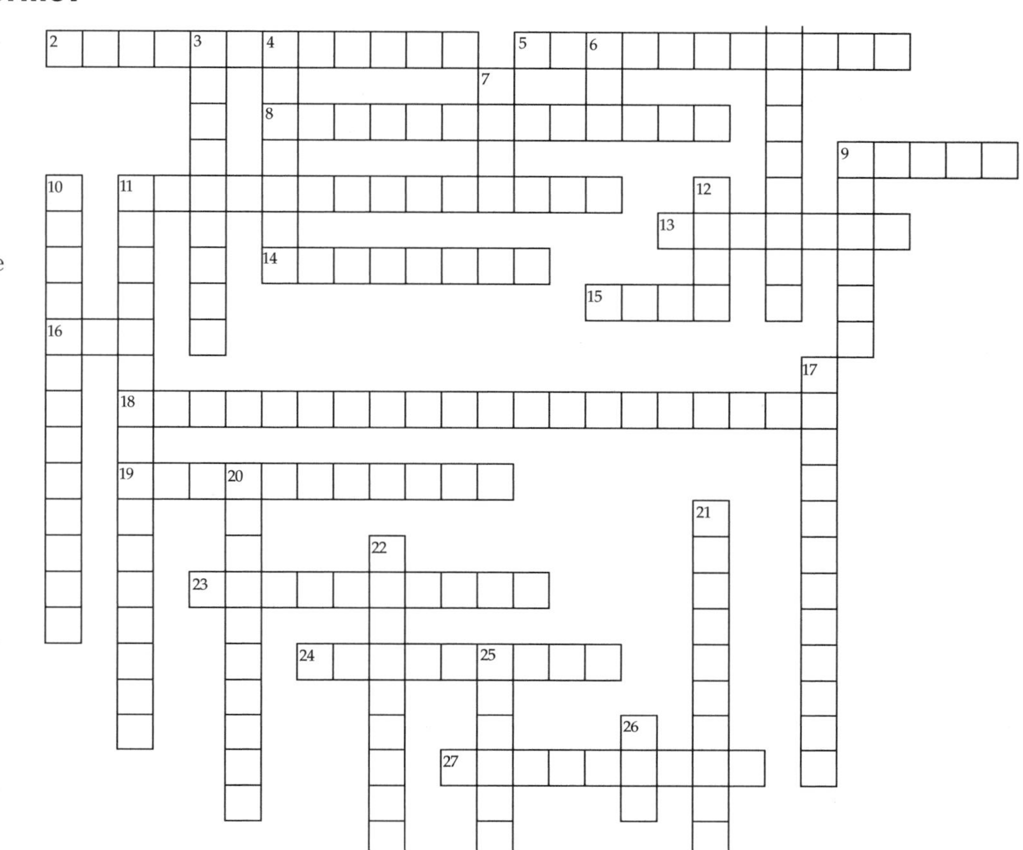

2. Technique that can pinpoint promoter locations.
5. Complex ________ produce two or more *different* mRNAs and polypeptides.
8. Proteins necessary for the activity of RNA Pol II are ________ factors.
9. "Accelerated evolution in a test tube."
11. DNA-directed enzymes that require a DNA template and ATP, GTP, UTP, and CTP. (2 words)
13. The newly synthesized RNA molecule is the ________ transcript.
14. As in film editing, cutting and ________ an RNA by one frame too few or too many can be devastating to the final product.
15. Molecule that carries the genetic message from the chromosome to the ribosomes. (abbr.)
16. Protein factor that signals termination. (word)
18. RNA-dependent DNA polymerase. (2 words)
19. DNA strand identical in sequence to the RNA transcript, except with Ts instead of Us.
23. Term describing the form of RNA polymerase that contains the σ subunit.
24. Areas of DNA containing consensus sequences.
27. A poison; why you shouldn't eat strange mushrooms.

DOWN

1. Type of enzymes; RNase P, for example.
3. —$(AAAAAAAAAAAAAAAAAAAA)_4$—OH(3′), at least. (2 words)
4. These are lacking in histone genes but present in almost all other vertebrate genes.
6. Structural analog of deoxythymidine. (abbr.)
7. Adapter molecule between mRNA and amino acid. (abbr.)
9. Not Saturday morning cartoon characters, but complexes containing RNA and protein and important in group III splicing reactions. (abbr.)
10. Actinomycin D and acridine, for example.
11. "Defective viruses, trapped in one cell."
12. Nucleic acid serving as a component of ribosomes. (abbr.)
17. Molecular proof that genetic information can flow "backward."
20. Enzyme that prevents the slow nibbling away of the ends of chromosomes.
21. Positive ________ form ahead of the transcription bubble.
22. Good genes gone bad; for example, the *src* gene.
25. Consensus sequence of the −10 region in *E. coli.*
26. Virus containing a tragically inaccurate reverse transcriptase. (abbr.)

Do You Know the Facts?

1. The processes of DNA replication and transcription of RNA are similar in some respects and different in others. Indicate whether each of the following describes replication only, transcription only, or both.

	Replication	Transcription	Both
(a) A specific region of the DNA is recognized and bound by the polymerase.	______	______	______
(b) The direction of polymerization is 5′ → 3′.	______	______	______
(c) The direction of enzyme movement on the template strand is 3′ → 5′.	______	______	______
(d) The process includes its own 3′ → 5′ exonuclease proofreading mechanism.	______	______	______
(e) The mechanism of reaction is attack by the 3′-OH group of the pentose on the α-phosphorus of an incoming nucleoside triphosphate.	______	______	______
(f) The process requires a primer.	______	______	______

2. Match each enzyme in the table with its function in eukaryotic RNA synthesis, from the list below. Indicate (with a check mark) which inhibitor(s) affect each enzyme.

		Inhibited by:		
Enzyme	**Principle Function**	**α-Amanitin**	**Rifampcin**	**Actinomycin D**
RNA polymerase I				
RNA polymerase II				
RNA polymerase III				

(a) Synthesizes mRNA.
(b) Synthesizes tRNAs.
(c) Synthesizes preribosomal RNA.

3. Consider the mRNA sequence: (5′) AAUGCAGCUUUAGCA (3′). The sequence of the *coding* strand of DNA is:
A. (5′) ACGATTTCGACGTAA (3′)
B. (3′) TTACGTCGAAATCGA (5′)
C. (5′) AATGCAGCTTTAGCA (3′)
D. (5′) AAUGCAGCUUUAGCA (3′)
E. (3′) AATGCAGCTTTAGCA (5′)

4. ________ is found *within* the transcription bubble.
A. Positive supercoils in the DNA
B. Negative supercoils in the DNA
C. An intact DNA duplex
D. A short DNA-RNA hybrid
E. An RNA primer

5. The σ^{70} subunit of the *E. coli* RNA polymerase:

A. acts as the catalytic site for polymerization.
B. recognizes promoters.
C. has a proofreading function.
D. increases the processivity of the enzyme.
E. recognizes termination signals.

6. In ρ-independent termination, the termination sequence is on the ________ and the hairpin structure is formed by the ________.

A. DNA; RNA
B. RNA; DNA
C. DNA; ρ protein
D. RNA; RNA
E. DNA; DNA-RNA hybrid

7. Indicate (yes/no) whether each of the following events can occur during the processing of eukaryotic mRNA transcripts.

________ **(a)** Attachment of a poly(A) tail to the 5′ end of the transcript.
________ **(b)** Methylation of all G residues.
________ **(c)** Excision of introns.
________ **(d)** Conversion of standard bases to modified bases such as inosine.
________ **(e)** Splicing together of exons.
________ **(f)** Differential cutting and splicing to produce two different proteins.

8. On Mars, it seems, the genetic material is double-stranded RNA rather than DNA. Martian RNA polymerase uses the double-stranded RNA as a template to make mRNA; it is an RNA-directed RNA polymerase. Assume all other aspects of transcription and translation and the functions and integrity of the enzymes involved are the same on Mars as on Earth (i.e., RNA has the same function, the same base pairs are complementary to each other, and chemistry in general is the same). Over time, the integrity of the genetic message in Martian RNA would: (Hint: Check Chapter 10 as well!)

A. decline more rapidly than that of DNA because of the lack of editing function of RNA polymerase.
B. be comparable with that of DNA.
C. decline more rapidly than that of DNA because the deamination of cytosine to uracil in RNA would confuse the message.
D. be greater than that of DNA because the message would be more direct: RNA → protein, with one less step in which mistakes could happen.
E. A and C are true.

9. Indicate whether each of the following is true of ribozymes only, protein enzymes only, or both.

	Ribozymes	Protein enzymes	Both
(a) Base-pairing reactions can help align an RNA substrate for catalysis.	________	________	________
(b) They can be denatured.	________	________	________
(c) They are saturable.	________	________	________
(d) They are recycled for use in the same reaction several to many times.	________	________	________
(e) They are subject to inhibition by substrate analogs.	________	________	________

10. What are the various functions of TFIIH?

Applying What You Know

1. The antibiotic *cordycepin,* an adenosine analog that lacks a 3′-OH group, inhibits bacterial RNA synthesis. What is the mechanism of this inhibition?

2. Ethidium bromide is an intercalator. It is used as a fluorescent dye specific for nucleic acids and is often used to stain DNA in agarose gels. When handling this substance, laboratory personnel should always wear gloves. Why?

3. Suppose you wished to separate out all the *mature* (fully processed) mRNAs from a particular eukaryotic cell lysate. Suggest a way to partition the mature mRNAs from the primary transcripts.

4. The only mature eukaryotic mRNAs that generally lack poly(A) tails are histone mRNAs.

(a) What could you guess about these mRNAs, given what is known about the function of poly(A) tails?

(b) Histone genes are among the very few vertebrate genes that lack introns. Can you suggest a reason why histones do not have introns?

5. β-Thalassemia is a class of diseases in which the β-globin gene of hemoglobin is barely expressed. One form of the disease appears to result from a single nucleotide change. If this base change does *not* result in an alteration in the encoded amino acid sequence, what other reasons can you think of for its having such severe consequences for the protein product?

ANSWERS

Do You Know the Terms?

ACROSS

2. footprinting
5. transcripts
8. transcription
9. SELEX
11. RNA polymerases
13. primary
14. splicing
15. mRNA
16. rho
18. reverse transcriptase
19. nontemplate
23. holoenzyme
24. promoters
27. α-amanitin

DOWN

1. ribozymes
3. poly(A) tail
4. introns
6. AZT
7. tRNA
9. snRNPs
10. intercalators
11. retrotransposons
12. rRNA
17. retroviruses
20. telomerase
21. supercoils
22. oncogenes
25. TATAAT
26. HIV

Do You Know the Facts?

1. **(a), (b), (c),** and **(e)** both; **(d)** and **(f)** replication only.

2.

Enzyme	Principle Function	Inhibited by:		
		α-Amanitin	Rifampcin	Actinomycin D
RNA polymerase I	(c)			✓
RNA polymerase II	(a)	✓		✓
RNA polymerase III	(b)	✓ (at high conc.)		✓

Note that rifampicin would not inhibit any of these enzymes; it specifically reacts with the β subunit of *prokaryotic* RNA polymerases.

3. C
4. D
5. B
6. A
7. **(a)** N; **(b)** N; **(c)** Y; **(d)** N; **(e)** Y; **(f)** Y
8. E
9. **(a)** ribozymes only; **(b)** both; **(c)** both; **(d)** in general, protein enzymes, but true for some ribozymes; **(e)** both
10. TFIIH:
- acts as a necessary part of the closed complex.
- acts as a helicase to unwind DNA.
- has a kinase activity that phosphorylates RNA Pol II, causing a conformational change that initiates transcription.
- interacts with damaged DNA and recruits the nucleotide-excision repair complex.

Applying What You Know

1. The addition of cordycepin to the 3′ end of a growing RNA strand, as will occur in 5′ → 3′ chain growth, prevents any further elongation because the incorporated antibiotic lacks a 3′-OH group and the next nucleoside triphosphate cannot be added.

2. Intercalating agents such as ethidium bromide inhibit RNA polymerases; actinomycin D and acridine are other examples. They tightly bind to duplex DNA and strongly inhibit transcription, presumably by interfering with the passage of the polymerases. Intercalators therefore cause inhibition of transcription—clearly to be avoided!

3. Affinity chromatography using poly(T) molecules as ligand would preferentially bind mature mRNAs by their poly(A) tails.

4. (a) Histone mRNAs have much shorter lifetimes than do other mRNAs: <30 min in the cytosol, compared with hours or days for other mRNAs. Although there are other factors involved in the rate of histone mRNA degradation, this is one piece of evidence that poly(A) tails provide RNAs with some protection (though not permanently) from degradation in the cytosol.

(b) Given the importance of histones' function and their relatively conserved sequences across species, perhaps the possibility of incorrect splicing has proven too great a danger.

5. The nucleotide change could be at an intron-exon junction, altering the removal of introns and/or splicing of exons. Another possibility is a defect in the promoter of the gene, leading to inefficient transcription. Premature termination is another feasible cause, due to a change in the nucleotide sequence producing an early signal for termination. All of these single-nucleotide effects would reduce β-globin gene expression.

chapter 27

Protein Metabolism

STEP-BY-STEP GUIDE

Major Concepts

The amino acid sequence of proteins is determined by the sequence of nucleotides in coding regions of DNA.

Regions of DNA that code for proteins are transcribed into a complementary **messenger RNA** (mRNA) molecule. Each amino acid is coded for by a sequence of three nucleotide residues in the mRNA. The code is degenerate in that more than one triplet, or **codon,** can be used to specify a given amino acid. For most codons, the third base tends to pair more loosely with the corresponding base of its tRNA anticodons (they "wobble"), and this has some interesting consequences for the code. The **initiation** and **termination** points of the protein-coding region are also specified by triplet codons. In general, regions of DNA containing an initiation codon followed by at least 50 codons specifying amino acids usually produce functional proteins. Codons do not usually overlap (though there are exceptions in viral systems) and there is no "punctuation" between codons.

Protein synthesis occurs in the cytosol on ribosomes.

Ribosomes are cytoplasmic complexes made up of proteins and **ribosomal RNA** (rRNA). After leaving the nucleus, mRNA molecules become associated with ribosomes. The molecule of mRNA serves as a template and the ribosome provides the structure and catalytic activity for protein synthesis.

The mRNA codons are recognized by transfer RNA (tRNA) anticodons.

Transfer RNA molecules contain a single strand of RNA that is folded (due to intramolecular base pairing) into a cloverleaf secondary structure. Codon recognition is the result of base pairing between the nucleotides of the mRNA strand and a three-nucleotide region **(anticodon)** on one "arm" of the tRNA molecule.

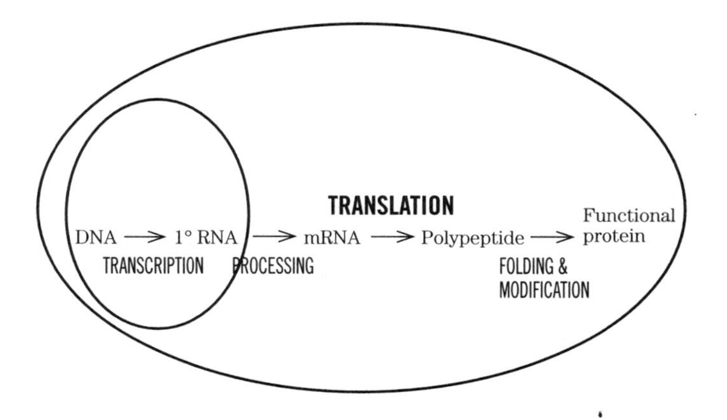

Because a given tRNA must recognize both a *specific* mRNA codon and a specific amino acid, tRNA molecules can be thought of as adapters linking two dissimilar entities.

The 3′ end of the tRNA molecule is the point of attachment for its amino acid.

tRNAs with the same anticodon sequence will bind the same amino acid at their 3′ end. Amino acids are first activated by ATP and then attached to their tRNA to form an **aminoacyl-tRNA.** This reaction, driven by the hydrolysis of pyrophosphate and essentially irreversible, is catalyzed by **aminoacyl-tRNA synthetases.** Each enzyme is specific for *both* the tRNA and the amino acid in order to ensure that the genetic information is faithfully translated into a specific amino acid sequence.

Polypeptide synthesis is initiated at the amino-terminal end of the protein and requires an AUG codon.

The AUG start condon, which codes for methionine, and a specific class of tRNA that binds *N*-formylmethionine (fMet) is required for **initiation.** Protein synthesis in bacteria begins with the formation of the 70S initiation complex, which contains the following: both the 30S and 50S ribosomal subunits, the mRNA molecule, the fMet-tRNA, three proteins called **initiation factors,** GTP, and Mg^{2+}. The fMet-tRNA is bound to one of two tRNA-binding sites on the initiation complex, referred to as the **P** (for peptidyl) **site.**

Polypeptide elongation involves the formation of peptide bonds between amino acids on adjacent tRNAs.

Elongation requires the 70S initiation complex, the tRNA specified by the next codon in the mRNA sequence, three proteins called **elongation factors,** and GTP. The incoming tRNA is first bound to the **A** (for aminoacyl) **site** of the ribosome complex. Next, a peptide bond is formed between the initial methionine and the amino acid on the next tRNA; this reaction is probably catalyzed by the 23S rRNA of the 50S ribosomal subunit. The dipeptide is then translocated to the P site of the ribosome complex and the uncharged tRNA is released from the E (for exit) site in preparation for a new elongation cycle. Proofreading is limited at this stage to the codon-anticodon pairing; no check of the amino acid attached to the tRNA is carried out during elongation.

Termination of polypeptide synthesis requires one of three termination codons and one of three termination factors.

In **termination,** each of the three termination codons is recognized by one of the three **termination** or **release factors.** Binding of the release factor causes hydrolysis of the polypeptide chain from the P-site tRNA, release of these components from the P site, and dissociation of the ribosomal complex into its subunits. Protein synthesis is energetically very expensive but understandably so; the cell must synthesize its molecular "Machinery", which includes the enzymes, transport proteins, defense systems, regulatory proteins, and all other polypeptides that direct a cell's metabolism.

Some antibiotics are naturally occurring inhibitors of prokaryotic protein synthesis.

They are useful clinically and as tools for basic research into the processes of protein synthesis.

Polypeptide chains undergo posttranslational modification.

Protein folding and processing occur following protein synthesis. **Protein folding** is largely determined by noncovalent interactions between amino acids in the polypeptide chain. **Protein modification** includes (1) removal of the amino-terminal Met residue; (2) acetylation of the carboxyl terminus; (3) covalent modification of individual amino acids in the chain, including phosphorylation, glycosylation, methylation, addition of isoprenyl groups, and addition of prosthetic groups; (4) proteolytic cleavage of precursor proteins into smaller, active forms; and (5) formation of disulfide bonds.

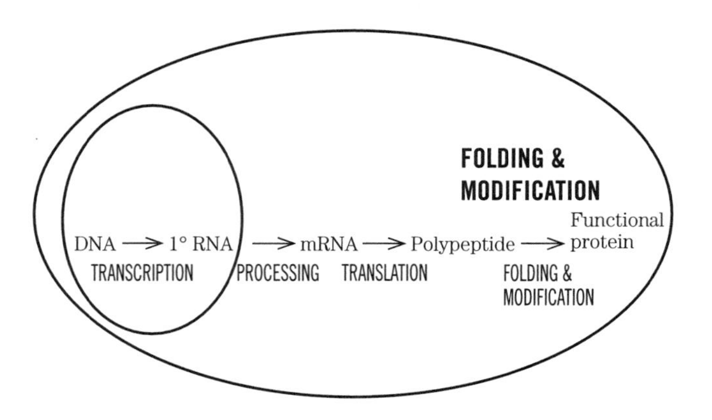

In addition, certain proteins are directed to specific regions in the cell by information contained in signal sequences.

These sequences determine whether a protein will be inserted into the membrane or translocated into the lumen of the endoplasmic reticulum (where certain of the posttranslational modification reactions occur in eukaryotes) for eventual secretion. Following translocation, signal sequences are proteolytically removed. In addition, there are nuclear localization signals, which are not removed, that direct some proteins to the nucleus.

The mechanisms triggering the degradation of proteins are poorly understood.

One possible signal may be the specific amino-terminal residue of a protein: certain amino-terminal amino acids appear to have a stabilizing effect (in terms of half-life of the protein) whereas others have a destabilizing effect. In eukaryotes, proteins are marked for destruction by the covalent attachment of one or more molecules of the protein **ubiquitin.** The mechanism of targeting by ubiquitin is unknown.

What to Review

Answering the following questions and reviewing the relevant concepts, which you have already studied, should make this chapter more understandable.

- Since proteins are composed of **amino acids** linked by **peptide bonds** it will be helpful to review the properties of individual amino acids and the details of peptide bond formation (Chapter 5, especially pp. 118–120, 126).

- The function of proteins is highly dependent on their structure, which is the result of protein folding and posttranslational modifications. Review the various factors that affect **protein folding** and review the **levels of protein structure** (Chapter 6).
- Protein folding and final structure are influenced by the presence of hydrophobic regions. Review the properties of soluble proteins compared with membrane-associated proteins (pp. 396–403).
- **Noncovalent interactions** between nucleic acids are critically important in the translation of mRNA sequences into proteins. Review the different types of interactions that affect base-pair formation (pp. 331–332), the formation of hybrid duplexes (p. 347), and the formation of secondary structures by DNA and RNA molecules (pp. 339–341).

Topics for Discussion

Answering each of the following questions, especially in the context of a study group discussion, should help you understand the important points of this chapter.

The Genetic Code

1. What are the essential elements of the protein synthetic machinery in cells?

2. Why is DNA not *directly* involved in this process?

The Genetic Code Was Cracked Using Artificial mRNA Templates

3. What pieces of evidence indicated that the genetic code was based on "three-letter words"?

4. What is the difference between overlapping and nonoverlapping codes?

5. What is meant by "reading frame" and why is it so critical to the production of a functional protein?

6. What technological developments have occurred since the breaking of the genetic code that would have made this accomplishment relatively straightforward? (Hint: See Chapter 12.)

7. Why are termination codons also called "nonsense" codons? Why isn't the initiation codon a "nonsense" codon?

8. What is the relationship between an "open reading frame" and a gene?

9. At which position in the codon is the genetic code *usually* degenerate? What are the exceptions?

Wobble Allows Some tRNAs to Recognize More than One Codon

10. What are the cellular advantages of having a "wobble" nucleotide in codons? (You should be able to come up with at least three from the information in the text.)

Overlapping Genes in Different Reading Frames Are Found in Some Viral DNAs

11. Most examples of functional proteins produced by frameshifting and "genes within genes" are found in viruses and bacteria. What are the probable reasons for this?

Protein Synthesis

The Ribosome Is a Complex Supramolecular Machine

12. How do RNA molecules contribute to the function of ribosomes?

13. What are the contributions of the protein components of ribosomes?

Transfer RNAs Have Characteristic Structural Features

14. What forces are responsible for the three-dimensional structure of tRNA?

15. Why does the anticodon arm always contain seven unpaired nucleotides?

Stage 1: Aminoacyl-tRNA Synthetases Attach the Correct Amino Acids to Their tRNAs

16. How are amino acids correctly paired with (added to) the tRNA that contains the appropriate anticodon?

17. What makes this process irreversible?

18. What is the nature of the bond that is formed?

Proofreading by Aminoacyl-tRNA Synthetases

19. Which aminoacyl-tRNA synthetases are *most* likely to require a proofreading capability? Which aminoacyl-tRNA synthetases probably do *not* have a proofreading capability?

20. Why is a proofreading capability more important for the enzymes involved in DNA replication than for aminoacyl-tRNA synthetases?

Interaction between an Aminoacyl-tRNA Synthetase and a tRNA: A "Second Genetic Code"

21. What constitutes a "codon" in the second genetic code?

Stage 2: A Specific Amino Acid Initiates Protein Synthesis

22. Which end of a protein is coded for by the 3′ end of an mRNA molecule? Which end is coded for by the 5′ end?

23. Why is an initiation codon critical to correct protein synthesis?

24. Suggest why there is only *one* initiation codon, but three termination codons.

25. Is fMet found in the proteins of vertebrate cells?

The Three Steps of Initiation

26. What additional information can you now add to your answer to the question: How do RNA molecules contribute to the function of ribosomes?

27. What is the role of GTP in this process?

28. What is the maximum number of tRNA molecules that can bind to a ribosome?

29. *Cell Map:* If you still have room, draw an initiation complex using different colors to indicate the different species of RNA involved.

Initiation in Eukaryotic Cells

30. How is the initial AUG distinguished from internal AUGs in eukaryotes?

Stage 3: Peptide Bonds Are Formed in the Elongation Stage

31. What additional elements are required for protein elongation, besides the 70S initiation complex?

Elongation Step 1: Binding of an Incoming Aminoacyl-tRNA

32. What elements are required for binding tRNA to the A site?

Elongation Step 2: Peptide Bond Formation

33. What additional information can you now add to your answer to the question: How do RNA molecules contribute to the function of ribosomes?

Elongation Step 3: Translocation

34. How is movement of mRNA through the ribosome accomplished?

35. What are the roles of ATP and GTP in the elongation process? How much energy (in terms of high-energy phosphate bonds) is expended in the formation of one peptide bond?

Proofreading on the Ribosome

36. What is the molecular basis for proofreading during elongation? In what way does this molecular mechanism support the concept of a second genetic code?

Stage 4: Termination of Polypeptide Synthesis Requires a Special Signal

37. How do termination factors "recognize" the UAA, UAG, or UGA termination codons?

Energy Cost of Fidelity in Protein Synthesis

38. At which point(s) is an energy input required in the process of translation?

Rapid Translation of a Single Message by Polysomes

39. Why are polysomes especially important to the process of protein synthesis in bacteria?

Stage 5: Newly Synthesized Polypeptide Chains Undergo Folding and Processing

40. Does protein folding require an input of energy?

41. Which amino acid residues can be modified following translation?

42. Are all posttranslational modifications permanent?

43. Give some of the reasons why proteins require posttranslational modification.

Protein Synthesis Is Inhibited by Many Antibiotics and Toxins

44. Some antibiotics selectively block bacterial protein synthesis, which is the basis for their therapeutic effects. What are some of the differences (encountered in this chapter) between bacterial and eukaryotic protein synthesis that may account for the selectivity of these agents?

Protein Targeting and Degradation

45. Which properties of amino acids could contribute to the targeting function of signal sequences?

Posttranslational Modification of Many Eukaryotic Proteins Begins in the Endoplasmic Reticulum

46. Signal sequences on proteins destined for the endoplasmic reticulum (ER) usually contain a core of hydrophobic amino acids. What does this suggest about the signal recognition particle?

47. Proteins destined to be secreted by cells must somehow make their way into the lumen of the ER. Why is a peptide translocation complex required for the successful production of these secretory proteins? How do you think this complex functions? (Hint: Consider what you have learned about the properties of proteins and cell membranes.)

Glycosylation Plays a Key Role in Protein Targeting

48. Most membrane proteins and secretory proteins are glycosylated in the lumen of the ER. What consequence does this have on the protein's topology?

49. Based on the information on hydrolase sorting, what appears to be the molecular basis for the use of oligosaccharides as targeting signals?

Proteins Are Targeted to Mitochondria and Chloroplasts by Similar Pathways

50. Why are both Tom and Tim needed to transport proteins into the mitochondrial matrix?

51. Which mitochondrial proteins would you expect to have a "stop transfer" signal?

Signal Sequences for Nuclear Transport Are Not Cleaved

52. Why can the NLS be located in the interior of a protein?

53. *Cell Map:* Look at the various cellular components that you have included on your map. Which ones required signal sequences?

Bacteria Also Use Signal Sequences for Protein Targeting

54. Suggest why a protein in its final, three-dimensional form would no longer be "translocation-competent."

Cells Import Proteins by Receptor-Mediated Endocytosis

55. How do cells "select" the specific extracellular items that they want to import via clathrin-coated pits? Is this mechanism foolproof?

Protein Degradation Is Mediated by Specialized Systems in All Cells

56. Which different mechanisms do cells use to ensure that only certain proteins (e.g., old or defective ones) and not other cellular components are degraded?

57. What aspect of protein structure influences the life span of cellular proteins?

SELF-TEST

Do You Know the Terms?

ACROSS

2. Cellular complex that increases the speed of protein synthesis.

5. Sometimes contains the nucleotide inosinate; portion of tRNA responsible for decoding the mRNA message.

6. "Start" codons are distinguished from other, identical codons occurring later in the mRNA sequence by ______-______ interactions between the Shine-Dalgarno sequence in the mRNA and stretches of rRNA near the P site.

11. "Id on otl ikeg reene ggsa ndh aml d on otl iket hemS am-I-Am" is analogous to a shift in the ______ of a gene. (2 words; apologies to Dr. Seuss)

13. Hydrolysis of this nucleotide triphosphate is required for translocation of an mRNA molecule during the elongation phase of protein synthesis. (abbr.)

14. Polynucleotide that contains the bases adenine, guanine, cytosine, and uracil and contains the code for a polypeptide. (abbr.)

16. "Word," in genetic terms.

18. Region of the ribosome that binds all charged tRNAs except $tRNA^{fMet}$.

22. "Start," in genetic terms.

23. Polynucleotide that contains the bases adenine, guanine, cytosine, and uracil and forms covalent associations with an amino acid. (abbr.)

24. Complex of proteins and ribonucleotides that has both structural and catalytic functions in protein synthesis.

25. Strong base pairing between the first two nucleotides in a codon and the complementary nucleotides in a tRNA anticodon contributes to the *specificity* of codon recognition; the weaker pairing, or ______, that occurs between the third pair of nucleotides influences the *speed* of protein synthesis.

26. Describe the incipient or initial protein formed on ribosomes.

28. Frame shifting is an example of ______ modification.

29. Covalent attachment of an amino acid to a specific tRNA requires energy in the form of ATP; the product of this reaction is commonly referred to as being ______.

DOWN

1. Proteins, such as the Na^+/K^+-ATPase, require a ______ sequence in the DNA code.

2. The "start" codon of an mRNA binds to ribosomes at the P, or ______, site.

3. "I do not like green eggs and ham I do not like them Sam-I-Am" is analogous to an ______ reading frame, in genetic terms.

4. Stage of protein synthesis in which an aminoacyl-tRNA binds to AUG on the A site of a ribosome.

7. Recognition and binding of a specific tRNA by a specific aminoacyl-tRNA synthetase is the basis for the ______ genetic code.

8. Stage of protein synthesis in which an aminoacyl-tRNA binds to AUG on the P site of a ribosome.

9. Cytosolic protein involved in targeting proteins to the endoplasmic reticulum. (abbr.)

10. The peptide ______ complex serves as both a docking protein and a transmembrane channel.

12. Posttranslational protein modification includes processes such as ______ and phosphorylation.

15. Scaffold:Building construction as ______:Protein synthesis. (abbr.)

17. Each codon specifies only one amino acid, meaning the genetic code is unambiguous; however, any one amino acid may have more than one codon, meaning the genetic code is ______.

19. Stage of protein synthesis in which the ester bond of peptidyl-tRNA is hydrated.

20. Protein occurring *ubiquitously* in cells; signals protein destruction.

21. Activated form of tRNA, known as ______-tRNA, that binds to the A site of ribosomes.

27. The process of protein synthesis requires a number of "accessory" proteins known as initiation, elongation, and releasing ______.

Do You Know the Facts?

1. Assuming the 5′ → 3′ connection of writing nucleotide sequences, indicate (yes/no) whether each of the following mRNA codons can be recognized by the tRNA anticodon ICG.

_____ **(a)** UGC
_____ **(b)** CGA
_____ **(c)** UGA
_____ **(d)** CGU
_____ **(e)** CGC

2. Indicate (yes/no) whether nascent insulin molecules are altered by each of the following posttranslational modifications.

_____ **(a)** Cleavage of a signal sequence.
_____ **(b)** Carboxylation of Asp residues.
_____ **(c)** Phosphorylation of Tyr residues.
_____ **(d)** Proteolytic cleavage of internal sequences.
_____ **(e)** Formation of disulfide cross-links.

3. The enzymatic process of "charging" a molecule of tRNA is similar to other enzyme catalyzed reactions you have encountered in previous chapters. Which of the following correctly describes such similarities?

A. Formation of an acyl-adenylate intermediate coupled to the hydrolysis of pyrophosphate is similar to reactions involved in fatty acid activation.
B. Proofreading and correcting abilities that prevent incorporation of the wrong molecule are analogous to the action of DNA polymerases.
C. The aminoacyl-tRNA synthetase is a relatively nonspecific enzyme (i.e., it can activate many different amino acids) and in this respect is similar to ribonucleotide reductase.
D. A and B
E. B and C

4. A new compound, vivekine, was recently discovered by a clever graduate student. It was isolated from bacteria found in deep sea-dwelling organisms. Vivekine inhibits protein synthesis in eukaryotes: Protein synthesis can initiate, but only dipeptides are formed and these remain bound to the ribosome. This toxin affects eukaryotic protein synthesis by blocking the:

A. binding of formylmethionyl-tRNA to ribosomes.
B. activity of elongation factors.
C. activation of amino acids.
D. recognition of stop signals.
E. formation of peptide bonds.

5. Which of the following describes the structure of transfer RNAs?

A. Their unusual bases are coded for by the DNA template.
B. Hydrophobic interactions between adjacent aromatic rings stabilize the three-dimensional structure.
C. The activated amino acid recognizes the codon.
D. All four "arms" have double-helical structures.
E. All of the above are true.

6. Indicate whether each of the following statements about prokaryotic translation is true or false.

_____ **(a)** An aminoacyl-tRNA synthetase catalyzes formation of an ester bond.
_____ **(b)** An mRNA molecule cannot be used to direct protein synthesis until it has been completely transcribed.
_____ **(c)** The positioning of fMet-tRNA on the A site defines the reading frame.
_____ **(d)** Incoming aminoacyl-tRNAs are first bound to the A site.
_____ **(e)** Formation of the 70S initiation complex requires an input of energy.
_____ **(f)** The carboxyl group of the amino acid on the aminoacyl-tRNA is transferred to the amino group of a peptidyl-tRNA.
_____ **(g)** Release factors cause the peptidyl transferase activity of the ribosome to use H_2O as a substrate.

7. Based on what you now know about DNA and RNA, suggest why cells don't use mRNA instead of DNA as the repository of genetic information, thus eliminating the need for a transcription step.

8. Which would you expect to be more critical to the production of a functional protein: the precise positioning of the initiation codon or the precise positioning of the termination codon?

Applying What You Know

1. Certain bacteria have suppressor tRNAs that suppress the effects of mutations in the DNA. In theory, in what *two* ways could suppressor tRNA differ from normal tRNA to account for this function?

2. One of the most clinically important missense mutations is that which causes sickle-cell disease: the alteration from a glutamate to a valine in the sixth position of the β chains of the hemoglobin tetramer. What would be the simplest explanation for how such a change occurred?

3. (a) You wish to set up a system for in vitro protein translation. What components would the system require? Assume you are working with a prokaryotic system.

(b) You repeat the experiment above using mRNA for insulin and partially purified eukaryotic components obtained from a *crude extract* (a common phrase used to denote the possible presence of undefined items in the mixture). You conclude that protein synthesis in eukaryotes must be different from that in prokaryotes because translation stops after a relatively short time. Examination of the components of your system reveals that a short (200 amino acid) peptide has been synthesized but is "stuck" on the ribosome, as is the mRNA molecule. Explain this result. What must you add to your in vitro system to permit complete protein synthesis?

ANSWERS

Do You Know the Terms?

ACROSS

2. polysome
5. anticodon
6. base-pair
11. reading frame
13. GTP
14. mRNA
16. codon
18. A site
22. AUG
23. tRNA
24. ribosome
25. wobble
26. nascent
28. translational
29. charged

DOWN

1. signal
2. peptidyl
3. open
4. elongation
7. second
8. initiation
9. SRP
10. translocation
12. glycosylation
15. rRNA
17. degenerate
19. termination
20. ubiquitin
21. aminoacyl
27. factors

Do You Know the Facts?

1. **(a)** N; **(b)** Y; **(c)** N; **(d)** Y; **(e)** Y
2. **(a)** Y; **(b)** N; **(c)** N; **(d)** Y; **(e)** Y
3. D
4. B
5. B
6. **(a)** T; **(b)** F; **(c)** F; **(d)** T; **(e)** T; **(f)** F; **(g)** T
7. Some organisms *do* use RNA for storing their genetic information (e.g., retroviruses, such as the virus that causes AIDS), but they do not sidestep the transcription requirement. DNA is used as the repository of genetic information by most organisms primarily because it is much more stable than RNA: DNA exists as a double-stranded molecule and lacks the reactive —OH group present on the ribose residues of RNA molecules. In addition, uracils formed by spontaneous deamination of cytosine cannot be recognized as "foreign" in RNA molecules and therefore cannot be "repaired," as they are in DNA. In fact, DNA is the only biomolecule for which repair systems exist—a critical property for a molecule with so much depending on the constancy of its sequences.
8. The precise positioning of the initiation codon is more critical to protein synthesis because the reading frame is determined by this position. If the reading frame is off by one or two codons the *entire* sequence of the protein will be altered; a misplaced start codon has a 66.6% chance of producing a garbled protein. The mispositioning of a termination codon, assuming that it remains somewhere near the end of the protein-coding sequence, means the loss (or gain) of some carboxyl-terminal amino acids. This would most likely have much less of an impact on protein function than a completely altered amino acid sequence.

Applying What You Know

1. Suppressor tRNAs could have complementary mutations in their anticodons that allow them to base pair with the mutated mRNA codon, negating the effect of the DNA mutation. Such mutations have been observed (see Box 27–3). In theory, it is also possible that a mutation in a particular aminoacyl-tRNA synthetase could change its binding specificity for either the tRNA or the amino acid, such that an amino acid now pairs with an anticodon that recognizes the mutated mRNA codon. The problem with this mechanism, and the reason it is probably not found in vivo, is that *all* of the cell's tRNAs with the anticodon for the mutant code would be improperly aminoacylated, and thus all other proteins in the cell would also contain inappropriate amino acids.
2. The codon for glutamate is GAA or GAG. A single substitution of a U for an A gives GUA or GUG, both of which are codons for valine. Note that a change of the second A in GAA to a U would not be as calamitous; it would alter the Glu to an Asp, which is also a negatively charged amino acid.

3. (a) The process of prokaryotic protein synthesis in a test tube would require the following:

Ribosomes (bacterial variety, 50S and 30S subunits)
mRNA containing both initiation and termination codons
All 20 of the amino acids
fMet-tRNA
At least 20 tRNAs that can bind the 20 amino acids
All the aminoacyl-tRNA synthetases
ATP, GTP, and Mg^{2+}
Initiation factors: IF-1, IF-2, and IF-3
Elongation factors: EF-Tu, EF-Ts, and EF-G
Termination (release) factors: RF_1, RF_2, and RF_3

(b) Synthesis stopped after about 200 amino acid residues because the polypeptide product of the insulin mRNA has a signal sequence to which a signal recognition particle (SRP) binds, halting translation. Synthesis will resume if you add vesicles composed of endoplasmic reticulum membrane on which the SRP-ribosomal complex can dock. Docking allows the SRP to detach and protein synthesis to resume.

Regulation of Gene Expression

STEP-BY-STEP GUIDE

Major Concepts

The regulation of gene expression determines which proteins are made by which cells and how much of each protein is produced.

The transcription of DNA to mRNA, catalyzed by RNA polymerases, is the most important point at which gene expression is regulated. Nevertheless, regulation can also occur at each of the steps involved in transforming the genetic message into functional proteins: RNA processing, translation, and protein (posttranslational) processing. The structural characteristics of proteins and polynucleotides provide the molecular basis for the various regulatory strategies used by cells.

Protein "factors" play key roles in the regulation of transcription.

In prokaryotes these proteins include **specificity factors,** which regulate the ability of RNA polymerase to bind to specific promoter regions on the DNA; **repressors,** which block polymerase binding; and **activators,** which enhance polymerase binding. In eukaryotes, gene activators are called **transcription factors.** These proteins have amino acid domains that bind DNA (some can recognize specific DNA sequences) and/or bind to other DNA-binding proteins. The mechanisms used to regulate gene expression are, even for simple organisms, highly complex, reflecting the complex compositions of the intracellular and extracellular environments to which cells must adapt.

Regulation of gene expression can be either positive or negative and often involves the interactions of multiple regulatory factors.

The precise control of gene expression is ensured by the use of **combinations** of regulators. This use of multiple regulators is an example of cellular economy: a relatively small pool of factors can be used in different combinations to regulate the expression of a large number of genes. In general, it appears that gene expression is not regulated in an "all-or-none" fashion. The use of multiple regulators makes it possible to *partially* turn on or turn off expression of a gene, making cells more responsive to fluctuations in their environment.

Prokaryotes and eukaryotes differ in some of their regulatory strategies.

Prokaryotes have smaller cell sizes than do eukaryotes and correspondingly smaller genomes. They compensate by clustering the genes for related proteins together in units called **operons** and by producing **polycistronic mRNA** that can be translated into a number of different proteins. Examples of operons that have been closely studied are the *lac, ara,* and *trp* operons of *E. coli.* The larger genomes of **eukaryotes** are highly condensed in order to fit into the nucleus. Consequently, in regions containing the genes to be expressed the DNA structure must be altered for transcription to occur. The regulation of gene expression in eukaryotes is usually through **activation** of transcription (i.e., positive regulation), as opposed to repression (negative regulation). Because of the complexity and large size of eukaryotic genomes, the possibility of nonspecific binding of transcription factors and inappropriate gene expression is increased. As a result, eukaryotic regulatory mechanisms are more complex and employ larger numbers of regulatory factors than those of prokaryotes. Finally, the physical separation of the genomic information in the nucleus from the protein-synthesizing machinery in the cytoplasm provides more points for regulation of gene expression than are available in prokaryotes, in which transcription and translation are tightly coupled.

What to Review

Answering the following questions and reviewing the relevant concepts, which you have already studied, should make this chapter more understandable.

- You have already been introduced to factors that alter cell function by affecting gene expression; these factors are the **steroid** and **thyroid hormones.** Review their structure and function (pp. 888–889, 892).
- The concept of **protein domains** is critical to an understanding of how proteins interact with other proteins and with DNA to regulate gene expression. Be sure that you are clear on the relationship between protein structure, protein function, and protein domains (Chapter 6).
- The use of **transcription terminators** is an important molecular mechanism in gene regulation schemes. Review the ways in which transcription can be terminated (p. 986; Fig. 26–7).
- DNA and RNA structure are crucial factors in the regulation of gene expression. Review the structure of DNA presented in Chapter 23, especially the way in which DNA is condensed in the nucleus (pp. 922–927).

Topics for Discussion

Answering each of the following questions, especially in the context of a study group discussion, should help you understand the important points of this chapter.

1. Why is the regulation of gene expression critical to the function of all organisms?

2. Make a flowchart listing the steps from the DNA code to the production of a functional protein. Show the points at which gene expression can be regulated.

Principles of Gene Regulation

3. What proteins have you encountered in your study of biochemistry that are likely to be coded for by housekeeping genes?

4. In addition to those cited in the text, what other proteins can you think of whose synthesis can be induced or repressed?

RNA Polymerase Binds to DNA at Promoters

5. What factors affect the binding of RNA polymerase to DNA?

6. How does RNA polymerase binding affect protein production?

7. Is the DNA binding by RNA polymerase an "all-or-none" process?

Transcription Initiation Is Regulated by Proteins That Bind to or near Promoters

8. What *molecular mechanisms* might account for the effects of specificity factors? Repressors? Activators?

Most Prokaryotic Genes Are Regulated in Units Called Operons

9. What are the different regions of DNA that make up an operon?

10. Why does it make biological sense for bacteria to have operons, whereas most higher eukaryotes do not?

The *lac* Operon Is Subject to Negative Regulation

11. What role does lactose play in the regulation of gene expression in *E. coli*?

12. How does the Lac repressor protein exert its effects?

Regulatory Proteins Have Discrete DNA-Binding Domains

13. What types of interactions can occur between DNA and proteins that would allow proteins to recognize specific nucleotide sequences?

14. Which region of a DNA molecule provides the greatest opportunity for such interactions to occur?

15. Why does a Gln residue only interact with adenine in DNA and not, for example, with guanine? (Hint: See Fig. 28–9.) Do other amino acid residues also interact with adenine?

Helix-Turn-Helix

16. The DNA-binding region of this type of regulator contains relatively short stretches of 7 to 9 amino acids that interact with only a portion of the DNA recognition sequence. These structures are usually unstable. How do such proteins function in vivo? (Hint: See Fig. 28–11 for an example.)

Zinc Finger

17. How is the structure of this type of DNA-binding protein stabilized?

18. Would you expect a protein with only one zinc finger to bind to DNA with more or with less specificity than a protein with 37 zinc fingers?

Homeodomain

19. In what way is the homeodomain related to the helix-turn-helix motif?

Regulatory Proteins Also Have Protein-Protein Interaction Domains

20. Why might it be important for regulatory proteins to be able to interact with RNA polymerase?

Leucine Zipper

21. What forces are responsible for the dimerization of leucine-zipper transcription factors?

Basic Helix-Loop-Helix

22. What molecular mechanisms are used by helix-loop-helix proteins to permit dimerization and DNA binding?

23. What is the significance of short stretches of basic amino acid residues in these proteins?

Subunit Mixing in Eukaryotic Regulatory Proteins

24. In what ways can protein-protein interactions contribute to the *specificity* of protein-DNA interactions? (Hint: Regulatory proteins bind to DNA as both *homo*dimers and *hetero*dimers.)

25. With which other cellular components might transcription factors interact in order to regulate gene expression? (Hint: You can find clues to this question from the section on the *lac* operon.)

Regulation of Gene Expression in Prokaryotes

The *lac* operon provides important insights into the molecular strategies used by cells to regulate gene expression. Keep in mind as you work through this section that the information gained from studies of prokaryotes suggests things to look for in eukaryotes.

The *lac* Operon Is Subject to Positive Regulation

26. Why do even relatively simple organisms such as bacteria use complex regulatory schemes to regulate gene expression?

27. How do bacteria respond to the presence of lactose (i.e., how do they transmit information about the presence of lactose to their DNA)?

28. How do bacteria respond to the presence of glucose?

29. Are these responses "all-or-none" phenomena?

30. Why is it useful for cells to have a few regulatory molecules control many different genes?

The *ara* Operon Undergoes Both Positive and Negative Regulation by a Single Regulatory Protein

31. What is meant by autoregulation?

32. How does DNA structure contribute to the regulation of gene expression?

33. Do you think the DNA looping mechanism of gene regulation would still work if the *ara*O_2 site were moved further to the left in Figure 28–20? What if it were moved to the right side of *ara*BAD?

Many Genes for Amino Acid Biosynthesis Are Regulated by Transcription Attenuation

34. In the case of the *trp* operon, how does DNA structure contribute to the regulation of gene expression?

35. The regulatory "leader protein" of the *trp* operon has *no* function in the cell other than as a "sensor" for tryptophan levels. What is its "sensory" mechanism?

36. Why is the tight coupling of transcription and translation in prokaryotes critical to the transcription attenuation mechanism?

37. In what ways is transcription attenuation similar to feedback inhibition of enzymes by their reaction product?

Induction of the SOS Response Requires Destruction of Repressor Proteins

38. How does the RecA protein act as a sensor of DNA damage in cells?

39. What is the molecular basis for RecA's signaling actions?

Synthesis of Ribosomal Proteins Is Coordinated with rRNA Synthesis

40. How do bacteria coordinate the rate of ribosomal protein synthesis with that of rRNA?

41. What intracellular signaling pathway is used to relay information about the levels of amino acids in bacterial cells?

42. How do lowered amino acid levels decrease translation in bacteria?

Some Genes Are Regulated by Genetic Recombination

43. How does genetic recombination cause a switch in the transcription of different flagellin genes in *Salmonella*?

44. How can this recombination event be so specific that it does not affect the expression of other genes necessary to the function of the organism?

45. Is this regulation an "all-or-none" response?

(Note: A similar strategy is used by the human immunodeficiency virus (HIV) to escape detection by the immune system. The development of a vaccine to protect against HIV has been greatly hindered by the virus's ability to continuously alter the conformation of proteins in its coat.)

Regulation of Gene Expression in Eukaryotes

46. Which major structural differences between eukaryotes and prokaryotes affect the way their gene expression is regulated?

47. Why does "positive regulation" of gene expression predominate in eukaryotes?

Transcriptionally Active Chromatin Is Structurally Distinct from Inactive Chromatin

48. What are "hypersensitive sites" and why is it likely that they correspond to areas of active gene transcription?

Modifications Increase the Accessibility of DNA

49. Why does it make sense that transcriptionally active DNA is deficient in histone H1?

50. What is the effect of methylation on DNA structure?

Chromatin Is Remodeled by Acetylation and Nucleosomal Displacements

51. Why does histone acetylation promote gene transcription?

Many Eukaryotic Promoters Are Positively Regulated

52. Why does the regulation of eukaryotic genes usually involve multiple regulatory sites (i.e., promoter and enhancer regions) and multiple regulatory proteins (i.e., general and specific transcription factors)?

53. In what ways is the initiation of gene transcription in cells analogous to opening a combination lock?

DNA-Binding Transactivators and Coactivators Facilitate Assembly of the General Transcription Factors

54. Enhancer elements that affect transcription of genes in eukaryotes are usually located 100 to 5,000 base pairs away from the gene they regulate, on either the 3′ *or* the 5′ side. Why is this possible? (Compare this with TATA boxes, which are located 25 to 30 base pairs "upstream," on the 5′ side of the gene.)

Three Classes of Proteins Are Involved in Transcriptional Activation

55. Where do each of the three classes of proteins bind and what is their role in gene activation?

The Genes Required for Galactose Metabolism in Yeast Are Subject to Both Positive and Negative Regulation

56. What are the positive and negative regulators in this system?

DNA-Binding Transactivators Have a Modular Structure

57. Can you think of a mechanism that would account for the *specific* activation of a gene by factors that bind relatively *nonspecifically* to DNA domains, such as the glutamine-rich, proline-rich, or acidic activation domains?

Eukaryotic Gene Expression Can Be Regulated by Intercellular and Intracellular Signals

58. What is the transducer that couples a steroid hormone signal to alterations in gene expression?

59. Is this mechanism of transcription activation an "all-or-none" phenomena?

Regulation Can Occur through Phosphorylation of Nuclear Transcription Factors

60. Why do nonsteroid hormones require second messengers to affect transcription?

61. What is the primary difference that distinguishes the effects of nonsteroid hormones from steroid hormones on transcriptional regulation?

Many Eukaryotic mRNAs Are Subject to Translational Repression

62. Why is translational repression an important regulatory mechanism in eukaryotes but not in prokaryotes?

63. What are the molecular mechanisms used by eukaryotes to achieve translational repression?

Development is Controlled by a Cascade of Regulatory Proteins

64. Suggest why *Drosophila* oocytes require a complement of *maternal* mRNA.

65. In what ways is the function of maternal genes such as *bicoid* in *Drosophila* oocytes similar to the developmental switches used by *Salmonella*?

66. The development of *Drosophila* embryos is thought to be regulated by a cascade of genes, many of which code for transcription factors. Can you think of a way in which such a cascade mechanism could cause the *differential* development of a single egg cell into the many different cell types that constitute the adult fly? (Hint: Use the anteriorly *localized bicoid* mRNA, which specifies the "head" end of the fly, and the posteriorly *localized nanos* mRNA, which specifies the "tail" end of the fly, as the top tier of the cascade.)

67. What are the different classes of regulatory genes involved in pattern formation during *Drosophila* embryogenesis and where do they originate?

SELF-TEST

Do You Know the Terms?

ACROSS

1. Region of transcription factor with a structure similar to that of the DNA-binding region of helix-loop-helix proteins.
4. Region of regulatory protein, the _____-_____-_____, contains two amphipathic α helices connected by extended amino acid chains of varying lengths.
5. Region of mRNA transcribed from the *trp* operon that "senses" the levels of tryptophan in *E. coli* cells.
8. Hairpin loops of nascent mRNA sometimes act as _____, abruptly terminating DNA transcription.
9. Transcription factor that regulates transcription of the *lac* operon genes and acts as a glucose sensor. (abbr.)
11. The ability to control yourself, for example.
13. Adenylate cyclase is to cAMP as _____ is to ppGpp. (2 words)
14. Molecule inactivated by binding of lactose to a regulatory protein, with consequent increase in expression of genes in the *lac* operon.
16. One of the gene products of the _____ operon is tryptophan synthetase.
19. Regions of DNA that usually include some or all of the promoter region.
21. Describes genes of the *lac* operon, whose gene products are needed only when [glucose] is low and [lactose] is high.
23. _____ factors regulate the formation of mRNA.
24. Describes transcription of *lac* genes in cells with mutations in the Lac repressor binding region.

DOWN

1. Class of genes used by all cells and essential to normal function; the genes for ribosomal proteins, for example.
2. In eukaryotes, a nucleic acid conformation that contributes to the interaction of enhancer and promoter elements. (2 words)
3. In the absence of glucose, CAP binds cAMP and undergoes a conformational change that allows it to bind to the *lac* promoter, resulting in the _____ of gene transcription; in the presence of glucose, cAMP levels fall and CAP is prevented from binding to DNA.
4. Describes sites in DNA that lack nucleosomes and tend to be found near transcription start sites; they are especially susceptible to the action of DNases.
5. Region of transcription factor; dimer held together by hydrophobic interactions between two α helices. (2 words)
6. Region of transcription factor; has one or more prominent loops of amino acids stabilized by interactions between cysteine (and sometimes histidine) and a zinc atom. (2 words)
7. These are transcribed into mRNA that is eventually packaged into unfertilized oocytes. (2 words)
10. Describes prokaryotic mRNA produced from gene clusters that are transcribed together.
12. During induction of the SOS response in a severely damaged cell, the RecA protein cleaves specific repressor proteins that normally prevent viruses from switching between the _____ life cycle to the lytic, or replication competent, form.

15. UASs are to yeast as _____ are to higher eukaryotes.

17. The DNA to which the *lac* repressor binds contains the following inverted repeat:

TGTGAGCGGATAACA
ACACTCGCCTATTGT

which is an almost perfect _____.

18. Class of segmentation genes transcribed in *Drosophila* embryos following fertilization; often their translation is induced by transcription factors produced from maternal mRNA.

19. Region of DNA that contains the binding sites for promoters, repressors, and RNA polymerases, and the coding regions for two or more proteins.

20. Describes amino acids that make up short stretches of the DNA-binding domains of many transcription factors.

Do You Know the Facts?

1. A cartoon of a "typical" eukaryotic cell is shown below. Identify each of the numbered steps involved in gene expression.

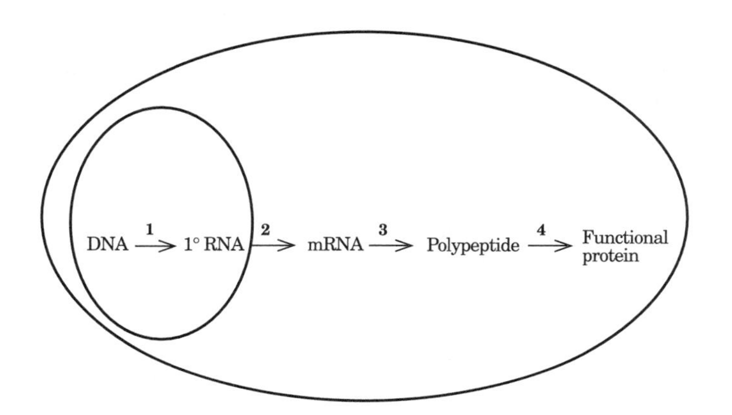

2. Provide a specific, molecular example of how regulation can occur at each step in question 1.

3. Indicate whether each statement describes transcription and/or translation in prokaryotes or in eukaryotes.

(a) Transcription and translation of the same gene occur almost simultaneously.

(b) Messenger RNAs undergo extensive processing.

(c) Transcription is both positively and negatively regulated.

(d) A single mRNA codes for several proteins.

(e) Transcription and translation occur in separate cellular compartments.

(f) The DNA code for mRNA is noncontiguous.

(g) Unwinding from histones is probably necessary for transcriptional activation of the DNA.

4. Indicate (yes/no) whether each type of bond or interaction contributes to the ability of transcription factors to recognize specific DNA sequences.

_____ **(a)** hydrophobic interactions

_____ **(b)** van der Waals interactions

_____ **(c)** hydrogen bonds

_____ **(d)** covalent bonds

_____ **(e)** phosphoanhydride bonds

5. Which of the following statements about enhancer elements is correct?

A. They can be located on either side of the gene they activate.
B. They will *not* function unless located at some distance from the gene they control.
C. They bind transcription factors.
D. An enhancer element activating a specific gene is not functional in all cell types.
E. All of the above are true.

6. The *lac* operon is shown diagrammatically below. Match the elements of the *lac* operon (listed below the diagram) with the appropriate features or functions (more than one may apply per element).

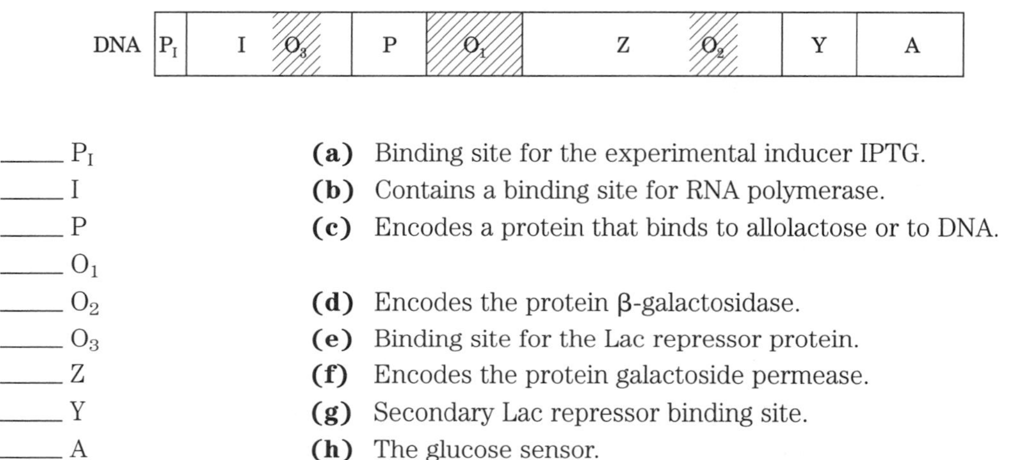

_____ P_I **(a)** Binding site for the experimental inducer IPTG.

_____ I **(b)** Contains a binding site for RNA polymerase.

_____ P **(c)** Encodes a protein that binds to allolactose or to DNA.

_____ O_1

_____ O_2 **(d)** Encodes the protein β-galactosidase.

_____ O_3 **(e)** Binding site for the Lac repressor protein.

_____ Z **(f)** Encodes the protein galactoside permease.

_____ Y **(g)** Secondary Lac repressor binding site.

_____ A **(h)** The glucose sensor.

_____ Lac repressor **(i)** Encodes the protein thiogalactoside transacetylase.

7. Indicate whether each of the following statements about the *lac* operon is true or false.

_____ **(a)** All seven genes are transcribed into mRNA, which is then translated into five different proteins.

_____ **(b)** The repressor binds to the structural genes, preventing their transcription.

_____ **(c)** Efficient binding of RNA polymerase to DNA occurs only in the presence of CRP and cAMP.

_____ **(d)** cAMP acts as an inducer by binding to the Lac repressor.

_____ **(e)** In the absence of lactose, the repressor binds the operator.

_____ **(f)** A mutation in the I gene such that no gene product is made leads to constitutive expression of the Z, Y, and A genes.

8. CRP has been found to contain a typical helix-turn-helix motif. Which of the following best describes the interactions between CRP and the *lac* operon?

A. Covalent bonds occur between the DNA helix and CRP helices.
B. Nonpolar interactions and hydrogen bonds contribute to the specificity of the CRP-DNA interactions.
C. The double helix of the operon DNA interacts with the double helix of the CRP DNA.
D. The base sequence of the DNA has no influence on the binding of CRP.
E. The specific amino acid sequence of CRP has no influence on DNA binding.

9. Transcriptional termination can be modulated in a variety of ways. Match each regulatory element or mechanism with the correct description from the list below.

_____ **(1)** GC-rich hairpin followed by several U residues
_____ **(2)** ρ protein
_____ **(3)** N protein of λ phage
_____ **(4)** attenuation

(a) Acts as an antiterminator.
(b) Recognizes certain RNA sequences and acts as an ATPase.
(c) Acts through translation of a leader peptide.
(d) Causes dissociation of RNA molecules from the DNA template and from RNA polymerase.

10. *Both* prokaryotic *and* eukaryotic promoters:

A. contain AT-rich regions.
B. interact with transcription factors.
C. are influenced by base sequences thousands of base pairs away, which increase their activity.
D. are recognized by three different types of RNA polymerases, depending on location in the genome and type of transcript.
E. have all of the above properties.

Applying What You Know

1. (a) The average *Homo sapiens* genome has approximately 3×10^9 base pairs, and it has been estimated that various human cell types make between 100 to 500,000 different proteins over their lifetime. If an "average" protein contains 400 amino acids, what is the maximum number of proteins that can be encoded by the human genome?

(b) What percentage of the total genome is actually used to encode proteins? (Assume for this calculation that human cells make an "average" of 250,000 different proteins over their lifetime, and that the average size of these proteins is again 400 amino acids.)

(c) How would you account (in molecular terms, of course) for the answers you obtained for (a) and (b)?

2. The regulation of gene expression that occurs during the development of multicellular organisms is an intricately orchestrated series of "molecular switches." These molecular switches are responsible for turning on developmental pathways that result in the differentiation of a single, pluripotent egg cell into the thousands of different cell types that make up an organism. Many of these switches are *transcription factors* that can turn on whole groups of genes needed to produce, say, a head structure. The maternal gene *bicoid* codes for just such a transcription factor in *Drosophila.* The *bicoid* gene product is localized at one end of the unfertilized egg due to the presence of a 3′ untranslated region (UTR) in the maternal mRNA that "tethers" mRNA molecules to some intracellular structure in this future "head" end. If a mutation occurred in the *bicoid* gene that deleted all of the 3′ UTR, what would be the effect on the phenotype of the adult fly?

Biochemistry on the Internet

* For the current URLs of the databases needed for the problem below, go to the Lehninger Web site at http://www.worthpublishers.com/lehninger.

The study of "gene cascades" involved in *Drosophila* development have provided new insights into the roles of nucleotide binding proteins and gene regulation. The genetic mechanisms for establishment of the anterior-posterior axes in the early stages of embryological development have been determined and a number of nucleotide binding proteins have been implicated.

1. One gene that has been implicated in the establishment of the posterior system is a maternal factor, *nanos* (*nos*). The *nanos* mRNA is localized in the posterior region of the embryo. Its gene product is a putative zinc-finger protein. These types of nucleotide-binding proteins have a protruding "finger" of amino acids that consists of about 30 amino acid residues, four of which (either four Cys or two Cys and two His) coordinate a single Zn^{2+} atom, thereby stabilizing the structure.

(a) At what level of gene expression does this protein function and how does its function contribute to the generation of a posterior axis?

(b) In the process of characterizing the functional domains of this protein you make a mutant strain in which all the Cys residues in this protein are replaced by acidic amino acid residues. How would you expect this substitution to affect the function of this protein?

(c) How many zinc fingers is this protein likely to have? (Hint: You can use FlyBase, the information source for research on *Drosophila*, or try the Interactive Fly at this same site.

2. The *Drosophila* gene *engrailed* encodes a protein product that is involved in determining the anterior-posterior axis in the wing of the fly. The name is derived from the fact that the mutant fly, which lacks this gene product, has a wing margin with a scalloped or engrailed edge. In the normal fly, this edge is typical of the anterior portion of the wing whereas the posterior region has a smoother edge. In the engrailed mutant, the posterior portion of the wing is replaced by a mirror image of the anterior wing structures, giving the wing an overall "engrailed" appearance.

(a) The *engrailed* gene product is a transcription factor. What kind of DNA-binding motif does it use to bind to DNA and alter transcription? (Hint: You can use FlyBase or try the SCOP database and do a keyword search.)

(b) What is the amino acid sequence of the DNA-binding region of the Engrailed protein? (Hint: Try using the 3Dinsight database and search tool for this exercise.)

(c) Use the Protein Explorer and PDB file 3HDD (a model of the Engrailed protein binding to DNA) to answer the following questions:

1. What is the structure of the DNA-binding region of the *engrailed* gene product?

2. Can you locate the amino acid residues in the Engrailed protein that hydrogen bond with the DNA? (Hint: Hydrogen bonds generally occur with intermolecular distances of between 2.7 and 3.3 Å.) The diagram below of the amino acid sequence for 3HDD may help you to keep track of the amino acids that are binding to the DNA:

1					|					|					|					|					|				30
E	K	R	P	R	T	A	F	S	S	E	Q	L	A	R	L	K	R	E	F	N	E	N	R	Y	L	T	E	R	R

31					|					|					|					|					|				60
R	Q	Q	L	S	S	E	L	G	L	N	E	A	Q	I	K	I	W	F	Q	N	K	R	A	K	I	K	K	S	T

3. A general consensus amino acid sequence for the homeodomain DNA-binding motif has been described. It was obtained by comparing over 300 homeobox proteins and is shown below. In this sequence, seven of the residues are absolutely conserved (marked with *), six are highly conserved (+), and the rest of the conserved positions have no more than 9 different amino acid residues.

```
    +          *   *     +             +    ++** * *  * +
xxxxRxxYxxxQLxxLExxFxxxxYLxxxxRxxLAxxLxLxxxQVKIWFQNRRxKxKxxx
```

Source: *Guidebook to the Homeobox Genes,* Dennis Duboule (Ed.), Oxford Univ. Press, 1994.

Which of these conserved residues correspond to those identified in part **(b)** of this exercise?

4. The consensus DNA-binding sequence for this protein is "TAATTA." Can you use the model to predict the identity of the residues that play the biggest role in recognizing a specific DNA sequence? How could you test your hypothesis?

ANSWERS

Do You Know the Terms?

ACROSS

1. homeodomain
4. helix-loop-helix
5. leader
8. attenuators
9. CRP
11. autoregulation
13. stringent factor
14. repressor
16. *trp*
19. operators
21. inducible
23. transcription
24. constitutive

DOWN

1. housekeeping
2. DNA looping
3. induction
4. hypersensitive
5. leucine zipper
6. zinc finger
7. maternal genes
10. polycistronic
12. lysogenic
15. enhancers
17. palindrome
18. gap
19. operon
20. basic

Do You Know the Facts?

1. The major steps linking the genetic information residing in DNA sequences to functional proteins and a functional organism are:

1 *Transcription* of DNA into RNA molecules in the nucleus.

2 *Posttranscriptional processing* of RNA and export to the cytosol.

3 *Translation* of the mRNA transcript to a polypeptide.

4 *Folding and posttranslational processing and modification* of the polypeptide.

2. The following are just a few examples of the types of regulation that can occur at each step in gene expression (this list is *not* exhaustive).

1 Transcriptional control

- Induction by activators and transcription factors; e.g., CRP in the regulation of *lac* operon transcription.
- Repression; e.g., the Lac repressor protein.
- Regulation by a form of feedback inhibition; e.g., the regulation of transcription of the *trp* operon by tryptophan.
- DNA structure can determine which genes are available for transcription; e.g., DNA-associated proteins such as histones affect DNA packing.
- Genetic recombination; e.g., used by *Salmonella* to regulate which of two genes for flagellin proteins are transcribed.

2 RNA processing
- Splicing of introns by snRNAs and self-splicing mechanisms.
- Differential (or alternative) splicing of complex transcripts; e.g., the production of calcitonin in the thyroid and calcitonin gene–related peptide in the brain from a common primary transcript.
- Base modification by methylation, deamination, or reduction in tRNAs.
- Addition of 5′ caps and 3′ poly(A) tails to mRNAs.
- Differential degradation of mRNAs; this type of regulation may be related to base modification.

3 Translational control
- A variety of initiation, elongation, and release factors control when and how much of a gene product is made; the activity of these factors is often modulated by protein phosphorylation or the availability of GTP.
- Regulation of the number of ribosomes available for translation in bacteria.
- Translation repressor proteins regulate the synthesis of ribosomal proteins in bacteria.

4 Protein processing
- Proteolytic processing; e.g., formation of a functional insulin dimer from a large, nonfunctional precursor molecule.
- Glycosylation; often targets proteins to specific cellular locations.
- Isoprenylation; also helps in localizing proteins (anchors them in membranes).
- Amino acid modification; e.g., reversible phosphorylation of Ser, Thr, or Tyr residues that activates or inactivates enzymes.
- Formation of disulfide bonds; e.g., in insulin, three disulfide bonds must be formed between the correct pairs of Cys residues to produce a functional protein.

3. **(a), (c), (d),** prokaryotes; **(b), (e), (f), (g),** eukaryotes; **(c)** also occurs in eukaryotes, but is not currently well-characterized.

4. **(a)** Y; **(b)** N; **(c)** Y; **(d)** N; **(e)** N

5. E

6. P_I (b); I (c); P (b), (g); O_1 (e); O_2 (g); O_3 (g); Z (d); Y (f); A (i); Lac repressor (a)

7. **(a)** F; **(b)** F; **(c)** T; **(d)** F; **(e)** T; **(f)** T

8. B

9. **(1)** (d); **(2)** (b); **(3)** (a); **(4)** (c)

10. A

Applying What You Know

1. **(a)** If the average protein contains 400 amino acids, the average size for a gene is 1,200 base pairs. Dividing the total number of base pairs (3×10^9) by 1,200 gives a maximum of **2.5×10^6 proteins** that can be encoded by the human genome.

(b) In actuality, only about **10% of the genome** is used to encode proteins. This figure is obtained by determining the number of base pairs (bp) needed to encode 250,000 proteins (250,000 proteins $\times$ 1,200 bp per protein) and dividing that number by the total number of base pairs in the genome (3×10^9 bp).

(c) The "extra" 90% is noncoding DNA containing a variety of types of sequences, including what some refer to as "junk" DNA. Also included in the extra DNA are regulatory regions (enhancers and promoters); spacer DNA; duplicated portions that are essentially redundant; pseudogenes; repetitious DNA; centromeric and telomeric regions; and other regions that remain to be identified. Much of this seemingly extraneous DNA is fodder for the evolutionary "mill."

2. The absence of the 3′ UTR in the maternal mRNA for the *bicoid* transcription factor would prevent the mRNA from binding to internal structures in one end of the egg. Consequently, the maternal message would *diffuse further* into the egg than in an egg with the unmutated (wild-type) message. Recall that the *bicoid* transcription factor is responsible for turning on a cascade of genes, ultimately resulting in the formation of head (anterior) structures. The presence of the factor in new locations in the egg would cause these head structures to form in inappropriate (ectopic) positions. The resulting phenotype would be a *Drosophila* with a *larger than normal, or inappropriately positioned,* "head" region.

Biochemistry on the Internet

1. **(a)** This morphogen is a translational repressor that binds to mRNA primarily in the posterior region of the embryo. It acts by suppressing production of proteins involved in the formation of anterior structures.

(b) The *nanos* gene product is a putative zinc-finger protein. The zinc-finger structure is thought to bind to specific DNA sequences and regulate transcription. Substitution of other amino acid residues for Cys residues in this protein would prevent the formation of the "finger" structure and interfere with this protein's ability to regulate gene expression.

(c) At FlyBase, perform a search for the gene *nanos.* Click on "*nos*" in the search output and then look up either of the Protein Accession numbers presented to get the amino acid sequence of the *nanos* gene product. There are relatively few Cys and His residues at the carboxyl end of this protein. Probably only a single zinc finger can be generated by this protein.

2. (a) This protein is identified as a homeodomain-containing DNA-binding protein.

(b) At the 3Dinsight database, search for Keyword "engrailed" and look at the information for 1ENH. Click on Amino Acid Analysis and get a List of the data available for this structure. You should come up with the following amino acid sequence:

RPRTAFSSEQLARLKREFNENRYLTER
RRQQLSSELGLNEAQIKIWFQNKRAKI

Most proteins that contain a homeobox domain can be further classified into three subfamilies based on their sequence characteristics. These are engrailed, antennapedia, and paired. Proteins currently known to belong to the engrailed subfamily include two *Drosophila* proteins (Engrailed and Invected); two homologous proteins in silk moths; E30 and E60 proteins in honeybee; G-En protein in grasshopper; En-1 and En-2 in mammals and birds; Eng-1, -2, -3 in zebrafish; SU-HB-en in sea urchin; Ht-En in leech; and ceh-16 in *C. elegans.*

(c) **1.** To examine the interactions between this homeodomain-containing transcription factor and DNA, use the Protein Explorer, which can be found at the Chime Resources Web site. (You will also need to have Chime installed on your computer.) Load the molecule of interest (2HDD) and **Explore Its Structure!** If you toggle off the **Option** to **Display hetero atoms** (click on the MDL logo in the lower right-hand corner of the molecule frame for the menu) and then **Color** by **Structure,** you will see three pink segments representing three α helices connected by short loops. One of these helices is clearly in contact with the DNA.

2. Switch the molecular display to one where the DNA is shown in a **Stick** format and Chain A of the homeodomain-containing polypeptides is in a **Spacefill** format. It will be easier to locate the interacting elements in this form. By changing the **Color** to display the different amino acid side-chain properties, it should become obvious that the residues that lie closest to the DNA are blue (representing positively-charged amino acids).

To identify the specific amino acid residues that form hydrogen bonds, go **Back** to the Explore! page and click on **Find noncovalent bonds. Select** Chain A using the MDL logo and then click on **Find noncovalent bonds.** It may be easier to see things if you click on the radio button that toggles on the **Spacefill: Yellow/White option.** Be sure the distance increment for finding interactions is set to 0.1 Å. It will be easier to identify a few interacting partners at a time than all at once.

Click **Find next,** the window that identifies the distance between found molecules as 2.5 Å. Hydrogen-bonding atoms on the DNA will be highlighted with a stippled surface and the bonding partner on the protein will be displayed in CPK color. (Hydrogen bonds with water are indicated in purple. You can eliminate these by typing "water" in the **Don't find** box.) A hydrogen on T211 in the DNA backbone (chain C) will be highlighted red, as will the interacting oxygen that carries the hydrogen on Arg^{5}. Clicking on the red oxygen will give you the identity of the residue; be sure the **Mouse Clicks:** window shows the function **Identify.**

No interactions between DNA and the protein are found between 2.5 and 2.7 Å. Click **Find** again to identify a base-pair-protein interaction between T382 (on DNA chain D) and Tyr^{25} that occurs within 2.8 Å. Click once more to find an interaction between A329 (D) and Lys^{57} at 2.9 Å. At 3.0 Å there is an interaction between T328 (D) and Arg^{53}. At 3.1 Å there are two interactions: one between A213 (C) and Thr^{6}, the other between T211 (C) and Arg^{5}. At 3.2 Å there is an interaction between A212 (C) and Arg^{5}; T211 (C) and Lys^{55}; and A213 (C) and Asn^{51}. Finally, at an interatomic distance of 3.3 Å (the upper limit for hydrogen bonds) there are interactions between A213 (C) and Asn^{51}, and between A329 (D) and Lys^{57}. A gray carbon is also highlighted at this distance. This probably represents a hydrophobic interaction between the DNA and the protein since this type of interaction usually occurs between 3.3 and 4.0 Å.

Click on the **Found** button to see a summary of all the found interactions. A summary in chart form of the amino acid residues that interact with DNA in this model is shown below (H indicates a potential hydrogen bond):

1				H					|					|
E	K	R	P	R	T	A	F	S	S	E	Q	L	A	R

				|				H					30
L	K	R	E	F	N	E	N	R	Y	L	T	E	R	R

31				|				|					|
R	Q	Q	L	S	S	E	L	G	L	N	E	A	Q	I

				|	H		H		H		H			60
K	I	W	F	Q	N	K	R	A	K	I	K	K	S	T

3. All six of the residues in the DNA-binding domain of Engrailed that are capable of forming hydrogen bonds with the DNA correspond to conserved amino acids in the consensus sequence for this motif as indicated below:

```
    +          *   *    +
xxxxRxxYxxxQLxxLExxFxxxxYLxxxxRx
    H                   H

       +    + +* *  *  * * +
xLAxxLxLxxxQVKIWFQNRRxKxKxxx
                  H H H H
```

4. You can alter the display of the DNA to show the double-helical structure and all the hydrogen bond interactions with the Engrailed protein. After finding all the bonds that occur between 2.5 and 3.3 Å, click the **Found** button to display them all. Click the MDL logo to select the DNA **(Select → Nucleic → DNA)** and display it in a stick format **(Display → Sticks).** Moving this model around with the mouse will allow you to view all the hydrogen bonds. A number of them are made with the DNA backbone. Since this is the same for all nucleotides, these interactions cannot impart any information as to the DNA sequence of this piece of DNA. Two of the amino acid residues, however, seem to insert into the grooves of the helix and form hydrogen bonds with specific nucleotides. These two residues, Arg^{5} and Asn^{51}, give the impression of a "clamp-like" structure. Changing the identity of these residues would alter the ability of this protein to recognize and bind to this specific DNA sequence.

Recombinant DNA Technology

STEP-BY-STEP GUIDE

Major Concepts

The development of recombinant DNA technology has revolutionized the way we study cells and their biomolecules.

Cloning technology provides researchers with the ability to "cut and paste" selected pieces of DNA and to insert them into many different types of cells. The basic tools are the following: (1) the "scissors" are **sequence-specific endonucleases,** or **restriction endonucleases;** (2) the "paste" is the enzyme **DNA ligase;** and (3) the vectors are self-replicating DNAs (**plasmids, bacteriophages,** and **cosmids**) that carry pieces of DNA into host cells. The **recombinant DNA,** then, is the vector DNA with a small piece of foreign DNA that has been inserted with the aid of restriction endonucleases and DNA ligase.

A variety of techniques can be used to insert vectors into a host cell, usually a strain of E. coli*, in a process called transformation.*

Once inside cells, the vector DNA replicates producing many copies of the inserted, foreign DNA along with the vector DNA. Not all cells are transformed by these procedures, however, and a process must be employed to **select** the cells that actually contain the recombinant DNA. The end result is a renewable source of a specific piece of DNA that can be replicated, transcribed, and translated in essentially unlimited quantities.

The source of DNA fragments is often a DNA library.

Genomic libraries contain the total genetic information of an organism in the form of small fragments that have been inserted into vectors. Each colony of bacteria in the library contains vectors with just *one* piece of the fragmented DNA. Fragments are randomly produced and thus may not contain the entire coding region for a given gene. Because they are derived from genomic DNA the fragments also contain DNA sequences other than the protein-coding sequences, such as promotor regions. **cDNA libraries,** on the other hand, contain the sequence information that is carried by mRNA molecules—that is, the entire coding sequences for individual proteins.

Recombinant DNA technology provides tools for use in research, medical therapies, and commercial production.

The ability to express cloned genes and obtain large quantities of the encoded protein is the basis for many of the applications of recombinant DNA technology (e.g., the production of human insulin). The ability to amplify specific DNA sequences using the polymerase chain reaction (PCR) makes it possible to detect as little as one copy of a DNA molecule (which is useful in the early detection of viral infections such as the human immunodeficiency virus). Mutagenesis techniques allow researchers to selectively alter genes, thereby manipulating protein functional domains and gaining insights into protein function (ultimately it may be possible to design proteins with specific functions that have commercial applications).

What to Review

Answering the following questions and reviewing the relevant concepts, which you have already studied, should make this chapter more understandable.

- Recombinant DNA technology makes use of many of the biochemical techniques already introduced in the text. A thorough understanding of the principles underlying these methods will make it easier for you to follow the discussion of recombinant

DNA technology. These techniques include (but are not limited to):

Gel electrophoresis (pp. 134–136, 351–353)
Chromatography techniques (pp. 130–133, 231, 318–319, 384–385)
Use of antibodies for isolation, detection, and localization of proteins (pp. 231–232)
Use of radioactivity to tag molecules for detection (pp. 334, 934–935)
DNA sequencing techniques (pp. 351–354)

- Restriction endonucleases cut DNA at specific nucleotide sequences. The enzyme-DNA interactions that allow this specific recognition to occur are similar to the interactions between transcription factors and DNA (pp. 1080–1085).
- Much of recombinant DNA technology relies upon the ability of DNA and RNA to form duplexes and hybrid duplexes. Be sure that you understand the molecular basis for base-pair formation between polynucleotide strands (pp. 331–332) and the concept of complementarity (pp. 336, 933–934).

Topics for Discussion

Answering each of the following questions, especially in the context of a study group discussion, should help you understand the important points of this chapter.

DNA Cloning: The Basics

1. What exactly is recombinant DNA?

2. What are the enzymes that allow researchers to cut, paste, and then duplicate specific pieces of DNA?

Restriction Endonucleases and DNA Ligase Yield Recombinant DNA

3. Endonucleases and ligases are normal constituents of some cells. What are the biological functions of these enzymes?

4. How do restriction endonucleases recognize specific DNA sequences? What other cellular components are capable of recognizing specific DNA sequences?

5. What makes "sticky ends" sticky?

6. What are the advantages of sticky ends for the insertion of DNA pieces into a vector? (Hint: Remember that translation proceeds only in the $5' \rightarrow 3'$ direction.)

7. Why is the efficiency of ligation of two blunt-cut ends less than that of ligation involving sticky ends?

Cloning Vectors Allow Amplification of Inserted DNA Segments

8. Given that plasmid DNA is self-replicating, why must it be inserted into a host cell?

9. What factors affect the transformation of bacterial cells? How can you determine whether or not a cell has been transformed?

10. How could you determine whether the plasmid in a transformed cell actually contains a foreign DNA fragment?

11. What factor must be considered when choosing the type of vector to use for cloning?

Isolating a Gene from a Cellular Chromosome

12. Specific proteins can be identified on the basis of their three-dimensional shape through the use of antibodies. What characteristics of genes can be used to unambiguously identify a specific gene?

Cloning a Gene Often Requires a DNA Library

13. How are the DNA pieces used to construct a genomic library obtained?

14. How are the DNA pieces used to construct a cDNA library generated?

15. What is absent from the cloned DNA in a cDNA library that is present in the DNA of a genomic library? (Hint: Consider the source!)

Specific DNA Sequences Can Be Amplified

16. What characteristic of DNA duplexes is critical to the successful application of the PCR technique?

Hybridization Allows the Detection of Specific Sequences

17. What is the molecular basis for the use of labeled probes to identify clones containing specific DNA sequences?

A Potent Weapon in Forensic Medicine

18. In DNA fingerprinting, a DNA probe is used to label a few of the thousands of DNA "bands" that are generated when genomic DNA is digested with restriction enzymes. When using this technique to match DNA samples, what should you do to increase the odds that the DNA in two different samples comes from the same source?

DNA Microarrays Provide Compact Libraries for Studying Genes and Their Expression

19. Why are DNA chips such an important technological development?

Applications of Recombinant DNA Technology

Cloned Genes Can Be Expressed

20. Which DNA sequences must be contained in an expression vector that are absent in a vector used only for cloning? (Hint: The next question may provide clues to part of the answer.)

21. The compound isopropylthiogalactoside, or IPTG, is often used to produce large amounts of foreign proteins from genes cloned into expression vectors. What is the role of IPTG in this system?

Cloned Genes Can Be Altered

22. What are the different strategies for generating specific mutations in cloned DNA sequences?

Yeast Is an Important Eukaryotic Host for Recombinant DNA

23. In what ways does cloning in yeast differ from cloning strategies in *E. coli*?

Very Large DNA Segments Can Be Cloned in Yeast Artificial Chromosomes

24. How do YACs differ from plasmids as cloning vectors?

Cloning in Plants Is Aided by a Bacterial Plant Parasite

25. In the transformation of plant cells, new genes are introduced using an engineered shuttle vector containing the gene to be inserted, a selectable marker, and two T-DNA repeats. Why is it also necessary to include an engineered Ti plasmid that no longer contains its T-DNA gene?

The Human Genome and Human Gene Therapy

26. Name some problems that must be overcome in order for gene therapy to be successful.

Cloning in Animal Cells Points the Way to Human Gene Therapy

27. In what way are retroviruses an important tool for use in gene therapy?

Recombinant DNA Technology Yields New Products and Choices

28. What are the major ethical hurdles to the practical use of recombinant DNA technologies? Do you think the potential benefits outweigh the potential problems?

SELF-TEST

Do You Know the Terms?

ACROSS

1. Vectors of choice for medium-size DNA fragments.
5. One approach to ______-______ mutagenesis is to use an oligonucleotide primer containing an altered base for the synthesis of a duplex DNA.
7. The new, improved *product* of two fused genes is a ______ protein.
8. Small, circular, extrachromosomal DNA molecule.
10. Viral ______ are often modified retroviruses.
12. A genomic ______ contains more information than its largest homolog at any university.
13. Synthetic DNA fragment with recognition sequences for several restriction endonucleases.
15. These cleave DNA at specific base sequences; the scalpels of molecular biology. (2 words)
16. In bacteria, event induced by a cold, calcium chloride bath followed by heat shock or a strong jolt of electricity.
18. Natural genetic engineer in plants. (2 words)
20. Synthetic DNA, complementary in sequence to an RNA template. (abbr.)
21. Describes ends of DNA fragments that have no overhang.
23. To ______ or not to ______ is no longer the question; everyone in biochemistry is doing it! (Hint: To make an identical copy of an organism, a cell, or a DNA segment.) (1 word)
24. Describes composite DNA molecules containing DNA from two or more species.

DOWN

2. A self-replicating DNA plasmid that contains a eukaryotic inducible promoter and a polylinker is called a(n) ______ vector.
3. Contains all the genetic information to make an organism.
4. Radioactive DNA fragment that can ferret out and bind to specific DNA sequences.
5. Self-replicating piece of DNA that is capable of surviving in both *E. coli* and *S. cerevisiae.* (2 words)
6. Shocking technique that makes cells transiently permeable to DNA.
9. It joins DNA strands; molecular "glue." (2 words)
10. ______ vectors are the equivalent of a molecular "syringe."
11. Differences in lengths and sequences of DNA fragments produced by random cutting of genomic DNA by restriction enzymes; vary from one individual to another. (abbr.)
13. Gene amplification technique that relies on the activity of a heat-stable polymerase (*Taq*I) isolated from hot-spring bacteria. (abbr.)
14. The hugely oversized mouse in Fig. 29–24 is a famous example of a ______ animal.
17. Restriction endonucleases that make staggered cuts produce ______. (2 words)
19. Following electrophoresis, proteins can be identified using antibodies in a Western blot, mRNA can be identified using DNA probes in a Northern blot, and DNA fragments can be identified using DNA probes in a ______ blot.
22. Used for cloning genomic DNA, these vectors are actually unstable when they contain inserts of less than 100,000 bp.
23. Although it is not used in computers, a DNA ______ contains an enormous amount of information.

Do You Know the Facts?

Questions 1–5 refer to the following illustration.

The illustration below shows a fragment of DNA that has been isolated from a genomic DNA library. The fragment contains a protein-coding region (indicated by the black box); the sites at which a number of restriction endonucleases cut (a restriction map) are indicated. Also shown is a plasmid that has been engineered to be used as an expression vector (the curved arrow indicates the direction of translation from the promoter); its restriction enzyme sites are indicated.

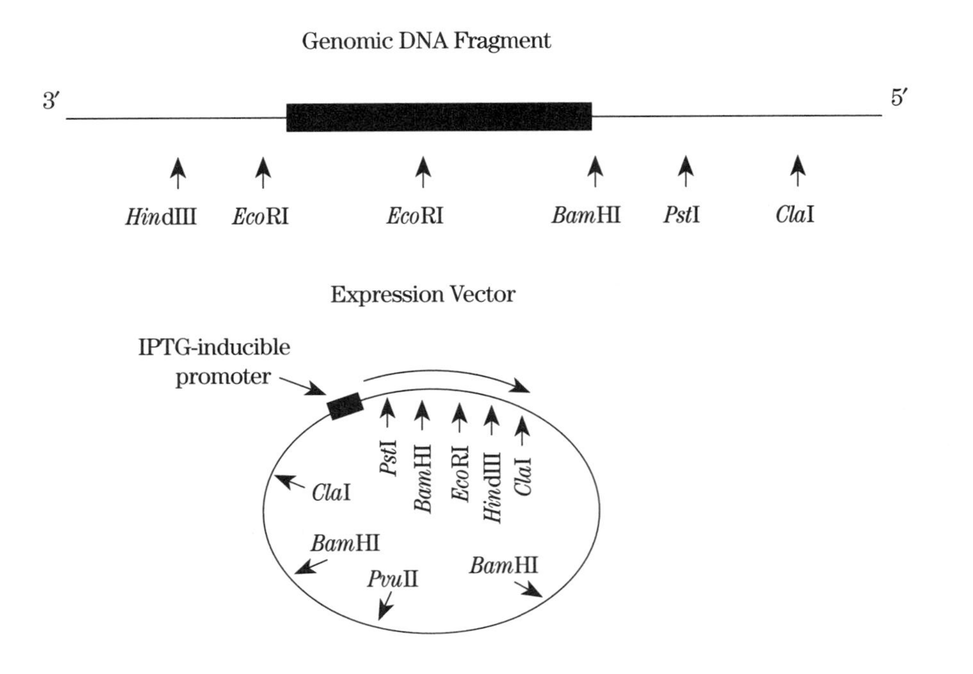

1. Name the region of the plasmid, near the promoter, where *Pst*I, *Bam*HI, *Eco*RI, *Hin*dIII, and *Cla*I have recognition sequences.

2. If you cut the vector with the restriction enzyme *Bam*HI, how many pieces will you get?

3. You wish to insert the coding region of the DNA fragment into the expression vector so that the gene will be expressed (i.e., produce the protein it codes for). Indicate where the gene fragment must be inserted into the vector and the orientation it must have.

4. If your goal is to obtain only a *portion* of the total protein for use in generating antibodies, which single enzyme could you use to cut both the vector and the DNA fragment so that the vector could produce all or part of the protein? Would *all* vectors with inserts make protein?

5. If your goal is to produce a *complete, functional* protein, could you still use a single restriction enzyme? Which enzyme or combination of enzymes could you use to get enough of the gene fragment into the vector to ensure formation of the entire gene product?

Applying What You Know

1. You have received from a colleague a plasmid containing a piece of DNA that is expressed only in neuronal cells in *Drosophila.* You suspect that this gene codes for some type of signaling peptide.

(a) What must you do to test your hypothesis? In other words, how do you get from a piece of DNA in a plasmid, to the nucleotide sequence, to a protein you can examine functionally?

(b) How could you find out whether the gene is present in other species (e.g., a laboratory rat) and determine its pattern of expression?

(c) Why would you or anyone care what proteins are present in *Drosophila* or in the lab rat?

2. How would you generate DNA probes for the following applications?

(a) You have isolated an uncharacterized protein and you want to find out which cells and/or tissues actually express this protein.

(b) You have isolated the mRNA for insulin from sheep pancreas. Your goal is to isolate a clone containing the insulin gene from a human cDNA library in order to produce authentic human insulin protein in *E. coli.*

ANSWERS

Do You Know the Terms?

ACROSS

1. bacteriophages
5. site directed
7. fusion
8. plasmid
10. vectors
12. library
13. polylinker
15. restriction enzymes
16. transformation
18. Ti plasmid
20. cDNA
21. blunt
23. clone
24. recombinant

DOWN

2. expression
3. genome
4. probe
5. shuttle vector
6. electroporation
9. DNA ligase
10. viral
11. RFLPs
13. PCR
14. transgenic
17. sticky ends
19. Southern
22. YACs
23. chip

Do You Know the Facts?

1. The polylinker region.
2. Three pieces.
3. The gene fragment must be inserted in the polylinker region, adjacent to and to the right of the promoter region. It must be oriented with its 5′ end to the left (at the beginning of the curved arrow) and its 3′ end to the right (at the arrowhead end): The DNA fragment will be flipped 180° relative to its orientation in the illustration.
4. You can use *Eco*RI to cut both the vector and the DNA fragment; this will linearize the circular plasmid so that you can insert the fragment. You will also obtain a piece of the coding region that includes the genes for the carboxyl-terminal half of the protein. Because all the ends are cut with the same restriction enzyme, all are compatible. This means that the inserted DNA can be oriented in either the 5′ → 3′ or the 3′ → 5′ direction. However, only those plasmids with inserts oriented in the 5′ → 3′ direction will be able to translate the entire piece of DNA.
5. You could cut both the vector and the DNA fragment with two restriction enzymes: *Pst*I and *Hin*dIII. When the fragment is ligated into the vector, its *Pst*I-cut 5′ end will be closest to the promoter (at the unique *Pst*I site in the polylinker) and its *Hin*dIII-cut 3′ end will be farthest from the promoter (inserting into the unique *Hin*dIII site in the polylinker). Note that you would not want to use *Bam*HI. It would insert the entire coding region into the vector in the correct orientation when used in concert with *Hin*dIII, but it also cuts at other sites within the vector.

Applying What You Know

1. **(a)** The first step is to determine the DNA sequence of the cloned gene. This can be done by purifying the plasmid vector DNA, excising the inserted DNA with restriction endonucleases, and using the dideoxy sequencing technique to sequence the excised fragment. If you obtained the DNA fragment from a genomic library, the sequence will likely include promoter regions, introns, exons, etc. Finding the initiation AUG sequence will provide you with the starting point of the open reading frame. If you obtained the DNA fragment from a cDNA library, determining the coding sequence will be more straightforward.

   The second step is to deduce the amino acid sequence from the DNA sequence, which is now usually done with computer programs. This information can then be compared with the sequences in various databases to see whether any similar genes have been sequenced and characterized. (This capability often proves an invaluable tool for researchers; someone else may already have cloned a homologous gene, and this information may give clues about the function of the gene you are studying.)

   Once you know the amino acid sequence, you can synthesize short segments of the protein for use in generating antibodies. These antibodies can localize the protein in situ, or they can be attached to column material for affinity-chromatography purification of the protein from homogenized *Drosophila*. An alternative approach would be to clone the DNA into a fusion protein vector and make the protein in *E. coli.*

Having obtained a functional protein, you are in a position to test whether it is a signaling peptide (neurotransmitter).

(b) You could look for homologous genes in other organisms using labeled DNA probes and tissue-specific cDNA libraries. If any of the clones in the libraries hybridizes to the probe, that clone most likely contains a gene similar to that from *Drosophila.* The source of the cDNA library will provide information about which tissues in rat, for example, actually express the homologous gene.

(c) You should now be acutely aware that the same basic molecular mechanisms are used in most biological systems. Cellular and molecular mechanisms that operate in *Drosophila*—and those that operate in the rat—are likely to have counterparts in other organisms, including *Homo sapiens.* The homeobox genes that play important roles in the development of *Drosophila* embryos, for example, are closely related to the *Hox* genes that regulate development in vertebrates.

2. (a) To generate DNA probes given only an uncharacterized protein, you first have to establish the amino acid sequence of the protein so as to deduce the nucleotide sequence of its gene. It is then possible to generate a synthetic complementary oligonucleotide containing sequences complementary to a portion of the deduced sequence. You can tag the oligonucleotide with a radioactive or fluorescent probe and perform in situ hybridization on fixed tissue slices to locate cells that contain mRNA that binds the probe. The results of this study will indicate which cells *might* contain the uncharacterized protein.

(b) Generating DNA probes when you have a supply of mRNA for a protein is relatively straightforward. Using reverse transcriptase and the mRNA as a template, you can make cDNA. This can be radioactively labeled and used to screen a genomic library. Once you have located a clone that contains the entire coding region, the foreign DNA fragment can be excised and cloned into an expression vector; it can then be used to produce quantities of the desired protein—in this case, insulin.

Solutions Manual

chapter

2

Cells

1. The Size of Cells and Their Components

(a) If you were to magnify a cell 10,000 fold (which is typical of the magnification that is achieved using a microscope), how big would it appear? Assume you are viewing a "typical" eukaryotic cell with a cellular diameter of 50 microns.

(b) If this cell were a muscle cell, how many molecules of actin could it hold assuming there were no other cellular components present? (Actin molecules are spherical with a diameter of 3.6 nm; assume the muscle cell is spherical, the volume of a sphere is 4/3 πr^3.)

(c) If this were a liver cell of the same dimensions, how many mitochondria could it hold assuming there were no other cellular components present? (Assume mitochondria are spherical with a diameter of 1.5 μm; assume the liver cell is spherical, the volume of a sphere is 4/3 πr^3.)

(d) Glucose is the major energy-yielding nutrient for most cells. Assuming it is present at a concentration of 1 mM, calculate how many molecules of glucose would be present in our hypothetical (and spherical) eukaryotic cell. (Avogadro's number, the number of molecules in 1 mol of a non-ionized substance, is 6.02×10^{23}.)

(e) Hexokinase is an important enzyme in the metabolism of glucose by cells (see Chapter 15). If the molar concentration of hexokinase in our eukaryotic cell is 20 μM, how many glucose molecules are available for each hexokinase enzyme molecule to metabolize?

Answer

(a) The magnified cell would have a diameter of 50×10^4 μm $= 500 \times 10^3$ μm $= 500$ mm, or 20 inches, which is about the diameter of a large pizza.

(b) The radius of a globular actin molecule is 3.6 nm/2 = 1.8 nm; the volume of the molecule, in cubic meters, is $4/3[3.14(1.8 \times 10^{-9}\text{ m})^3] = 2.44 \times 10^{-26}\text{ m}^3$.

The total number of actin molecules that could fit inside the cell is found by dividing the muscle cell volume by the actin molecule volume. Cell volume $= 4/3[3.14(25 \times 10^{-6}\text{ m})^3] = 6.5 \times 10^{-14}\text{ m}^3$. Thus the number of actin molecules in the hypothetical muscle cell is

$$(6.5 \times 10^{-14}\text{ m}^3)/(2.44 \times 10^{-26}\text{ m}^3) = 2.66 \times 10^{12} \approx 2.7 \times 10^{12}\text{ molecules*}$$

This is approximately 2.7 trillion actin molecules.

(c) The radius of the mitochondrion is 1.5 μm/2 = 0.75 μm, therefore the volume of a spherical mitochondrion is $4/3[3.14(0.75 \times 10^{-6}\text{ m})^3] = 1.77 \times 10^{-18}\text{ m}^3$. The number of

**Significant figures* In multiplication and division, the answer can be expressed with no more significant figures than the least precise value in the calculation. Since some of the data in these problems are derived from measured values, we must round off the calculated answer to reflect this. In this case, the cell volume ($6.5 \times 10^{-14}\text{ m}^3$) has two significant figures, so the final answer (2.7×10^{12} molecules) can be expressed with no more than two significant figures. It will be standard practice in these expanded answers to first show the calculated answer, then to round it off to the proper number of significant figures.

mitochondria in the hypothetical liver cell is

$$(6.5 \times 10^{-14}\ \text{m}^3)/(1.77 \times 10^{-18}\ \text{m}^3) = 36{,}723 \approx 37{,}000 \text{ mitochondria}$$

(d) The volume of the eukaryotic cell is $6.5 \times 10^{-14}\ \text{m}^3$ as determined above, which is $6.5 \times 10^{-8}\ \text{cm}^3$ or 6.5×10^{-8} mL. One liter of a 1 mM solution has (0.001 mol/1000 mL) (6.02×10^{23} molecules/mol) = 6.02×10^{17} molecules/mL. The total number of glucose molecules is the product of the cell volume and glucose concentration:

$$\begin{aligned}(6.5 \times 10^{-8}\ \text{mL})(6.02 \times 10^{17}\ \text{molecules/mL}) &= 3.91 \times 10^{10}\ \text{molecules}\\ &\approx 3.9 \times 10^{10}\ \text{molecules}\end{aligned}$$

Thus there are approximately 39 billion glucose molecules in this cell.

(e) The concentration ratio of glucose/hexokinase is 0.001 M/0.00002 M, or 50/1, meaning that each enzyme molecule would have about 50 molecules of glucose available as substrate.

2. Components of *E. coli* *E. coli* cells are rod-shaped, about 2 μm long and 0.8 μm in diameter. The volume of a cylinder is $\pi r^2 h$, where h is the height of the cylinder.

(a) If the average density of *E. coli* (mostly water) is 1.1×10^3 g/L, what is the mass of a single cell?

(b) The protective cell wall of *E. coli* is 10 nm thick. What percentage of the total volume of the bacterium does the wall occupy?

(c) *E. coli* is capable of growing and multiplying rapidly because of the inclusion of some 15,000 spherical ribosomes (diameter 18 nm) in each cell, which carry out protein synthesis. What percentage of the total cell volume do the ribosomes occupy?

Answer

(a) The volume of a single *E. coli* cell can be calculated from $\pi r^2 h$:

$$\pi(0.4)^2(2) = 3.14(4 \times 10^{-5}\ \text{cm})^2(2 \times 10^{-4}\ \text{cm}) = 1.0 \times 10^{-12}\ \text{cm}^3 = 1 \times 10^{-15}\ \text{L}$$

Density (g/L) multiplied by volume (L) gives the weight of a single cell:

$$(1.1 \times 10^3\ \text{g/L})(10^{-15}\ \text{L}) = 1.1 \times 10^{-12}\ \text{g} \approx 1\ \text{pg}$$

(b) The diameter of the *E. coli* cell contained *within* the cell wall (i.e., excluding the thickness of the cell wall) is 0.8 − 0.02 = 0.78 μm; the height is 2 − 0.02 = 1.98 μm. The volume of the *E. coli* cell that does *not* include the cell wall (using πr^2h) is

$$3.14[(7.8 \times 10^{-5})/(2)\ \text{cm}]^2(1.98 \times 10^{-4}\ \text{cm}) = 9.5 \times 10^{-13}\ \text{cm}^3$$

Thus the *E. coli* volume within the cell wall occupies $(9.5 \times 10^{-13})/(1 \times 10^{-12}) = 0.95$ or 95% of the total volume. This means that the cell wall must account for 5% of the total volume of this bacterium.

(c) Ribosomal radius is 9 nm = 0.9×10^{-6} cm. Volume of a single ribosome is

$$4/3\ \pi r^3 = 1.33(3.14)(0.9 \times 10^{-6}\ \text{cm})^3 = 3.0 \times 10^{-18}\ \text{cm}^3$$

This volume, multiplied by the total number of ribosomes, gives the total volume occupied by ribosomes:

$$15{,}000(3.0 \times 10^{-18}\ \text{cm}^3) = 4.6 \times 10^{-14}\ \text{cm}^3 = 0.046 \times 10^{-12}\ \text{cm}^3$$

Given a cell volume of $9.5 \times 10^{-13}\ \text{cm}^3$, this represents

$$(4.6 \times 10^{-14}\ \text{cm}^3)/(9.5 \times 10^{-13}\ \text{cm}^3) = 0.048 \text{ or } 4.8\% \text{ of the total } \textit{E. coli} \text{ cell volume.}$$

3. Genetic Information in *E. Coli* DNA The genetic information contained in DNA consists of a linear sequence of successive coding units, known as codons. Each codon is a specific sequence of three nucleotides (three nucleotide pairs in double-stranded DNA), and each codon codes for a single amino acid unit in a protein. The molecular weight of an *E. coli* DNA molecule is about 3.1×10^9. The average molecular weight of a nucleotide pair is 660, and each nucleotide pair contributes 0.34 nm to the length of DNA.

(a) Calculate the length of an *E. coli* DNA molecule. Compare the length of the DNA molecule with the cell dimensions (see Problem 2). How does the DNA molecule fit into the cell?

(b) Assume that the average protein in *E. coli* consists of a chain of 400 amino acids. What is the maximum number of proteins that can be coded by an *E. coli* DNA molecule?

Answer

(a) The number of nucleotide pairs in the DNA molecule can be calculated by dividing the molecular weight of DNA by that of a single pair:

$$(3.1 \times 10^9)/(0.66 \times 10^3) = 4.7 \times 10^6 \text{ pairs}$$

Multiplying the number of pairs by the length per pair gives

$$(4.7 \times 10^6 \text{ pairs})(0.34 \text{ nm/pair}) \approx 1.6 \times 10^6 \text{ nm} = 1.6 \text{ mm}$$

The length of the cell is 2.0 μm (from Problem 2), or 0.002 mm, which means that the DNA is 1.6 mm/0.002 mm or 800 times longer than the cell. The DNA must be tightly coiled to fit into the cell.

(b) Since the DNA molecule has 4.7×10^6 nucleotide pairs, as calculated in (a), it must have 1/3 this number of triplet codons:

$$(4.7 \times 10^6)/3 = 1.57 \times 10^6 \text{ codons}$$

If each protein has an average of 400 amino acids, each requiring one codon, the number of proteins that can be coded by *E. coli* DNA is

$$(1.57 \times 10^6)/400 = 3{,}930 \approx 3{,}900 \text{ proteins per cell}$$

4. The High Rate of Bacterial Metabolism Bacterial cells have a much higher rate of metabolism than animal cells. Under ideal conditions some bacteria will double in size and divide in 20 min, whereas most animal cells under rapid growth conditions require 24 h. The high rate of bacterial metabolism requires a high ratio of surface area to cell volume.

(a) Why would the surface-to-volume ratio have an effect on the maximum rate of metabolism?

(b) Calculate the surface-to-volume ratio for the spherical bacterium *Neisseria gonorrhoeae* (diameter 0.5 μm), responsible for the disease gonorrhea. Compare it with the surface-to-volume ratio for a globular amoeba, a large eukaryotic cell of diameter 150 μm. The surface area of a sphere is $4\pi r^2$.

Answer

(a) Metabolic rate is limited by diffusion of fuels into the cell and waste products out of the cell. This diffusion in turn is limited by the surface area of the cell. As the ratio of surface area to volume decreases, the rate of diffusion cannot keep up with the rate of metabolism within the cell.

(b) For a sphere, surface area $= 4\pi r^2$ and volume $= 4/3\ \pi r^3$. The ratio of the two is the surface-to-volume ratio, S/V, which equals $3/r$ or $6/D$, where D = diameter. Thus, rather than calculating S and V separately for each cell, one can rapidly calculate and compare S/V ratios for cells of different diameters.

$$S/V \text{ for } N.\ gonorrhoeae = 6/(0.5\ \mu\text{m}) = 12\ \mu\text{m}^{-1} = 1.2 \times 10^7\ \text{m}^{-1}$$

$$S/V \text{ for amoeba} = 6/(150\ \mu\text{m}) = 0.04\ \mu\text{m}^{-1} = 4 \times 10^4\ \text{m}^{-1}$$

$$\text{Thus } \frac{S/V \text{ for bacterium}}{S/V \text{ for amoeba}} = \frac{12\ \mu\text{m}^{-1}}{0.04\ \mu\text{m}^{-1}} \approx 300$$

5. A Strategy to Increase the Surface Area of Cells Certain cells whose function is to absorb nutrients, such as the cells lining the small intestine or the root hair cells of a plant, are optimally adapted to their role because their exposed surface area is increased by microvilli. Consider a spherical epithelial cell (diameter 20 μm) in the lining of the small intestine. Given that only a part of the cell surface faces the interior of the intestine, assume that a "patch" corresponding to 25% of the cell area is covered with microvilli. Furthermore, assume that the microvilli are cylinders 0.1 μm in diameter, 1.0 μm long, and spaced in a regular grid 0.2 μm on center.

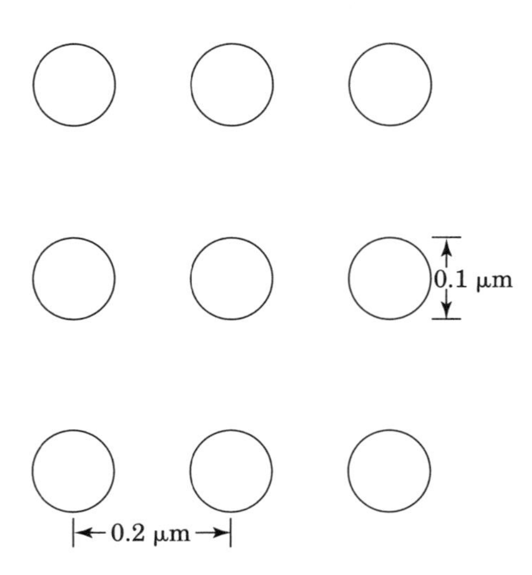

Arrangement of microvilli on the "patch"

(a) Calculate the number of microvilli on the patch.
(b) Calculate the surface area of the patch, assuming it has no microvilli.
(c) Calculate the surface area of the patch, assuming it does have microvilli.
(d) What percentage improvement in absorptive capacity (reflected by the surface-to-volume ratio) does the presence of microvilli provide?
(e) What other organelles and organ systems utilize this strategy to improve absorptive capacity?

Answer

(a) The total surface area of an epithelial cell is
$4\pi r^2 = 4(3.14)(10\ \mu m)^2 = 1{,}256\ \mu m^2$
25% of this is exposed, or $0.25(1{,}256\ \mu m^2) = 314\ \mu m^2$. Taking the square root of this number, the patch can be viewed as a square, 17.7 μm on a side. The number of microvilli on a side can be calculated by dividing the side dimension by the spacing:
$17.7\ \mu m/0.2\ \mu m = 88.6$
Squaring this value gives the total number of microvilli = 7,850 ≈ 7,900.

(b) The surface area as calculated in (a) $= 314\ \mu m^2$
$= 3.14 \times 10^2(10^{-6}\ m)^2$
$= 3.14 \times 10^{-10}\ m^2 \approx 3.1 \times 10^{-10}\ m^2$

(c) The surface area of one microvillus is $\pi r^2 + \pi Dh$, where D = diameter of a cylinder:
$(3.14)(0.05\ \mu m)^2 + (3.14)(0.1)(1)\ \mu m^2 \approx 0.314\ \mu m^2$
The surface area of 7,850 microvilli = $7{,}850(0.314)\ \mu m^2 = 2{,}465\ \mu m^2$.
The total area of the patch is the area covered by microvilli *plus* the area not covered by microvilli.

Area covered by microvilli $= 7{,}850 \times$ (surface area of one end of microvillus)
$= 7{,}850(3.14)(0.05)^2\ \mu m^2 = 61.6\ \mu m^2$

Area not covered $=$ total area of patch − area covered by microvilli
$= 3.14 \times 10^{-10}\ m^2 - 61.6\ \mu m^2$
$= (314 - 62)\ \mu m^2$
$= 252\ \mu m^2$

Total surface area of the microvilli-covered patch

$$= \text{surface area of microvilli} + \text{area of patch not covered}$$
$$= (2{,}465 + 252)\ \mu\text{m}^2 = 2{,}717\ \mu\text{m}^2 \approx 2.7 \times 10^{-9}\ \text{m}^2$$

(d) The surface area of the patch without microvilli = 314 μm^2, and the surface area with microvilli = 2,717 μm^2. Therefore, the improvement in surface area, expressed as a percentage, is

$$\frac{(2{,}717 - 314)\ \mu\text{m}^2\ (100)}{314\ \mu\text{m}^2} = 765\%$$

(e) A similar strategy for increasing surface area is used in mitochondria, chloroplasts, the gill lamellae of fish, and the root hairs of plants.

6. Fast Axonal Transport Neurons possess long, thin processes called axons, which are structures specialized for conducting signals throughout an organism. Some axonal processes can be as long as 2 m—for example, the axons that originate in the spinal cord and terminate in the muscles of your toes. Small membrane-enclosed vesicles carrying materials essential to axonal function move along microtubules from the cell body to the tip of axons by a kinesin-dependent "fast axonal transport."

(a) If the average velocity of a vesicle is 1 μm/s, how long does it take a vesicle to move from a cell body in the spinal cord to the axonal tip in the toes?

(b) Movement of large molecules in cells by diffusion occurs relatively slowly in cells. (For example, hemoglobin diffuses at a rate of approximately 5 μm/s.) However, the diffusion of sucrose in an aqueous solution occurs at a rate approaching that of fast transport mechanisms (about 4 μm/s). What are some advantages to a cell or an organism of fast, directed transport mechanisms, compared to what a cell could do relying on diffusion alone?

(c) Some of the studies which originally determined the velocity of vesicular movement were performed on microtubules *in vitro* (in a dish). In order to isolate the microtubules for these studies, intact neurons were initially homogenized (broken) in the presence of 0.2 M sucrose to prevent osmotic swelling and bursting of intracellular organelles. Why is this an important consideration in studies involving cell fractionation?

Answer

(a) Transport time is equal to distance traveled/velocity, or

$$(2 \times 10^6\ \mu\text{m})/(1\ \mu\text{m/s}) = 2 \times 10^6\ \text{s, or about 23 days!}$$

(b) Diffusion occurs randomly in all directions, but fast transport mechanisms allow for unidirectional motion along microtubular "tracks." Fast, directed transport mechanisms therefore allow cells to have larger sizes than they could achieve if they relied solely on diffusional mechanisms. Because the process requires energy, it can be regulated by the level of ATP available and in response to other cellular signals. In addition, such transport mechanisms allow cells to localize cellular functions to discrete regions.

For example, neurons have long axonal processes that can approach 2 m in length. These cells rely on fast transport mechanisms for constructing and maintaining axons that are distant from the cell body, where the necessary components are made. Additionally, neurons release chemicals that transmit information to other neurons at discrete sites along these axonal processes. Fast transport is used to move these chemicals, which are typically synthesized and packaged into vesicles in the cell body, to distant release sites in sufficient quantities for use in neuronal communication.

(c) Breakage of lysosomes would release lytic (cleaving) enzymes that break down proteins and other macromolecules; peroxisomes would release enzymes and chemicals that cause oxidation, either inactivating enzymes or potentiating them for further breakdown. The result would be damage to the cellular component or biomolecule—in this case, the microtubule—that one is attempting to isolate.

chapter

3 Biomolecules

1. **Vitamin C: Is the Synthetic Vitamin as Good as the Natural One?** A claim sometimes put forth by purveyors of health foods is that vitamins obtained from natural sources are more healthful than those obtained by chemical synthesis. For example, pure L-ascorbic acid (vitamin C) extracted from rose hips is better for you than pure L-ascorbic acid manufactured in a chemical plant. Are the vitamins from the two sources different? Can the body distinguish a vitamin's source?

   Answer The properties of the vitamin—like any other compound—are determined by its chemical structure. Because vitamin molecules from the two sources are structurally identical, their properties are identical, and no organism can distinguish between them. If, however, different vitamin preparations contain different impurities, the biological effects of the *mixtures* may vary with the source. The ascorbic acid in such preparations, however, will be identical.

2. **Identification of Functional Groups** Figures 3–5 and 3–6 show some common functional groups of biomolecules. Because the properties and biological activities of biomolecules are largely determined by their functional groups, it is important to be able to identify them. In each of the compounds below, circle and identify by name each constituent functional group.

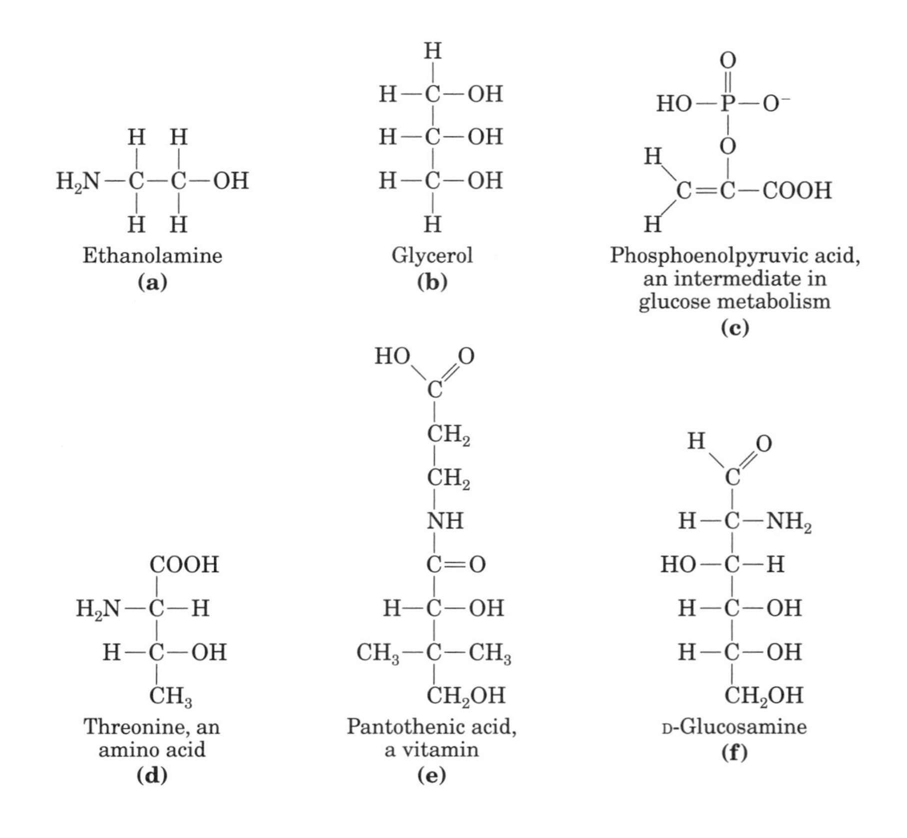

Answer

(a) —NH_2 = amino; —OH = hydroxyl

(b) —OH = hydroxyl (three)

(c) —$P(OH)O_2^-$ = phosphoryl (in its ionized form); —COOH = carboxyl

(d) —COOH = carboxyl; —NH_2 = amino; —OH = hydroxyl; —CH_3 = methyl (two)

(e) —COOH = carboxyl; —CO—NH— = amino; —OH = hydroxyl (two);
—CH_3 = methyl (two)

(f) —CHO = aldehyde; —NH_2 = amino; —OH = hydroxyl (four)

3. Drug Activity and Stereochemistry The quantitative differences in biological activity between the two enantiomers of a compound are sometimes quite large. For example, the D isomer of the drug isoproterenol, used to treat mild asthma, is 50 to 80 times more effective as a bronchodilator than the L isomer. Identify the chiral center in isoproterenol. Why do the two enantiomers have such radically different bioactivity?

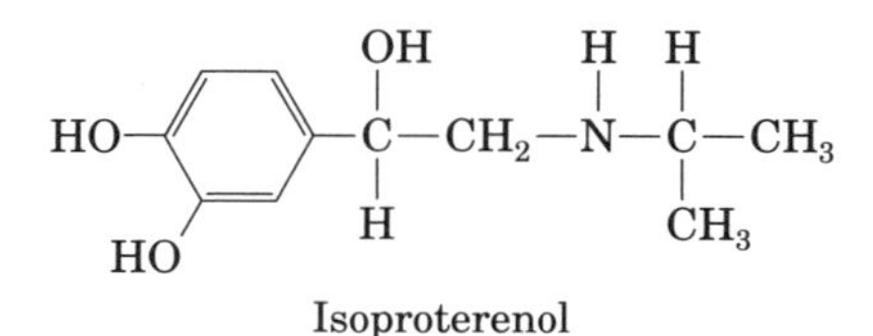

Isoproterenol

Answer A chiral center or chiral carbon is a carbon atom that is bonded to four different groups. One consequence of a single chiral center is that the molecule has two enantiomers, usually designated as D and L (or in the RS system, *S* and *R*). In isoproterenol, only one carbon has four different groups around it; this is the chiral center:

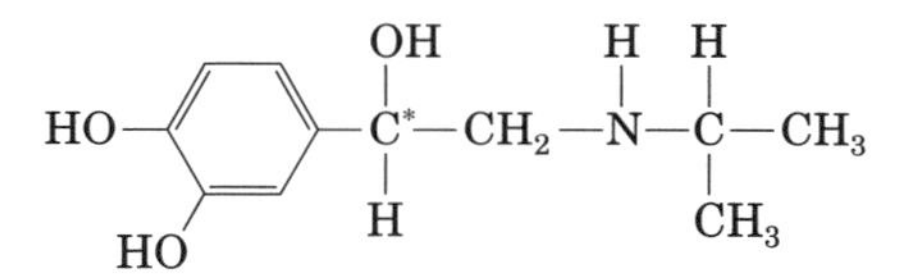

The bioactivity of a drug is the result of interaction with a biological "receptor," a protein molecule with a binding site that is also chiral and stereospecific. As a result, the interaction of the D isomer of a drug with a chiral receptor site will differ from the interaction of the L isomer with that site.

4. Drug Action and Shape of Molecules Some years ago two drug companies marketed a drug under the trade names Dexedrine and Benzedrine. The structure of the drug is shown below.

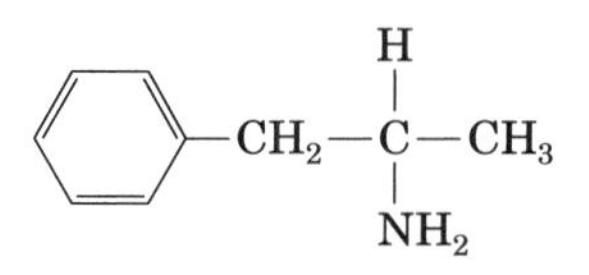

The physical properties (C, H, and N analysis, melting point, solubility, etc.) of Dexedrine and Benzedrine were identical. The recommended oral dosage of Dexedrine (which is still available) was 5 mg/day, but the recommended dosage of Benzedrine (no longer available) was twice that. Apparently it required considerably more Benzedrine than Dexedrine to yield the same physiological response. Explain this apparent contradiction.

Answer Only one of the two enantiomers of the drug molecule (which has a chiral center) is physiologically active, for reasons described in the answer to Problem 3 (interaction with a stereospecific receptor site). Dexedrine, as manufactured, consists of the single enantiomer (D-amphetamine) recognized by the receptor site. In contrast, Benzedrine was a racemic mixture (having equal amounts of D and L isomers). Thus, a much larger dose of Benzedrine was required to obtain the same effect.

5. Components of Complex Biomolecules Figure 3–24 shows the major components of complex biomolecules. For each of the three important biomolecules below (shown in their ionized forms at physiological pH), identify the constituents.

(a) Guanosine triphosphate (GTP), an energy-rich nucleotide that serves as a precursor to RNA.
(b) Phosphatidylcholine, a component of many membranes.
(c) Methionine enkephalin, the brain's own opiate.

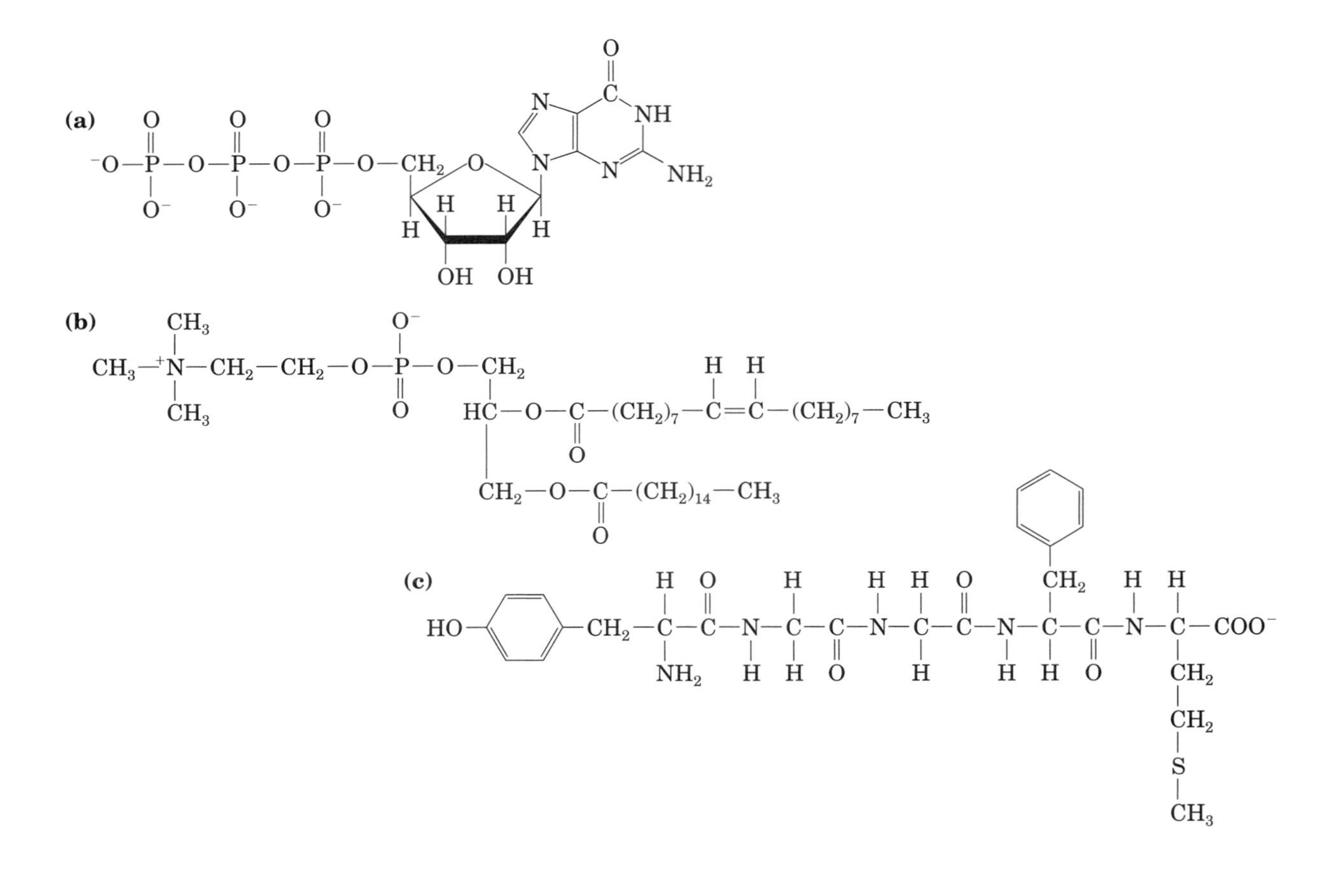

Answer

(a) Three phosphoric acid molecules (linked by two anhydride bonds), esterified to an α-D-ribose (at the 5′ position), which is attached at C-1 to guanine.
(b) Choline esterified to a phosphoric acid group, which is esterified to glycerol, which is esterified to two fatty acids, oleic acid and palmitic acid.
(c) Tyrosine, two glycines, phenylalanine, and methionine, all linked by peptide bonds.

6. Determination of the Structure of a Biomolecule An unknown substance, X, was isolated from rabbit muscle. Its structure was determined from the following observations and experiments. Qualitative analysis showed that X was composed entirely of C, H, and O. A weighed sample of X was completely oxidized, and the H_2O and CO_2 produced were measured; this quantitative analysis revealed that X contained 40.00% C, 6.71% H, and 53.29% O by weight. The molecular mass of X, determined by a mass spectrometer, was 90.00 amu. An infrared spectrum showed that X contained one double bond. X dissolved readily in water to give an acidic solution; the solution demonstrated optical activity when tested in a polarimeter.

(a) Determine the empirical and molecular formula of X.

(b) Draw the possible structures of X that fit the molecular formula and contain one double bond. Consider *only* linear or branched structures and disregard cyclic structures. Note that oxygen makes very poor bonds to itself.

(c) What is the structural significance of the observed optical activity? Which structures in (b) does this observation eliminate? Which structures are consistent with the observation?

(d) What is the structural significance of the observation that a solution of X is acidic? Which structures in (b) are now eliminated? Which structures are consistent with the observation?

(e) What is the structure of X? Is more than one structure consistent with all the data?

Answer

(a) From the C, H, and O analysis, and knowing the mass of X is 90.00 amu, we can calculate the relative atomic proportions by dividing the weight percents by the atomic weights:

Atom	Relative atomic proportion	No. of atoms relative to O
C	90.00 amu(40.00/100)/12 amu = 3	3/3 = 1
H	90.00 amu(6.71/100)/1.008 amu = 6	6/3 = 2
O	90.00 amu(53.29/100)/16.0 amu = 3	3/3 = 1

Thus, the empirical formula is CH_2O, with a formula weight of 12 + 2 + 16 = 30. The molecular formula, based on X having a mass of 90.00 amu, must be three times this empirical formula, or $C_3H_6O_3$.

(b) Structures **1** through **5** can be eliminated because they are unstable enol isomers of the corresponding carbonyl derivatives. Structures **9, 10,** and **12** can also be eliminated on the basis of their instability: they are hydrated carbonyl derivatives (vicinal diols).

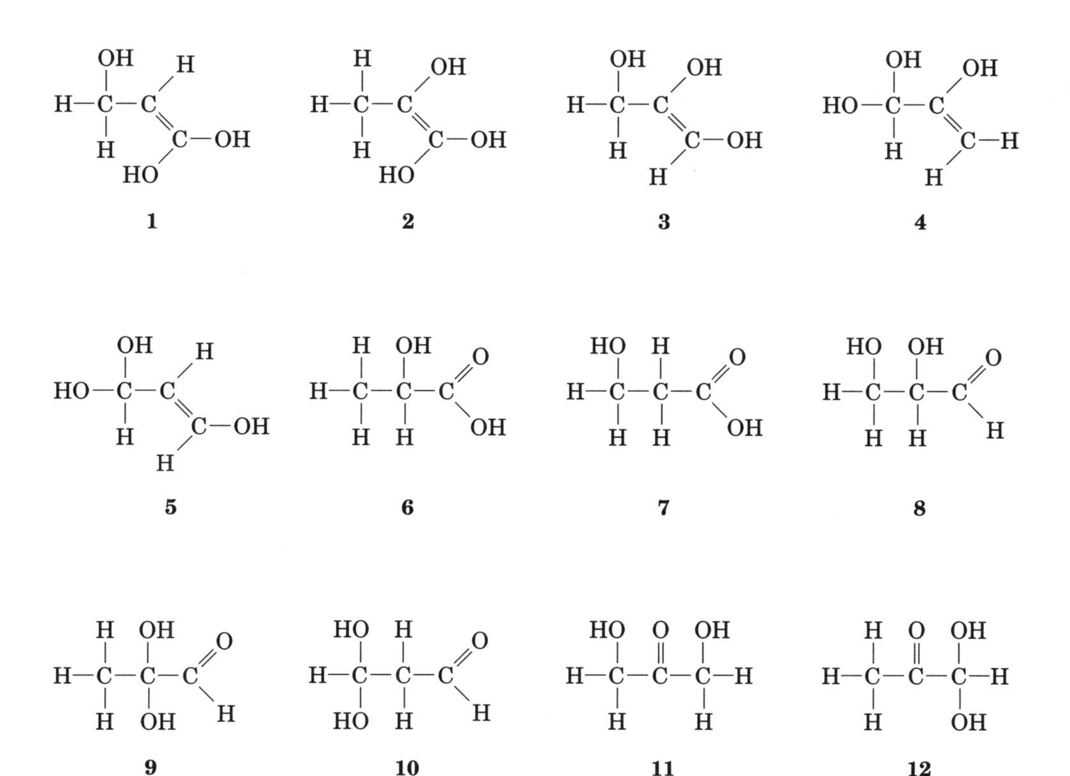

(c) The presence of optical activity eliminates all structures that lack a chiral center (a C atom surrounded by four different groups). Only structures **6** and **8** have chiral centers.

(d) X contains an acidic functional group, which structure **8** does not. Structure **6** contains a carboxyl group.

(e) Structure **6** is substance X. This compound exists in two enantiomeric forms that cannot be distinguished, even by measuring specific rotation. One could determine absolute stereochemistry by x-ray crystallography.

7. Separating Biomolecules In laboratory biochemistry, it is first necessary to separate the molecule of interest from other biomolecules in the sample—that is, to *purify* the protein, nucleic acid, carbohydrate, or lipid. Specific purification techniques will be addressed later in the text. However, just by looking at the monomeric subunits from which the larger biomolecules are made, you should have some ideas as to what characteristics of those biomolecules would allow you to separate them from one another.

(a) What characteristics of amino acids and fatty acids would allow them to be easily separated from each other?

(b) How might nucleotides be separated from glucose molecules?

Answer

(a) Both amino acids and fatty acids have carboxyl groups, whereas only the amino acids have amino groups. Thus a technique that separates molecules on the basis of the properties (charge or binding affinity) of amino groups could be used. Also, fatty acids have long hydrocarbon chains and therefore are less soluble in water than amino acids. And, the size and shape of these two types of molecules are quite different. Either of these properties may provide ways to separate the two types of compounds.

(b) A nucleotide molecule has three components: a nitrogenous organic base, a five-carbon sugar, and phosphate. Glucose is a six-carbon sugar; it is smaller than a nucleotide. The size difference could be used to separate the molecules. Alternatively, the nitrogenous bases and/or the phosphate groups characteristic of the nucleotides could be used for separation (based on differences in solubility, charge) from glucose.

8. Silicon-Based Life? Silicon is in the same group of the periodic table as carbon and, like carbon, can form up to four single bonds. Many science fiction stories have been written based on the premise of silicon-based life. Is this realistic? What characteristics of silicon make it less well adapted to performing as the central organizing element for life? To answer this question, use the information in this chapter about carbon's bonding versatility, and refer to a beginning inorganic chemistry textbook for silicon's bonding properties.

Answer

It is improbable that silicon could serve as the central organizing element for life on Earth, for several reasons. Long chains of silicon atoms are not readily synthesized, and thus the polymeric macromolecules necessary for more complex functions would not readily form. Also, oxygen disrupts bonds between two silicon atoms, which would cause problems for silicon-based life forms in an oxygen-containing atmosphere. Once formed, the bonds between silicon and oxygen are extremely stable and difficult to break, which would prevent the breaking and making (degradation and synthesis) of biomolecules that is essential to the processes of living organisms.

chapter

4

Water

1. **Simulated Vinegar** One way to make vinegar (*not* the preferred way) is to prepare a solution of acetic acid, the sole acid component of vinegar, at the proper pH (see Fig. 4–13) and add appropriate flavoring agents. Acetic acid (M_r 60) is a liquid at 25 °C with a density of 1.049 g/mL. Calculate the volume that must be added to distilled water to make 1 L of simulated vinegar (see Fig. 4–14).

Answer From Fig. 4–13, the pH of vinegar is 3.0; from Fig. 4–14, the pK_a of acetic acid is 4.76. Using the Henderson-Hasselbalch equation

$$\text{pH} = \text{p}K_a + \log \frac{[\text{A}^-]}{[\text{HA}]}$$

and the fact that dissociation of HA gives equimolar [H^+] and [A^-] (where HA is CH_3COOH, and A^- is CH_3COO^-), we can write

$$\begin{aligned} 3.00 &= 4.76 + \log ([\text{A}^-]/[\text{HA}]) \\ 3.00 - 4.76 &= -1.76 = \log ([\text{A}^-]/[\text{HA}]) \\ [\text{HA}]/[\text{A}^-] &= 10^{1.76} = 58 \end{aligned}$$

At pH 3, [H^+] = [A^-] = 10^{-3}. Substituting this value into the denominator gives

$$[\text{HA}] = 5.8 \times 10^{-2}\ \text{M} = 0.058\ \text{mol/L}$$

Dividing density (g/mL) by molecular weight (g/mol) for acetic acid gives

$$\frac{1.049\ \text{g/mL}}{60\ \text{g/mol}} = 0.0175\ \text{mol/mL}$$

Dividing this into 0.058 mol/L gives the volume of acetic acid needed to prepare 1.0 L of a 0.058 M solution:

$$\frac{0.058\ \text{mol/L}}{0.0175\ \text{mol/mL}} = 3.32\ \text{mL per liter of solution}$$

2. **Acidity of Gastric HCl** In a hospital laboratory, a 10.0 mL sample of gastric juice, obtained several hours after a meal, was titrated with 0.1 M NaOH to neutrality; 7.2 mL of NaOH was required. The patient's stomach contained no ingested food or drink, thus assume that no buffers were present. What was the pH of the gastric juice?

Answer Multiplying volume (mL) by molar concentration (mol/L) gives the number of moles in that volume of solution. If x is the concentration of gastric HCl (mol/L),

$$\begin{aligned} (10\ \text{mL})x &= (7.2\ \text{mL})(0.1\ \text{mol/L}) \\ x &= 0.072\ \text{M gastric HCl} \end{aligned}$$

Given that pH = $-\log [H^+]$ and that HCl is a strong acid,

$$pH = -\log (7.2 \times 10^{-2}) = 1.1$$

3. Measurement of Acetylcholine Levels by pH Changes The concentration of acetylcholine (a neurotransmitter) in a sample can be determined from the pH changes that accompany its hydrolysis. When the sample is incubated with the enzyme acetylcholinesterase, acetylcholine is quantitatively converted into choline and acetic acid, which dissociates to yield acetate and a hydrogen ion:

$$\underset{\text{Acetylcholine}}{CH_3{-}\overset{O}{\overset{\|}{C}}{-}O{-}CH_2{-}CH_2{-}{}^+N(CH_3)_3} \xrightarrow{H_2O} \underset{\text{Choline}}{HO{-}CH_2{-}CH_2{-}{}^+N(CH_3)_3} + \underset{\text{Acetate}}{CH_3{-}\underset{\underset{O}{\|}}{C}{-}O^-} + H^+$$

In a typical analysis, 15 mL of an aqueous solution containing an unknown amount of acetylcholine had a pH of 7.65. When incubated with acetylcholinesterase, the pH of the solution decreased to 6.87. Assuming that there was no buffer in the assay mixture, determine the number of moles of acetylcholine in the 15 mL sample.

Answer Given that pH = $-\log [H^+]$, we can calculate $[H^+]$ at the beginning and at the end of the reaction:

$$\text{At pH 7.65, } \log [H^+] = -7.65 \qquad [H^+] = 10^{-7.65} = 2.24 \times 10^{-8}\ \text{M}$$

$$\text{At pH 6.87, } \log [H^+] = -6.87 \qquad [H^+] = 10^{-6.87} = 1.35 \times 10^{-7}\ \text{M}$$

The difference in $[H^+]$ is

$$(1.35 - 0.22) \times 10^{-7}\ \text{M} = 1.13 \times 10^{-7}\ \text{M}$$

For a volume of 15 mL, or 0.015 L, multiplying volume by molarity gives

$$(0.015\ \text{L})(1.13 \times 10^{-7}\ \text{mol/L}) = 1.7 \times 10^{-9}\ \text{mol of acetylcholine}$$

4. Osmotic Balance in a Marine Frog The crab-eating frog of Southeast Asia, *Rana cancrivora*, is born and matures in fresh water but searches for its food in coastal mangrove swamps (80% to full-strength seawater). Consequently, when the frog moves from its freshwater home to seawater it experiences a large change in the osmolarity of its environment (from hypotonic to hypertonic).

(a) Eighty percent seawater contains 460 mM NaCl, 10 mM KCl, 10 mM $CaCl_2$, and 50 mM $MgCl_2$. What are the concentrations of the various ionic species in this seawater? Assuming that these salts account for nearly all the solutes in seawater, what is the osmolarity of the seawater?

(b) The chart below lists the cytoplasmic concentrations of ions in *Rana cancrivora*. Ignoring dissolved proteins, amino acids, nucleic acids, and other small metabolites, what is the osmolarity of the frog's cells based solely on the ionic concentrations given below?

	Na^+ (mM)	K^+ (mM)	Cl^- (mM)	Ca^{2+} (mM)	Mg^{2+} (mM)
Rana cancrivora	122	10	100	2	1

(c) Like all frogs, the crab-eating frog can exchange gases through its permeable skin, allowing it to stay underwater for long periods of time without breathing. How does the high permeability of frog skin affect the frog's cells when it moves from fresh water to seawater?

(d) The crab-eating frog uses two mechanisms to maintain its cells in osmotic balance with its environment. First, it allows the Na^+ and Cl^- concentrations in its cells to slowly increase as the ions diffuse down their concentration gradients. Second, like many elasmobranchs (sharks), it retains the waste product urea in its cells. The addition of both NaCl and urea increases the osmolarity of the cytosol to a value that is nearly equal to that of the surrounding environment.

$$H_2N-C(=O)-NH_2$$

Urea CH_4N_2O

Assuming the volume of water in a typical frog is 100 mL, how many grams of NaCl (formula weight (FW 58.44) does the frog need to take up in order to make its tissues isotonic with seawater?

(e) How many grams of urea (FW 60) must it retain to accomplish the same thing?

Answer

(a)

	Na^+ (mM)	K^+ (mM)	Cl^- (mM)	Ca^{2+} (mM)	Mg^{2+} (mM)
80% seawater	460	10	(460 + 10 + 20 + 100) = 590	10	50

The osmolarity of the 80% seawater of the mangrove swamp is the sum of the concentration of all dissolved species. Given that the listed salts are the primary solutes and that they completely dissociate in solution, the osmolarity is 460 + 10 + 590 + 10 + 50 = 1120 mosmol/L.

(b) The osmolarity of the cells is the sum of the concentration of all dissolved species in the cytosol. Assuming that the ions listed are the primary solutes, the osmolarity is 122 + 10 + 100 + 2 + 1 = 235 mosmol/L.

(c) The high permeability of the frog's skin, which permits effective O_2 and CO_2 exchange between its tissues and the surrounding water, also permits diffusion of water and ions down their concentration gradients. Because seawater has a higher osmolarity than the cytosol, ions (notably Na^+ and Cl^-) tend to diffuse down their concentration gradients, into the cells. At the same time, water tends to move down its concentration gradient, out of the cells. The immediate result is severe loss of water from the frog's cells (dehydration).

(d) If the frog uses only NaCl to adjust its osmolarity, it will need (1120 − 235) mosmol = 885 mosmol/L to become isotonic with the seawater. Because NaCl dissociates into two species in solution, a 1 M solution contains 2 osmol/L. The volume of water in the frog is assumed to be 100 mL, so the frog needs (0.885 osmol/L)(1/2 mol/osmol NaCl)(0.1 L)(58.44 g/mol) = 2.59 g of NaCl (or about 1/2 teaspoon).

(e) A 1 M solution of urea (a nondissociating molecule) contains 1 osmol/L. To increase the osmolarity of its cells by 885 mosmol/ L using urea, so as to become isotonic with the seawater, the frog needs (0.885 osmol/L)(1 mol/osmol urea)(0.1 L)(60 g/mol) = 5.31 g of urea.

5. Properties of a Buffer The amino acid glycine is often used as the main ingredient of a buffer in biochemical experiments. The amino group of glycine, which has a pK_a of 9.6, can exist either in the protonated form ($-NH_3^+$) or as the free base ($-NH_2$) because of the reversible equilibrium

$$R-NH_3^+ \rightleftharpoons R-NH_2 + H^+$$

(a) In what pH range can glycine be used as an effective buffer due to its amino group?

(b) In a 0.1 M solution of glycine at pH 9.0, what *fraction* of glycine has its amino group in the $-NH_3^+$ form?

(c) How much 5 M KOH must be added to 1.0 L of 0.1 M glycine at pH 9.0 to bring its pH to exactly 10.0?

(d) When 99% of the glycine is in its $-NH_3^+$ form, what is the numerical relation between the pH of the solution and the pK_a of the amino group?

Answer

(a) In general, a buffer functions best in the zone from about one pH unit below to one pH unit above its pK_a. Thus glycine is a good buffer (through ionization of its amino group) between pH 8.6 and pH 10.6.

(b) Using the Henderson-Hasselbalch equation

$$\text{pH} = \text{p}K_a + \log \frac{[A^-]}{[HA]}$$

we can write

$$9.0 = 9.6 + \log \frac{[A^-]}{[HA]}$$

$$\frac{[A^-]}{[HA]} = 10^{-0.6} = 0.25$$

which corresponds to a ratio of 1/4. This indicates that the amino group of glycine is about 1/5 deprotonated and about 4/5 protonated at pH 9.0.

(c) From (b) we know that the amino group is about 1/5 or 20% deprotonated at pH 9.0. Thus in moving from pH 9.0 to pH 9.6 (at which, by definition, the amino group is 50% deprotonated), 30% or 0.3 of the glycine is titrated. We can now calculate from the Henderson-Hasselbalch equation the percentage protonation at pH 10.0:

$$10.0 = 9.6 + \log \frac{[A^-]}{[HA]}$$

$$\frac{[A^-]}{[HA]} = 10^{0.4} = 2.5 = 5/2$$

This ratio indicates that glycine is 5/7 or 71% deprotonated at pH 10.0, an additional 21% or 0.21 deprotonation above that (50% or 0.5) at the pK_a. Thus the total fractional deprotonation in moving from pH 9.0 to 10.0 is 0.30 + 0.21 = 0.51, which corresponds to

$$0.51 \times 0.1 \text{ mol} = 0.051 \text{ mol of KOH}$$

Thus the volume of 5 M KOH solution required is about 0.01 L or 10 mL.

(d) From the Henderson-Hasselbalch equation

$$\begin{aligned} \text{pH} &= \text{p}K_a + \log([-NH_2]/[-NH_3^+]) \\ &= \text{p}K_a + \log(0.01/0.99) \\ &= \text{p}K_a + (-2) = \text{p}K_a - 2 \end{aligned}$$

In general, any group with an ionizable hydrogen is almost completely protonated at a pH at least two pH units below its pK_a value.

6. The Effect of pH on Solubility The strongly polar hydrogen-bonding nature of water makes it an excellent solvent for ionic (charged) species. By contrast, nonionized, nonpolar organic molecules, such as benzene, are relatively insoluble in water. In principle, the aqueous solubility of any organic acid or base can be increased by conversion of the molecules to charged species. For example, the solubility of benzoic acid in water is low. The addition of sodium bicarbonate to a mixture of water and benzoic acid raises the pH and deprotonates the benzoic acid to form benzoate ion, which is quite soluble in water.

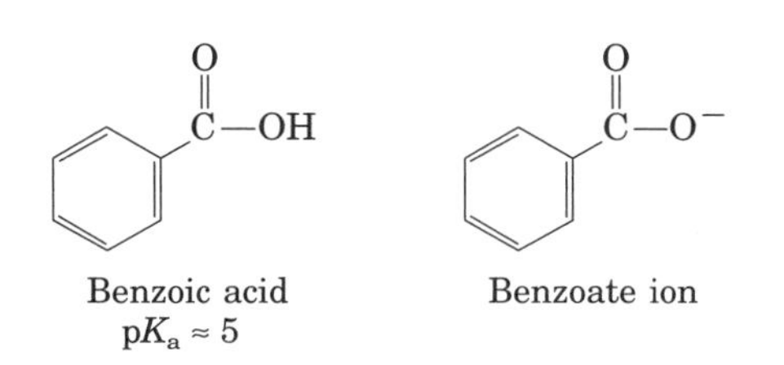

Are the following compounds more soluble in an aqueous solution of 0.1 M NaOH or 0.1 M HCl?

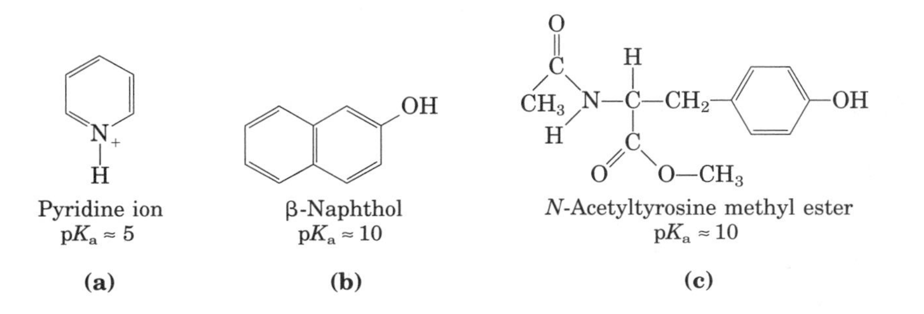

Answer

(a) Pyridine is ionic in its protonated form and therefore more soluble at the lower pH, in 0.1 M HCl.

(b) β-Naphthol is ionic when *de*protonated and thus more soluble at the higher pH, in 0.1 M NaOH.

(c) *N*-Acetyltyrosine methyl ester is ionic when *de*protonated and thus more soluble in 0.1 M NaOH.

7. Treatment of Poison Ivy Rash The components of poison ivy and poison oak that produce the characteristic itchy rash are catechols substituted with long-chain alkyl groups.

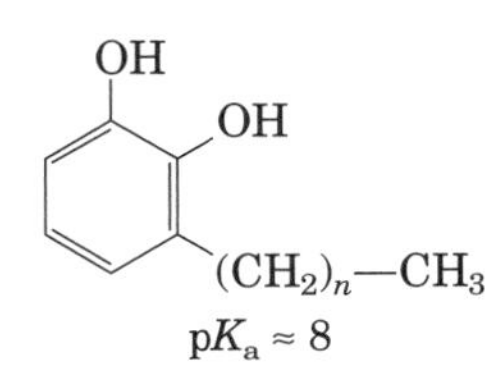

If you were exposed to poison ivy, which of the treatments below would you apply to the affected area? Justify your choice.

(a) Wash the area with cold water.

(b) Wash the area with dilute vinegar or lemon juice.

(c) Wash the area with soap and water.

(d) Wash the area with soap, water, and baking soda (sodium bicarbonate).

Answer Soap helps to emulsify and dissolve the hydrophobic alkyl group of an alkylcatechol. Given that the pK_a of an alkylcatechol is about 8, in a mildly alkaline solution of bicarbonate ($NaHCO_3$) its —OH group ionizes to $—O^-$, making the compound much more water-soluble. A neutral aqueous or an acidic solution, as in (a) or (b), would not be effective. Thus (d) is the best choice.

8. pH and Drug Absorption Aspirin is a weak acid with a pK_a of 3.5.

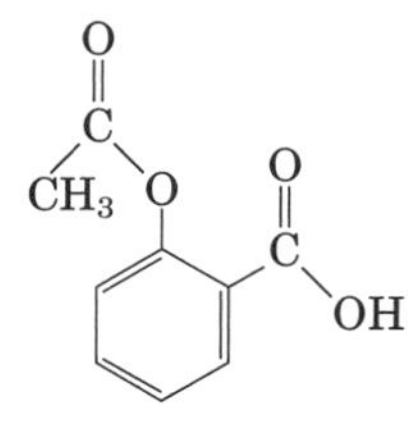

It is absorbed into the blood through the cells lining the stomach and the small intestine. Absorption requires passage through the plasma membrane, the rate of which is determined by the polarity of the molecule: charged and highly polar molecules pass slowly, whereas neutral hydrophobic ones pass rapidly. The pH of the stomach contents is about 1.5, and the pH of the contents of the small intestine is about 6. Is more aspirin absorbed into the bloodstream from the stomach or from the small intestine? Clearly justify your choice.

Answer With a pK_a of 3.5, aspirin is in its protonated (neutral) form at pH below 2.5. At higher pH, it becomes increasingly deprotonated (anionic). Thus aspirin is better absorbed in the more acidic environment of the stomach.

9. Preparation of Standard Buffer for Calibration of a pH Meter The glass electrode used in commercial pH meters gives an electrical response proportional to the concentration of hydrogen ion. To convert these responses into pH, glass electrodes must be calibrated against standard solutions of known H^+ concentration. Determine the weight in grams of sodium dihydrogen phosphate ($NaH_2PO_4 \cdot H_2O$; FW 138.01) and disodium hydrogen phosphate (Na_2HPO_4; FW 141.98) needed to prepare 1 L of a standard buffer at pH 7.00 with a total phosphate concentration of 0.100 M (see Fig. 4–14).

Answer In solution, the two salts ionize as indicated below.

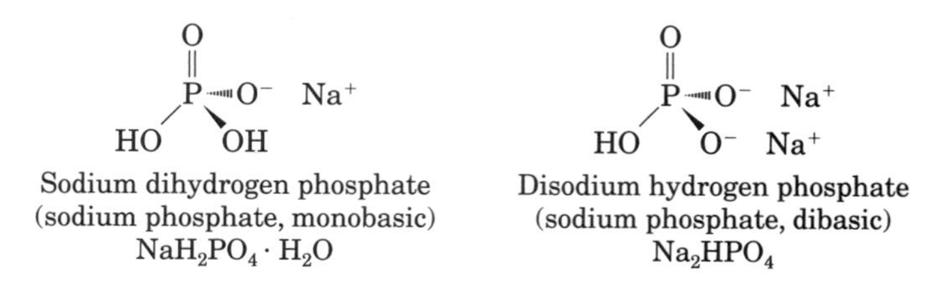

The buffering capacity of the solution is determined by the concentration ratio of proton acceptor (A^-) to proton donor (HA), or [disodium phosphate]/[dihydrogen phosphate]. From Figure 4–14, the pK_a for the dissociation of the ionizable hydrogen of dihydrogen phosphate

$$H_2PO_4^- \rightleftharpoons HPO_4^{2-} + H^+$$

is 6.86. Using the Henderson-Hasselbalch equation

$$\text{pH} = \text{p}K_a + \log \frac{[\text{A}^-]}{[\text{HA}]}$$

$$7.00 - 6.86 = \log \frac{[\text{A}^-]}{[\text{HA}]}$$

$$\frac{[\text{A}^-]}{[\text{HA}]} = 10^{0.14} = 1.38$$

This ratio is approximately 7/5; that is, 7 parts Na_2HPO_4 to 5 parts $NaH_2PO_4 \cdot H_2O$. Because $[HPO_4^{2-}] + [H_2PO_4^-] = 0.100$ M, $[H_2PO_4^-] = 0.100\text{ M} - [HPO_4^{2-}]$.
We can now calculate the amount of each salt required for a 0.100 M solution.

$$\frac{[\text{HPO}_4^{2-}]}{0.100\text{ M} - [\text{HPO}_4^{2-}]} = 1.38$$

$$[\text{HPO}_4^{2-}] = \frac{0.138}{2.38}\text{ M} = 0.058\text{ M} = 0.058\text{ mol/L}$$

$$[\text{H}_2\text{PO}_4^-] = 0.100\text{ M} - 0.058\text{ M} = 0.042\text{ M} = 0.042\text{ mol/L}$$

The amount needed for 1 L of solution = FW × (mol/L).

For $NaH_2PO_4 \cdot H_2O$: (138.01 g/mol)(0.042 mol/L) = 5.80 g/L

For Na_2HPO_4: (141.98 g/mol)(0.058 mol/L) = 8.23 g/L

10. Control of Blood pH by Respiration Rate

(a) The partial pressure of CO_2 in the lungs can be varied rapidly by the rate and depth of breathing. For example, a common remedy to alleviate hiccups is to increase the concentration of CO_2 in the lungs. This can be achieved by holding one's breath, by very slow and shallow breathing (hypoventilation), or by breathing in and out of a paper bag. Under such conditions, the partial pressure of CO_2 in the air space of the lungs rises above normal. Qualitatively explain the effect of these procedures on the blood pH.

(b) A common practice of competitive short-distance runners is to breathe rapidly and deeply (hyperventilation) for about half a minute to remove CO_2 from their lungs just before running in, say, a 100 M dash. Their blood pH may rise to 7.60. Explain why the blood pH increases.

(c) During a short-distance run the muscles produce a large amount of lactic acid ($CH_3CH(OH)COOH$, $K_a = 1.38 \times 10^{-4}$) from their glucose stores. In view of this fact, why might hyperventilation before a dash be useful?

Answer

(a) Blood pH is controlled by the carbon dioxide–bicarbonate buffer system, as shown by the net equation

$$CO_2 + H_2O \rightleftharpoons H^+ + HCO_3^-$$

During *hypoventilation* the concentration of CO_2 in the lungs and arterial blood increases, driving the equilibrium to the right and raising the $[H^+]$; that is the pH is lowered.

(b) During *hyperventilation* the concentration of CO_2 in the lungs and arterial blood falls. This drives the equilibrium to the left, which requires the consumption of hydrogen ions, reducing $[H^+]$ and increasing pH.

(c) Lactate is a moderately strong acid (pK_a = 3.86) that completely dissociates under physiological conditions:

$$CH_3CH(OH)COOH \rightleftharpoons CH_3CH(OH)COO^- + H^+$$

This lowers the pH of the blood and muscle tissue. Hyperventilation is useful because it removes hydrogen ions, raising the pH of the blood and tissues in anticipation of the acid build up.

chapter

5 Amino Acids, Peptides, and Proteins

1. Absolute Configuration of Citrulline The citrulline isolated from watermelons has the structure shown below. Is it a D- or L-amino acid? Explain.

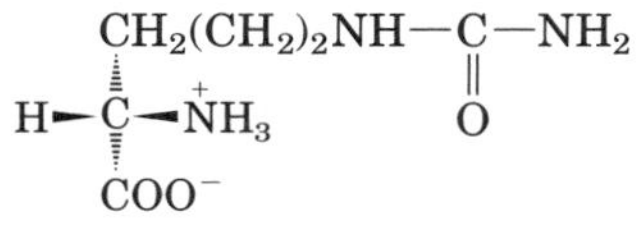

$$\begin{array}{l} \quad CH_2(CH_2)_2NH-\overset{\displaystyle O}{\overset{\|}{C}}\!-NH_2 \\ H-C-\overset{+}{N}H_3 \\ \quad COO^- \end{array}$$

Answer Rotating the structural representation by 180° in the plane of the page puts the most highly oxidized group—the carboxyl group—at the top, in the same position as the —CHO group of glyceraldehyde in Figure 5–4. In this orientation, the amino group is on the left and thus the absolute configuration of the citrulline is L.

2. Relationship between the Titration Curve and the Acid-Base Properties of Glycine A 100 mL solution of 0.1 M glycine at pH 1.72 was titrated with 2 M NaOH solution. The pH was monitored and the results were plotted on a graph, as shown below. The key points in the titration are designated I to V. For each of the statements (a) to (o), *identify* the appropriate key point in the titration and *justify* your choice.

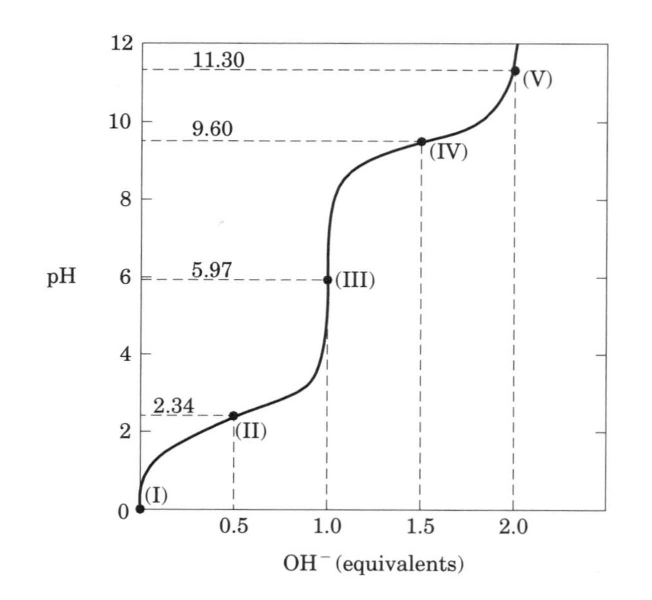

Note: Before considering statements (a) through (o) refer to Figure 5–10. The three species involved in the titration of glycine can be considered in terms of a useful physical analogy. Each ionic species can be viewed as a different floor of a building, each with a different net charge:

$^+H_3N—CH_2—COOH$	+1
$^+H_3N—CH_2—COO^-$	0 (zwitterion)
$H_2N—CH_2—COO^-$	−1

The floors are connected by steep stairways, and each stairway has a landing halfway between the floors. A titration curve traces the path one would follow between the different floors, as the pH changes in response to added OH^-. Recall that the pK_a of an acid represents the pH at which half of the acid is deprotonated. The isoelectric point (pI) is the pH at which the average net charge is zero.

(a) Glycine is present predominantly as the species $^+H_3N—CH_2—COOH$.
(b) The *average* net charge of glycine is +1/2.
(c) Half of the amino groups are ionized.
(d) The pH is equal to the pK_a of the carboxyl group.
(e) The pH is equal to the pK_a of the protonated amino group.
(f) Glycine has its maximum buffering capacity.
(g) The *average* net charge of glycine is zero.
(h) The carboxyl group has been completely titrated (first equivalence point).
(i) Glycine is completely titrated (second equivalence point).
(j) The predominant species is $^+H_3N—CH_2—COO^-$.
(k) The *average* net charge of glycine is −1.
(l) Glycine is present predominantly as a 50:50 mixture of $^+H_3N—CH_2—COOH$ and $^+H_3N—CH_2—COO^-$.
(m) This is the isoelectric point.
(n) This is the end of the titration.
(o) These are the *worst* pH regions for buffering power.

Answer

(a) I; maximum protonation occurs at the lowest pH (the highest [H^+]).
(b) II; at the first pK_a (2.34), half of the protons are removed from the α-carboxyl group (i.e., it is half deprotonated), changing its charge from 0 to −1/2. The net charge is (−1/2) + 1 = 1/2.
(c) IV; the α-amino group is half-deprotonated at its pK_a (9.6).
(d) II; from the Henderson-Hasselbalch equation, pH = pK_a + log ([A^-]/[HA]). If [A^-]/[HA] = 1 or [A^-] = [HA], then pH = pK_a. (Recall that log 1 = 0.)
(e) IV; see answers (c) and (d).
(f) II and IV; in the pK_a regions, acid donates protons to or base abstracts protons from glycine, with minimal pH changes.
(g) III; this occurs at the isoelectric point; pI = (pK_1 + pK_2)/2 = (2.34 + 9.60)/2 = 5.97.
(h) III; the pH at which 1.0 equivalent of OH^- has been added, pH 5.97 (2.6 pH units away from either pK_a).
(i) V; pH 11.3 (nearly two pH units above pK_2).
(j) III; at pI (5.97) the carboxyl group is fully negatively charged (deprotonated) and the amino group is fully positively charged (protonated).
(k) V; both groups are fully deprotonated, with a neutral amino group and a negatively charged carboxyl group.
(l) II; the carboxyl group is half-ionized at pH = pK_1.

(m) III; see answers to (g) and (j).

(n) V; glycine is fully titrated after 2.0 equivalents of OH^- have been added.

(o) I, III, and V; each is several pH units removed from either pK_a, where the best pH buffering action occurs.

3. How Much Alanine Is Present as the Completely Uncharged Species? At a pH equal to the isoelectric point of alanine, the *net* charge on alanine is zero. Two structures can be drawn that have a net charge of zero, but the predominant form of alanine at its pI is zwitterionic.

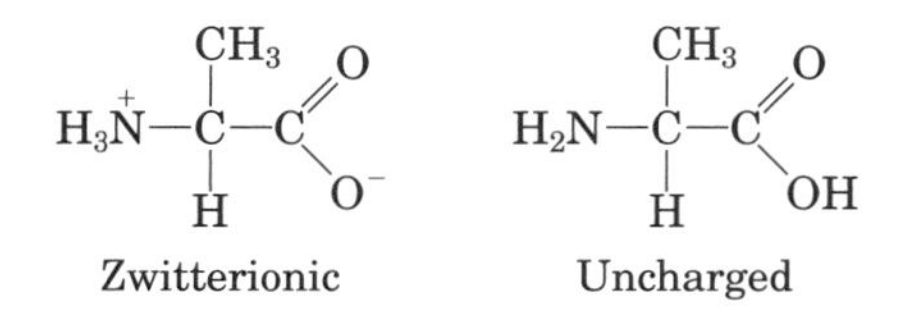

(a) Why is alanine predominantly zwitterionic rather than completely uncharged at its pI?

(b) What fraction of alanine is in the completely uncharged form at its pI? Justify your assumptions.

Answer

(a) The pI of alanine occurs at a pH well above the pK_a of the α-carboxyl group and well below the pK_a of the α-amino group. Hence, both groups are present predominantly in their charged (ionized) forms.

(b) The pI of alanine is 6.01, midway between the two pK_a values, 2.34 and 9.69. From the Henderson-Hasselbach equation, $\text{pH} - pK_a = \log([A^-]/[HA])$.

$$\log \frac{[A^-]}{[HA]} = 3.67$$

$$\frac{[A^-]}{[HA]} = 10^{-3.67} = \frac{1}{4.68 \times 10^3}$$

That is, one molecule in 4,680 is still in the form —COOH. Similarly, at pH = pI, one molecule in 4,680 is in the form $—NH_2$. Thus, the fraction of molecules with both groups uncharged (—COOH *and* $—NH_2$) is 1 in 4,680 × 4,680, or 1 in 2.19×10^7.

4. Ionization State of Amino Acids Each ionizable group of an amino acid can exist in one of two states, charged or neutral. The electric charge on the functional group is determined by the relationship between its pK_a and the pH of the solution. This relationship is described by the Henderson-Hasselbalch equation.

(a) Histidine has three ionizable functional groups. Write the equilibrium equations for its three ionizations and assign the proper pK_a for each ionization. Draw the structure of histidine in each ionization state. What is the net charge on the histidine molecule in each ionization state?

(b) Draw the structures of the predominant ionization state of histidine at pH 1, 4, 8, and 12. Note that the ionization state can be approximated by treating each ionizable group independently.

(c) What is the net charge of histidine at pH 1, 4, 8, and 12? For each pH, will histidine migrate toward the anode (+) or cathode (−) when placed in an electric field?

Answer

(a)

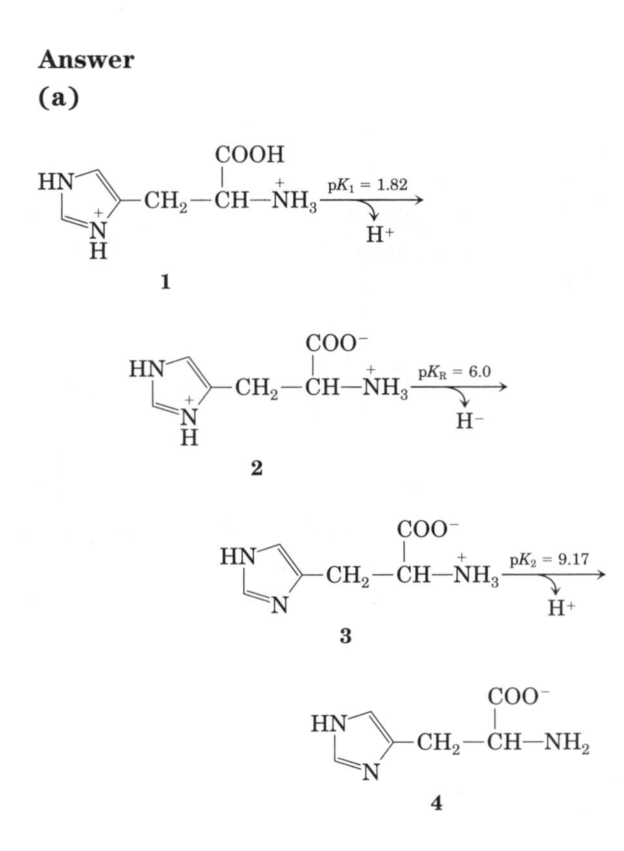

Starting with the most highly protonated species (that found at the most acidic pH), find the pK_a for each group in Table 5–1. As base is added, the group with the lowest pK_a loses its proton first, followed by the group with the next lowest pK_a, then the group with the highest pK_a. (In the following table, R = —CH_2-imidazole.)

	Structure	Total charge
1	$^{+}H_3N$—CH(RH^{+})—COOH	+2
2	$^{+}H_3N$—CH(RH^{+})—COO^{-}	+1
3	$^{+}H_3N$—CH(R)—COO^{-}	0
4	H_2N—CH(R)—COO^{-}	−1

(b) and **(c)** See the structures in **(a)**.

pH	Structure	Net charge	Migrates toward
1	**1**	+2	Cathode (−)
4	**2**	+1	Cathode (−)
8	**3**	0	Does not migrate
12	**4**	−1	Anode (+)

5. **Separation of Amino Acids by Ion-Exchange Chromatography** Mixtures of amino acids are analyzed by first separating the mixture into its components through ion-exchange chromatography. Amino acids placed on a cation-exchange resin containing sulfonate groups (see Fig. 5–18) flow down the column at different rates because of two factors that influence their movement: (1) ionic attraction between the $—SO_3^-$ residues on the column and positively charged functional groups on the amino acids, and (2) hydrophobic interactions between amino acid side chains and the strongly hydrophobic backbone of the polystyrene resin. For each pair of amino acids listed, determine which will be eluted first from an ion-exchange column using a pH 7.0 buffer.
   (a) Asp and Lys
   (b) Arg and Met
   (c) Glu and Val
   (d) Gly and Leu
   (e) Ser and Ala

Answer See Table 5–1 for pK_a values for the amino acid side chains. At $\mathrm{pH} < \mathrm{pI}$, an amino acid has a net positive charge; at $\mathrm{pH} > \mathrm{pI}$, it has a net negative charge. For any pair of amino acids, the more negatively charged one passes through the sulfonated resin faster. For two neutral amino acids, the less polar one passes through more slowly because of its stronger hydrophobic interactions with the polystyrene.

	pI values	Net charge (pH 7)	Elution order	Basis for separation
(a) Asp, Lys	2.8, 9.7	−1, +1	Asp, Lys	Charge
(b) Arg, Met	10.8, 5.7	+1, 0	Met, Arg	Charge
(c) Glu, Val	3.2, 6.0	−1, 0	Glu, Val	Charge
(d) Gly, Leu	5.97, 5.98	0, 0	Gly, Leu	Polarity
(e) Ser, Ala	5.68, 6.01	0, 0	Ser, Ala	Polarity

6. **Naming the Stereoisomers of Isoleucine** The structure of the amino acid isoleucine is

$$
\begin{array}{c}
\mathrm{COO^-} \\
| \\
\mathrm{H_3\overset{+}{N}—C—H} \\
| \\
\mathrm{H—C—CH_3} \\
| \\
\mathrm{CH_2} \\
| \\
\mathrm{CH_3}
\end{array}
$$

   (a) How many chiral centers does it have?
   (b) How many optical isomers?
   (c) Draw perspective formulas for all the optical isomers of isoleucine.

Answer

(a) Two; at C-2 and C-3 (the α and β carbons).

(b) Four; the two chiral centers permit four possible diastereoisomers: (*S*,*S*), (*S*,*R*), (*R*,*R*), and (*R*,*S*).

(c)

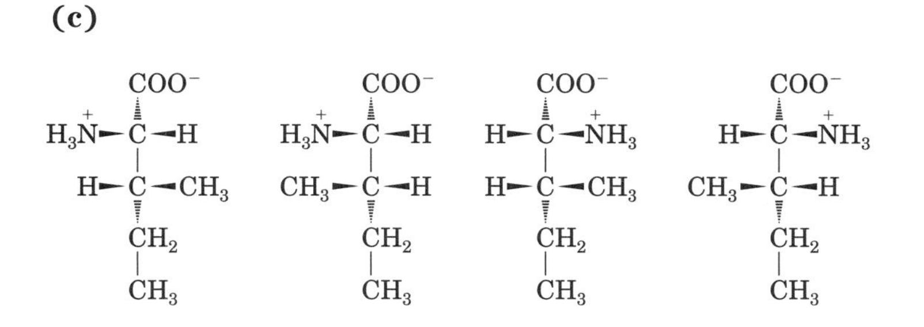

7. Comparing the pK_a Values of Alanine and Polyalanine The titration curve of alanine shows the ionization of two functional groups with pK_a values of 2.34 and 9.69, corresponding to the ionization of the carboxyl and the protonated amino groups, respectively. The titration of di-, tri-, and larger oligopeptides of alanine also shows the ionization of only two functional groups, although the experimental pK_a values are different. The trend in pK_a values is summarized in the table.

Amino acid or peptide	pK_1	pK_2
Ala	2.34	9.69
Ala–Ala	3.12	8.30
Ala–Ala–Ala	3.39	8.03
Ala–(Ala)$_n$–Ala, $n \geq 4$	3.42	7.94

(a) Draw the structure of Ala–Ala–Ala. Identify the functional groups associated with pK_1 and pK_2.
(b) Why does the value of pK_1 *increase* with each addition of an Ala residue to the Ala oligopeptide?
(c) Why does the value of pK_2 *decrease* with each addition of an Ala residue to the Ala oligopeptide?

Answer

(a) The structure at pH 7 is:

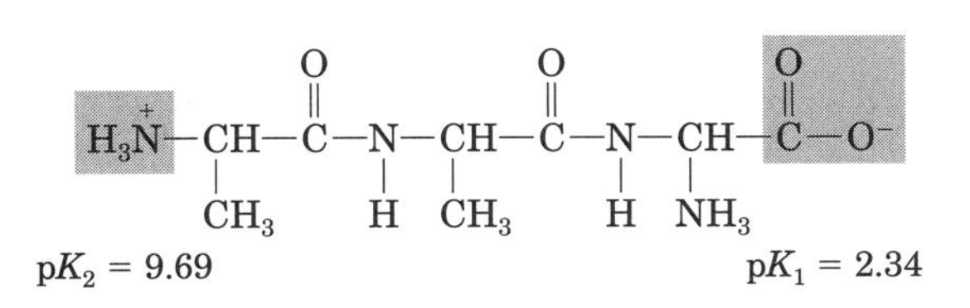

Note that only the amino- and carboxyl-terminal groups ionize.

(b) As the length of poly-Ala increases, the two terminal groups are moved farther apart, separated by an intervening sequence with an "insulating" nonpolar structure. The carboxyl group becomes a weaker acid, as reflected in its higher pK_a, because the electrostatic repulsion between the carboxyl proton and the positive charge on the NH_3^+ group diminishes as the groups become more distant.

(c) The negative charge on the terminal carboxyl group has a stabilizing effect on the positively charged (protonated) terminal amino group. With increasing numbers of intervening Ala residues, this stabilizing effect is diminished and the NH_3^+ group loses it proton more easily. The lower pK_2 indicates that the terminal amino group has become a weaker base (stronger acid).

8. The Size of Proteins What is the approximate molecular weight of a protein with 682 amino acid residues in a single polypeptide chain?

Answer Assuming that the average M_r per residue is 110 (corrected for loss of water in formation of the peptide bond), a protein containing 682 residues has an M_r of approximately $682 \times 110 = 75{,}020$.

9. The Number of Tryptophan Residues in Bovine Serum Albumin A quantitative amino acid analysis reveals that bovine serum albumin (BSA) contains 0.58% tryptophan (M_r 204) by weight.

(a) Calculate the *minimum* molecular weight of BSA (i.e., assuming there is only one tryptophan residue per protein molecule).

(b) Gel filtration of BSA gives a molecular weight estimate of 70,000. How many tryptophan residues are present in a molecule of serum albumin?

Answer

(a) The M_r of a Trp residue must be adjusted to account for the removal of water during peptide bond formation: $M_r = 204 - 18 = 186$. The molecular weight of BSA can be calculated using the following proportionality, where n is the number of Trp residues in the protein:

$$\frac{\text{wt Trp}}{\text{wt BSA}} = \frac{n(M_r\ \text{Trp})}{M_r\ \text{BSA}}$$

$$\frac{0.58\ \text{g}}{100\ \text{g}} = \frac{n(186)}{M_r\ \text{BSA}}$$

A *minimum* molecular weight can be found by assuming only one Trp residue per BSA molecule ($n = 1$).

$$\frac{(100\ \text{g})(186)(1)}{0.58\ \text{g}} = 32{,}100$$

(b) Given that the M_r of BSA is approximately 70,000, then BSA has $\sim 70{,}000/32{,}100 = 2$ Trp residues per molecule. (The remainder from this division suggests that the estimate of M_r 70,000 for BSA is somewhat high.)

10. Net Electric Charge of Peptides A peptide has the sequence

Glu–His–Trp–Ser–Gly–Leu–Arg–Pro–Gly

(a) What is the net charge of the molecule at pH 3, 8, and 11? (Use pK_a values for side chains and terminal amino and carboxyl groups as given in Table 5–1.)

(b) Estimate the pI for this peptide.

Answer

(a) When pH > pK_a, ionizing groups lose their protons. The pK_a values of importance here are those of the amino-terminal (2.3) and carboxyl-terminal (9.6) groups and those of the R groups of the Glu (4.3), His (6.0), and Arg (12.5) residues.

pH	^+H_3N	Glu	His	Arg	COO^-	Net charge
3.0	+1	0	+1	+1	−1	+2
8.0	+1	−1	0	+1	−1	0
11	0	−1	0	+1	−1	−1

(b) Two different methods can be used to estimate pI. Average the pK_a values for the two side chains that ionize near pH 8, the amino-terminal α-amino group of Glu and the His imidazole group:

$$pI = \frac{9.6 + 6.0}{2} = 7.8$$

Alternatively, plot the calculated net charges as a function of pH, and determine graphically the pH at which the net charge is zero on the vertical axis. More data points are needed to use this method accurately.

Note: Although at any instant an individual amino acid molecule will have an integral charge, it is possible for a population of amino acid molecules in solution to have a fractional charge. For example, at pH 1.0 glycine exists entirely as the form $^{+}H_3N—CH_2—COOH$ with a net positive charge of 1.0. However, at pH 2.34, where there is an equal mixture of amino acids in the forms $^{+}H_3N—CH_2—COOH$ and $^{+}H_3N—CH_2—COO^{-}$, the average or net charge on the population of amino acids is 0.5 (see the discussion on p. 125). You can use the Henderson-Hasselbalch equation to calculate the exact ratio of charged and uncharged species at equilibrium at various pH values.

11. Isoelectric Point of Pepsin. Pepsin is the name given to several digestive enzymes that are secreted (as larger precursor proteins) by glands that line the stomach. These glands also secrete hydrochloric acid, which dissolves the particulate matter in food, allowing pepsin to enzymatically cleave individual protein molecules. The resulting mixture of food, HCl, and digestive enzymes is known as chyme and has a pH near 1.5. What pI would you predict for the pepsin proteins? What functional groups must be present to confer this pI on pepsin? Which amino acids in the proteins would contribute to such groups?

Answer Pepsin proteins have a relatively low pI (near the pH of gastric juice) in order to remain soluble and thus functional in the stomach. (Pepsin has a pI of ~1.) As pH increases, pepsins acquire a net charge and undergo ionic interactions with oppositely charged molecules (such as dissolved salts), causing the pepsin proteins to precipitate. Pepsin is active only in the stomach. In the relatively high pH of the intestine, pepsin proteins precipitate and become inactive.

A low pI requires large numbers of negatively charged (low pK_a) groups. These are contributed by the carboxylate groups of Asp and Glu residues.

12. The Isoelectric Point of Histones Histones are proteins found in eukaryotic cell nuclei, tightly bound to DNA, which has many phosphate groups. The pI of histones is very high, about 10.8. What amino acid residues must be present in relatively large numbers in histones? In what way do these residues contribute to the strong binding of histones to DNA?

Answer Large numbers of positively charged (high pK_a) groups in a protein give it a high pI. In histones, the positively charged R groups of Lys, Arg, and His residues interact strongly with the negatively charged phosphate groups of DNA through ionic interactions.

13. Solubility of Polypeptides One method for separating polypeptides makes use of their differential solubilities. The solubility of large polypeptides in water depends upon the relative polarity of their R groups, particularly on the number of ionized groups: the more ionized groups there are, the more soluble the polypeptide. Which of each pair of polypeptides below is more soluble at the indicated pH?

(a) $(Gly)_{20}$ or $(Glu)_{20}$ at pH 7.0
(b) $(Lys–Ala)_3$ or $(Phe–Met)_3$ at pH 7.0
(c) $(Ala–Ser–Gly)_5$ or $(Asn–Ser–His)_5$ at pH 6.0
(d) $(Ala–Asp–Gly)_5$ or $(Asn–Ser–His)_5$ at pH 3.0

Answer

(a) $(Glu)_{20}$; it is highly negatively charged (polar) at pH 7. $(Gly)_{20}$ is uncharged except for the amino- and carboxyl-terminal groups.

(b) $(Lys–Ala)_3$; this is highly positively charged (polar) at pH 7. $(Phe–Met)_3$ is much less polar and hence less soluble.

(c) $(Asn–Ser–His)_5$; both polymers have polar Ser side chains, but $(Asn–Ser–His)_5$ also has the polar Asn side chains and partially protonated His side chains.

(d) $(Asn–Ser–His)_5$; at pH 3, the carboxylate side chains of Asp are partially protonated and neutral, whereas the imidazole groups of His are fully protonated and positively charged.

14. Purification of an Enzyme A biochemist discovers and purifies a new enzyme, generating the purification table below.

Procedure	Total protein (mg)	Activity (units)
1. Crude extract	20,000	4,000,000
2. Precipitation (salt)	5,000	3,000,000
3. Precipitation (pH)	4,000	1,000,000
4. Ion-exchange chromatography	200	800,000
5. Affinity chromatography	50	750,000
6. Size-exclusion chromatography	45	675,000

(a) From the information given in the table, calculate the specific activity of the enzyme solution after each purification procedure.

(b) Which of the purification procedures used for this enzyme is most effective (i.e., gives the greatest relative increase in purity)?

(c) Which of the purification procedures is least effective?

(d) Is there any indication based on the results shown in the table that the enzyme after step 6 is now pure? What else could be done to estimate the purity of the enzyme preparation?

Answer

(a) From the percentage recovery of activity (units), we can calculate percentage yield and specific activity (units/mg).

Procedure	Protein (mg)	Activity (units)	% Yield	Specific activity (units/mg)	Purification factor
1.	20,000	4,000,000	(100)	200	(1.0)
2.	5,000	3,000,000	75	600	× 3.0
3.	4,000	1,000,000	25	250	× 1.25
4.	200	800,000	20	4,000	× 20
5.	50	750,000	19	15,000	× 75
6.	45	675,000	17	15,000	× 75

(b) Step 4, ion-exchange chromatography, which gives the greatest increase in specific activity (an index of purity and degree of increase in purification).

(c) Step 3, pH precipitation, in which two-thirds of the total activity from the previous step was lost.

(d) Yes. The specific activity did not increase further after step 5. SDS-polyacrylamide gel electrophoresis is an excellent, standard way of checking homogeneity and purity.

15. Sequence Determination of the Brain Peptide Leucine Enkephalin A group of peptides that influence nerve transmission in certain parts of the brain has been isolated from normal brain tissue. These peptides are known as opioids, because they bind to specific receptors that also bind opiate drugs, such as morphine and naloxone. Opioids thus mimic some of the properties of opiates. Some researchers consider these peptides to be the brain's own pain killers. Using the information below, determine the amino acid sequence of the opioid leucine enkephalin. Explain how your structure is consistent with each piece of information.

(a) Complete hydrolysis by 6 M HCl at 110 °C followed by amino acid analysis indicated the presence of Gly, Leu, Phe, and Tyr, in a 2:1:1:1 molar ratio.

(b) Treatment of the peptide with 1-fluoro-2,4-dinitrobenzene followed by complete hydrolysis and chromatography indicated the presence of the 2,4-dinitrophenyl derivative of tyrosine. No free tyrosine could be found.

(c) Complete digestion of the peptide with pepsin followed by chromatography yielded a dipeptide containing Phe and Leu, plus a tripeptide containing Tyr and Gly in a 1:2 ratio.

Answer

(a) The empirical composition is (2Gly, Leu, Phe, Tyr)$_n$.

(b) Tyr is the amino-terminal residue, so the sequence is Tyr–(2Gly, Leu, Phe).

(c) As shown in Table 5–7, pepsin cleaves on the amino side of aromatic residues (Phe, Tyr, Trp). Given that the peptide has only two such residues, one of which (Tyr) is amino-terminal, the dipeptide containing Phe and Leu must have been cleaved from the tripeptide at Phe. Thus the sequence must be

Tyr–(2Gly)–Phe–Leu = Tyr–Gly–Gly–Phe–Leu

16. Structure of a Peptide Antibiotic from *Bacillus brevis* Extracts from the bacterium *Bacillus brevis* contain a peptide with antibiotic properties. This peptide forms complexes with metal ions and apparently disrupts ion transport across the cell membranes of other bacterial species, killing them. The structure of the peptide has been determined from the following observations.

(a) Complete acid hydrolysis of the peptide followed by amino acid analysis yielded equimolar amounts of Leu, Orn, Phe, Pro, and Val. Orn is ornithine, an amino acid not present in proteins but present in some peptides. It has the structure

$$H_3\overset{+}{N}-CH_2-CH_2-CH_2-\underset{^{+}NH_3}{\overset{H}{\underset{|}{\overset{|}{C}}}}-COO^-$$

(b) The molecular weight of the peptide was estimated as about 1,200.

(c) The peptide failed to undergo hydrolysis when treated with the enzyme carboxypeptidase. This enzyme catalyzes the hydrolysis of the carboxyl-terminal residue of a polypeptide unless that residue is Pro or does not contain a free carboxyl group for some reason.

(d) Treatment of the intact peptide with 1-fluoro-2,4-dinitrobenzene, followed by complete hydrolysis and chromatography, yielded only free amino acids and the following derivative:

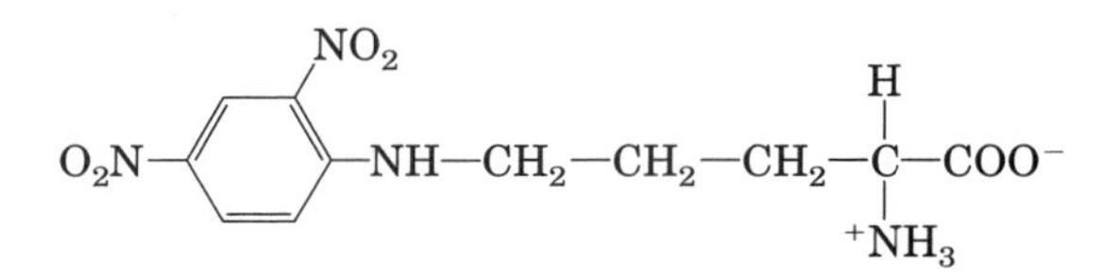

(Hint: Note that the 2,4-dinitrophenyl derivative involves the amino group of a side chain rather than the α-amino group.)

(e) Partial hydrolysis of the peptide followed by chromatographic separation and sequence analysis yielded the following di- and tripeptides (the amino-terminal amino acid is always at the left):

Leu–Phe Phe–Pro Orn–Leu Val–Orn

Val–Orn–Leu Phe–Pro–Val Pro–Val–Orn

Given the above information, deduce the amino acid sequence of the peptide antibiotic. Show your reasoning. When you have arrived at a structure, demonstrate that it is consistent with *each* experimental observation.

Answer The information obtained from each experiment is as follows.

(a) The simplest empirical formula for the peptide is (Leu, Orn, Phe, Pro, Val).

(b) Assuming an average residue M_r of 110, the minimum molecular weight for the peptide is 550. Since 1,200/550 ≈ 2, the empirical formula is (Leu, Orn, Phe, Pro, Val)$_2$.

(c) Failure of carboxypeptidase to cleave the peptide could result from Pro at the carboxyl-terminus or the absence of a carboxyl-terminal residue—as in a cyclic peptide.

(d) Failure of FDNB to derivatize an α-amino group indicates either the absence of a free amino-terminal group or that Pro (an imino acid) is at the amino-terminal position. (The derivative formed is 2,4 dinitrophenyl-ε-Orn.)

(e) The presence of Pro at an internal position in the peptide Phe–Pro–Val indicates that it is *not* at the amino or carboxyl terminus. The information from these experiments suggest that the peptide is cyclic. The alignment of overlapping sequences is

```
Leu–Phe
    Phe–Pro
    Phe–Pro–Val
        Pro–Val–Orn
            Val–Orn
                Orn–(Leu)
```

Thus the peptide is a cyclic dimer of Leu–Phe–Pro–Val–Orn:

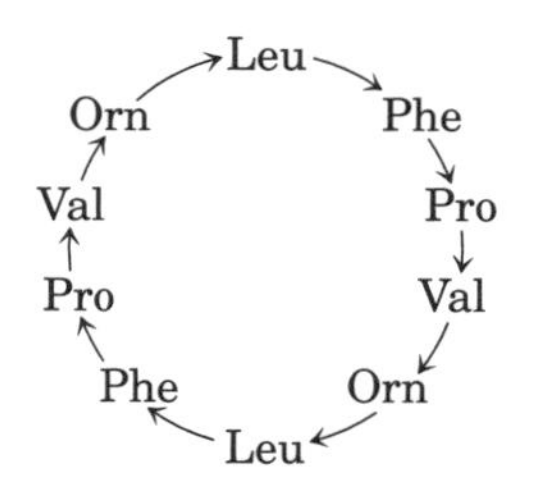

where the arrows indicate the –CO–NH– or C → N direction of the peptide bonds. This structure is consistent with all the data.

17. Efficiency in Peptide Sequencing A peptide with the primary structure Lys–Arg–Pro–Leu–Ile–Asp–Gly–Ala is sequenced by the Edman procedure. If each Edman cycle were 96% efficient, what percentage of the amino acids liberated in the fourth cycle would be leucine? Do the calculation a second time, but assume a 99% efficiency for each cycle.

Answer 88%, 97%. The formula for calculating the percentage of correct amino acid liberated after sequencing cycle n, given an efficiency x, is x^n/x, or x^{n-1}. If the efficiency is 0.96, the fraction of correct amino acid liberated in the fourth cycle is $(0.96)^3 = 0.88$. If the efficiency is 0.99, the fraction is $(0.99)^3 = 0.97$.

The $n - 1$ term can be explained by considering what happens in each cycle. In the first cycle, a Lys residue is liberated from 96% of the ends, so that 96% of the termini now have Arg and the remaining 4% still have Lys. However, 100% of the residues actually removed in this cycle are Lys. In the second cycle, Arg is removed from 96% of the ends that contain it, or from $0.96 \times 0.96 = 0.922 = 92.2\%$ of the ends. Lys is removed from $0.96 \times 0.04 = 0.38 = 3.8\%$ of the ends. However, the fraction of liberated residues in this cycle that are Arg is not 92.2%, because only 96% of the ends had residues removed. Hence, the fraction of residues liberated as Arg in the second cycle is $0.922/0.96 = 0.96 = 96\%$, and so forth.

18. Biochemistry Protocols: Your First Protein Purification As the newest and least experienced student in a biochemistry research lab, your first few weeks are spent washing glassware and labeling test tubes. You then graduate to making buffers and stock solutions for use in various laboratory procedures. Finally, you are given responsibility for purifying a protein. It is a citric acid cycle enzyme, citrate synthase, located in the mitochondrial matrix. Following a protocol for the purification, you proceed through the steps below. As you work, a more experienced student questions you about the rationale for each procedure. Supply the answers. (Hint: See Chapter 2 for information on separation of organelles from cells, and Chapter 4 for information about osmolarity.)

(a) You pick up 20 kg of beef hearts from a nearby slaughterhouse. You transport the hearts on ice, and perform each step of the purification in a walk-in cold room or on ice. You homogenize the beef heart tissue in a high-speed blender in a medium containing ~0.2 M sucrose, buffered to a pH of 7.2. *Why do you use beef heart tissue, and in such large quantity? What is the purpose of keeping the tissue cold and suspending it in 0.2 M sucrose, at pH 7.2? What happens to the tissue when it is homogenized?*

(b) You subject the resulting heart homogenate, which is dense and opaque, to a series of differential centrifugation steps. *What does this accomplish?*

(c) You proceed with the purification using the supernatant fraction that contains mostly intact mitochondria. Next you osmotically lyse the mitochondria. The lysate, which is less dense than the homogenate, but still opaque, consists primarily of mitochondrial membranes and internal mitochondrial contents. To this lysate you add ammonium sulfate, a highly soluble salt, to a specific concentration. You centrifuge the solution, decant the supernatant, and discard the pellet. To the supernatant, which is clearer than the lysate, you add *more* ammonium sulfate. Once again, you centrifuge the sample, but this time you save the pellet because it contains the protein of interest. *What is the rationale for the two-step addition of the salt?*

(d) You solubilize the ammonium sulfate pellet containing the mitochondrial proteins and dialyze it overnight against large volumes of buffered (pH 7.2) solution. *Why isn't ammonium sulfate included in the dialysis buffer? Why do you use the buffer solution instead of water?*

(e) You run the dialyzed solution over a size-exclusion chromatographic column. Following the protocol, you collect the *first* protein fraction that exits the column, and discard the rest of the fractions that elute from the column later. You detect the protein by measuring UV absorption (at 280 nm) in the fractions. *What does the instruction to collect the first fraction tell you about the protein? Why is UV absorption at 280 nm a good way to monitor for the presence of protein in the eluted fractions?*

(f) You place the fraction collected in (e) on a cation-exchange chromatographic column. After discarding the initial solution that exits the column (the flowthrough), you add a washing solution of higher pH to the column and collect the protein fraction that immediately elutes. *Explain what you are doing.*

(g) You run a small sample of your fraction, now very reduced in volume and quite clear (though tinged pink), on an isoelectric focusing gel. When stained, the gel shows three sharp bands. According to the protocol, the protein of interest is the one with the pI of 5.6, but you decide to do one more assay of the protein's purity. You cut out the pI 5.6 band and subject it to SDS-polyacrylamide gel electrophoresis. The protein resolves as a single band. *Why were you unconvinced of the purity of the "single" protein band on your IEF gel? What did the results of the SDS gel tell you? Why is it important to do the SDS gel electrophoresis* after *the isoelectric focusing?*

Answer

(a) *Why do you use beef heart tissue, and why do you need so much of it?* The protein you are to purify (citrate synthase, CS) is found in mitochondria, which are abundant in cells with high metabolic activity such as heart muscle cells. Beef hearts are relatively cheap and easy to get at the local slaughterhouse. You begin with a large amount of tissue because cells contain thousands of different proteins, and no single protein is present in high concentration—that is, the specific activity of CS is low. To purify a significant quantity, you must start with a large excess of tissue.

Why do you need to keep the tissues cold? The cold temperatures inhibit the action of lysosomal enzymes that would destroy the sample.

Why is the tissue suspended in 0.2 M sucrose, at pH 7.2? Sucrose is used in the homogenization buffer to create a medium that is isotonic with the organelles. This prevents diffusion of water into the organelles, causing them to swell, burst, and spill their contents. A pH of 7.2 helps to decrease the activity of lysosomal enzymes and maintain the native structure of proteins in the sample.

What happens to the tissue when it is homogenized? Homogenization breaks open the heart muscle cells, releasing the organelles and cytosol.

(b) *What does differential centrifugation accomplish?* Organelles differ in size and therefore sediment at different rates during centrifugation. Larger organelles and pieces of cell debris sediment first, and progressively smaller cellular components can be isolated in a series of centrifugation steps at increasing speed. The contents of each fraction can be determined microscopically or by enzyme assay.

(c) *What is the rationale for the two-step addition of ammonium sulfate?* Proteins have characteristic solubilities at different salt concentrations, depending on the functional groups in the protein. In a concentration of ammonium sulfate just below the precipitation point of CS, some unwanted proteins can be precipitated (salted out). The ammonium sulfate concentration is then increased so that CS is salted out. It can then be recovered by centrifugation.

(d) *Why is a buffer solution without ammonium sulfate used for the dialysis step?* Osmolarity (as well as pH and temperature) affects the conformation and stability of proteins. To solubilize and renature the protein, the ammonium sulfate must be removed. In dialysis against a buffered solution containing no ammonium sulfate, the ammonium sulfate in the sample moves into the buffer until its concentration is equal in both solutions. By dialyzing against large volumes of buffer that are changed frequently, the concentration of ammonium sulfate in the sample can be reduced to almost zero. This procedure usually takes a long time (typically overnight). The dialysate must be buffered to keep the pH (and ionic strength) of the sample in a range that promotes the native conformation of the protein.

(e) *What does the instruction to collect the first fraction tell you about the protein?* The CS molecule is larger than the pore size of the chromatographic gel. Size-exclusion columns retard the flow of smaller molecules, which enter the pores of the column matrix material. Larger molecules flow around the matrix, taking a direct route through the column.

Why is UV absorption at 280 nm a good way to monitor for the presence of protein in the eluted fractions? The aromatic side chains of Tyr and Trp residues strongly absorb at 280 nm.

(f) *Explain the procedure at the cation-exchange chromatography column.* CS has a positive charge (at the pH of the separation) and binds to the negatively charged beads of the cation-exchange column, while negatively charged and neutral proteins pass through. CS is displaced from the column by raising the pH of the mobile phase and thus altering the charge on the CS molecules.

(g) *Why were you unconvinced of the purity of the "single" protein band on your IEF gel?* Several different proteins, all with the same pI, could be focused in the "single" band. SDS-polyacrylamide gel electrophoresis separates on the basis of mass and therefore would separate any polypeptides in the pI 5.6 band.

Why is it important to do the SDS gel electrophoresis after the isoelectric focusing? SDS is a highly negatively charged detergent that binds tightly and uniformly along the length of a polypeptide. Removing SDS from a protein is difficult, and a protein with only traces of SDS no longer has its native acid-base properties, including it native pI.

chapter

6 The Three-Dimensional Structure of Proteins

1. **Properties of the Peptide Bond** In x-ray studies of crystalline peptides, Linus Pauling and Robert Corey found that the C—N bond in the peptide link is intermediate in length (1.32 Å) between a typical C—N single bond (1.49 Å) and a C=N double bond (1.27 Å). They also found that the peptide bond is planar (all four atoms attached to the C—N group are located in the same plane) and that the two α-carbon atoms attached to the C—N are always trans to each other (on opposite sides of the peptide bond):

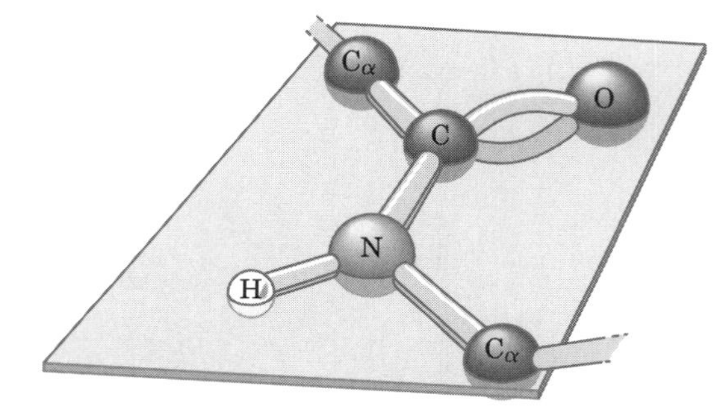

(a) What does the length of the C—N bond in the peptide linkage indicate about its strength and its bond order (i.e., whether it is single, double, or triple)?

(b) What do the observations of Pauling and Corey tell us about the ease of rotation about the C—N peptide bond?

Answer

(a) The higher the bond order (double or triple vs. single), the shorter and stronger are the bonds. Thus bond length is an indication of bond order. For example, the C=N bond is shorter (1.27 Å) and has a higher order ($n = 2.0$) than a typical C—N bond (length = 1.49 Å, $n = 1.0$). The length of the C—N linkage in the peptide bond (1.32 Å) indicates that it is intermediate in strength and bond order between a single and double bond.

(b) Rotation about a double bond is generally impossible at physiological temperatures, and the steric relationship of the groups attached to the two atoms involved in the double bond is spatially "fixed." Since the peptide bond has considerable double-bond character, there is essentially no rotation, and the C=O and N—H groups are fixed in the trans configuration.

2. **Structural and Functional Relationships in Fibrous Proteins** William Astbury discovered that the x-ray pattern of wool shows a repeating structural unit spaced about 5.2 Å along the direction of the wool fiber. When he steamed and stretched the wool, the x-ray pattern showed a new repeating

structural unit at a spacing of 7.0 Å. Steaming and stretching the wool and then letting it shrink gave an x-ray pattern consistent with the original spacing of about 5.2 Å. Although these observations provided important clues to the molecular structure of wool, Astbury was unable to interpret them at the time.

(a) Given our current understanding of the structure of wool, interpret Astbury's observations.

(b) When wool sweaters or socks are washed in hot water or heated in a dryer, they shrink. Silk, on the other hand, does not shrink under the same conditions. Explain.

Answer

(a) The principal structural units in the wool fiber polypeptides are successive turns of the α helix, which are spaced at 5.4 Å intervals. The intrinsic stability of the helix (and thus the fiber) results from *intra* chain hydrogen bonds (see Fig. 6–4b). Steaming and stretching the fiber yields an extended polypeptide chain with the β conformation, in which the distance between adjacent R groups is about 7.0 Å. Upon resteaming, the polypeptide chains assume the less-extended α helix conformation, and the fiber shrinks.

(b) Wool freshly sheared from a sheep is primarily in its α-keratin (α-helical coiled coil) form (see Fig. 6–11). Because raw wool is crimped or curly, it is combed and stretched to straighten it before it is spun into fibers for clothing. This processing converts the wool from its native α-helical conformation to a more extended β form. Moist heat triggers a conformational change back to the native α-helical structure, which shrinks both the fiber and the clothing. Under conditions of mechanical tension and moist heat, wool can be stretched back to a fully extended form. In contrast, the polypeptide chains of silk have a very stable β-pleated sheet structure, fully extended along the axis of the fiber (see Fig. 6–7), and have small, closely packed amino acid side chains (see Fig. 6–14). These characteristics make silk resistant to stretching or shrinking.

3. Rate of Synthesis of Hair α-Keratin Hair grows at a rate of 15 to 20 cm/yr. All this growth is concentrated at the base of the hair fiber, where α-keratin filaments are synthesized inside living epidermal cells and assembled into ropelike structures (see Fig. 6–11). The fundamental structural element of α-keratin is the α helix, which has 3.6 amino acid residues per turn and a rise of 5.4 Å per turn (see Fig. 6–4b). Assuming that the biosynthesis of α-helical keratin chains is the rate-limiting factor in the growth of hair, calculate the rate at which peptide bonds of α-keratin chains must be synthesized (peptide bonds per second) to account for the observed yearly growth of hair.

Answer

Since there are 3.6 amino acids (AAs) per turn and the rise is 5.4 Å/turn, the length per AA of the α helix is

$$\frac{5.4\ \text{Å/turn}}{3.6\ \text{AA/turn}} = 1.5\ \text{Å/AA} = 1.5 \times 10^{-10}\ \text{m/AA}$$

A growth rate of 20 cm/yr is equivalent to

$$\frac{20\ \text{cm/year}}{(365\ \text{days/yr})(24\ \text{h/day})(60\ \text{min/h})(60\ \text{s/min})} = 6.4 \times 10^{-7}\ \text{cm/s} = 6.4 \times 10^{-9}\ \text{m/s}$$

Thus the rate at which AAs are added is

$$\frac{6.4 \times 10^{-9}\ \text{m/s}}{1.5 \times 10^{-10}\ \text{m/AA}} \approx 43\ \text{AA/s} \approx 43\ \text{peptide bonds per second}$$

4. The Effect of pH on the Conformation of α-Helical Secondary Structures The unfolding of the α helix of a polypeptide to a randomly coiled conformation is accompanied by a large decrease in a property called its specific rotation, a measure of a solution's capacity to rotate plane-polarized light.

Polyglutamate, a polypeptide made up of only L-Glu residues, has the α-helical conformation at pH 3. However, when the pH is raised to 7, there is a large decrease in the specific rotation of the solution. Similarly, polylysine (L-Lys residues) is an α helix at pH 10, but when the pH is lowered to 7, the specific rotation also decreases, as shown by the following graph.

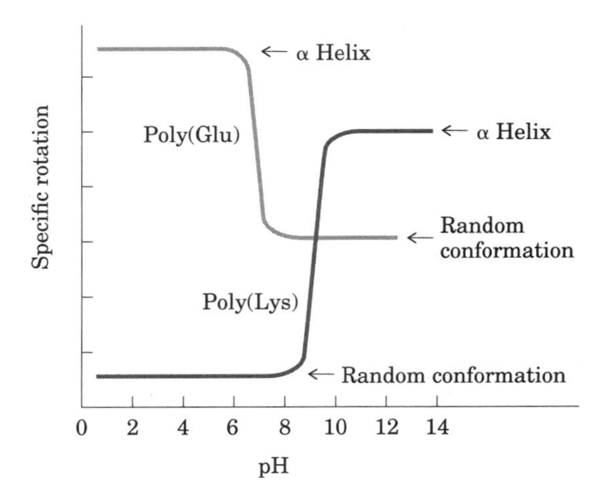

What is the explanation for the effect of the pH changes on the conformations of poly(Glu) and poly(Lys)? Why does the transition occur over such a narrow range of pH?

Answer At pH values above 6, deprotonation of the carboxylate side chains of poly(Glu) leads to repulsion between adjacent negatively charged groups, which destabilizes the α helix and results in unfolding. Similarly, at pH 7 protonation of the amino-group side chains of poly(Lys) causes repulsion between positively charged groups, which leads to unfolding.

5. **Disulfide Bonds Determine the Properties of Many Proteins** A number of natural proteins are very rich in disulfide bonds, and their mechanical properties (tensile strength, viscosity, hardness, etc.) are correlated with the degree of disulfide bonding. For example, glutenin, a wheat protein rich in disulfide bonds, is responsible for the cohesive and elastic character of dough made from wheat flour. Similarly, the hard, tough nature of tortoise shell is due to the extensive disulfide bonding in its α-keratin.

(a) What is the molecular basis for the correlation between disulfide-bond content and mechanical properties of the protein?

(b) Most globular proteins are denatured and lose their activity when briefly heated to 65 °C. However, globular proteins that contain multiple disulfide bonds often must be heated longer at higher temperatures to denature them. One such protein is bovine pancreatic trypsin inhibitor (BPTI), which has 58 amino acid residues in a single chain and contains three disulfide bonds. On cooling a solution of denatured BPTI, the activity of the protein is restored. What is the molecular basis for this property?

Answer

(a) Disulfide bonds are covalent bonds, which are much stronger than the noncovalent interactions (hydrogen bonds, hydrophobic interactions, van der Waals interactions) that stabilize the three-dimensional structure of most proteins. Disulfide bonds serve to cross-link protein chains, increasing stiffness, hardness, and mechanical strength.

(b) As the temperature is raised, the increased thermal motion of the polypeptide chains and vibrational motions of hydrogen bonds ultimately lead to thermal denaturation (unfolding) of a protein. Cystine residues (disulfide bridges), depending on their location in the protein structure, can prevent or restrict the movement of folded protein domains, block access of solvent water to the interior of the protein, and prevent the complete unfolding of the protein. Refolding to the native structure from a random conformation is seldom spontaneous, owing to the very large number of conformations possible. Disulfide bonds limit the number of conformations by allowing only a few minimally unfolded structures, and hence allow the protein to return to its native conformation more easily upon cooling.

6. Amino Acid Sequence and Protein Structure Our growing understanding of how proteins fold allows researchers to make predictions about protein structure based on primary amino acid sequence data.

1	2	3	4	5	6	7	8	9	10
Ile –	Ala –	His –	Thr –	Tyr –	Gly –	Pro –	Phe –	Glu –	Ala –

11	12	13	14	15	16	17	18	19	20
Ala –	Met –	Cys –	Lys –	Trp –	Glu –	Ala –	Gln –	Pro –	Asp –

21	22	23	24	25	26	27	28
Gly –	Met –	Glu –	Cys –	Ala –	Phe –	His –	Arg

(a) Based on the amino acid sequence above, where would you predict that bends or β turns would occur?

(b) Where might intrachain disulfide cross-linkages be formed?

(c) Assuming that this sequence is part of a larger globular protein, indicate the probable location (the external surface or interior of the protein) of the following amino acid residues: Asp, Ile, Thr, Ala, Gln, Lys. Explain your reasoning. (Hint: See the hydropathy index in Table 5–1.)

Answer

(a) Bends or turns are most likely to occur at residues 7 and 19, since Pro residues are often (but not always) found at bends in globular folded proteins. Other candidates are Ser, Thr, and Ile.

(b) Intrachain disulfide cross-linkages can only form between residues 13 and 24 (Cys residues).

(c) Amino acids with ionic (charged) or strongly polar neutral groups (for example Asp, Gln, and Lys in this protein) are located on the external surface, where they interact optimally with solvent water. In contrast, residues with nonpolar side chains (such as Ala and Ile) are situated in the interior, where they escape the polar solvent environment. Thr (and also Ser) are of intermediate polarity and can be found either in the interior or on the exterior surface (see Table 5–1).

7. Bacteriorhodopsin in Purple Membrane Proteins Under the proper environmental conditions, the salt-loving bacterium *Halobacterium halobium* synthesizes a membrane protein (M_r 26,000) known as bacteriorhodopsin, which is purple because it contains retinal. Molecules of this protein aggregate into "purple patches" in the cell membrane. Bacteriorhodopsin acts as a light-activated proton pump that provides energy for cell functions. X-ray analysis of this protein reveals that it consists of seven parallel α-helical segments, each of which traverses the bacterial cell membrane (thickness 45 Å). Calculate the minimum number of amino acids necessary for one segment of α helix to traverse the membrane completely. Estimate the fraction of the bacteriorhodopsin protein that is involved in membrane-spanning helices. (Use an average amino acid residue weight of 110.)

Answer If there are seven helices in a protein of M_r 26,000 with an average residue M_r of 110, assuming that all residues are involved in helices, the number of amino acids per helix would be

$$(26{,}000)/(7 \text{ helices/protein})(110 \text{ AA}) = 34 \text{ AA/helix (maximum)}$$

Using the parameters from Problem 3 (3.6 AA/turn, 5.4 Å/turn), we can calculate that there are 0.67 AA/Å along the axis of the helix. Thus in 45 Å of membrane, a minimum of (45 Å)(0.67 AA/Å) = 30 AAs are required to span the membrane per segment of helix. Thus the fraction of residues actually involved in membrane-spanning helices is 30/34 = 0.88 = 88%.

8. Pathogenic Action of Bacteria That Cause Gas Gangrene The highly pathogenic anaerobic bacterium *Clostridium perfringens* is responsible for gas gangrene, a condition in which animal tissue structure is destroyed. This bacterium secretes an enzyme that efficiently catalyzes the hydrolysis of the peptide bond indicated by an asterick in the sequence:

$$\text{—X}\overset{*}{\text{—}}\text{Gly—Pro—Y—} \xrightarrow{H_2O} \text{—X—COO}^- + H_3\overset{+}{N}\text{—Gly—Pro—Y—}$$

where X and Y are any of the 20 standard amino acids. How does the secretion of this enzyme contribute to the invasiveness of this bacterium in human tissues? Why does this enzyme not affect the bacterium itself?

Answer Collagen is distinctive in its amino acid composition having a very high proportion of Gly (35%) and Pro residues. The enzyme (a collagenase) secreted by the bacterium destroys the connective-tissue barrier (skin, hide, etc.) of the host, allowing the bacterium to invade the host tissues. Bacteria do not contain collagen and thus are unaffected by collagenase.

9. The Number of Polypeptide Chains in a Multisubunit Protein A sample (660 mg) of an oligomeric protein of M_r 132,000 was treated with an excess of 1-fluoro-2,4-dinitrobenzene (Sanger's reagent) under slightly alkaline conditions until the chemical reaction was complete. The peptide bonds of the protein were then completely hydrolyzed by heating it with concentrated HCl. The hydrolysate was found to contain 5.5 mg of the following compound:

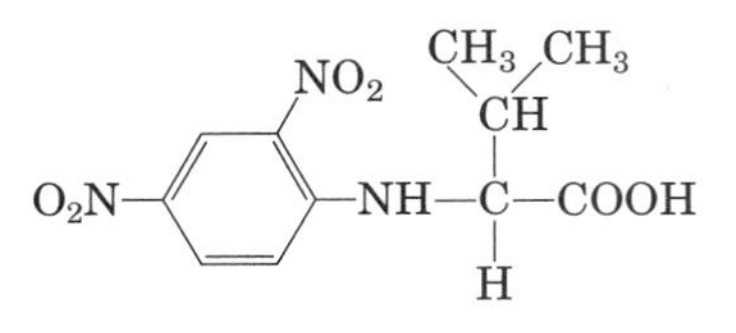

However, 2,4-dinitrophenyl derivatives of the α-amino groups of other amino acids could not be found.

(a) Explain how this information can be used to determine the number of polypeptide chains in an oligomeric protein.

(b) Calculate the number of polypeptide chains in this protein.

(c) What other protein analysis technique could you employ to determine whether the polypeptide chains in this protein are similar or different?

Answer

(a) Since only a single 2,4-dinitrophenyl (DNP) amino acid derivative is found, there is only one kind of amino acid at the amino terminus, i.e., all of the polypeptide chains have the same amino-terminal residue. Comparing the number of moles of this derivative to the number of moles of protein gives the number of polypeptide chains.

(b) The amount of protein = (0.66 g)/(132,000 g/mol) = 5×10^{-6} mol.
Since M_r for DNP-Val ($C_{11}H_{13}O_6N_3$) = 283,
the amount of DNP-Val = (0.0055 g)/(283 g/mol) = 19.4×10^{-6} mol.
The ratio of moles of DNP-Val to moles of protein gives the number of amino-terminal residues and thus the number of chains per oligomer:

$$\frac{1.9 \times 10^{-5} \text{ mol DNP-Val}}{5 \times 10^{-6} \text{ mol protein}} = 3.8 \approx 4 \text{ polypeptide chains}$$

An alternative approach to the problem is through the proportionality:

$$\frac{n(283 \text{ g/mol})}{132{,}000 \text{ g/mol}} = \frac{5.5 \text{ mg}}{660 \text{ mg}}$$

$$n = \frac{(5.5 \text{ mg})(132{,}000 \text{ g/mol})}{(660 \text{ mg})(283 \text{ g/mol})} = 3.89 \approx 4$$

(c) Polyacrylamide gel electrophoresis in the presence of a detergent (such as sodium dodecylsulfate, SDS) and an agent that prevents the formation of disulfide bonds (such as β-mercaptoethanol) would provide information on subunit structure of a protein. In the example here, the 132,000 Da oligomeric protein would dissociate into identical subunits that would appear on a gel as a single band. This band would have an apparent M_r of approximately 33,000. Calculate the number of subunits by dividing the apparent M_r of the oligomer by the apparent M_r of the individual subunits (132,000/33,000 = 4). If the protein were made up of different polypeptide subunits, they would likely appear as multiple discrete bands on the gel.

Biochemistry on the Internet

10. Protein Modeling on the Internet A group of patients suffering from Crohn's disease (an inflammatory bowel disease) underwent biopsies of their intestinal mucosa in an attempt to identify the causative agent. A protein was identified that was expressed at higher levels in patients with Crohn's disease than in patients with an unrelated inflammatory bowel disease or in unaffected controls. The protein was isolated and the following *partial* amino acid sequence was obtained:

EAELCPDRCI HSFQNLGIQC VKKRDLEQAI SQRIQTNNNP FQVPIEEQRG DYDLNAVRLC
FQVTVRDPSG RPLRLPPVLP HPIFDNRAPN TAELKICRVN RNSGSCLGGD EIFLLCDKVQ
KEDIEVYFTG PGWEARGSFS QADVHRQVAI VFRTPPYADP SLQAPVRVSM QLRRPSDREL
SEPMEFQYLP DTDDRHRIEE KRKRTYETFK SIMKKSPFSG PTDPRPPPRR IAVPSRSSAS VPKPAPQPYP

(a) You can identify this protein using a protein database on the Internet. Some good places to start include PIR-International Protein Sequence Database, Structural Classification of Proteins (SCOP), and Prosite. For the current URLs of these and other protein database sites, use an Internet search engine or go to the *Principles of Biochemistry,* 3/e site at http://www.worthpublishers.com/lehninger. At these sites, follow links to locate the sequence comparison "engine." Enter about 30 residues from the sequence of the protein in the appropriate search field and submit it for analysis. What does this analysis tell you about the identity of the protein?

(b) Try using different portions of the protein amino acid sequence. Do you always get the same result?

(c) A variety of Web sites provide information about the three-dimensional structure of proteins. Find information about the protein's secondary, tertiary, and quaternary structure using database sites such as the Protein Data Bank (PDB) or SCOP.

(d) In the course of your Web searches try to find information about the cellular function of the protein.

Answer

(a) The following is one example of a sequence similarity search using the PIR-International Protein Sequence Database. Go to their site at http://pir.georgetown.edu/pirwww/dbinfo/pirpsd.html, click on **Search Databases,** and choose a **BLAST** search. You can then paste your amino acid sequence (the first 30 amino acids were used for this analysis) into the appropriate field and submit your sequence for comparison. The output will look something like the following:

Accession # Database ID	Title	Sequence similarity score	Statistical significance (e-value)
PIR: I15379	NF-kappa-B transcription factor subunit - human	64.4	1e-11
PIR: A40851	transforming protein (rel) homolog - human	64.4	1e-11
PIR: A42017	transcription factor NF-kappa-B chain p65 - human	64.4	1e-11
PIR: A37932	nuclear factor kappa-B chain p65 - mouse	59.7	3e-10
PIR: JC2004	nuclear factor-kappa B p65 chain - chicken	54.9	9e-09
PIR: A38631	transforming protein (rel) homolog - African clawed ...	52.3	5e-08
PIR: S10893	transforming protein (rel) - chicken	43.5	2e-05
PIR: TVTK	transforming protein rel - turkey (fragment)	43.5	2e-05
PIR: TVCHRL	transforming protein rel - chicken	43.5	2e-05
PIR: VCVDA	transforming protein rel - avian ...	43.5	2e-05

This information provides the names of the various proteins that contain this amino acid sequence, the sequence similarity score (in terms of "bits"), and the statistical significance of the match. The number before the name of each protein is its ID number, which allows you to find the relevant entry in the database.

(b) Sequence matching based on only 30 amino acids will occasionally generate "hits" that are not closely related to the actual protein, depending on the portion of the sequence that you use for your search. For example, this amino acid sequence is for the NF-κ-B protein from humans, but the sequence is similar to that of transforming protein (rel) from chicken. The more sequence information you have, the more likely you are to pinpoint the identity of your protein; nevertheless, you should be impressed by the fact that this search engine generated the correct match (NF-κ-B) based on a limited number of amino acid residues.

(c) The following example of an Internet-based analysis of protein structure uses the Protein Data Bank at http://www.rcsb.org/pdb/. On the home page click on **SearchLite** to access the PDB SearchLite utility.

Type in the name of the protein you identified from your sequence search in part **(a),** NF-κ-B transcription factor subunit - human. (Note: You may have to try different methods of inputting the keyword information. For example, you will not find any matches using "NF-kappa-B transcription factor subunit - human" or "NF-kappa-B." However, you will be successful in finding the entry for this protein if you make your query less specific. For example, use only "Nuclear Factor" or "NF-kappa."

You will be presented with a list of possible matches to your query. Find the protein you are looking for from this list of possible matches and click on **Explore.** For this analysis, see if you can find the entry with PDB ID 1NFK, the Nuclear

Factor-κ-B Fragment: P50 subunit. The information page that is presented will contain links to a number of protein analysis utilities. Click on the **View Structure** link and then click on **Quick PDB** to view the backbone structure of the protein. For 1NFK you will see a complex of two polypeptide chains representing a homodimer of the p50 subunit of the nuclear factor-κ-B. You will also see the primary structure for each subunit. Clicking on an amino acid residue in the primary structure will highlight that residue in the protein backbone model. Selecting **Secondary Structure** will color code β conformations (blue) and α helices (red) in the protein model.

Alternatively you can view this structure using the 3-D viewers Rasmol or Chime. (For instructions on how to download these free programs, go to the *Principles of Biochemistry* Web site at http://www.worthpublishers.com/lehninger.) Clicking on **Rasmol** will cause the 1NFK PDB file to be downloaded onto your computer. You will be able to open and view this file if the Rasmol program has already been installed on your hard drive.

Clicking on **Chime** will cause the 1NFK file to be opened automatically within the Internet window. The link also provides information on how to download the Chemscape Chime plug-in if you need to do so. Both the Chime and Rasmol viewers provide a three-dimensional view of the transcription factor as well as the DNA helix with which this protein interacts.

(d) The NF-κ-B protein shown in the 1NFK model is a homodimer of two identical protein subunits (p50 subunits). The biologically active form of this molecule is probably a heterodimer composed of one p50 protein subunit (visualized in the model discussed above) and a related protein called p65. NF-κ-B is a transcription factor that binds specific DNA sequences. A consequence of this binding is the enhancement of transcription of nearby genes. One such gene is the immunoglobulin κ light chain, from which the factor gets its name.

chapter

7 Protein Function

1. **Relationship between Affinity and Dissociation Constant** Protein A has a binding site for ligand X with a K_d of 10^{-6} M. Protein B has a binding site for ligand X with a K_d of 10^{-9} M. Which protein has a higher affinity for ligand X? Explain your reasoning. Convert the K_d to K_a for both proteins.

   Answer Protein B has a higher affinity for ligand X. The lower K_d indicates that protein B will be half-saturated with bound ligand X at a much lower concentration of X than will protein A. Protein A has a K_a of 10^6 M^{-1}. Protein B has a K_a of 10^9 M^{-1}.

2. **Negative Cooperativity** Which of the following situations would produce a Hill plot with $n_H < 1.0$? Explain your reasoning in each case.
   (a) The protein has multiple subunits, each with a single ligand-binding site. Binding of ligand to one site decreases the binding affinity of other sites for the ligand.
   (b) The protein in a single polypeptide with two ligand-binding sites, each having a different affinity for the ligand.
   (c) The protein is a single polypeptide with a single ligand-binding site. The protein preparation is heterogeneous, containing some protein molecules that are partially denatured and thus have a lower binding affinity for the ligand.

   Answer All three situations would produce a Hill coefficient, n_H, of less than 1.0. An $n_H <$ 1.0 generally suggests that situation (a)—the classic case of negative cooperativity—applies. However, closer examination of the properties of a protein exhibiting apparent negative cooperativity in ligand binding often reveals situation (b) or (c). When two or more types of ligand-binding sites with different affinities for the ligand are present on the same or different proteins in the same solution, apparent negative cooperativity is observed. In situation (b), the higher-affinity ligand-binding sites bind the ligand first. As the ligand concentration is increased, binding to the lower-affinity sites produces an $n_H < 1.0$, even though binding to the two ligand-binding sites is completely independent. Even more common is situation (c) in which the protein preparation is heterogeneous. Unsuspected proteolytic digestion by contaminating proteases and partial denaturation of the protein under certain solvent conditions are common artifacts of protein purification. There are few well-documented cases of true negative cooperativity.

3. **Affinity for Oxygen in Myoglobin and Hemoglobin** What is the effect of the following changes on the O_2 affinity of myoglobin and hemoglobin? (a) A drop in the pH of blood plasma from 7.4 to 7.2. (b) A decrease in the partial pressure of CO_2 in the lungs from 6 kPa (holding one's breath) to 2 kPa (normal). (c) An increase in the BPG level from 5 mM (normal altitudes) to 8 mM (high altitudes).

Answer The affinity of hemoglobin for O_2 is regulated by the binding of the ligands H^+, CO_2, and BPG. The binding of each ligand shifts the O_2-saturation curve to the right—that is, the O_2 affinity of hemoglobin in the presence of the ligand is reduced. In contrast, the O_2 affinity of myoglobin is *not* affected by the presence of these ligands. Hence, the changing conditions only affect the binding of oxygen to hemoglobin: (a) decreases the affinity; (b) increases the affinity; (c) decreases the affinity.

4. Cooperativity in Hemoglobin Under appropriate conditions, hemoglobin dissociates into its four subunits. The isolated α subunit binds oxygen, but the O_2-saturation curve is hyperbolic rather than sigmoid. In addition, the binding of oxygen to the isolated α subunit is not affected by the presence of H^+, CO_2, or BPG. What do these observations indicate about the source of the cooperativity in hemoglobin?

Answer These observations indicate that the cooperative behavior—the sigmoid O_2-binding curve and the positive cooperativity in ligand binding—of hemoglobin arises from interaction between subunits.

5. Comparison of Fetal and Maternal Hemoglobins Studies of oxygen transport in pregnant mammals have shown that the O_2-saturation curves of fetal and maternal blood are markedly different when measured under the same conditions. Fetal erythrocytes contain a structural variant of hemoglobin, HbF, consisting of two α and two γ subunits ($\alpha_2\gamma_2$), whereas maternal erythrocytes contain HbA ($\alpha_2\beta_2$).

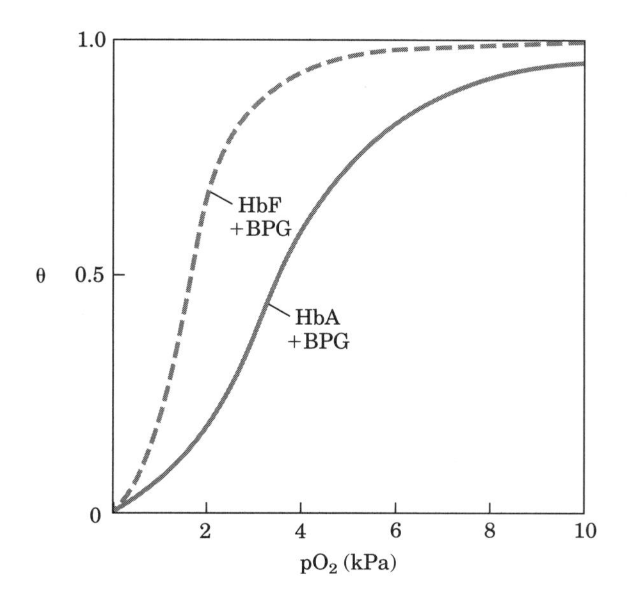

(a) Which hemoglobin has a higher affinity for oxygen under physiological conditions, HbA or HbF? Explain.

(b) What is the physiological significance of the different O_2 affinities?

(c) When all the BPG is carefully removed from samples of HbA and HbF, the measured O_2-saturation curves (and consequently the O_2 affinities) are displaced to the left. However, HbA now has a greater affinity for oxygen than does HbF. When BPG is reintroduced, the O_2-saturation curves return to normal, as shown in the figure. What is the effect of BPG on the O_2 affinity of hemoglobin? How can this information explain the different O_2 affinities of fetal and maternal hemoglobin?

Answer

(a) The observation that hemoglobin A (HbA; maternal) is only 33% saturated when the pO_2 is 4 kPa, while hemoglobin F (HbF; fetal) is 58% saturated under the same physiological conditions, indicates that the O_2 affinity of HbF is higher than that of HbA. In other words, at identical O_2 concentrations, HbF binds more oxygen than does HbA. Thus HbF must bind oxygen more tightly (with higher affinity) than HbA under physiological conditions.

(b) The higher O_2 affinity of HbF assures that oxygen will flow from maternal blood to fetal blood in the placenta. For maximal oxygen transport, the oxygen pressure at which fetal blood approaches full saturation must be in the region where the O_2 affinity of HbA is low. This is indeed the case.

(c)

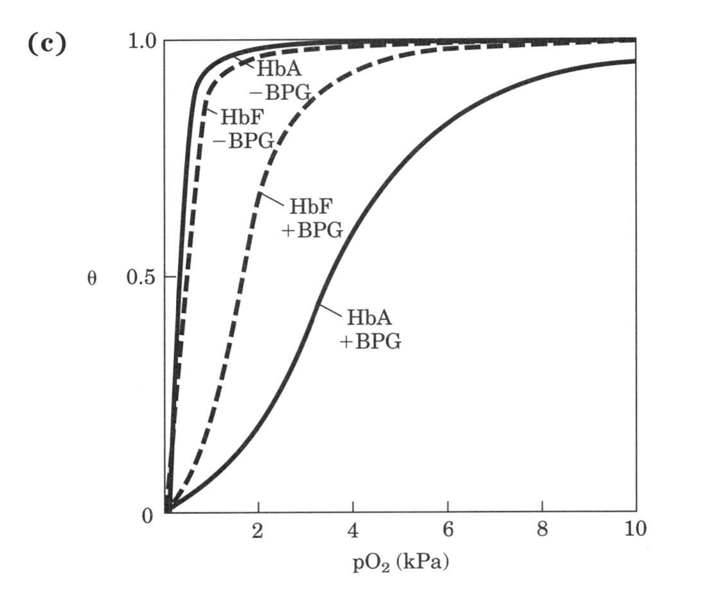

Binding of BPG to hemoglobin reduces the affinity of hemoglobin for O_2, as shown in the figure. The O_2-saturation curve for HbA shifts far to the right when BPG binds (solid curves)—that is, the O_2 affinity is dramatically lowered. The O_2-saturation curve for HbF also shifts to the right when BPG binds (dashed curves), but not as far. The observation that the O_2-saturation curve of HbA undergoes a larger shift on BPG binding than does that of HbF suggests that HbA binds BPG more tightly than does HbF. Differential binding of BPG to the two hemoglobins may determine the difference in their O_2 affinities.

6. Hemoglobin Variants There are almost 500 naturally occurring variants of hemoglobin. Most are the result of a single amino acid substitution in a globin polypeptide chain. Some variants produce clinical illness, though not all variants have deleterious effects. A brief sample is presented below:

HbS (sickle-cell Hb): substitutes a Val for a Glu on the surface
Hb Cowtown: eliminates an ion pair involved in T-state stabilization
Hb Memphis: substitutes one uncharged polar residue for another of similar size on the surface
Hb Bibba: substitutes a Pro for a Leu involved in an α helix
Hb Milwaukee: substitutes a Glu for a Val
Hb Providence: substitutes an Asn for a Lys that normally projects into the central cavity of the tetramer
Hb Philly: substitutes a Phe for a Tyr, disrupting hydrogen bonding at the $\alpha_1\beta_1$ interface

Explain your choices for each of the following:

(a) The Hb variant *least* likely to cause pathological symptoms.

(b) The variant(s) most likely to show pI values different from that of HbA when run on an isoelectric focusing gel.

(c) The variant(s) most likely to show a decrease in BPG binding and an increase in the overall affinity of the hemoglobin for oxygen.

Answers

(a) Hb Memphis has a conservative substitution that is unlikely to have a significant effect on function.

(b) HbS, Hb Milwaukee, and Hb Providence all have substitutions that alter the net charge on the protein, which will change the pI. The loss of an ion pair in Hb Cowtown may indicate loss of a charged residue, which would also change the pI, but there is not enough information to be sure.

(c) Hb Providence has an Asn residue in place of a Lys that normally projects into the central cavity of hemoglobin. Loss of the positively charged Lys that normally interacts with the negative charges on BPG results in Hb Providence having lower affinity for BPG and thus higher affinity for O_2.

7. Reversible (but Tight) Binding to an Antibody An antibody binds to an antigen with a K_d of 5×10^{-8} M. At what concentration of antigen will θ be (a) 0.2, (b) 0.5, (c) 0.6, (d) 0.8?

Answer (a) 1.25×10^{-8} M, **(b)** 5×10^{-8} M, **(c)** 7.5×10^{-8} M, **(d)** 2×10^{-7} M. Note that a rearrangement of Eqn 7–8 gives $[L] = \theta K_d/(1 - \theta)$.

8. Using Antibodies to Probe Structure-Function Relationships in Proteins A monoclonal antibody binds to G-actin but not to F-actin. What does this tell you about the epitope recognized by the antibody?

Answer The epitope is likely to be a structure that is buried when G-actin polymerizes to form F-actin.

9. The Immune System and Vaccines A host organism needs time, often days, to mount an immune response against a new antigen, but memory cells permit a rapid response to pathogens previously encountered. A vaccine to protect against a particular viral infection often consists of weakened or killed virus or isolated proteins from a viral protein coat. When injected into a human patient, the vaccine generally does not cause an infection and illness, but it effectively "teaches" the immune system what the viral particles look like, stimulating the production of memory cells. On subsequent infection, these cells can bind to the virus and trigger a rapid immune response. Some pathogens, including HIV, have developed mechanisms to evade the immune system, making it difficult or impossible to develop effective vaccines against them. What strategy could a pathogen use to evade the immune system? Assume that antibodies and/or T-cell receptors are available to bind to any structure that might appear on the surface of a pathogen and that, once bound, the pathogen is destroyed.

Answer A variety of pathogens, including HIV, have evolved mechanisms by which they can repeatedly alter the surface proteins to which immune system components initially bind. Thus the host organism regularly faces new antigens and requires time to mount an immune response to each one. As the immune system responds to one variant, new variants are created. Some molecular mechanisms used to vary viral surface proteins are described in Part IV of this text. HIV uses an additional strategy to evade the immune system: it actively infects and destroys immune system cells.

10. How We Become a "Stiff" When a higher vertebrate dies, its muscles stiffen as they are deprived of ATP, a state called rigor mortis. Explain the molecular basis of the rigor state.

Answer Binding of ATP to myosin triggers dissociation of myosin from the actin thin filament. In the absence of ATP, actin and myosin bind tightly to each other.

11. Sarcomeres from Another Point of View The symmetry of thick and thin filaments in a sarcomere is such that six thin filaments ordinarily surround each thick filament in a hexagonal array. Draw a cross section (transverse cut) of a myofibril at the following points: (a) at the M line; (b) through the I band; (c) through the dense region of the A band; (d) through the less dense region of the A band, adjacent to the M line (see Fig. 7–31b).

Answer

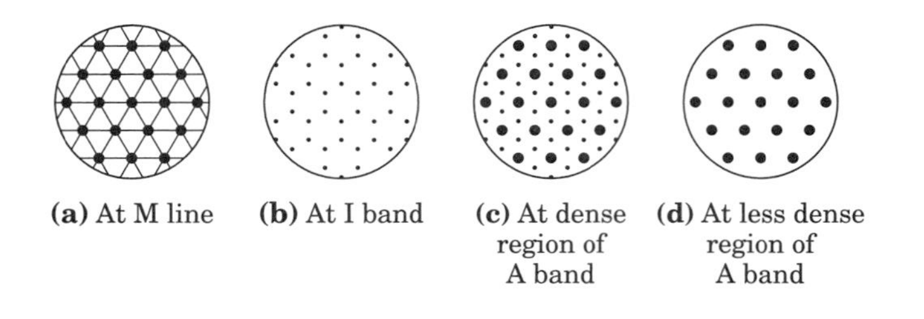

The less dense region of the A band, also known as the H zone, is the region in which the myosin thick filaments do not overlap the actin thin filaments. When the sarcomere contracts, the H zone and the I band decrease in width.

Biochemistry on the Internet

12. Lysozyme and Antibodies To fully appreciate how proteins function in a cell, it is helpful to have a three-dimensional view of how proteins interact with other cellular components. Fortunately, this is possible using the Internet and on-line protein databases. Go to the biochemistry site at http://www.worthpublishers.com/lehninger to learn how to use the Chemscape® Chime® three-dimensional molecular viewing utility. You can then use the Protein Data Bank and Chemscape Chime to investigate the interactions between antibodies and antigens in more detail.

To examine the interactions between the enzyme lysozyme (Chapter 6) and the Fab portion of the anti-lysozyme antibody, go to the Protein Data Bank Website. Use the PDB identifier 1FDL to retrieve the data page for the IgG1 Fab Fragment-Lysozyme Complex (antibody-antigen complex). Open the structure using Chemscape Chime, and use the different viewing options to answer the following questions:

(a) Which chains in the three-dimensional model correspond to the antibody fragment and which correspond to the antigen, lysozyme?

(b) What secondary structure predominates in this Fab fragment?

(c) How many amino acid residues are in the heavy and light chains of the Fab fragment and in lysozyme? Estimate the percentage of the lysozyme that interacts with the antigen-binding site of the antibody fragment.

(d) Identify the specific amino acid residues in lysozyme and in the variable regions of the heavy and light chains that appear to be situated at the antigen-antibody interface. Are the residues contiguous in the primary sequence of the polypeptide chain?

Answer

(a) Chain L is the light chain and chain H is the heavy chain of the Fab fragment of this antibody molecule. Chain Y is lysozyme.

(b) Using Chemscape Chime to view the secondary structure of this protein complex, you should be able to identify the β structures in the variable and constant regions of both the light and heavy chains. The β strands form a supersecondary structure called the immunoglobulin-like β sandwich. A similar IgG structure is shown in Figure 7–23.

(c) The heavy chain of the Fab fragment has 218 amino acid residues, the light chain fragment has 214, and lysozyme has 129. Viewing the structure in the spacefill mode shows that less than 15% of the total lysozyme molecule is in contact with the combined V_L and V_H domains of the antibody fragment.

(d) While viewing the structure in spacefill mode, with each chain a different color, click quickly on the residues that appear to be in contact. In the H chain these residues include Gly^{31},Tyr^{32}, Arg^{99}, Asp^{100}, and Tyr^{101}. In the L chain these residues include Tyr^{32}, Tyr^{49}, Tyr^{50}, and Trp^{92}. In lysozyme, residues Asn^{19}, Gly^{22}, Tyr^{23}, Ser^{24}, Lys^{116}, Gly^{117}, Thr^{118}, Asp^{119}, Gln^{121}, and Arg^{125} appear to be situated at the antigen-antibody interface. Not all these residues are adjacent in the primary structure. In any antibody, the residues in the V_L and V_H domains that come into contact with the antigen are located primarily in the loops connecting the β strands of the β-sandwich supersecondary structure. Folding of the polypeptide chain into higher levels of structure brings the nonconsecutive residues together to form the antigen-binding site.

chapter

8 Enzymes

1. **Keeping the Sweet Taste of Corn** The sweet taste of freshly picked corn (maize) is due to the high level of sugar in the kernels. Store-bought corn (several days after picking) is not as sweet because about 50% of the free sugar is converted into starch within one day of picking. To preserve the sweetness of fresh corn, the husked ears can be immersed in boiling water for a few minutes ("blanched") then cooled in cold water. Corn processed in this way and stored in a freezer maintains its sweetness. What is the biochemical basis for this procedure?

   Answer After an ear of corn has been removed from the plant, the enzyme-catalyzed conversion of sugar to starch continues. Inactivation of these enzymes slows down the conversion to an imperceptible rate. One of the simplest techniques for inactivating enzymes is heat denaturation. Freezing the corn lowers any remaining enzyme activity to an insignificant level.

2. **Intracellular Concentration of Enzymes** To approximate the actual concentration of enzymes in a bacterial cell, assume that the cell contains equal concentrations of 1,000 different enzymes in solution in the cytosol and that each protein has a molecular weight of 100,000. Assume also that the bacterial cell is a cylinder (diameter 1 μm, height 2.0 μm), that the cytosol (specific gravity 1.20) is 20% soluble protein by weight, and that the soluble protein consists entirely of enzymes. Calculate the *average* molar concentration of each enzyme in this hypothetical cell.

   Answer There are three different ways to approach this problem.

   (i) The concentration of total protein in the cell is

   $$\frac{(1.2 \text{ g/mL})(0.2)}{100{,}000 \text{ g/mol}} = 0.24 \times 10^{-5} \text{ mol/mL} = 2.4 \times 10^{-3} \text{ M}$$

   Thus, for one enzyme in 1,000

   $$\text{the enzyme concentration} = \frac{2.4 \times 10^{-3} \text{ M}}{(1000)} = 2.4 \times 10^{-6} \text{ mol/L} \approx 2 \ \mu\text{M}$$

   (ii) $$\text{The average molar concentration} = \frac{\text{moles of each enzyme in cell}}{\text{volume of cell in liters}}$$

   $$\text{Volume of bacterial cytosol} = \pi r^2 h = (3.14)(0.5)^2(2) \ \mu\text{m}^3 = 1.57 \ \mu\text{m}^3 = 1.57 \times 10^{-15} \text{ L}$$

   Amount (in moles) of each enzyme in cell is

   $$\frac{(0.20)(1.2 \text{ g/cm}^3)(1.57 \ \mu\text{m}^3)(10^{-12} \text{cm}^3/\mu\text{m}^3)}{(100{,}000 \text{ g/mol})(100{,}000)} = 3.77 \times 10^{-21} \text{ mol}$$

   $$\text{Average molar concentration} = \frac{3.77 \times 10^{-21} \text{ mol}}{1.57 \times 10^{-15} \text{ L}} = 2.4 \times 10^{-6} \text{ mol/L} \approx 2 \ \mu\text{M}$$

(iii) Volume of bacterial cytosol $= \pi r^2 h$

$= (3.14)(0.5\ \mu\text{m})^2(2\ \mu\text{m}) = 1.57\ \mu\text{m}^3 = 1.57 \times 10^{-12}$ mL

Weight of cytosol = (specific gravity)(volume)

$= (1.2\ \text{g/mL})(1.57 \times 10^{-12}\ \text{mL}) = 1.88 \times 10^{-12}$ g

Average weight of each protein (1 in 1000, 20% wt/wt protein)

$= (1.88 \times 10^{-12}\ \text{g})(0.2)/(1000) = 3.77 \times 10^{-16}$ g

Average molar concentration of each protein

= (average weight)/(M_r)(volume)

$= (3.77 \times 10^{-16}\ \text{g})/(10^5\ \text{g/mol})(1.57 \times 10^{-12}\ \text{mL})(1\ \text{L}/1000\ \text{mL})$

$= 2.4 \times 10^{-6}\ \text{mol/L} \approx 2\ \mu\text{M}$

3. Rate Enhancement by Urease The enzyme urease enhances the rate of urea hydrolysis at pH 8.0 and 20 °C by a factor of 10^{14}. If a given quantity of urease can completely hydrolyze a given quantity of urea in 5 min at 20 °C and pH 8.0, how long would it take for this amount of urea to be hydrolyzed under the same conditions in the absence of urease? Assume that both reactions take place in sterile systems so that bacteria cannot attack the urea.

Answer

Time to hydrolyze urea

$$= \frac{(5\ \text{min})(10^{14})}{(60\ \text{min/h})(24\ \text{h/day})(365\ \text{days/yr})}$$

$= 9.5 \times 10^8$ yr

= 950 million years!

4. Protection of an Enzyme against Denaturation by Heat When enzyme solutions are heated, there is a progressive loss of catalytic activity over time due to denaturation of the enzyme. A solution of the enzyme hexokinase incubated at 45 °C lost 50% of its activity in 12 min, but when incubated at 45 °C in the presence of a very large concentration of one of its substrates, it lost only 3% of its activity in 12 min. Suggest why thermal denaturation of hexokinase was retarded in the presence of one of its substrates.

Answer One possibility is that the ES complex is more stable than the free enzyme. This implies that the ground state for the ES complex is at a lower energy level than that for the free enzyme, thus *increasing the height of the energy barrier* to be crossed in passing from the native to the denatured or unfolded state.

An alternative view is that enzymes undergo unfolding in two stages involving reversible conversion of native, active enzyme (N) to an unfolded, inactive state (U) followed by conversion to the irreversibly inactivated enzyme (I):

$$\text{N} \rightleftharpoons \text{U} \longrightarrow \text{I}$$

If substrate binds only to N, saturation with S to form NS complexes means that less free N is available for conversion to U or I, as the N $\rightleftharpoons$ U equilibrium is perturbed toward N. If N (free enzyme) but not NS complexes are converted to U or I, this will cause stabilization.

5. Requirements of Active Sites in Enzymes Carboxypeptidase, which sequentially removes carboxyl-terminal amino acid residues from its peptide substrates, is a single polypeptide of 307 amino acids. The two essential catalytic groups in the active site are furnished by Arg^{145} and Glu^{270}.

(a) If the carboxypeptidase chain were a perfect α helix, how far apart (in Ångstroms) would Arg^{145} and Glu^{270} be? (Hint: See Fig. 6–4b.)

(b) Explain how two amino acids, separated by this distance, can catalyze a reaction occurring in the space of a few Ångstroms.

Answer

(a) Arg^{145} is separated from Glu^{270} by (270 − 145) = 125 amino acid (AA) residues. From Figure 6–4b we see that the α helix has 3.6 AA/turn and increases in length along the major axis by 5.4 Å/turn. Thus, the distance between the two residues is

$$\frac{(125 \text{ AA})(5.4 \text{ Å/turn})}{3.6 \text{ AA/turn}} = 188 \text{ Å}$$

(b) Three-dimensional folding of the enzyme brings the two amino acid residues into close proximity.

6. Quantitative Assay for Lactate Dehydrogenase The muscle enzyme lactate dehydrogenase catalyzes the reaction

$$\underset{\text{Pyruvate}}{CH_3\text{—CO—COO}^-} + \text{NADH} + H^+ \longrightarrow \underset{\text{Lactate}}{CH_3\text{—CH(OH)—COO}^-} + \text{NAD}^+$$

NADH and NAD^+ are the reduced and oxidized forms, respectively, of the coenzyme NAD. Solutions of NADH, but *not* NAD^+, absorb light at 340 nm. This property is used to determine the concentration of NADH in solution by measuring spectrophotometrically the amount of light absorbed at 340 nm by the solution. Explain how these properties of NADH can be used to design a quantitative assay for lactate dehydrogenase.

Answer The reaction rate can be measured by following the decrease in NADH absorption at 340 nm as the reaction proceeds. Three pieces of information are required to develop a good quantitative assay for lactate dehydrogenase:

(i) Determine K_m values (see Box 8–1);

(ii) Measure the initial rate at several known concentrations of enzyme with saturating concentrations of NADH and pyruvate; and

(iii) Plot the initial rates as a function of [E]. This plot should be linear, with a slope that provides a measure of lactate dehydrogenase concentration.

7. Relation between Reaction Velocity and Substrate Concentration: Michaelis-Menten Equation

(a) At what substrate concentration will an enzyme with a k_{cat} of 30 s^{-1} and a K_m of 0.005 M show one-quarter of its maximum rate?

(b) Determine the fraction of V_{max} that would occur at the following substrate concentrations: $[S] = \frac{1}{2}K_m$, $2K_m$, and $10K_m$.

Answer

(a) Since $V_0 = V_{max}[S]/(K_m + [S])$, and $V_0 = 0.25(30\ s^{-1})$, $= 7.5\ s^{-1}$, we can substitute into the Michaelis-Menten equation to give

$$7.5\ s^{-1} = (30\ s^{-1}\ [S]/(5\ \text{mM} + [S])$$

$$[S] = 1.7\ \text{mM} = 1.7 \times 10^{-3}\ \text{M}$$

(b) We can rearrange the Michaelis-Menten equation into the form

$$V_0/V_{max} = [S]/(K_m + [S])$$

Substituting $[S] = \frac{1}{2}K_m$ into this equation gives

$$V_0/V_{max} = 0.33$$

Substituting [S] = $2K_m$ into the equation gives

$$V_0/V_{max} = 0.67$$

Finally, substituting [S] = $10K_m$ into the equation gives

$$V_0/V_{max} = 0.91$$

8. Estimation of V_{max} and K_m by Inspection Although graphical methods are available for accurate determination of the V_{max} and K_m of an enzyme-catalyzed reaction (see Box 8–1), sometimes these quantities can be quickly estimated by inspecting values of V_0 at increasing [S]. Estimate the approximate V_{max} and K_m of the enzyme-catalyzed reaction for which the following data were obtained:

[S] (M)	V_0 (μM/min)	[S] (M)	V_0 (μM/min)
2.5×10^{-6}	28	4×10^{-5}	112
4.0×10^{-6}	40	1×10^{-4}	128
1×10^{-5}	70	2×10^{-3}	139
2×10^{-5}	95	1×10^{-2}	140

Answer Notice how little the velocity changes as the substrate concentration increases by fivefold from 2 to 10 mM. Thus, we can estimate a V_{max} of 140 μM/min. K_m is defined as the substrate concentration that produces a velocity of $\frac{1}{2}V_{max}$, or 70 μM/min. Inspection of the table indicates that this V_0 occurs at [S] = 1×10^{-5} M ≈ K_m.

9. Properties of an Enzyme of Prostaglandin Synthesis Prostaglandins are a class of eicosanoids, fatty acid derivatives with a variety of extremely potent actions on vertebrate tissues, whose structure and action will be discussed further in Chapters 11 and 21. Prostaglandins are responsible for producing fever and inflammation and its associated pain. They are derived from the 20-carbon fatty acid arachidonic acid in a reaction catalyzed by the enzyme prostaglandin endoperoxide synthase. This enzyme, a cyclooxygenase, uses oxygen to convert arachidonic acid to PGG_2, the immediate precursor of many different prostaglandins (Chapter 21).

(a) The kinetic data given below are for the reaction catalyzed by prostaglandin endoperoxide synthase. Focusing here on the first two columns, determine the V_{max} and K_m of the enzyme.

[Arachidonic acid] (mM)	Rate of formation of PGG_2 (mM/min)	Rate of formation of PGG_2 with 10 mg/mL ibuprofen (mM/min)
0.5	23.5	16.67
1.0	32.2	25.25
1.5	36.9	30.49
2.5	41.8	37.04
3.5	44.0	38.91

(b) Ibuprofen is an inhibitor of prostaglandin endoperoxide synthase. By inhibiting the synthesis of prostaglandins, ibuprofen reduces inflammation and pain. Using the data in the first and third columns of the table, determine the type of inhibition that ibuprofen exerts on prostaglandin endoperoxide synthase.

Answer

(a) Calculate the reciprocal values for the data, as in parentheses below, and prepare a double-reciprocal plot to determine the kinetic parameters.

[S] (mM) (1/[S] (mM^{-1}))	V_0 (mM/min) (1/V_0 (min/mM))	V_0 with 10 mg/mL ibuprofen (mM/min) (1/V_0 (min/mM))
0.5 (2.0)	23.5 (0.043)	16.67 (0.06)
1.0 (1.0)	32.2 (0.031)	25.25 (0.0396)
1.5 (0.67)	36.9 (0.027)	30.49 (0.0328)
2.5 (0.4)	41.8 (0.024)	37.04 (0.027)
3.5 (0.27)	44.0 (0.023)	38.91 (0.0257)

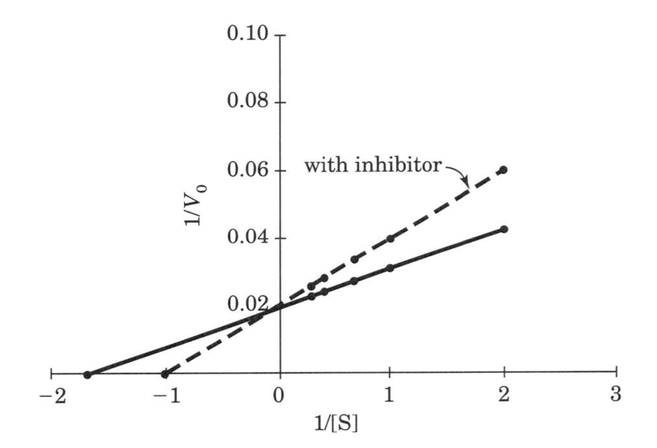

From the graph:

$$V_{max} = 51.55 \text{ mM/min}$$

$$K_m = 0.598 \text{ mM}$$

(b) Ibuprofen acts as a competitive inhibitor. The double-reciprocal plot (with inhibitor) shows that, in the presence of ibuprofen, the V_{max} of the reaction is unchanged (the intercept on the $1/V_0$ axis is the same) and K_m is increased ($-1/K_m$ is closer to the origin).

10. Graphical Analysis of V_{max} and K_m The following experimental data were collected during a study of the catalytic activity of an intestinal peptidase with the substrate glycylglycine:

$$\text{Glycylglycine} + H_2O \longrightarrow 2 \text{ glycine}$$

[S] (mM)	Product formed (μmol/min)	[S] (mM)	Product formed (μmol/min)
1.5	0.21	4.0	0.33
2.0	0.24	8.0	0.40
3.0	0.28	16.0	0.45

Use graphical analysis (see Box 8–1) to determine the K_m and V_{max} for this enzyme preparation and substrate.

Answer As described in Box 8–1, the standard method is to use V_0 versus [S] data to calculate $1/V_0$ and 1/[S].

V_0 (mg/min)	$1/V_0$ (min/mg)	[S] (mM)	1/[S] (mM^{-1})
0.21	4.76	1.5	0.67
0.24	4.17	2.0	0.50
0.28	3.57	3.0	0.33
0.33	3.03	4.0	0.25
0.40	2.50	8.0	0.13
0.45	2.22	16.0	0.06

Graphing these values gives a Lineweaver-Burk plot. From the best straight line through the data, the intercept on the horizontal axis $= -1/K_m$ and the intercept on the vertical axis $= 1/V_{max}$. From these values, we can calculate K_m and V_{max}:

$$K_m = 2.2 \text{ mM}$$

$$V_{max} = 0.51 \ \mu\text{mol/min}$$

11. The Eadie-Hofstee Equation One transformation of the Michaelis-Menten equation is the Lineweaver-Burk, or double-reciprocal, equation. Multiplying both sides of the Lineweaver-Burk equation by V_{max} and rearranging gives the Eadie-Hofstee equation:

$$V_0 = (-K_m)\frac{V_0}{[S]} + V_{max}$$

A plot of V_0 vs. $V_0/[S]$ for an enzyme-catalyzed reaction is shown on the next page. The curve with a slope of $-K_m$ (the "normal" curve) was obtained in the absence of inhibitor. Which of the other curves (A, B, or C) shows the enzyme activity when a competitive inhibitor is added to the reaction mixture?

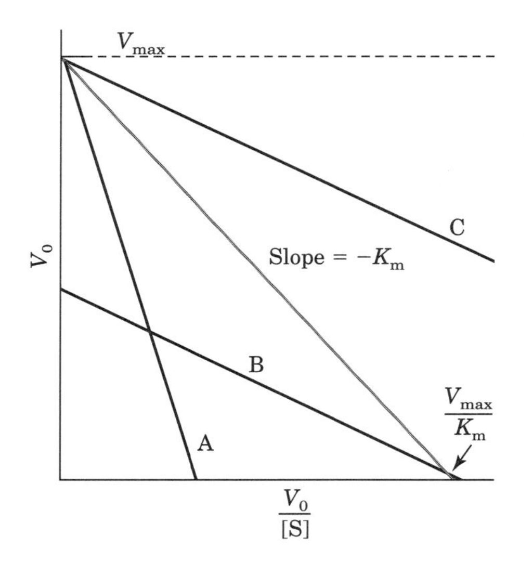

Answer Curve A shows competitive inhibition. V_{max} for A is the same as for the normal curve, as seen by the identical intercepts on the V_0 axis. And, for every value of [S] (until maximal velocity is reached at saturating substrate levels), V_0 is lower for curve A than for the normal curve, indicating competitive inhibition. Note that as [S] increases, V_0/[S] decreases, so that V_{max}—that is, the V_0 at the highest (saturating) [S]—is found at the intersection of the curve at the y-axis. Curve C, while also having an identical V_{max}, shows higher V_0 values for every [S] (and for every V_0/[S]) than for the normal reaction, which is not indicative of inhibition.

12. The Turnover Number of Carbonic Anhydrase Carbonic anhydrase of erythrocytes (M_r 30,000) has one of the highest turnover numbers among known enzymes. It catalyzes the reversible hydration of CO_2:

$$H_2O + CO_2 \rightleftharpoons H_2CO_3$$

This is an important process in the transport of CO_2 from the tissues to the lungs. If 10 μg of pure carbonic anhydrase catalyzes the hydration of 0.30 g of CO_2 in 1 min at 37 °C at V_{max}, what is the turnover number (k_{cat}) of carbonic anhydrase (in units of min^{-1})?

Answer The turnover number of an enzyme is the number of substrate molecules transformed per unit time by a single enzyme molecule (or a single catalytic site) when the enzyme is saturated with substrate:

$$k_{cat} = V_{max}/E_t$$

where E_t = total moles of active sites.

We can convert the values given in the problem into a turnover number (min^{-1}) by converting the weights of enzyme and substrate to molar amounts:

$$V_{max} \text{ (moles of } CO_2\text{/min)} = \frac{0.3 \text{ g/min}}{44 \text{ g/mol}} = 6.8 \times 10^{-3} \text{ mol/min}$$

$$\text{Amount of enzyme (moles)} = \frac{(10\ \mu\text{g})(1 \text{ g}/10^6\ \mu\text{g})}{30{,}000 \text{ g/mol}} = 3.3 \times 10^{-10} \text{ mol}$$

The turnover number is obtained by dividing moles of CO_2/min by moles of enzyme:

$$k_{cat} = \frac{6.8 \times 10^{-3} \text{ mol/min}}{3.3 \times 10^{-10} \text{ mol}} = 2.0 \times 10^7 \text{ min}^{-1}$$

13. Deriving a Rate Equation for Competitive Inhibition The rate equation for an enzyme subject to competitive inhibition is

$$V_0 = \frac{V_{max}[S]}{\alpha K_m + [S]}$$

Beginning with a new definition of total enzyme as

$$[E]_t = [E] + [EI] + [ES]$$

and the definitions of α and K_I on p. 266, derive the rate equation above. Use the derivation of the Michaelis-Menten equation in the text as a guide.

Answer The basic assumptions used to derive the Michaelis-Menten equation still hold. The reaction is at steady state, and the overall rate is determined by:

$$V_0 = k_2\,[ES] \qquad \text{(a)}$$

With the competitive inhibitor, I, now to be added, the goal again is to describe V_0 in terms of the measurable quantities $[E_t]$, [S], and [I]. In the presence of inhibitor,

$$[E_t] = [ES] + [E] + [EI] \qquad \text{(b)}$$

We first solve for [EI].

As we have seen,

$$K_I = \frac{[E][I]}{[EI]}; \text{ so } [EI] = \frac{[E][I]}{K_I}$$

substituting into (b), we get:

$$[E_t] = [ES] + [E] + \frac{[E][I]}{K_I} \qquad \text{(c)}$$

Simplifying gives:

$$[E_t] = [ES] + [E]\left(1 + \frac{[I]}{K_I}\right) = [ES] + [E]\alpha \qquad \text{(d)}$$

The term α describes the effect of the competitive inhibitor. The term [E] in the absence of inhibitor can be obtained from a rearrangement of Eqn. 8–19 (remembering that $[E_t]$ = [ES] + [E]), giving:

$$[E] = \frac{[ES]\,K_M}{[S]} \qquad \text{(e)}$$

Substituting (e) into (d) gives

$$[E_t] = [ES] + \left(\frac{[ES]\,K_M}{[S]}\right)\alpha \qquad \text{(f)}$$

Rearranging and solving for [ES] gives:

$$[ES] = \frac{[E_t][S]}{K_M\alpha + [S]} \qquad \text{(g)}$$

Substituting (g) into (a), and defining $k_2[E_t] = V_{max}$, gives the final equation for velocity in the presence of a competitive inhibitor:

$$V_0 = \frac{V_{max}\,[S]}{K_M\alpha + [S]} \qquad \text{(h)}$$

14. Irreversible Inhibition of an Enzyme Many enzymes are inhibited irreversibly by heavy-metal ions such as Hg^{2+}, Cu^{2+}, or Ag^{+}, which can react with essential sulfhydryl groups to form mercaptides:

$$\text{Enz—SH} + Ag^{+} \longrightarrow \text{Enz—S—Ag} + H^{+}$$

The affinity of Ag^{+} for sulfhydryl groups is so great that Ag^{+} can be used to titrate —SH groups quantitatively. To 10 mL of a solution containing 1.0 mg/mL of a pure enzyme, an investigator added just enough $AgNO_3$ to completely inactivate the enzyme. A total of 0.342 μmol of $AgNO_3$ was required. Calculate the *minimum* molecular weight of the enzyme. Why does the value obtained in this way give only the minimum molecular weight?

Answer An equivalency exists:

$$0.342 \times 10^{-3} \text{ mmol} = \frac{(10 \text{ mg/mL})(10 \text{ mL})}{(\text{minimum } M_r)\ (\text{mg/mmol})}$$

Thus the minimum $M_r = \dfrac{(10 \text{ mg/mL})(10 \text{ mL})}{0.342 \times 10^{-3} \text{ mmol}} = 2.9 \times 10^{4} = 29{,}000$

This calculation assumes that the enzyme contains only one titratable —SH group per molecule.

15. Clinical Application of Differential Enzyme Inhibition Human blood serum contains a class of enzymes known as acid phosphatases, which hydrolyze biological phosphate esters under slightly acidic conditions (pH 5.0):

$$\text{R—O—}PO_3^{2-} + H_2O \longrightarrow \text{R—OH} + \text{HO—}PO_3^{2-}$$

Acid phosphatases are produced by erythrocytes, the liver, kidney, spleen, and prostate gland. The enzyme from the prostate gland is clinically important because its increased activity in the blood is frequently an indication of prostate cancer. The phosphatase from the prostate gland is strongly inhibited by the tartrate ion, but acid phosphatases from other tissues are not. How can this information be used to develop a specific procedure for measuring the activity of the acid phosphatase of the prostate gland in human blood serum?

Answer First measure the *total* acid phosphatase activity in a blood sample in units of μmol of phosphate ester hydrolyzed per mL of serum. Then remeasure this activity in the presence of tartrate ion at a concentration sufficient to completely inhibit the enzyme from the prostate gland. The difference between the two activities represents the activity of acid phosphatase from the prostate gland.

16. Inhibition of Carbonic Anhydrase by Acetazolamide Carbonic anhydrase is strongly inhibited by the drug acetazolamide, which is used as a diuretic (to increase the production of urine) and to treat glaucoma (to reduce excessively high pressure in the eye due to accumulation of intraocular fluid). Carbonic anhydrase plays an important role in these and other secretory processes because it participates in regulating the pH and bicarbonate content of a number of body fluids. The experimental curve of initial reaction velocity (as percentage of V_{max}) versus [S] for the carbonic anhydrase reaction is illustrated on the next page (upper curve). When the experiment is repeated in the presence of acetazolamide, the lower curve is obtained. From an inspection of the curves and your knowledge of the kinetic properties of competitive and mixed enzyme inhibitors, determine the nature of the inhibition by acetazol-amide. Explain.

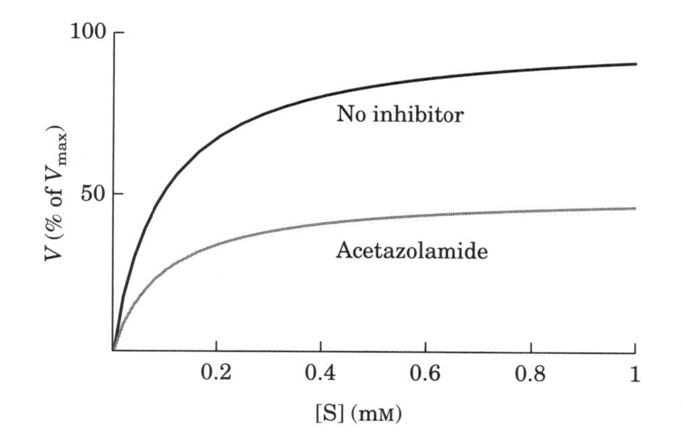

Answer The graph gives us several pieces of information. First, the inhibitor prevents the enzyme from achieving the same V_{max} as in the absence of inhibitor. Second, the overall shape of the two curves is very similar: at any [S] the ratio of the two velocities (±inhibitor) is the same. Third, the velocity does not change very much above [S] = 1 mM, so at much higher [S] the observed velocity is essentially V_{max} for each curve. Fourth, if we estimate the [S] at which $\frac{1}{2}V_{max}$ is achieved, this value is nearly identical for both curves. Noncompetitive inhibition, a special form of mixed inhibition that is rarely observed, alters the V_{max} of enzymes but leaves K_m unchanged. Thus, acetazolamide acts as a noncompetitive (mixed) inhibitor of carbonic anhydrase.

17. pH Optimum of Lysozyme The active site of lysozyme contains two amino acid residues essential for catalysis: Glu^{35} and Asp^{52}. The pK_a values of the carboxyl side chains of these two residues are 5.9 and 4.5, respectively. What is the ionization state (protonated or deprotonated) of each residue at pH 5.2, the pH optimum of lysozyme? How can the ionization states of these residues explain the pH-activity profile of lysozyme shown below?

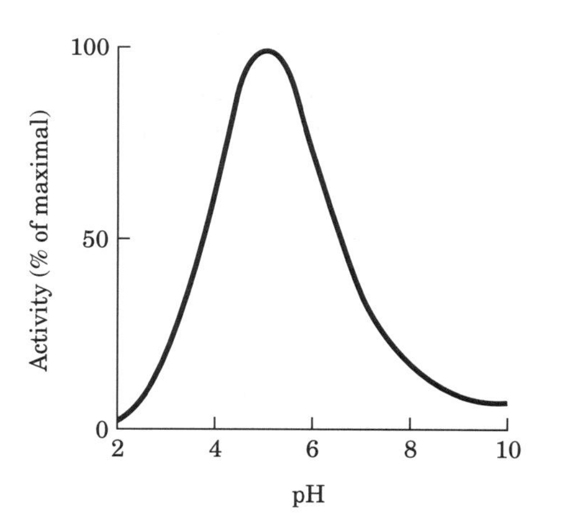

Answer At a pH midway between the two pK_a values (pH 5.2), the side-chain carboxyl group of Asp^{52}, with the lower pK_a (4.5), is mainly deprotonated (—COO^-), whereas Glu^{35}, with the higher pK_a (5.9; the stronger base) is protonated (—COOH). At pH values below 5.2, Asp^{52} becomes protonated and the activity decreases. Similarly, at pH values above 5.2, Glu^{35} becomes deprotonated and the activity also decreases. The pH-activity profile suggests that maximum catalytic activity occurs at a pH midway between the pK_a values of the two acidic groups, when Glu^{35} is protonated and Asp^{52} is deprotonated.

chapter

9 Carbohydrates and Glycobiology

1. **Determination of an Empirical Formula** An unknown substance containing only C, H, and O was isolated from goose liver. A 0.423 g sample produced 0.620 g of CO_2 and 0.254 g of H_2O after complete combustion in excess oxygen. Is the empirical formula of this substance consistent with its being a carbohydrate? Explain.

Answer The stoichiometry of the reaction must first be calculated. The reaction can be written

$C_?H_?O_?$	+	$?O_2$	$\longrightarrow$	CO_2	+	H_2O
(0.423 g)		(?)		(0.620 g)		(0.254 g)
(? mol)		(? mol)		(1.41×10^{-2} mol)		(1.41×10^{-2} mol)

where each ? represents a different unknown value. Because CO_2 and H_2O are produced in equal amounts (in moles), for every C produced the reaction also produces 2 H and 3 O. The simplest equation to describe this stoichiometry is

$$CH_2O + O_2 \longrightarrow CO_2 + H_2O$$

The empirical formula CH_2O is characteristic of a carbohydrate.

2. **Sugar Alcohols** In the monosaccharide derivatives known as sugar alcohols, the carbonyl oxygen is reduced to a hydroxyl group. For example, D-glyceraldehyde can be reduced to glycerol. However, this sugar alcohol is no longer designated D or L. Why?

Answer With reduction of the carbonyl oxygen to a hydroxyl group, the chemistry at C-1 and C-3 is the same; the glycerol molecule is not chiral.

3. **Melting Points of Monosaccharide Osazone Derivatives** Many carbohydrates react with phenylhydrazine ($C_6H_5NHNH_2$) to form bright yellow crystalline derivatives known as osazones:

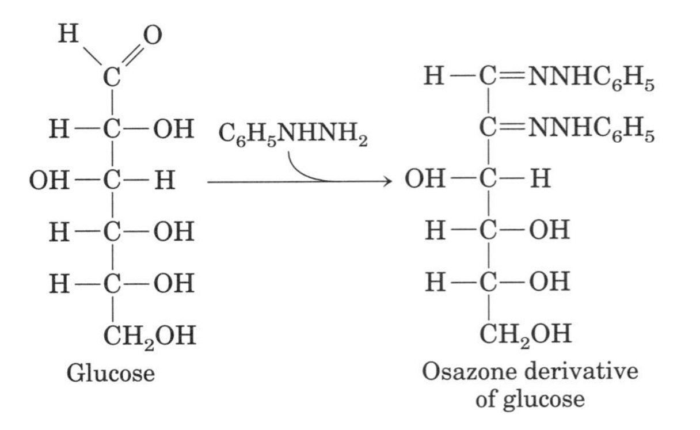

The melting temperatures of these derivatives are easily determined and are characteristic for each osazone. This information was used to help identify monosaccharides before the development of HPLC or gas-liquid chromatography. Listed below are the melting points (MPs) of some aldose-osazone derivatives:

Monosaccharide	MP of anhydrous monosaccharide (°C)	MP of osazone derivative (°C)
Glucose	146	205
Mannose	132	205
Galactose	165–168	201
Talose	128–130	201

As the table shows, certain pairs of derivatives have the same melting points, although the underivatized monosaccharides do not. Why do glucose and mannose, and galactose and talose, form osazone derivatives with the same melting points?

Answer Osazone formation destroys the configuration around C-2 of aldoses, so aldoses differing only at the C-2 configuration give the same derivative, with the same melting point. Glucose and mannose are C-2 epimers, as are galactose and talose.

4. Interconversion of D-Galactose Forms A solution of one stereoisomer of a given monosaccharide rotates plane-polarized light to the left (counterclockwise) and is called the levorotatory isomer, designated (−); the other stereoisomer rotates plane-polarized light to the same extent but to the right (clockwise) and is called the dextrorotatory isomer, designated (+). An equimolar mixture of the (+) and (−) forms does not rotate plane-polarized light.

The optical activity of a stereoisomer is expressed quantitatively by its *optical rotation,* the number of degrees by which plane-polarized light is rotated on passage through a given path length of a solution of the compound at a given concentration. The *specific rotation* $[\alpha]_D^{25°C}$ of an optically active compound is defined thus:

$$[\alpha]_D^{25°C} = \frac{\text{observed optical rotation (°)}}{\text{optical path length (dm)} \times \text{concentration (g/mL)}}$$

The temperature and the wavelength of the light employed (usually the D line of sodium, 589 nm) must be specified in the definition.

A freshly prepared solution of α-D-galactose (1 g/mL in a 10 cm cell) shows an optical rotation of +150.7°. Over time, the observed rotation of the solution gradually decreases and reaches an equilibrium value of +80.2°. In contrast, a freshly prepared solution (1 g/mL) of β-D-galactose shows an optical rotation of only +52.8°. Moreover, the rotation increases over time to an equilibrium value of +80.2°, identical to that reached by α-D-galactose.

(a) Draw the Haworth perspective formulas of the α and β forms of D-galactose. What feature distinguishes the two forms?

(b) Why does the optical rotation of a freshly prepared solution of the α form gradually decrease with time? Why do solutions of the α and β forms (at equal concentrations) reach the same optical rotation at equilibrium?

(c) Calculate the percentage composition of the two forms of D-galactose at equilibrium.

Answer

(a)

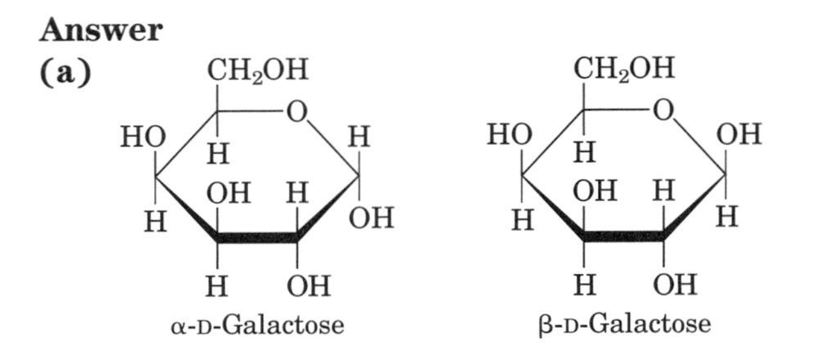

The two forms are distinguished by the configuration of the OH group on C-1, the anomeric carbon.

(b) A fresh solution of the α form of galactose undergoes *mutarotation* to an equilibrium mixture containing both the α and β forms. The same applies to a fresh solution of the β form.

(c) The change in optical rotation of a solution in changing from 100% α form (150.7°) to 100% β form (52.8°) is 97.9°. For an equilibrium mixture that rotates light 80.2°, the fraction of D-galactose in the α form is

$$\frac{80.2° - 52.8°}{150.7° - 52.8°} = \frac{27.4°}{97.9°} = 0.279 \approx 28\%$$

Thus the mixture contains 28% α-D-galactose and 72% β-D-galactose.

5. A Taste of Honey The fructose in honey is mainly in the β-D-pyranose form. This is one of the sweetest substances known, about twice as sweet as glucose. The β-D-furanose form of fructose is much less sweet. The sweetness of honey gradually decreases at a high temperature. Also, high-fructose corn syrup (a commercial product in which much of the glucose in corn syrup is converted to fructose) is used for sweetening *cold* but not *hot* drinks. What chemical property of fructose could account for both of these observations?

Answer Straight-chain fructose can cyclize to yield either the pyranose or the furanose structure. Increasing the temperature shifts the equilibrium in the direction of the furanose form, reducing the sweetness of the monosaccharide. The higher the temperature, the less sweet is the fructose solution.

6. Glucose Oxidase in Determination of Blood Glucose The enzyme glucose oxidase isolated from the mold *Penicillium notatum* catalyzes the oxidation of β-D-glucose to D-glucono-δ-lactone. This enzyme is highly specific for the β anomer of glucose and does not affect the α anomer. In spite of this specificity, the reaction catalyzed by glucose oxidase is commonly used in a clinical assay for total blood glucose—that is, for solutions consisting of a mixture of β-and α-D-glucose. How is this possible? Aside from allowing the detection of smaller quantities of glucose, what advantage does glucose oxidase offer over Fehling's reagent for the determination of blood glucose?

Answer The rate of mutarotation (interconversion of the α and β anomers) is sufficiently high that as the enzyme consumes β-D-glucose, more α-D-glucose is converted to the β form. Eventually all of the glucose is oxidized. Glucose oxidase is specific for glucose and does not detect other reducing sugars such as galactose that react with Fehling's reagent.

7. Invertase "Inverts" Sucrose The hydrolysis of sucrose (specific rotation +66.5°) yields an equimolar mixture of D-glucose (specific rotation +52.5°) and D-fructose (specific rotation −92°).

(a) Suggest a convenient way to determine the rate of hydrolysis of sucrose by an enzyme preparation extracted from the lining of the small intestine.

(b) Explain why an equimolar mixture of D-glucose and D-fructose formed by hydrolysis of sucrose is called invert sugar in the food industry.

(c) The enzyme invertase (its preferred name is now sucrase) is allowed to act on a solution of sucrose until the optical rotation of the solution becomes zero. What fraction of the sucrose has been hydrolyzed?

Answer

(a) When an equimolar mixture of D-glucose and D-fructose is completely hydrolyzed, its optical rotation is $52.5° + (-92°) = -39.5°$. Enzyme (sucrase) activity can be assayed by observing the change in optical rotation of a solution of 100% sucrose (optical rotation $+66.5°$) as it is converted to a 1:1 mixture of D-glucose and D-fructose.

(b) The optical rotation of the mixture is negative (inverted) relative to that of the sucrose solution.

(c) When sucrose is completely hydrolyzed, the change in optical rotation (from $+66.5°$ to $-39.5°$) is 106°. At zero optical rotation (0°), the change is 66.5°. Thus the fraction hydrolyzed $= 66.5°/106° = 0.63$.

8. Manufacture of Liquid-Filled Chocolates The manufacture of chocolates containing a liquid center is an interesting application of enzyme engineering. The flavored liquid center consists largely of an aqueous solution of sugars rich in fructose to provide sweetness. The technical dilemma is the following: the chocolate coating must be prepared by pouring hot melted chocolate over a solid (or almost solid) core, yet the final product must have a liquid, fructose-rich center. Suggest a way to solve this problem. (Hint: Sucrose is much less soluble than a mixture of glucose and fructose.)

Answer Prepare the core as a semi-solid slurry of sucrose and water. Add a small amount of sucrase, and quickly coat the semi-solid mixture with chocolate. After the chocolate coat has cooled and hardened, the sucrase will hydrolyze enough of the sucrose to reduce the viscosity of the mixture of glucose and fructose, converting it to a more nearly liquid consistency.

9. Anomers of Sucrose? Although lactose exists in two anomeric forms, no anomeric forms of sucrose have been reported. Why?

Answer Unlike lactose, sucrose has no free anomeric carbon to undergo mutarotation.

10. Physical Properties of Cellulose and Glycogen The almost pure cellulose obtained from the seed threads of *Gossypium* (cotton) is tough, fibrous, and completely insoluble in water. In contrast, glycogen obtained from muscle or liver disperses readily in hot water to make a turbid solution. Although they have markedly different physical properties, both substances are composed of (1→4)-linked D-glucose polymers of comparable molecular weight. What structural features of these two polysaccharides underlie their different physical properties? Explain the biological advantages of their respective properties.

Answer Native cellulose consists of glucose units linked by (β1→4) glycosidic bonds. The β linkages force the polymer chain into an extended conformation (see Fig. 9–17). Parallel series of these extended chains can form intermolecular hydrogen bonds, thus aggregating into long, tough, insoluble fibers. Glycogen consists of glucose units linked by (α1→4) glycosidic bonds. The α linkages cause bends in the chain and prevent the formation of long fibers. In addition, glycogen is highly branched and, because many of the glucose hydroxyl groups are exposed to water, is highly hydrated and therefore very water-soluble. It can be extracted as a dispersion in hot water.

The physical properties of the two polymers are well suited to their biological roles. Cellulose serves as a structural material in plants, consistent with its side-by-side aggregation into tough, insoluble fibers. Glycogen is a storage fuel in animals. The highly hydrated glycogen granules with their abundance of free, nonreducing ends can be rapidly hydrolyzed by glycogen phosphorylase to release glucose 1-phosphate.

11. Growth Rate of Bamboo The stems of bamboo, a tropical grass, can grow at the phenomenal rate of 0.3 m/day under optimal conditions. Given that the stems are composed almost entirely of cellulose fibers oriented in the direction of growth, calculate the number of sugar residues per second that must be added enzymatically to growing cellulose chains to account for the growth rate. Each D-glucose unit in the cellulose molecule is about 0.45 nm long.

Answer First calculate the growth per second:

$$\frac{0.3 \text{ m/day}}{(24 \text{ h/day})(60 \text{ min/h})(60 \text{ s/min})} = 3.47 \times 10^{-6} \text{ m/s}$$

Given that each glucose residue is 0.45 nm (0.45×10^{-9} m) long, the number of residues added per second is

$$\frac{3.47 \times 10^{-6} \text{ m/s}}{0.45 \times 10^{-9} \text{ m/residue}} = 7{,}716 \text{ residues/s} \approx 7{,}700 \text{ residues/s}$$

12. Glycogen as Energy Storage: How Long Can a Game Bird Fly? Since ancient times it has been observed that certain game birds, such as grouse, quail, and pheasants, are easily fatigued. The Greek historian Xenophon wrote, "The bustards . . . can be caught if one is quick in starting them up, for they will fly only a short distance, like partridges, and soon tire; and their flesh is delicious." The flight muscles of game birds rely almost entirely on the use of glucose 1-phosphate for energy, in the form of ATP (Chapter 15). In game birds, glucose 1-phosphate is formed by the breakdown of stored muscle glycogen, catalyzed by the enzyme glycogen phosphorylase. The rate of ATP production is limited by the rate at which glycogen can be broken down. During a "panic flight," the game bird's rate of glycogen breakdown is quite high, approximately 120 μmol/min of glucose 1-phosphate produced per gram of fresh tissue. Given that the flight muscles usually contain about 0.35% glycogen by weight, calculate how long a game bird can fly.

Answer The average molecular weight of a glucose unit in glycogen is 160. The amount of usable glucose (as glycogen) in 1 g of tissue is

$$\frac{0.0035 \text{ g}}{160 \text{ g/mol}} = 2.19 \times 10^{-5} \text{ mol}$$

In 1 min, 120 μmol of glucose 1-phosphate is produced, so 120 μmol of glucose is hydrolyzed. Thus depletion of the glycogen would occur in

$$\frac{2.19 \times 10^{-5} \text{ mol } (60 \text{ s/min})}{120 \times 10^{-6} \text{ mol/min}} = 10.9 \text{ s} \approx 11 \text{ s}$$

13. Volume of Chondroitin Sulfate in Solution One critical function of chondroitin sulfate is to act as a lubricant in skeletal joints by creating a gel-like medium that is resilient to friction and shock. This function appears to be related to a distinctive property of chondroitin sulfate: the volume occupied by the molecule is much greater in solution than in the dehydrated solid. Why is the volume occupied by the molecule so much larger in solution?

Answer The negative charges of chondroitin sulfate repel each other and force the molecule into an extended conformation. The polar molecule also attracts many water molecules (water of hydration), further increasing the molecular volume.

14. Heparin Interactions Heparin, a highly negatively charged glycosaminoglycan, is used clinically as an anticoagulant. It acts by binding several plasma proteins, including antithrombin III, an inhibitor of blood clotting. The 1:1 binding of heparin to antithrombin III appears to cause a conformational change in the protein that greatly increases its ability to inhibit clotting. What amino acid residues of antithrombin III are likely to interact with heparin?

Answer Positively charged amino acid residues would be the best candidates to bind to the highly negatively charged groups on heparin. In fact, Lys residues of anithrombin III interact with heparin.

15. Information Content of Oligosaccharides The carbohydrate portion of some glycoproteins may serve as a cellular recognition site. In order to perform this function, the oligosaccharide moiety of glycoproteins must have the potential to exist in a large variety of forms. Which can produce a greater variety of structures: oligopeptides composed of five different amino acid residues or oligosaccharides composed of five different monosaccharide residues? Explain.

Answer Oligosaccharides; their monosaccharide subunits can be combined in more ways than the amino acid subunits of oligopeptides. Each of the several hydroxyl groups of each monosaccharide can participate in glycosidic bonds, and the configuration of each glycosidic bond can be either α or β. Furthermore, the polymer can be linear or branched. Oligopeptides are unbranched polymers, with all amino acid units linked through identical peptide bonds.

16. Determination of the Extent of Branching in Amylopectin The extent of branching (number of (α1→6) glycosidic bonds) in amylopectin can be determined by the following procedure. A sample of amylopectin is exhaustively treated with a methylating agent (methyl iodide) that replaces all the hydrogens of the sugar hydroxyls with methyl groups, converting —OH to $—OCH_3$. All the glycosidic bonds in the treated sample are then hydrolyzed in aqueous acid. The amount of 2,3-di-*O*-methylglucose in the hydrolyzed sample is determined.

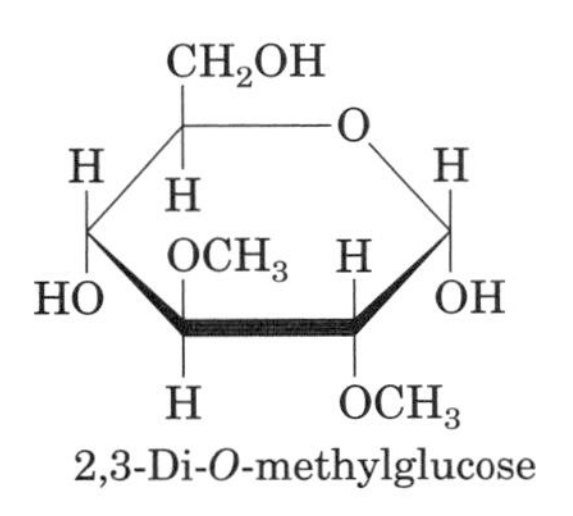

2,3-Di-*O*-methylglucose

(a) Explain the basis of this procedure for determining the number of (α1→6) branch points in amylopectin. What happens to the unbranched glucose residues in amylopectin during the methylation and hydrolysis procedure?

(b) A 258 mg sample of amylopectin treated as described above yielded 12.4 mg of 2,3-di-*O*-methylglucose. Determine what percentage of the glucose residues in amylopectin contain an (α1→6) branch.

Answer

(a) Glucose units at branch points are protected from complete methylation because the hydroxyl of C-6 is involved in branch formation. During methylation and subsequent hydrolysis, the branch-point residues yield 2,3-di-*O*-methylglucose and the unbranched residues yield 2,3,6-tri-*O*-methylglucose.

(b) The average molecular weight of glucose in amylopectin is 162. Thus 258 mg of amylopectin contains

$$\frac{258 \times 10^{-3}\ \text{g}}{162\ \text{g/mol}} = 1.59 \times 10^{-3}\ \text{mol of glucose}$$

The 12.4 mg yield of 2,3-di-*O*-methylglucose (M_r 208) is equivalent to

$$\frac{12.4 \times 10^{-3}\ \text{g}}{208\ \text{g/mol}} = 5.96 \times 10^{-5}\ \text{mol}$$

Thus the percentage of glucose residues in amylopectin that yield 2,3-di-*O*-methylglucose is

$$\frac{5.96 \times 10^{-5}\ \text{mol}\ (100)}{1.59 \times 10^{-3}\ \text{mol}} = 3.75\%$$

17. Structural Analysis of a Polysaccharide A polysaccharide of unknown structure was isolated, subjected to exhaustive methylation, and hydrolyzed. Analysis of the products revealed three methylated sugars in the ratio 20:1:1. The sugars were 2,3,4-tri-*O*-methyl-D-glucose; 2,4-di-*O*-methyl-D-glucose; and 2,3,4,6-tetra-*O*-methyl-D-glucose. What is the structure of the polysaccharide?

Answer Because the predominant product is 2,3,4-tri-*O*-methyl-D-glucose, the predominant linkage must be (1→6). The formation of 2,4-di-*O*-methyl-D-glucose indicates that the branch point occurs through C-3. The ratio of methylated sugars indicates that a branch occurs at a frequency of once every 20 residues. The 2,3,4,6-tetra-*O*-methyl-D-glucose is derived from nonreducing chain ends, which compose about 1/20 or 5% of the residues, also consistent with a high degree of branching. Thus the polysaccharide has chains of (1→6)-linked D-glucose residues with occasional (1→3)-linked branches, about one branch for every 20 residues.

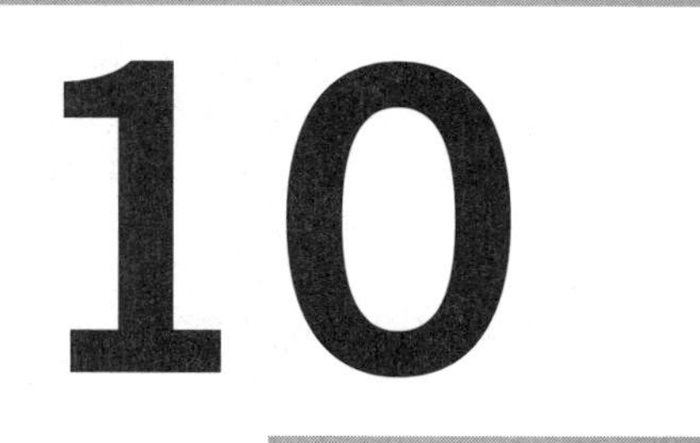

Nucleotides and Nucleic Acids

1. **Determination of Protein Concentration in a Solution Containing Proteins and Nucleic Acids** The concentration of protein or nucleic acid in a solution containing both can be estimated by using their different light absorption properties: proteins absorb most strongly at 280 nm and nucleic acids at 260 nm. Their respective concentrations in a mixture can be estimated by measuring the absorbance (A) of the solution at 280 nm and 260 nm and using the table below, which gives $R_{280/260}$, the ratio of the absorbances at 280 and 260 nm; the percentage of total mass that is nucleic acid; and a factor, F, that corrects the A_{280} reading and gives a more accurate protein estimate. The protein concentration (in mg/ml) = $F \times A_{280}$ (assuming the cuvette is 1 cm wide). Calculate the protein concentration in a solution of $A_{280} = 0.69$ and $A_{260} = 0.94$.

$R_{280/260}$	Proportion of nucleic acid (%)	F
1.75	0.00	1.116
1.63	0.25	1.081
1.52	0.50	1.054
1.40	0.75	1.023
1.36	1.00	0.994
1.30	1.25	0.970
1.25	1.50	0.944
1.16	2.00	0.899
1.09	2.50	0.852
1.03	3.00	0.814
0.979	3.50	0.776
0.939	4.00	0.743
0.874	5.00	0.682
0.846	5.50	0.656
0.822	6.00	0.632
0.804	6.50	0.607
0.784	7.00	0.585
0.767	7.50	0.565
0.753	8.00	0.545
0.730	9.00	0.508
0.705	10.00	0.478
0.671	12.00	0.422
0.644	14.00	0.377
0.615	17.00	0.322
0.595	20.00	0.278

Answer The ratio $R_{280/260} = 0.69/0.94 = 0.73$, so F is 0.508. The concentration of protein is $F \times A_{280} = 0.508 \times 0.69 = 0.35$ mg/ml. Note: The table applies to mixtures of proteins, such as might be found in a crude cellular extract, and reflects the absorption properties of average proteins. For a purified protein, the values of F would have to be altered to reflect the unique extinction coefficient of that protein.

2. **Nucleotide Structure** Which positions in a purine ring of a purine nucleotide in DNA have the potential to form hydrogen bonds but are not involved in Watson-Crick base pairing?

Answer All purine ring nitrogens (N-1, N-3, N-7, and N-9) have the potential to form hydrogen bonds (see Figs 10–1, 10–11, and 4–3). However, N-1 is involved in Watson-Crick hydrogen bonding with a pyrimidine, and N-9 is involved in the *N*-glycosidic linkage with deoxyribose and has very limited hydrogen-bonding capacity. Thus, N-3 and N-7 are available to form further hydrogen bonds.

3. **Base Sequence of Complementary DNA Strands** One strand of a double-helical DNA has the sequence (5′)GCGCAATATTTCTCAAAATATTGCGC(3′). Write the base sequence of the complementary strand. What special type of sequence is contained in this DNA segment? Does the double-stranded DNA have the potential to form any alternative structures?

Answer The complementary strand is (5′)GCGCAATATTTTGAGAAATATTGCGC(3′). (Note that the single strand sequence is always written in the 5′→3′ direction.) This sequence has a palindrome, an inverted repeat with twofold symmetry:

(5′)<u>GCGCAATATTT</u>CTCA<u>AAATATTGCGC</u>(3′)
(3′)<u>CGCGTTATAAA</u>GAGT<u>TTTATAACGCG</u>(5′)

Because this sequence is self-complementary, the individual strands have the potential to form hairpin structures. The two strands together may also form a cruciform.

4. **DNA of the Human Body** Calculate the weight in grams of a double-helical DNA molecule stretching from the earth to the moon (~320,000 km). The DNA double helix weighs about 1×10^{-18} g per 1,000 nucleotide pairs; each base pair extends 3.4 Å. For an interesting comparison, your body contains about 0.5 g of DNA!

Answer
The length of the DNA is

$$(320{,}000 \text{ km})\ (10^{12} \text{ nm/km})\ (10 \text{ Å/nm}) = 3.2 \times 10^{18} \text{ Å}$$

The number of base pairs (bp) is

$$\frac{3.2 \times 10^{18} \text{ Å}}{3.4 \text{ Å/bp}} = 9.4 \times 10^{17} \text{ bp}$$

Thus, the weight of the DNA molecule is

$$(9.4 \times 10^{17} \text{ bp})(1 \times 10^{-18} \text{ g}/10^{3} \text{ bp}) = 9.4 \times 10^{-4} \text{ g} = 0.00094 \text{ g}$$

5. **DNA Bending** Assume that a poly(A) tract five base pairs long produces a bend in a DNA strand of about 20°. Calculate the total (net) bend produced in a DNA if the center base pairs (the third of five) of two successive $(dA)_5$ tracts are located (a) 10 base pairs apart; (b) 15 base pairs apart. Assume 10 base pairs per turn in the DNA double helix.

Answer When bending elements are repeated in phase with the helix turn (that is, every 10 base pairs) as in **(a)**, the total bend is additive; when bending elements are repeated out of phase by one half a helix turn as in **(b)**, they cancel each other out. Thus the net bend is **(a)** 40°; **(b)** 0°.

6. Distinction between DNA Structure and RNA Structure Hairpins may form at palindromic sequences in single strands of either RNA or DNA. How is the helical structure of a long and fully base-paired (except at the end) hairpin in RNA different from that of a similar hairpin in DNA?

Answer The RNA helix assumes the A conformation; the DNA helix generally assumes the B conformation. (The presence of the 2′-OH group on ribose makes it sterically impossible for double-helical RNA to assume the B-form helix.)

7. Nucleotide Chemistry The cells of many eukaryotic organisms have highly specialized systems that specifically repair G–T mismatches in DNA. The mismatch is repaired to form a G≡C (not A═T) base pair. This G–T mismatch repair mechanism occurs in addition to a more general system that repairs virtually all mismatches. Can you suggest why cells might require a specialized system to repair G–T mismatches?

Answer Many C residues of CpG sequences in eukaryotic DNA are methylated at the 5′ position to 5-methylcytosine. (About 5% of all C residues are methylated.) Spontaneous deamination of 5-methylcytosine yields thymine, T, and a G–T mismatch resulting from spontaneous deamination of 5-methylcytosine in a G≡C base pair is one of the most common mismatches in eukaryotic cells. The specialized repair mechanism to convert G–T back to G≡C is directed at this common class of mismatch.

8. Nucleic Acid Structure Explain why the absorption of UV light increases (hyperchromic effect) when double-stranded DNA is denatured.

Answer The double-helical structure is stabilized by hydrogen bonding between complementary bases on opposite strands and base stacking between adjacent bases on the same strand. Base stacking in nucleic acids causes a decrease in the absorption of UV light. On denaturation of DNA, the base stacking is lost and UV absorption increases.

9. Base Pairing in DNA In samples of DNA isolated from two unidentified species of bacteria, adenine makes up 32% and 17%, respectively, of the total bases. What relative proportions of adenine, guanine, thymine, and cytosine would you expect to find in the two DNA samples? What assumptions have you made? One of these species was isolated from a hot spring (64 °C). Suggest which DNA came from this thermophilic bacterium. What is the basis for your answer?

Answer For any double-helical DNA, A = T and G = C. Because the G≡C base pair involves three hydrogen bonds and the A═T base pair involves only two, the higher the G + C content of a DNA molecule, the higher the melting temperature. The sample with 32% A must contain 32% T, 18% G, and 18% C. The sample with 17% A must contain 17% T, 33% G, and 33% C. This calculation is based on the assumption that both DNA molecules are double-stranded. The second DNA sample, with the higher G + C content (66%), most likely came from the thermophilic bacterium; it has a higher melting temperature and thus is more stable at the temperature of the hot spring.

10. DNA Sequencing The following DNA fragment was sequenced by the Sanger method. The asterisk indicates a fluorescent label.

```
*5′ ———— 3′-OH
 3′ ———— ATTACGCAAGGACATTAGAC---5′
```

A sample of the DNA was reacted with DNA polymerase and each of the nucleotide mixtures (in an appropriate buffer) listed below. Dideoxynucleotides (ddNTPs) were added in relatively small amounts.

1. dATP, dTTP, dCTP, dGTP, ddTTP
2. dATP, dTTP, dCTP, dGTP, ddGTP
3. dATP, dCTP, dGTP, ddTTP
4. dATP, dTTP, dCTP, dGTP

The resulting DNA was separated by electrophoresis on an agarose gel, and the fluorescent bands on the gel were located. The band pattern resulting from nucleotide mixture 1 is shown. Assuming that all mixtures were run on the same gel, what did the remaining lanes of the gel look like?

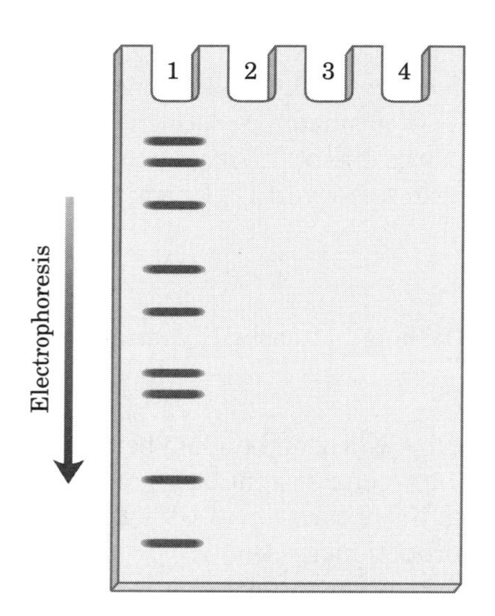

Answer
The reaction mixture that generated the fragments in **lane 1** included all the deoxynucleotides, plus dideoxythymidine. The fragments are of various lengths, all terminating where a ddTTP was substituted for a dTTP. For a small portion of the strands synthesized in the experiment, ddTTP would not be inserted and the strand would thus extend to the final G. Thus the products are

5′-primer-TAATGCGTTCCTGTAATCTG
5′-primer-TAATGCGTTCCTGTAATCT
5′-primer-TAATGCGTTCCTGTAAT
5′-primer-TAATGCGTTCCTGT
5′-primer-TAATGCGTTCCT
5′-primer-TAATGCGTT
5′-primer-TAATGCGT
5′-primer-TAAT

5′-primer-T

Similarly, **lane 2** would have the following four fragments, each terminating where ddGTP was inserted in place of dGTP:

5′-primer-TAATGCGTTCCTGTAATCTG

5′-primer-TAATGCGTTCCTG

5′-primer-TAATGCG

5′-primer-TAATG

Because mixture 3 lacked dTTP, every fragment was terminated immediately after the primer as ddTTP was inserted, to form 5′-primer-T. The result in **lane 3** is a thick band near the bottom of the gel. Finally, when all the deoxynucleotides were provided, as in mixture 4, a single labeled product formed: 5′-primer-TAATGCGTTCCTGTAATCTG. In **lane 4** this would be seen as a thick band at the top of the gel.

11. Solubility of the Components of DNA Draw the following structures and rate their relative solubilities in water (most soluble to least soluble): deoxyribose, guanine, phosphate. How are these solubilities consistent with the three-dimensional structure of double-stranded DNA?

Answer

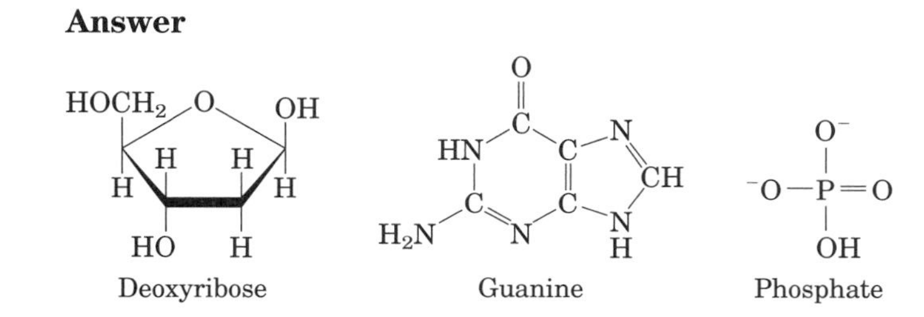

Solubilities: phosphate > deoxyribose > guanine. The negatively charged phosphate is the most water-soluble; the deoxyribose, with several hydroxyl groups, is quite water-soluble; and guanine, a hydrophobic base, is relatively insoluble in water. The polar phosphate groups and sugars are on the outside of the DNA double helix, exposed to water. The hydrophobic bases are located in the interior of the double helix, away from water.

12. Snake Venom Phosphodiesterase An exonuclease is an enzyme that sequentially cleaves nucleotides from the end of a polynucleotide strand. Snake venom phosphodiesterase, which hydrolyzes nucleotides from the 3′ end of any oligonucleotide with a free 3′-hydroxyl group, cleaves between the 3′ hydroxyl of the ribose or deoxyribose and the phosphoryl group of the next nucleotide. It acts on single-stranded DNA or RNA and has no base specificity. This enzyme was used in sequence determination experiments before the development of modern nucleic acid sequencing techniques. What are the products of partial digestion by snake venom phosphodiesterase of an oligonucleotide with the sequence (5′)GCGCCAUUGC(3′)—OH?

Answer When snake venom phosphodiesterase cleaves a nucleotide from a nucleic acid strand, it leaves the phosphoryl group attached to the 5′ position of the released nucleotide and a free 3′-OH group on the remaining strand. Partial digestion of the oligonucleotide gives a mixture of fragments of all lengths:

(5′)P—GCGCCAUUG(3′)—OH
(5′)P—GCGCCAUU(3′)—OH
(5′)P—GCGCCAU(3′)—OH
(5′)P—GCGCCA(3′)—OH
(5′)P—GCGCC(3′)—OH
(5′)P—GCGC(3′)—OH
(5′)P—GCG(3′)—OH
(5′)P—GC(3′)—OH

and the released nucleoside 5′-phosphates.

13. Preserving DNA in Bacterial Endospores Bacterial endospores form when the environment is no longer conducive to active cell metabolism. The soil bacterium *Bacillus subtilis,* for example, begins the process of sporulation when one or more nutrients are depleted. The end product is a small, metabolically dormant structure that can survive almost indefinitely with no detectable metabolism. Spores have mechanisms to prevent accumulation of potentially lethal mutations in their DNA over periods of dormancy that can exceed 1,000 years. *Bacillus subtilis* spores are much more resistant than the organism's growing cells to heat, UV radiation, and oxidizing agents, all of which promote mutations.

(a) One factor that prevents potential DNA damage in spores is their greatly decreased water content. How would this affect some types of mutations?

(b) Endospores have a category of proteins called small acid-soluble proteins (SASPs) that bind to their DNA, preventing formation of cyclobutane-type dimers. What causes cyclobutane dimers, and why do bacterial endospores need mechanisms to prevent their formation?

Answer

(a) Water is a participant in most biological reactions, including those that cause mutations. The low water content in endospores reduces the activity of mutation-causing enzymes and slows the rate of nonenzymatic depurination reactions, which are hydrolysis reactions.

(b) UV light induces the condensation of adjacent pyrimidine bases to form cyclobutane pyrimidine dimers. The spores of *B. subtilis,* a soil organism, are at constant risk of being lofted to the top of the soil or into the air, where they are subject to UV exposure, possibly for prolonged periods. Protection from UV-induced mutation is critical to spore DNA integrity.

Biochemistry on the Internet

14. The Structure of DNA Elucidation of the three-dimensional structure of DNA helped researchers to understand how this molecule conveys information that can be faithfully replicated from one generation to the next. To see the secondary structure of double-stranded DNA, go to the Protein Data Bank Web site. Use the PDB identifiers listed below to retrieve the data pages for the two forms of DNA. Open the structures using Chime and use the different viewing options to complete the following exercises.

(a) Obtain the file for 141D, a highly conserved DNA sequence from an HIV-1 long terminal repeat region. Display the molecule as a stick or ball-and-stick structure. Identify the sugar-phosphate backbone for each strand of the DNA duplex. Locate and identify individual bases. Which is the 5′ end of this molecule? Locate the major and minor grooves. Is this a right-or left-handed helix?

(b) Obtain the file for 145D, DNA in the Z conformation. Display the molecule as a stick or ball-and-stick structure. Identify the sugar-phosphate backbone for each strand of the DNA duplex. Is this a right-or left-handed helix?

(c) To fully appreciate the secondary structure of DNA, turn on the Stereo option in the viewer. You will see two images of the DNA molecule. Sit with your nose approximately 10 inches from the monitor and focus on the tip of your nose. In the background you should see three images of the DNA helix.

Shift your focus from the tip of your nose to the middle image, which should appear three-dimensional. For additional tips, see the Study Guide or *http://www.worthpublishers.com/lehninger*.

Answer

(a) The DNA fragment modeled in file 141D, from the human immunodeficiency virus, is in the B form, the standard Watson-Crick structure. To identify the sugar-phosphate backbone, turn on the **Backbone** display. To identify individual bases, return to the ball-and-stick structure and click on the relevant portion of the molecule; the corresponding base will be identified at the bottom of the frame. Alternatively, go the **Options** menu and click on **Labels** to display the names and numbers of every base in the structure. This fragment has an adenine at the 5′ end and a guanine at the 3′ end; click on the bases at each end of the helix to identify which is the 5′ end. When the helix is oriented so that the 5′ adenine is at the upper left-hand side of the model, the *minor groove* is in the center of the model. Rotating the model so that the 5′ adenine is at the upper right-hand side positions the *major groove* in the center. The spiral of this helix runs upward in a counterclockwise direction, so this is a right-handed helix.

(b) The model of DNA in the Z conformation includes a number of water molecules and several ligands. Note the shell of water molecules around the helix. Turn off the display of these water molecules in Chime by toggling the **Display Hetero Atoms** tool. You'll also find it helpful to toggle off the display of the hydrogen atoms. The backbone of DNA in the Z conformation is very different from that in the B conformation. Turn on the **Backbone** display to fully appreciate why this is referred to as the "Z" conformation. The spiral of the helix runs upward in a clockwise direction, so this is a left-handed helix.

(c) Viewing these structures in stereo takes a bit of practice, but perseverance will be rewarded! Here are some tips for successful three-dimensional viewing:

(1) Turn off or lower the room lighting.

(2) Sit directly in front of the screen.

(3) Use a ruler to make sure you are 10–11 inches from the screen.

(4) Position your head so that when you focus on the tip of your nose, the screen images are on either side of the tip (that is, look down your nose at the structures).

(5) Move your head slightly closer to or farther away from the screen to bring the middle image into focus. Don't look directly at the middle image as you try to bring it into focus.

(6) If you find it is uncomfortable to focus on the tip of your nose, try using the tip of a finger (positioned just beyond the tip of your nose) instead.

(7) Relax as you attempt to view the three-dimensional image.

Lipids

1. **Operational Definition of Lipids** How is the definition of "lipid" different from the types of definitions used for other biomolecules that we have considered, such as amino acids, nucleic acids, and proteins?

Answer The term "lipid" does not specify a particular chemical structure. Whereas one can write a general formula for an amino acid, nucleic acid, or protein, lipids are much more chemically diverse. Compounds are categorized as lipids based on their solubility: a greater solubility in organic solvents than in water.

2. **Melting Points of Lipids** The melting points of a series of 18-carbon fatty acids are stearic acid, 69.6 °C; oleic acid, 13.4 °C; linoleic acid, −5 °C; and linolenic acid, −11 °C.

(a) What structural aspect of these 18-carbon fatty acids can be correlated with the melting point? Provide a molecular explanation for the trend in melting points.

(b) Draw all of the possible triacylglycerols that can be constructed from glycerol, palmitic acid, and oleic acid. Rank them in order of increasing melting point.

(c) Branched-chain fatty acids are found in some bacterial membrane lipids. Would their presence increase or decrease the fluidity of the membranes (that is, make them have a lower or higher melting point)? Why?

Answer

(a) The number of cis double bonds. Each cis double bond causes a bend in the hydrocarbon chain, and bent chains are less well packed than straight chains in a crystal lattice. The lower the extent of packing, the lower is the melting temperature.

(b) Six different triacylglycerols are possible: one with glycerol and only palmitic acid (PPP); one with glycerol and only oleic acid (OOO); and four with glycerol and a mixture of oleic and palmitic acids. Four mixed triacylglycerols are possible because the three carbons of glycerol are not equivalent: thus OOP and OPO are positional isomers, as are POP and OPP. The greater the content of saturated fatty acid (P), the higher the melting point. Thus the order of melting points is OOO < OOP = OPO < POP = OPP < PPP. See Table 11–1 and Figure 11–2 for information on how to draw the triacylglycerols.

(c) Branched-chain fatty acids will increase the fluidity of membranes (that is, lower their melting point) because they decrease the extent of packing possible within the membrane. The effect of branches is similar to that of bends caused by double bonds.

3. Preparation of Béarnaise Sauce During the preparation of béarnaise sauce, egg yolks are incorporated into melted butter to stabilize the sauce and avoid separation. The stabilizing agent in the egg yolks is lecithin (phosphatidylcholine). Suggest why this works.

Answer Lecithin is an emulsifying agent, its amphipathic molecule solubilizing the fat (triacylglycerols) in butter. Lecithin is such a good emulsifying agent that it can be used to create a stable emulsion in a mixture that contains up to 75% oil. Mayonnaise is also an emulsion created with egg yolks, with an oil:vinegar mixture in a 3:1 ratio.

4. Hydrophobic and Hydrophilic Components of Membrane Lipids A common structural feature of membrane lipid molecules is their amphipathic nature. For example, in phosphatidylcholine, the two fatty acid chains are hydrophobic and the phosphocholine head group is hydrophilic. For each of the following membrane lipids, name the components that serve as the hydrophobic and hydrophilic units: (a) phosphatidylethanolamine; (b) sphingomyelin; (c) galactosylcerebroside; (d) ganglioside; (e) cholesterol.

Answer

	Hydrophobic unit(s)	Hydrophilic unit(s)
(a)	2 Fatty acids	Phosphoethanolamine
(b)	1 Fatty acid and the hydrocarbon chain of sphingosine	Phosphocholine
(c)	1 Fatty acid and the hydrocarbon chain of sphingosine	D-Galactose
(d)	1 Fatty acid and the hydrocarbon chain of sphingosine	Several sugar molecules
(e)	Steroid nucleus and acyl side chain	Alcohol group

5. Alkali Lability of Triacylglycerols A common procedure for cleaning the grease trap in a sink is to add a product that contains sodium hydroxide. Explain why this works.

Answer Triacylglycerols, a component of grease (consisting largely of animal fats), are hydrolyzed by NaOH to form glycerol and the sodium salts of free fatty acids, a process known as saponification. The fatty acids form micelles, which are more water-soluble than triacylglycerols.

6. The Action of Phospholipases The venom of the Eastern diamondback rattler and the Indian cobra contains phospholipase A_2, which catalyzes the hydrolysis of fatty acids at the C-2 position of glycerophospholipids. The phospholipid breakdown product of this reaction is lysolecithin (lecithin is phosphatidylcholine). At high concentrations, this and other lysophospholipids act as detergents, dissolving the membranes of erythrocytes and lysing the cells. Extensive hemolysis may be life-threatening.

(a) Detergents are amphipathic. What are the hydrophilic and hydrophobic portions of lysolecithin?

(b) The pain and inflammation caused by a snake bite can be treated with certain steroids. What is the basis of this treatment?

(c) Though high levels of phospholipase A_2 can be deadly, this enzyme is necessary for a variety of normal metabolic processes. What are these processes?

Answer

(a) The free OH group on C-2 and the phosphorylcholine head group on C-3 are the hydrophilic portions; the fatty acid on C-1 of the lysolecithin is the hydrophobic portion.

(b) Certain steroids such as prednisone inhibit the action of phospholipase A_2, the enzyme that releases the fatty acid arachidonic acid from the C-2 position. Arachidonic acid is converted to a variety of eicosanoids, some of which cause inflammation and pain.

(c) Phospholipase A_2 is necessary to release arachidonic acid, which is a precursor of other eicosanoids that have vital protective functions in the body. The enzyme is also important in digestion, breaking down dietary glycerophospholipids.

7. Intracellular Messengers from Phosphatidylinositols When the hormone vasopressin stimulates cleavage of phosphatidylinositol 4,5-bisphosphate by hormone-sensitive phospholipase C, two products are formed. Compare their properties and solubilities in water, and predict whether either would diffuse readily through the cytosol.

Answer Phosphatidylinositol 4,5-bisphosphate is a membrane lipid. The two products of cleavage are diacylglycerol and inositol 1,4,5-trisphosphate. Diacylglycerol is not water-soluble and remains in the membrane, acting as a second messenger. The inositol 1,4,5-trisphosphate is highly polar and very soluble in water; it readily diffuses in the cytosol, acting as a soluble second messenger.

8. Storage of Fat-Soluble Vitamins In contrast to water-soluble vitamins, which must be a part of our daily diet, fat-soluble vitamins can be stored in the body in amounts sufficient for many months. Suggest an explanation for this difference based on solubilities.

Answer Unlike water-soluble compounds, lipid-soluble compounds are not readily mobilized—that is, do not readily pass into aqueous solution. The body's lipids provide a reservoir for storage of lipid-soluble vitamins. Water-soluble vitamins cannot be stored and are rapidly removed from the blood by the kidneys.

9. Hydrolysis of Lipids Name the products of mild hydrolysis with dilute NaOH of **(a)** 1-stearoyl-2,3-dipalmitoylglycerol; **(b)** 1-palmitoyl-2-oleoylphosphatidylcholine

Answer Mild hydrolysis cleaves the ester linkages between glycerol and fatty acids, forming **(a)** glycerol and the sodium salts of palmitic and stearic acids; **(b)** D-glycerol 3-phosphorylcholine and the sodium salts of palmitic and oleic acids.

10. Effect of Polarity on Solubility Rank the following in order of increasing solubility in water: a triacylglycerol, a diacylglycerol, and a monoacylglycerol, all containing only palmitic acid.

Answer Solubilities: monoacylglycerol > diacylglycerol > triacylglycerol. Increasing the number of palmitic acid moieties increases the proportion of the molecule that is hydrophobic.

11. Chromatographic Separation of Lipids A mixture of lipids is applied to a silica gel column, and the column is then washed with increasingly polar solvents. The mixture consists of: phosphatidylserine, cholesteryl palmitate (a sterol ester), phosphatidylethanolamine, phosphatidylcholine, sphingomyelin, palmitate, *n*-tetradecanol, triacylglycerol, and cholesterol. In what order do you expect the lipids to elute from the column?

Answer Because silica gel is polar, the most hydrophobic lipids elute first, the most hydrophilic last. The neutral lipids elute first: cholesteryl palmitate and triacylglycerol. Cholesterol and *n*-tetradecanol, neutral but somewhat more polar, elute next. The neutral phospholipids phosphatidylcholine and phosphatidylethanolamine follow. Sphingomyelin, neutral but slightly more polar, elutes after the neutral phospholipids. The negatively charged phosphatidylserine and palmitate elute last, phosphatidylserine first because it is larger and has a lower charge-to-mass ratio.

12. Identification of Unknown Lipids Johann Thudichum, who practiced medicine in London about 100 years ago, also dabbled in lipid chemistry in his spare time. He isolated a variety of lipids from neural tissue, and characterized and named many of them. His carefully sealed and labeled vials of isolated lipids were rediscovered many years later.

(a) How would you confirm, using techniques not available to Thudichum, that the vials labeled "sphingomyelin" and "cerebroside" actually contain these compounds?

(b) How would you distinguish sphingomyelin from phosphatidylcholine by chemical, physical, or enzymatic tests?

Answer

(a) First, create an acid hydrolysate of each compound. Sphingomyelin yields sphingosine, fatty acids, phosphocholine, choline, and phosphate. Cerebroside yields sphingosine, fatty acids and sugars, but no phosphate. Subject each hydrolysate to chromatography (GLC or silica gel TLC) and compare the result with known standards.

(b) On strong alkaline hydrolysis, sphingomyelin yields sphingosine, fatty acid, and phosphocholine; phosphatidylcholine yields glycerol, fatty acids, and phosphocholine. The distinguishing features are the presence of *sphingosine* in sphingomyelin and *glycerol* in phosphatidylcholine, which can be detected on thin-layer chromatograms of each hydrolysate compared against known standards. The hydrolysates could also be distinguished by their reaction with the Sanger reagent (2,4-DNP): only the sphingosine in the sphingomyelin hydrolysate has a primary amine that would react with 2,4-DNP to form a colored product. Alternatively, enzymatic treatment of the two samples with phospholipase A_1 or A_2 would release free fatty acids from phosphatidylcholine, but not from sphingomyelin.

13. Ninhydrin to Detect Lipids on TLC Plates Ninhydrin reacts specifically with primary amines to form a purplish-blue product. A thin-layer chromatogram of rat liver phospholipids is sprayed with ninhydrin, and the color is allowed to develop. Which phospholipids can be detected this way?

Answer Phosphatidylethanolamine and phosphatidylserine; they are the only phospholipids that have primary amine groups that can react with ninhydrin.

chapter

12 Biological Membranes and Transport

1. **Determining the Cross-Sectional Area of a Lipid Molecule** When phospholipids are layered gently onto the surface of water, they orient at the air-water interface with their head groups in the water and their hydrophobic tails in the air. An experimental apparatus **(a)** has been devised that reduces the surface area available to a layer of lipids. By measuring the force necessary to push the lipids together, it is possible to determine when the molecules are packed tightly in a continuous monolayer; when that area is approached, the force needed to further reduce the surface area increases sharply **(b).** How would you use this apparatus to determine the average area occupied by a single lipid molecule in the monolayer?

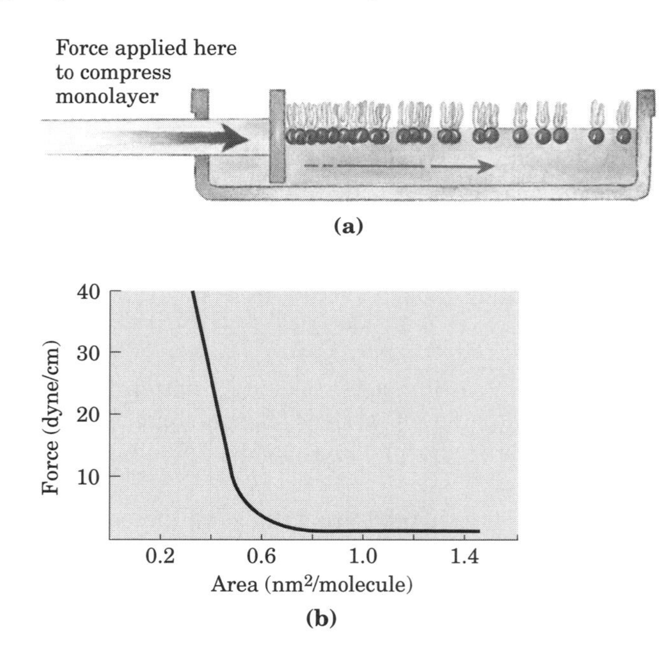

Answer Determine the surface area of the water at which the pressure increases sharply. Divide this surface area by the number of lipid molecules on the surface, which is calculated by multiplying the number of moles (calculated from the concentration and the molecular weight) by Avogadro's number.

2. **Evidence for Lipid Bilayer** In 1925, E. Gorter and F. Grendel used an apparatus like that described in Problem 1 to determine the surface area of a lipid monolayer formed by lipids extracted from erythrocytes of several animal species. They used a microscope to measure the dimensions of individual cells, from which they calculated the average surface area of one erythrocyte. They obtained the data shown on the next page. Were these investigators justified in concluding that "chromocytes [erythrocytes] are covered by a layer of fatty substances that is two molecules thick" (i.e., a lipid bilayer)?

Animal	Volume of packed cells (mL)	Number of cells (per mm^3)	Total surface area of lipid monolayer from cells (m^2)	Total surface area of one cell (μm^2)
Dog	40	8,000,000	62	98
Sheep	10	9,900,000	6.0	29.8
Human	1	4,740,000	0.92	99.4

Source: Data from Gorter, E. & Grendel, F. (1925) On bimolecular layers of lipoids on the chromocytes of the blood. *J. Exp. Med.* **41,** 439–443.

Answer The conclusions are justified for dog erythrocytes but not for sheep or human erythrocytes. The table provides the total surface area of the lipid monolayer. To determine the monolayer surface area per cell, first calculate the total number of cells. For example for dog erythrocytes, the number of cells is

$$8 \times 10^6 \text{ per mm}^3 = 8 \times 10^9 \text{ per cm}^3 \text{ (or per mL).}$$

In 40 mL, there is a total of (40 mL)(8×10^9 cells/mL) $= 3.2 \times 10^{11}$ cells. From the table, this number of cells yielded a monolayer surface area of 62 m^2 $= 6.2 \times 10^5$ cm^2. Dividing the surface area by the number of cells gives

$$\frac{6.2 \times 10^5 \text{ cm}^2}{3.2 \times 10^{11} \text{ cells}} = 1.9 \times 10^{-6} \text{ cm}^2\text{/cell}$$

Comparing this number to the total surface area of one erythrocyte (98 μm^2 $= 0.98 \times 10^{-6}$ cm^2), we find a 2 to 1 relationship. This result justifies the investigators' conclusion of a lipid bilayer in dog erythrocytes. Similar calculations for the sheep and human erythrocytes reveal a 1 to 1 relationship, suggesting a lipid monolayer. However, there were significant experimental errors in these early experiments; recent, more accurate measurements support a bilayer in all cases.

3. **Number of Detergent Molecules per Micelle** When a small amount of sodium dodecyl sulfate ($Na^+CH_3(CH_2)_{11}OSO_3^-$) is dissolved in water, the detergent ions enter solution as monomeric species. As more detergent is added, a concentration is reached (the critical micelle concentration) at which the monomers associate to form micelles. The critical micelle concentration of SDS is 8.2 mM. An examination of the micelles shows that they have an average particle weight (the sum of the molecular weights of the constituent monomers) of 18,000. Calculate the number of detergent molecules in the average micelle.

Answer The molecular weight of sodium dodecyl sulfate is 288. Given an average micelle particle weight of 18,000, there are 18,000/288 = 63 SDS molecules per micelle.

4. **Properties of Lipids and Lipid Bilayers** Lipid bilayers formed between two aqueous phases have this important property: they form two-dimensional sheets, the edges of which close upon each other and undergo self-sealing to form liposomes.

   (a) What properties of lipids are responsible for this property of bilayers? Explain.

   (b) What are the consequences of this property with regard to the structure of biological membranes?

Answer

(a) Lipids that form bilayers are amphipathic molecules: they contain a hydrophilic and a hydrophobic unit. In order to minimize the hydrophobic area that is exposed to the water surface, these lipids form two-dimensional sheets with the hydrophilic units exposed to water and the hydrophobic units buried in the interior of the sheet. Furthermore, to avoid exposing the hydrophobic edges of the sheet to water, lipid bilayers close upon themselves. Similarly, if the sheet is perforated, the hole will seal because the membrane is semifluid.

(b) These sheets form the closed membrane surfaces that envelop cells and compartments within cells (organelles).

5. Length of a Fatty Acid Molecule The carbon–carbon bond distance for single-bonded carbons such as those in a saturated fatty acyl chain is about 1.5 Å. Estimate the length of a single molecule of palmitate in its fully extended form. If two molecules of palmitate were placed end to end, how would their total length compare with the thickness of the lipid bilayer in a biological membrane?

Answer Given that the C–C bond length is 0.15 nm and that the bond angle of tetrahedral carbon is 109°, the distance between the first and third carbons in an acyl chain is about 0.26 nm. Thus the distance between two adjacent carbons is about 0.13 nm. For palmitic acid, the length is $16 \times 0.13 \text{ nm} \approx 2.1 \text{ nm}$. Thus two palmitic acids end to end (as in a bilayer) would extend 4.2 nm. This is slightly more than the thickness of a lipid bilayer (4 nm).

6. Temperature Dependence of Lateral Diffusion The experiment described in Figure 12–7 was performed at 37 °C. If the experiment were carried out at 10 °C, what effect would you expect on the rate of cell-cell fusion and the rate of membrane protein mixing? Why?

Answer When the temperature drops, the fluidity of a membrane decreases. This is caused by a decrease in the rate of diffusion of lipids and of the proteins associated with the lipids. Consequently, all processes depending on diffusion, such as cell-cell fusion and protein mixing, would slow down.

7. Synthesis of Gastric Juice: Energetics Gastric juice (pH 1.5) is produced by pumping HCl from blood plasma (pH 7.4) into the stomach. Calculate the amount of free energy required to concentrate the H^+ in 1 L of gastric juice at 37 °C. Under cellular conditions, how many moles of ATP must be hydrolyzed to provide this amount of free energy? The free-energy change for ATP hydrolysis under cellular conditions is about −58 kJ/mol (as we will explain in Chaper 14).

Answer Given that $\text{pH} = -\log[H^+]$, then $[H^+] = 10^{-\text{pH}} = -\text{antilog pH}$

At pH 1.5, $[H^+] = 10^{-1.5} = 3.16 \times 10^{-2}\ \text{M}$

At pH 7.4, $[H^+] = 10^{-7.4} = 3.98 \times 10^{-8}\ \text{M}$

$$\Delta G_t = RT \ln (C_2/C_1)$$

$$= (8.315 \text{ J/mol·K})(310 \text{ K}) \ln \left(\frac{3.16 \times 10^{-2}\ \text{M}}{3.98 \times 10^{-8}\ \text{M}}\right) = 35 \text{ kJ/mol}$$

The amount of ATP required to provide 35 kJ is

$$\frac{35 \text{ kJ}}{58 \text{ kJ/mol}} = 0.6 \text{ mol}$$

8. Energetics of the Na^+K^+ ATPase The concentration of Na^+ inside a vertebrate cell is about 12 mM, and that in blood plasma is about 145 mM. For a typical cell with a transmembrane potential of −0.07 V (inside negative relative to outside), what is the free-energy change for transporting 1 mol of Na^+ out of the cell and into the blood at 37 °C?

Answer

$$\Delta G_t = RT \ln (C_2/C_1) + Z\mathcal{F}\Delta\psi$$
$$= (8.315 \text{ J/mol·K})(310 \text{ K}) \ln \left(\frac{145 \text{ mM}}{12 \text{ mM}}\right) + (1)(96{,}480 \text{ J/V·mol})(0.07 \text{ V})$$
$$= 6.42 \text{ kJ/mol} + 6.75 \text{ kJ/mol} \approx 13.2 \text{ kJ/mol}$$

Note that 6.75 kJ/mol is the membrane potential portion.

9. Action of Ouabain on Kidney Tissue Ouabain specifically inhibits the Na^+K^+ ATPase activity of animal tissues but is not known to inhibit any other enzyme. When ouabain is added to thin slices of living kidney tissue, it inhibits oxygen consumption by 66%. Why? What does this observation tell us about the use of respiratory energy by kidney tissue?

Answer Oxidative phosphorylation to supply the cell with ATP accounts for the majority of oxygen consumption. A decrease in oxygen consumption by 66% on addition of ouabain indicates that consumption by the Na^+K^+ ATPase accounts for a large percentage of ATP produced, and thus the energy consumed, by kidney tissue.

10. Energetics of Symport Suppose that you determined experimentally that a cellular transport system for glucose, driven by symport of Na^+, could accumulate glucose to concentrations 25 times greater than in the external medium, while the external $[Na^+]$ was only ten times greater than the intracellular $[Na^+]$. Would this violate the laws of thermodynamics? If not, how could you explain this observation?

Answer No; the symport may be able to transport more than one equivalent of glucose per Na^+.

11. Location of a Membrane Protein The following observations are made on an unknown membrane protein, X. It can be extracted from disrupted erythrocyte membranes into a concentrated salt solution, and the isolated X can be cleaved into fragments by proteolytic enzymes. Treatment of erythrocytes with proteolytic enzymes followed by disruption and extraction of membrane components yields intact X. However, treatment of erythrocyte "ghosts" (which consist of only membranes, produced by disrupting the cells and washing out the hemoglobin) with proteolytic enzymes followed by disruption and extraction yields extensively fragmented X. What do these observations indicate about the location of X in the plasma membrane? Do the properties of X resemble those of an integral or peripheral membrane protein?

Answer Because protein X can be removed by salt treatment, it must be a peripheral membrane protein. Inability to digest the protein with proteases unless the membrane has been disrupted indicates that protein X is located internally, bound to the inner surface of the erythrocyte plasma membrane.

12. Membrane Self-Sealing Cellular membranes are self-sealing—if they are punctured or disrupted mechanically, they quickly and automatically reseal. What properties of membranes are responsible for this important feature?

Answer Hydrophobic interactions are the driving force for membrane formation. Because these forces are noncovalent and reversible, after membranes are disrupted they can be readily reannealed.

13. Lipid Melting Temperatures Membrane lipids in tissue samples obtained from different parts of the leg of a reindeer have different fatty acid compositions. Membrane lipids from tissue near the hooves contain a larger proportion of unsaturated fatty acids than those from tissue in the upper leg. What is the significance of this observation?

Answer The temperature of body tissues at the extremities, such as near the hooves, is generally lower than that of tissues closer to the center of the body. To maintain membrane fluidity, as required by the fluid-mosaic model, membranes at lower temperatures must contain a higher percentage of polyunsaturated fatty acids: unsaturated fatty acids lower the melting point of lipid mixtures.

14. Flip-Flop Diffusion The inner face (monolayer) of the human erythrocyte membrane consists predominantly of phosphatidylethanolamine and phosphatidylserine. The outer face consists predominantly of phosphatidylcholine and sphingomyelin. Although the phospholipid components of the membrane can diffuse in the fluid bilayer, this sidedness is preserved at all times. How?

Answer The energy required to flip a charged polar head group through a hydrophobic lipid bilayer is prohibitively high.

15. Membrane Permeability At pH 7, tryptophan crosses a lipid bilayer about 1,000 times more slowly than does the closely related substance indole:

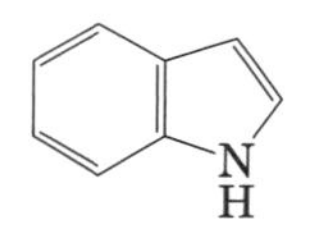

Suggest an explanation for this observation.

Answer At pH 7 tryptophan exists as a zwitterion (having both a positive and negative charge), whereas indole is uncharged. The movement of the less polar indole through a hydrophobic membrane is more energetically favorable.

16. Water Flow through an Aquaporin Each human erythrocyte has about 2×10^5 AQP-1 monomers. If water flows through the plasma membrane at a rate of 5×10^8 H_2O molecules per AQP-1 tetramer per second, and the volume of an erythrocyte is 5×10^{-11} mL, how rapidly could an erythrocyte halve its volume as it encounters the high osmolarity (1 M) in the interstitial fluid of the renal medulla?

Answer The volume of a red blood cell is 5×10^{-11} ml. If H_2O is 55 M, the amount of water in the cell can be calculated:

$(5 \times 10^{-11}$ml/cell$)(6.02 \times 10^{20}$ molecules of water/mmol$)(55$ mmol H_2O/ml$)$
$= 17 \times 10^{11}$ molecules H_2O/cell.

Half of these molecules (8.5×10^{11}) have to leave to reduce the cell to half its original volume, and they move through the 2×10^5 aquaporin molecule tetramers in a single cell.

Each tetramer of aquaporin passes 2×10^9 H_2O molecules per second, so the total flux through the aquaporin of one cell per second is

$(2 \times 10^9\ H_2O\ \text{molecules/s/aquaporin})(2 \times 10^5\ \text{aquaporin tetramers/cell})$
$= 4 \times 10^{14}\ H_2O\ \text{molecules/s}.$

To remove half the volume of water it takes $(8.5 \times 10^{11})/(4 \times 10^{14}\ H_2O/s) = 2.2 \times 10^{-3}$ s

17. Labeling the Galactoside Transporter The galactoside transporter of a bacterium, which is highly specific for its substrate lactose, contains a Cys residue that is essential to its transport activity. Covalent reaction of *N*-ethylmaleimide (NEM) with this Cys residue irreversibly inactivates the transporter. A high concentration of lactose in the medium prevents inactivation by NEM, presumably by sterically protecting the Cys residue, which is in or near the lactose binding site. You know nothing else about the transporter protein. Suggest an experiment that might allow you to determine the M_r of the Cys-containing transporter polypeptide.

Answer Treat a suspension of the bacteria as follows: Add lactose at a concentration well above the K_t, so that virtually every molecule of galactoside transporter binds lactose. Next add nonradiolabeled NEM and allow it to react with all available —SH groups on the cell surface. Remove excess lactose by centrifuging and resuspending the cells, then add radiolabeled NEM. The only Cys residues now available to react with NEM are those in the transporter protein. Dissolve the membrane proteins in sodium dodecylsulfate (SDS), and separate them on the basis of size by SDS gel electrophoresis. The M_r of the labeled band should represent that of the galactoside transporter.

18. Predicting Membrane Protein Topology from Sequence You have cloned the gene for a human erythrocyte protein, which you suspect is a membrane protein. From the nucleotide sequence of the gene, you know the amino acid sequence. From this sequence alone, how would you evaluate the possibility that your protein is an integral protein? Suppose the protein proves to be an integral protein, either type I or type II. Suggest biochemical or chemical experiments that might allow you to determine which type it is.

Answer Construct and analyze a hydropathy plot for the protein. You can assume that any hydrophobic regions of more than 20 consecutive residues are transmembrane segments. To determine whether the external domain is carboxyl- or amino-terminal, treat intact erythrocytes with a membrane-impermeant reagent known to react with primary amines (see Fig. 12–9) and determine whether the protein reacts. If it does, the amino terminus is on the external surface of the erythrocyte membrane and this is a type I protein (see Fig. 12–14).

19. Intestinal Uptake of Leucine You are studying the uptake of L-leucine by epithelial cells of the mouse intestine. Measurements of the rate of uptake of L-leucine and several of its analogs, with and without Na^+ in the assay buffer, yield the following results. What can you conclude about the properties and mechanism of the leucine transporter? Would you expect L-leucine uptake to be inhibited by ouabain?

Substrate	Uptake in presence of Na^+		Uptake in absence of Na^+	
	V_{max}	K_t (mM)	V_{max}	K_t (mM)
L-Leucine	420	0.24	23	0.24
D-Leucine	310	4.7	5	4.7
L-Valine	225	0.31	19	0.31

Answer The similar K_t values for L-leucine and L-valine indicate that the transporter binding site can accommodate the side chains of both amino acids equally well; it is probably a hydrophobic pocket of suitable size for either R group. The 20-fold higher K_t for D- than for L-leucine indicates that the binding site recognizes differences of configuration about the α carbon. Based on the lower V_{max} in the absence of Na^+, we know that Na^+ entry is essential for amino acid uptake; the transporter acts by symport of leucine (or valine) and Na^+.

20. Effect of an Ionophore on Active Transport Consider the leucine transporter described in Problem 20. Would V_{max} and/or K_t change if you added a Na^+ ionophore to the assay solution containing Na^+? Explain.

Answer By dissipating the transmembrane Na^+ gradient, the ionophore would prevent symport of L-leucine and reduce the rate of uptake, measured as V_{max}. K_t, a measure of the transporter's affinity for the substrate (L-leucine), should not change. Valinomycin (the likely ionophore here) does not resemble L-leucine in structure and almost certainly would not bind the transporter to affect K_t.

21. Surface Density of a Membrane Protein *E. coli* can be induced to make about 10,000 copies of the galactoside transporter (M_r 31,000) per cell. Assume that *E. coli* is a cylinder 1 μm in diameter and 2 μm long. What fraction of the plasma membrane surface is occupied by the galactoside transporter molecules? Explain how you arrived at this conclusion.

Answer The surface of a cylinder is $2\pi r^2 + \pi dh$, where r = radius, d = diameter, and h = height. For a cylinder 2 μm high and 1 μm in diameter, the surface area is $2.5\pi\ \mu m^2 = 7.9\ \mu m^2$. This is the *E. coli* surface area.

To estimate the cross-sectional area of a globular protein of M_r 31,000 we can use the dimensions for hemoglobin (M_r = 68,000; diameter = 5.5 nm; see text p. 210); thus a protein of M_r 31,000 has a diameter of about 3 nm, assuming the proteins have the same density. The cross-sectional area of a sphere of diameter 3 nm (0.003 μm), or of a single transporter molecule, is

$$\pi r^2 = 3.14(1.5 \times 10^{-3}\ \mu m)^2 = 7.1 \times 10^{-6}\ \mu m^2$$

and the total area of 10,000 transporter molecules is $7.1 \times 10^{-2}\ \mu m^2$. Thus the fraction of an *E. coli* cell surface covered by transporter molecules is

$$(7.1 \times 10^{-2}\ \mu m^2)/(7.9\ \mu m^2) = 0.009, \text{ or about } 1\%$$

This answer is clearly an approximation, given the method of estimating the diameter of the transporter molecule, but it is certainly of the right order of magnitude.

Biochemistry on the Internet

22. Membrane Protein Topology The receptor for the hormone epinephrine in animal cells is an integral membrane protein (M_r 64,000) that is believed to have seven membrane-spanning regions.

(a) Show that a protein of this size is capable of spanning the membrane seven times.

(b) Given the amino acid sequence of this protein, how would you predict which regions of the protein form the membrane-spanning helices?

(c) Go to the Protein Data Bank Web site. Use the PDB identifier 1DEP to retrieve the data page for a portion of the β-adrenergic receptor (one type of epinephrine receptor) from turkey. Using Chime to explore the structure, predict where this portion of the receptor is located: within the membrane or at the membrane surface. Explain.

(d) Retrieve the data for a portion of another receptor, the acetylcholine receptor of neurons and myocytes, using the PDB identifier 1A11. As in (c), predict where this portion of the receptor is located and explain your answer.

If you have not used the PDB or Chemscape Chime, you can find instructions at http://www.worthpublishers.com/lehninger.

Answer

(a) Assume that the transmembrane portion of the peptide is an α helix. The rise per amino acid (AA) residue of an α helix is 1.5 Å/AA = 0.15 nm/AA. The lipid bilayer is approximately 4 nm thick; to span this, (4 nm)/(0.15 nm/AA) = 27 AA are needed. Thus seven spans requires 7 × 27 = 189 residues. A protein of M_r 64,000 (if the average M_r of an amino acid in a peptide is 110) has approximately 64,000/110 = 580 AA residues.

(b) To identify potential transmembrane regions, hydropathy plots can be used (see Fig. 12-17 in the text). The most hydrophobic (hydropathic) stretches are those most likely to pass through the apolar lipid bilayer.

(c) This portion of the epinephrine receptor represents an intracellular loop that connects two adjacent membrane-spanning regions of the protein. You can predict that this α helix is not membrane-associated based on the properties of its amino acids. Using the Chime viewer first **Display** the structure in **Backbone** format and then **Select** the **Protein, Charged** residues and **Display** them in **Ribbon** format. It should be apparent that roughly half of the α helix consists of charged residues. It would be unlikely that such a structure would be stable within a lipid bilayer. In addition, this structure would be unlikely to span the lipid bilayer since there are only 15 amino acid residues (choose **Options → Labels** in Chime or look at the amino acid sequence using PDB summary), which are theoretically not sufficient to span the bilayer.

(d) This portion of the acetylcholine receptor represents one of the membrane-spanning regions of the protein. You can predict that this α helix is in fact membrane-associated based on the properties of its amino acids. Using the Chime viewer **Select** the **Protein → Hydrophobic** residues and **Display** them in **Ribbon** format. It should be apparent that the majority of the helix is composed of hydrophobic residues, which would be expected to be thermodynamically stable in a lipid environment. Displaying the charged residues shows that these are located on each end of the helical structure. In addition, if you choose **Options → Labels,** it is clear that there are enough hydrophobic amino acids between these charged residues (20 amino acids between the lysine in position 4 and arginine in position 25) in this structure to span a lipid bilayer.

chapter

13 Biosignaling

1. **Therapeutic Effects of Albuterol** The respiratory symptoms of asthma result from constriction of the bronchi and bronchioles of the lungs due to contraction of the smooth muscle of their walls. This constriction can be reversed by raising the [cAMP] in the smooth muscle. Explain the therapeutic effects of albuterol, a β-adrenergic agonist taken (by inhalation) for asthma. Would you expect this drug to have any side effects? How might one design a better drug that did not have these effects?

   Answer By mimicking the actions of epinephrine on smooth muscle, albuterol raises [cAMP], leading to relaxation and enlargement of the bronchi. Because β-adrenergic receptors control many processes other than smooth muscle relaxation in bronchi, drugs that act as β-adrenergic agonists generally have undesirable side effects. To minimize such effects, the goal is to find an agonist that is specific for the subtype of β-adrenergic receptors found in the smooth muscle of bronchi.

2. **Amplification of Hormonal Signals** Describe all the sources of amplification in the insulin receptor system.

   Answer The amplification results from catalysts activating catalysts. Two molecules of insulin can activate an insulin receptor dimer for a finite period, during which it can phosphorylate many molecules of IRS-1. One molecule of Raf phosphorylates many molecules of MEK, each of which phosphorylates and activates many molecules of MAPK. Each activated MAPK phosphorylates several molecules of transcription factor, and each of these molecules stimulates the transcription of multiple copies of mRNA for the genes under their control. Each mRNA can direct the synthesis of many copies of the protein it encodes.

3. **Termination of Hormonal Signals** Signals carried by hormones must eventually be terminated. Describe several different mechanisms for signal termination.

   Answer The hormone can be degraded by extracellular enzymes (such as acetylcholinesterase). The GTP bound to a G protein can be hydrolyzed to GDP. The second message can be degraded (for cAMP and cGMP), further metabolized (IP_3), or resequestered (Ca^{2+}, in the endoplasmic reticulum). The receptor may be desensitized (the acetylcholine receptor/channel), phosphorylated, bound to arrestin, removed from the surface (β-adrenergic receptor or rhodopsin).

4. Specificity of a Signal for a Single Cell Type Discuss the validity of the following proposition. A signaling molecule (hormone, growth factor, or neurotransmitter) elicits identical responses in different types of target cells if they contain identical receptors.

Answer The proposition is invalid because two cells expressing the same surface receptor for a given hormone may have different complements of target proteins for phosphorylation by protein kinases, resulting in different physiological and biochemical responses in the two cells.

5. Resting Membrane Potential A variety of unusual invertebrates, including giant clams, mussels, and polychaete worms live on the fringes of hydrothermal vents on the ocean bottom, where the temperature is 60 °C.

(a) The adductor muscle of a deep-sea giant clam has a resting membrane potential of −95 mV. Given the intracellular and extracellular ionic compositions shown below, would you have predicted this membrane potential? Why or why not?

Ion	Concentration (mM)	
	Intracellular	Extracellular
Na^+	50	440
K^+	400	20
Cl^-	21	560
Ca^{2+}	0.4	10

(b) Assuming that the adductor muscle membrane is permeable to only one of the ions listed above, which ion could determine the V_m?

Answer

(a) In most muscle cells at rest, the plasma membrane is permeable primarily to only K^+ ions. V_m is a function of the distribution of K^+ ions across the membrane. If V_m in the clam adductor muscle is determined primarily by K^+, then V_m at rest will be predicted by the Nernst equation using the values for $[K^+]$ given in the chart.

$$E_{ion} = (RT/Z\mathcal{F}) \ln (C_{out}/C_{in})$$
$$\begin{aligned} E_{K^+} &= [(8.315 \text{ J/mol·K})(333 \text{ K})/(1)(96{,}480 \text{ J/V·mol})] \ln (20 \text{ mM}/400 \text{ mM}) \\ &= [(2{,}769)/(96{,}480/\text{V})] \ln 0.05 \\ &= 0.0287 \text{ V} \times (-3.0) \\ &= -0.086 \text{ V or } -86 \text{ mV} \end{aligned}$$

Because the experimentally observed V_m is −95 mV, the plasma membrane in the adductor muscle must be permeable to some other ion or combination of ions.

(b) If the membrane is permeable to only one ion species, you can use the Nernst equation to calculate E for each ion above. The ion whose E is closest to the membrane potential will be the permeable ion that influences V_m.

$E_{K^+} = -86$ mV (see above)

$E_{Na^+} = 0.0287 \text{ V} \times \ln (440/50) = 0.062$ V or 62 mV

$E_{Cl^-} = (0.0287 \text{ V}/-1) \times \ln (560/21) = -0.094$ V or −94 mV
(Z for Cl^- is −1)

$E_{Ca^{2+}} = (0.0287 \text{ V}/2) \times \ln (10/0.4) = 0.046$ V or 46 mV
(Z for Ca^{2+} is 2)

$E_{Cl^-} = -94$ mV, very close to the resting V_m of −95 mV.

Since the calculated equilibrium potential for chloride is the same as the experimentally determined resting potential, it is likely that the membrane is permeable only to chloride ions at rest. You could verify this experimentally by changing the extracellular concentration of chloride ions and then measuring the effect on resting membrane potential. If this potential does depend only on chloride ions, the Nernst Equation should predict how the membrane potential will change.

6. **Membrane Potentials in Frog Eggs** Fertilization of a frog oocyte by a sperm cell triggers ionic changes similar to those observed in neurons (during movement of the action potential) and initiates the events that result in cell division and development of the embryo. Oocytes can be stimulated to divide without fertilization by suspending them in 80 mM KCl (normal pond water contains 9 mM KCl).
   (a) How does the change in extracellular [KCl] affect the resting membrane potential of the oocyte? (Hint: Assume the oocyte contains 120 mM K^+ and is permeable *only* to K^+.) Assume a temperature of 20 °C.
   (b) When the experiment is repeated in Ca^{2+}-free water, elevated [KCl] has no effect. What does this suggest about the mechanism of the KCl effect?

Answer

(a) $$V_m = \frac{RT}{Z\mathcal{F}} \ln \left(\frac{[K^+]_{out}}{[K^+]_{in}} \right)$$
$$= [(8.315 \text{ J/mol·K})(293 \text{ K})/(1)(96{,}480 \text{ J/V·mol})] \ln [K^+]_{out}/[K^+]_{in}$$
$$= (0.025 \text{ V}) \ln [K^+]_{out}/[K^+]_{in}$$
V_m in pond water $= 0.025$ V ln (9 mM/120 mM) $= -0.065$ V or -65 mV
V_m in high [KCl] $= 0.025$ V ln (80 mM/120 mM) $= -0.010$ V or -10 mV

The membrane of the oocyte has been *depolarized* by exposure to elevated extracellular $[K^+]$.

(b) This observation suggests that the effect of KCl depends on an influx of Ca^{2+} from the extracellular medium. KCl treatment must depolarize the oocyte sufficiently to open voltage-dependent Ca^{2+} channels in the oocyte membrane.

7. **Excitation Triggered by Hyperpolarization** In most neurons, membrane *depolarization* leads to the opening of voltage-dependent ion channels, generation of an action potential, and ultimately an influx of Ca^{2+}, which causes release of neurotransmitter at the axon terminus. Devise a cellular strategy by which *hyperpolarization* in rod cells could produce excitation of the visual pathway and passage of visual signals to the brain. (Hint: The neuronal signaling pathway in higher organisms consists of a series of neurons that relay information to the brain (see Fig. 13–20). The signal released by one neuron can be either excitatory or inhibitory to the following, postsynaptic neuron.)

Answer Hyperpolarization of rod cells in the retina occurs when the membrane potential, V_m, becomes negative. The negative V_m results in the closing of voltage-dependent Ca^{2+} channels in the presynaptic region of the rod cell. The resulting decrease in the intracellular levels of Ca^{2+} cause a corresponding decrease in the release of neurotransmitter by exocytosis. The neurotransmitter released by rod cells is actually an inhibitory neurotransmitter which results in the supression of activity in the next neuron of the visual circuit. When this inhibition is removed in response to a light stimulus the circuit becomes active and visual centers in the brain are excited.

8. Hormone Experiments in Cell-Free Systems In the 1950s, Earl W. Sutherland, Jr., and his colleagues carried out pioneering experiments to elucidate the mechanism of action of epinephrine and glucagon. Given what you have learned about hormone action in this chapter, interpret each of the experiments described below. Identify substance X and indicate the significance of the results.

(a) Addition of epinephrine to a homogenate of normal liver resulted in an increase in the activity of glycogen phosphorylase. However, if the homogenate was first centrifuged at a high speed and epinephrine or glucagon was added to the clear supernatant fraction that contains phosphorylase, no increase in the phosphorylase activity was observed.

(b) When the particulate fraction from the centrifugation in (a) was treated with epinephrine, substance X was produced. The substance was isolated and purified. Unlike epinephrine, substance X activated glycogen phosphorylase when added to the clear supernatant fraction of the centrifuged homogenate.

(c) Substance X was heat-stable; that is, heat treatment did not affect its capacity to activate phosphorylase. (Hint: Would this be the case if substance X were a protein?) Substance X was nearly identical to a compound obtained when pure ATP was treated with barium hydroxide. (Figure 10–6 will be helpful.)

Answer

(a) Because adenylate cyclase is a membrane-bound protein, centrifugation sediments it into the particulate fraction.

(b) Epinephrine stimulates the production of cAMP, a soluble substance that stimulates glycogen phosphorylase.

(c) Cyclic AMP is the heat-stable substance; it can be prepared by treating ATP with barium hydroxide.

9. Effect of Cholera Toxin on Adenylyl Cyclase The gram-negative bacterium *Vibrio cholerae* produces a protein, cholera toxin (M_r 90,000), that is responsible for the characteristic symptoms of cholera: extensive loss of body water and Na^+ through continuous, debilitating diarrhea. If body fluids and Na^+ are not replaced, severe dehydration results; untreated, the disease is often fatal. When the cholera toxin gains access to the human intestinal tract it binds tightly to specific sites in the plasma membrane of the epithelial cells lining the small intestine, causing adenylyl cyclase to undergo prolonged activation (hours or days).

(a) What is the effect of cholera toxin on [cAMP] in the intestinal cells?

(b) Based on the information above, suggest how cAMP normally functions in intestinal epithelial cells.

(c) Suggest a possible treatment for cholera.

Answer **(a)** It increases the level of cAMP. **(b)** The observations suggest that cAMP regulates Na^+ permeability. **(c)** Replace lost body fluids and electrolytes.

10. Effect of Dibutyryl cAMP versus cAMP on Intact Cells The physiological effects of epinephrine should in principle be mimicked by the addition of cAMP to the target cells. In practice, addition of cAMP to intact target cells elicits only a minimal physiological response. Why? When the structurally related derivative dibutyryl cAMP (shown below) is added to intact cells, the expected physiological response is readily apparent. Explain the basis for the difference in cellular response to these two substances. Dibutyryl cAMP is widely used in studies of cAMP function.

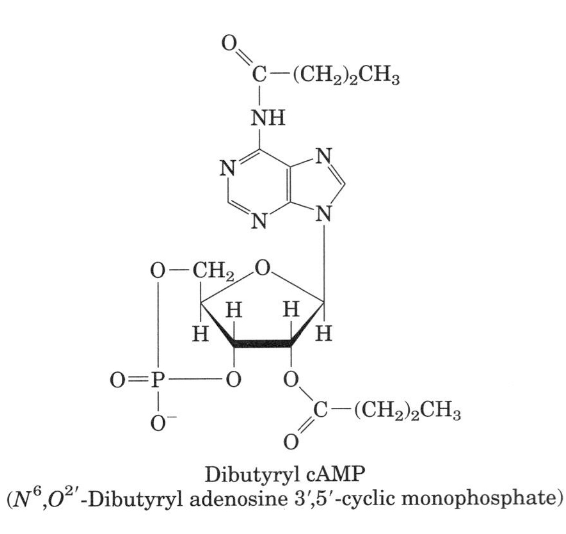

Dibutyryl cAMP
(N^6,$O^{2'}$-Dibutyryl adenosine 3′,5′-cyclic monophosphate)

Answer Unlike cAMP, dibutyryl-cAMP passes readily through the cell membrane.

11. Nonhydrolyzable GTP Analogs Many enzymes can hydrolyze GTP between the β and γ phosphates. The GTP analog β,γ-imidoguanosine 5′-triphosphate Gpp(NH)p, shown below, cannot be hydrolyzed between the β and γ phosphates. Predict the effect of microinjecting Gpp(NH)p into a myocyte on the cell's response to β-adrenergic stimulation.

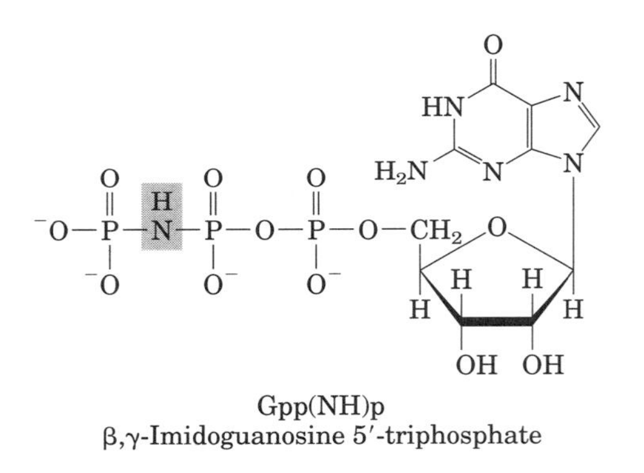

Gpp(NH)p
β,γ-Imidoguanosine 5′-triphosphate

Answer Nonhydrolyzable analogs of GTP have the effect of keeping the stimulatory G protein (G_s) in its activated form once it has encountered the receptor-hormone complex; it cannot shut itself off by converting the bound GTP (analog) to GDP. Injection of the analog would therefore be expected to prolong the effect of epinephrine on the injected cell.

12. G Protein Differences Compare the G proteins G_s, which act in transducing the signal from β-adrenergic receptors, and Ras. What properties do they share? How do they differ? What is the functional difference between G_s and G_i?

Answer Shared properties of Ras and G_s: both can bind either GDP or GTP; both are activated by GTP; both can, when active, activate a downstream enzyme; both have intrinsic GTPase activity that shuts them off after a short period of activation. Differences between Ras and G_s: Ras is a small, monomeric protein; G_s is trimeric. Functional differences: G_s activates adenylate cyclase, G_i inhibits it.

13. EGTA Injection EGTA (ethylene glycol-bis(β-aminoethyl ether)-N,N,N′,N′-tetraacetic acid) is a chelating agent with high affinity and specificity for Ca^{2+}. By microinjecting a cell with an appropriate Ca^{2+}-EDTA solution, an experimenter can prevent cytosolic [Ca^{2+}] from rising above 10^{-7} M. How would EGTA microinjection affect a cell's response to vasopressin (see Table 13–5)? To glucagon?

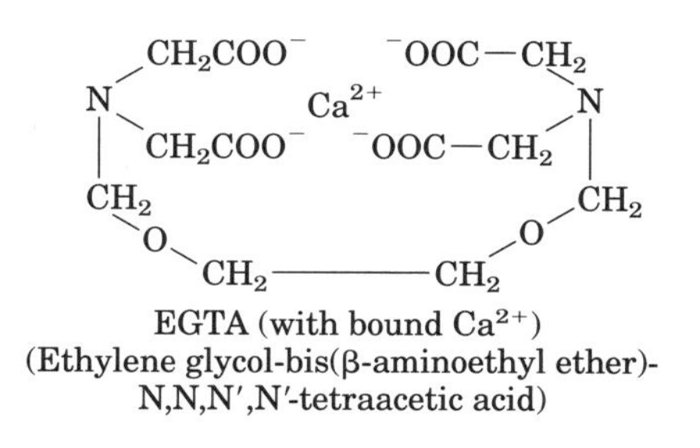

EGTA (with bound Ca^{2+})
(Ethylene glycol-bis(β-aminoethyl ether)-
N,N,N′,N′-tetraacetic acid)

Answer Vasopressin acts through a PLC-coupled serpentine receptor. The IP_3 released by PLC normally elevates cytosolic [Ca^{2+}] to 10^{-6} M, activating (with diacylglycerol) protein kinase C. Preventing this elevation of [Ca^{2+}] by using EGTA to "buffer" the internal [Ca^{2+}] will block vasopressin action, but should not directly affect the response to glucagon, which uses cAMP, *not* Ca^{2+}, as its intracellular messenger.

14. Visual Desensitization Oguchi's disease is an inherited form of night blindness. Affected individuals are slow to recover vision after a flash of bright light against a dark background, such as the headlights of a car on the freeway. Suggest what the molecular defect(s) might be in Oguchi's disease. Explain in molecular terms how this defect accounts for the night blindness.

Answer Some individuals with Oguchi disease have a defect in rhodopsin kinase that slows the process of recycling rhodopsin after its conversion to the all-trans form after illumination. This defect leaves retinal rod and cone cells insensitive for some time after a bright flash. Others have genetic defects in arrestin which prevent it from interacting with phosphorylated rhodopsin to trigger the process of internalization that leads to replacement of all-*trans*-retinal with 11-*cis*-retinal.

15. Mutations in PKA Explain how mutations in the R or C subunit of cAMP-dependent protein kinase (PKA) might lead to (a) a constantly active PKA or (b) a constantly inactive PKA.

Answer (a) If the mutation in the R subunit makes it unable to bind to the C subunit, the C subunit is never inhibited; it is constantly active. (b) If the mutation prevents the binding of cAMP to the R subunit but still allows normal R-C interaction, the inhibition of C by R cannot be relieved by elevated cAMP, and the enzyme is constantly inactive.

16. Mechanisms for Regulating Protein Kinases Identify eight general types of protein kinases found in eukaryotic cells, and explain what factor is *directly* responsible for activating each type.

Answer PKA (cAMP); PKG (cGMP); PKC (Ca^{2+}, DAG); Ca^{2+}/CaM kinase (Ca^{2+}, CaM); cyclin-dependent kinase (cyclin); protein tyrosine kinase (ligand for the receptor, such as insulin); MAP kinase (Raf); Raf (Ras); glycogen phosphorylase kinase (PKA).

17. Mutations in Tumor Suppressor Genes and Oncogenes Explain why mutations in tumor suppressor genes are recessive (both copies of the gene must be defective for the regulation of cell division to be defective) whereas mutations in oncogenes are dominant.

Answer An oncogene is a gene that normally encodes a cellular regulator protein, which, when defective, leads to unregulated cell growth and division. The role of the normal oncogene product is to send the signal to divide, but only when external or internal signals (such as growth factors) call for cell division. The mutant version of the oncogene product constantly sends the signal to divide, whether or not growth factors are present. A tumor suppressor gene in its normal cellular form encodes a protein that restrains cell division. Mutant forms of the protein fail to suppress cell division, but if either of the two alleles of the gene present encodes a normal protein, the normal function will continue. Only if both alleles of the tumor suppressor gene are defective will the suppression of cell division fail, leading to unregulated cell division.

18. Retinoblastoma in Children Explain why some children with retinoblastoma develop multiple tumors of the retina in both eyes, while others have a single tumor in only one eye.

Answer Children who develop multiple tumors in both eyes were born with every cell in the retina already having one defective copy of the *Rb* gene. Early in their lives, as retinal cells divide, there is a finite chance that one or several cells will each independently suffer a second mutation that damages the one remaining good copy of the *Rb* gene. Each of these cells will develop into a tumor. The later onset, single-tumor form of the disease occurs in children who at birth had two good copies of the *Rb* gene in every cell. They get a tumor because in a single cell, a mutation knocks out one copy of the *Rb* gene, and then a second mutation knocks out the second copy. Two mutations in the same gene in the same cell are so unlikely that this never happens twice in the same person, and these children therefore develop only one tumor, and in only one eye.

19. Mutations in *ras* How does a mutation in the *ras* gene that leads to a Ras protein with no GTPase activity affect a cell's response to insulin?

Answer When active, Ras activates the protein kinase Raf, which starts the cascade of protein kinases that includes the MAP kinases and leads to the phosphorylation of nuclear proteins. A mutation in *ras* that inactivates its GTPase activity creates a protein that, once activated by the binding of GTP, continues to give the signal to divide; it cannot shut itself off.

Principles of Bioenergetics

1. Entropy Changes during Egg Development Consider a system consisting of an egg in an incubator. The white and yolk of the egg contain proteins, carbohydrates, and lipids. If fertilized, the egg is transformed from a single cell to a complex organism. Discuss this irreversible process in terms of the entropy changes in the system, surroundings, and universe. Be sure that you first clearly define the system and surroundings.

Answer Consider the developing chick as the system. The nutrients, egg shell, and outside world are the surroundings. Transformation of the single cell into a chick drastically reduces the entropy of the system (increases the order). Initially, the parts of the egg outside the embryo (within the surroundings) contain complex fuel molecules (low entropy). During incubation, some of these complex molecules are converted into large numbers of CO_2 and H_2O molecules (high entropy). This increase in entropy of the surroundings is larger than the decrease in entropy of the chick (the system). Thus the entropy of the universe (the system + surroundings) increases.

2. Calculation of $\Delta G'^\circ$ from Equilibrium Constants Calculate the standard free-energy changes of the following metabolically important enzyme-catalyzed reactions at 25 °C and pH 7.0 from the equilibrium constants given.

(a) Glutamate + oxaloacetate $\xrightleftharpoons{\text{aspartate aminotransferase}}$ aspartate + α-ketoglutarate $\quad K'_{eq} = 6.8$

(b) Dihydroxyacetone phosphate $\xrightleftharpoons{\text{triose phosphate isomerase}}$ glyceraldehyde 3-phosphate $\quad K'_{eq} = 0.0475$

(c) Fructose 6-phosphate + ATP $\xrightleftharpoons{\text{phosphofructokinase}}$ fructose 1,6-bisphosphate + ADP $\quad K'_{eq} = 254$

Answer
$\Delta G = \Delta G'^\circ + RT \ln K'_{eq}$
At equilibrium $\Delta G = 0$ and $\Delta G'^\circ = -RT \ln K'_{eq}$
where $R = 8.315$ J/ mol·K and
$T = 25$ °C $= 298$ K, $RT = 2.479$ kJ/mol. We can now calculate the $\Delta G'^\circ$ values from the K'_{eq} for each reaction.
(a) $\Delta G'^\circ = -(2.479 \text{ kJ/mol}) \ln 6.8 = -4.75 \text{ kJ/mol}$
(b) $\Delta G'^\circ = -(2.479 \text{ kJ/mol}) \ln 0.0475 = 7.6 \text{ kJ/mol}$
(c) $\Delta G'^\circ = -(2.479 \text{ kJ/mol}) \ln 254 = -13.7 \text{ kJ/mol}$

3. Calculation of Equilibrium Constants from $\Delta G'^{\circ}$ Calculate the equilibrium constants K'_{eq} for each of the following reactions at pH 7.0 and 25 °C, using the $\Delta G'^{\circ}$ values of Table 14–4:

(a) Glucose 6-phosphate + $H_2O \xrightleftharpoons{\text{glucose 6-phosphatase}}$ glucose + P_i

(b) Lactose + $H_2O \xrightleftharpoons{\beta\text{-galactosidase}}$ glucose + galactose

(c) Malate $\xrightleftharpoons{\text{fumarase}}$ fumarate + H_2O

Answer

$\Delta G = \Delta G'^{\circ} + RT \ln K'_{eq}$

At equilibrium $\Delta G = 0$, so $\ln K'_{eq} = -\Delta G'^{\circ} /RT$

or $K'_{eq} = e^{-(\Delta G'^{\circ}/RT)}$ where $RT = 2.479$ kJ/mol at 25 °C (see Problem 2).

From these relationships, we can calculate K'_{eq} for each reaction using the values given for $\Delta G'^{\circ}$ in Table 14–4.

(a) For G 6-phosphatase:

$$\Delta G'^{\circ} = -13.8 \text{ kJ/mol}$$
$$\ln K'_{eq} = (-13.8 \text{ kJ/mol})/(2.479 \text{ kJ/mol}) = 5.567$$
$$K'_{eq} = 262 \text{ M}$$

(b) For β-galactosidase:

$$\Delta G'^{\circ} = -15.9 \text{ kJ/mol}$$
$$\ln K'_{eq} = (-15.9 \text{ kJ/mol})/(2.479 \text{ kJ/mol}) = 6.414$$
$$K'_{eq} = 610 \text{ M}$$

(c) For fumarase:

$$\Delta G'^{\circ} = 3.1 \text{ kJ/mol}$$
$$\ln K'_{eq} = -(3.1 \text{ kJ/mol})/(2.479 \text{ kJ/mol}) = -1.251$$
$$K'_{eq} = 0.29$$

4. Experimental Determination of K'_{eq} and $\Delta G'^{\circ}$ If a 0.1 M solution of glucose 1-phosphate is incubated with a catalytic amount of phosphoglucomutase, the glucose 1-phosphate is transformed to glucose 6-phosphate. At equilibrium, the concentrations of the reaction components are

$$\underset{4.5 \times 10^{-3}\text{ M}}{\text{Glucose 1-phosphate}} \rightleftharpoons \underset{9.6 \times 10^{-2}\text{ M}}{\text{glucose 6-phosphate}}$$

Calculate K'_{eq} and $\Delta G'^{\circ}$ for this reaction at 25 °C.

Answer

$$K'_{eq} = [\text{G6P}]/[\text{G1P}] = (9.6 \times 10^{-2} \text{ M})/(4.5 \times 10^{-3} \text{ M}) = 21.3 \approx 21$$
$$\Delta G'^{\circ} = -RT \ln K'_{eq} = -(2.479 \text{ kJ/mol})(\ln 21.3) = -7.6 \text{ kJ/mol}$$

5. Experimental Determination of $\Delta G'^\circ$ for ATP Hydrolysis A direct measurement of the standard free-energy change associated with the hydrolysis of ATP is technically demanding because the minute amount of ATP remaining at equilibrium is difficult to measure accurately. The value of $\Delta G'^\circ$ can be calculated indirectly, however, from the equilibrium constants of two other enzymatic reactions having less favorable equilibrium constants:

$$\text{Glucose 6-phosphate} + H_2O \longrightarrow \text{glucose} + P_i \qquad K'_{eq} = 270$$

$$\text{ATP} + \text{glucose} \longrightarrow \text{ADP} + \text{glucose 6-phosphate} \qquad K'_{eq} = 890$$

Using this information, calculate the standard free energy of hydrolysis of ATP at 25 °C.

Answer The reactions, if coupled together, constitute a "futile cycle" that results in the net hydrolysis of ATP:

(1) $\text{G6P} + H_2O \longrightarrow \text{glucose} + P_i$
(2) $\text{ATP} + \text{glucose} \longrightarrow \text{ADP} + \text{G6P}$

Sum: $\text{ATP} + H_2O \longrightarrow \text{ADP} + P_i$

$$\Delta G'^\circ = -RT \ln K'_{eq}$$

$$\Delta G_1'^\circ = (-2.479 \text{ kJ/mol})(\ln 270) = -13.9 \text{ kJ/mol}$$

$$\Delta G_2'^\circ = (-2.479 \text{ kJ/mol})(\ln 890) = -16.8 \text{ kJ/mol}$$

$$\Delta G'^\circ_{sum} = \Delta G_1'^\circ + \Delta G_2'^\circ = -30.7 \text{ kJ/mol}$$

6. Difference between $\Delta G'^\circ$ and ΔG Consider the following interconversion, which occurs in glycolysis (Chapter 15):

$$\text{Fructose 6-phosphate} \rightleftharpoons \text{glucose 6-phosphate} \qquad K'_{eq} = 1.97$$

(a) What is $\Delta G'^\circ$ for the reaction (assuming that the temperature is 25 °C)?

(b) If the concentration of fructose 6-phosphate is adjusted to 1.5 M and that of glucose 6-phosphate is adjusted to 0.5 M, what is ΔG?

(c) Why are $\Delta G'^\circ$ and ΔG different?

Answer

(a) At equilibrium,

$$\Delta G'^\circ = -RT \ln K'_{eq} = -(2.479 \text{ kJ/mol}) \ln 1.97 = -1.68 \text{ kJ/mol} \approx -1.7 \text{ kJ/mol}$$

(b) By definition, $\Delta G = \Delta G'^\circ + RT \ln K'_{eq}$

$$K'_{eq} = [\text{G6P}]/[\text{F6P}] = 0.5 \text{ M}/1.5 \text{ M} = 0.33$$

$$\Delta G = -1.68 \text{ kJ/mol} + (2.479 \text{ kJ/mol}) \ln 0.33 = -4.4 \text{ kJ/mol}$$

(c) $\Delta G'^\circ$ for any reaction is a fixed parameter, since it is defined for standard conditions of temperature (25 °C = 298 K) and concentration (both F6P and G6P at 1 M). In contrast, ΔG is a variable and can be calculated for any set of product and reactant concentrations. ΔG is defined as $\Delta G'^\circ$ (standard conditions) plus whatever differences occur in this parameter upon moving to nonstandard conditions.

7. Dependence of ΔG on pH The free energy released by the hydrolysis of ATP under standard conditions at pH 7.0 is −30.5 kJ/mol. If ATP is hydrolyzed under standard conditions but at pH 5.0, is more or less free energy released? Why?

Answer Less; the overall equation for ATP hydrolysis can be approximated as:

$$ATP^{4-} + H_2O \rightleftharpoons ADP^{3-} + HPO_4^{2-} + H^+$$

Because H^+ ions are produced in the reaction, if the pH at which this reaction is carried out is lower—that is, as $[H^+]$ increases—the equilibrium is shifted to the left (toward the reactants). As a result, at lower pH values the reaction does not proceed as far toward products, and less free energy is released.

8. The $\Delta G'^\circ$ for Coupled Reactions Glucose 1-phosphate is converted into fructose 6-phosphate in two successive reactions:

$$\text{Glucose 1-phosphate} \longrightarrow \text{glucose 6-phosphate}$$
$$\text{Glucose 6-phosphate} \longrightarrow \text{fructose 6-phosphate}$$

Using the $\Delta G'^\circ$ values in Table 14–4, calculate the equilibrium constant, K'_{eq}, for the sum of the two reactions at 25 °C:

$$\text{Glucose 1-phosphate} \longrightarrow \text{fructose 6-phosphate}$$

Answer

$$\begin{array}{lll}
(1) & \text{G1P} \longrightarrow \text{G6P} & \Delta G_1'^\circ = -7.3 \text{ kJ/mol} \\
(2) & \text{G6P} \longrightarrow \text{F6P} & \Delta G_2'^\circ = 1.7 \text{ kJ/mol} \\
\hline
\text{Sum:} & \text{G1P} \longrightarrow \text{F6P} & \Delta G_{sum}'^\circ = -5.6 \text{ kJ/mol}
\end{array}$$

$$\begin{aligned}
\ln K'_{eq} &= -\Delta G'^\circ/RT \\
&= -(-5.6 \text{ kJ/mol})/(2.479 \text{ kJ/mol}) \\
&= 2.259 \\
K'_{eq} &= 9.57 \approx 9.6
\end{aligned}$$

9. Strategy for Overcoming an Unfavorable Reaction: ATP-Dependent Chemical Coupling The phosphorylation of glucose to glucose 6-phosphate is the initial step in the catabolism of glucose. The direct phosphorylation of glucose by P_i is described by the equation

$$\text{Glucose} + P_i \longrightarrow \text{glucose 6-phosphate} + H_2O \qquad \Delta G'^\circ = 13.8 \text{ kJ/mol}$$

(a) Calculate the equilibrium constant for the above reaction. In the rat hepatocyte the physiological concentrations of glucose and P_i are maintained at approximately 4.8 mM. What is the equilibrium concentration of glucose 6-phosphate obtained by the direct phosphorylation of glucose by P_i? Does this reaction represent a reasonable metabolic step for the catabolism of glucose? Explain.

(b) In principle, at least, one way to increase the concentration of glucose 6-phosphate is to drive the equilibrium reaction to the right by increasing the intracellular concentrations of glucose and P_i. Assuming a fixed concentration of P_i at 4.8 mM, how high would the intracellular concentration of glucose have to be to give an equilibrium concentration of glucose 6-phosphate of 250 μM (normal physiological concentration)? Would this route be physiologically reasonable, given that the maximum solubility of glucose is less than 1 M?

(c) The phosphorylation of glucose in the cell is coupled to the hydrolysis of ATP; that is, part of the free energy of ATP hydrolysis is utilized to effect the endergonic phosphorylation of glucose:

(1)	Glucose + $P_i \longrightarrow$ glucose 6-phosphate + H_2O	$\Delta G'° = 13.8$ kJ/mol
(2)	ATP + $H_2O \longrightarrow$ ADP + P_i	$\Delta G'° = -30.5$ kJ/mol
Sum:	Glucose + ATP $\longrightarrow$ glucose 6-phosphate + ADP	

Calculate K'_{eq} for the overall reaction. For the ATP-dependent phosphorylation of glucose, what concentration of glucose is needed to achieve a 250 μM intracellular concentration of glucose 6-phosphate when the concentrations of ATP and ADP are 3.38 and 1.32 mM, respectively? Does this coupling process provide a feasible route, at least in principle, for the phosphorylation of glucose in the cell? Explain.

(d) Although coupling ATP hydrolysis to glucose phosphorylation makes thermodynamic sense, how this coupling is to take place has not been specified. Given that coupling requires a common intermediate, one conceivable route is to use ATP hydrolysis to raise the intracellular concentration of P_i and thus drive the unfavorable phosphorylation of glucose by P_i. Is this a reasonable route? (Think about the solubility products of metabolic intermediates.)

(e) The ATP-coupled phosphorylation of glucose is catalyzed in hepatocytes by the enzyme glucokinase. This enzyme binds ATP and glucose to form a glucose-ATP-enzyme complex, and the phosphoryl group is transferred directly from ATP to glucose. Explain the advantages of this route.

Answer

(a)
$$\Delta G'° = -RT \ln K'_{eq}$$
$$\ln K'_{eq} = -\Delta G'°/RT = -(13.8 \text{ kJ/mol})/(2.479 \text{ kJ/mol})$$
$$K'_{eq} = e^{-5.567} = 3.8 \times 10^{-3}\ \text{M}^{-1}$$

Note: this value has units M^{-1} because the expression for K'_{eq} from the chemical equilibrium includes H_2O (see below).

$$K'_{eq} = \frac{[\text{G6P}]}{[\text{Glc}][\text{P}_i]}$$
$$[\text{G6P}] = K'_{eq}[\text{Glc}][\text{P}_i] = (3.8 \times 10^{-3}\ \text{M}^{-1})(4.8 \times 10^{-3}\ \text{M})(4.8 \times 10^{-3}\ \text{M}) = 8.8 \times 10^{-8}\ \text{M}$$

This would not be a reasonable route for glucose catabolism because the cellular [G6P] is likely be much higher than 8.8×10^{-8} M, and the reaction would be unfavorable.

(b) Because
$$K'_{eq} = \frac{[\text{G6P}]}{[\text{Glc}][\text{P}_i]}$$
then
$$[\text{Glc}] = \frac{[\text{G6P}]}{K'_{eq}[\text{P}_i]} = \frac{250 \times 10^{-6}\ \text{M}}{(3.8 \times 10^{-3}\ \text{M}^{-1})(4.8 \times 10^{-3}\ \text{M})} = 14\ \text{M}$$

This would not be a reasonable route because the maximum solubility of glucose is less than 1 M.

(c)

(1)	Glc + $P_i \longrightarrow$ G6P + H_2O	$\Delta G'^{\circ}_1 = 13.8$ kJ/mol
(2)	ATP + $H_2O \longrightarrow$ ADP + P_i	$\Delta G'^{\circ}_2 = -30.5$ kJ/mol
Sum:	Glc + ATP $\longrightarrow$ G6P + ADP	$\Delta G'^{\circ}_{sum} = -16.7$ kJ/mol

$$\ln K'_{eq} = -\Delta G'^\circ/RT$$
$$= -(-16.7 \text{ kJ/mol})/(2.479 \text{ kJ/mol})$$
$$= 6.74$$
$$K'_{eq} = 843$$
$$\text{Since, } K'_{eq} = \frac{[\text{G6P}][\text{ADP}]}{[\text{Glc}][\text{ATP}]}$$
$$\text{then [Glc]} = \frac{[\text{G6P}][\text{ADP}]}{(K'_{eq})[\text{ATP}]}$$
$$= \frac{(250 \times 10^{-6} \text{ M})(1.32 \times 10^{-3} \text{ M})}{(843)(3.38 \times 10^{-3} \text{ M})}$$
$$= 1.2 \times 10^{-7} \text{ M}$$

This route is feasible because the [Glc] is reasonable.

(d) No; this is not reasonable. When glucose is at its physiological level, the required P_i concentration would be so high that phosphate salts of divalent cations would precipitate out.

(e) Direct transfer of the phosphoryl group from ATP to glucose takes advantage of the high phosphoryl group transfer potential of ATP and does not demand that the concentration of intermediates be very high, unlike the mechanism proposed in (d). In addition, the usual benefits of enzymatic catalysis apply, including binding interactions between the enzyme and its substrates; induced fit leading to the exclusion of water from the active site, so that only glucose is phosphorylated; and stabilization of the transition state.

10. Calculations of $\Delta G'^\circ$ for ATP-Coupled Reactions From data in Table 14–6 calculate the $\Delta G'^\circ$ value for the reactions

(a) Phosphocreatine + ADP ⟶ creatine + ATP

(b) ATP + fructose ⟶ ADP + fructose 6-phosphate

Answer

(a) The $\Delta G'^\circ$ value for the overall reaction is calculated from the sum of the $\Delta G'^\circ$ values for the coupled reactions.

(1)	Phosphocreatine + H_2O ⟶ creatine + P_i	$\Delta G_1'^\circ = -43.0$ kJ/mol
(2)	ADP + P_i ⟶ ATP + H_2O	$\Delta G_2'^\circ = 30.5$ kJ/mol
Sum:	Phosphocreatine + ADP ⟶ creatine + ATP	$\Delta G_{sum}'^\circ = -12.5$ kJ/mol

(b)

(1)	ATP + H_2O ⟶ ADP + P_i	$\Delta G_1'^\circ = -30.5$ kJ/mol
(2)	Fructose + P_i ⟶ F6P + H_2O	$\Delta G_2'^\circ = 15.9$ kJ/mol
Sum:	ATP + fructose ⟶ ADP + F6P	$\Delta G_{sum}'^\circ = -14.6$ kJ/mol

11. Coupling ATP Cleavage to an Unfavorable Reaction To explore the consequences of coupling ATP hydrolysis under physiological conditions to a thermodynamically unfavorable biochemical reaction, consider the hypothetical transformation X ⟶ Y, for which $\Delta G'^\circ = 20$ kJ/mol.

(a) What is the ratio [Y]/[X] at equilibrium?

(b) Suppose X and Y participate in a sequence of reactions during which ATP is hydrolyzed to ADP and P_i. The overall reaction is

$$\text{X} + \text{ATP} + \text{H}_2\text{O} \longrightarrow \text{Y} + \text{ADP} + \text{P}_i$$

Calculate [Y]/[X] for this reaction at equilibrium. Assume that the concentrations of ATP, ADP, and P_i are all 1 M when the reaction is at equilibrium.

(c) We know that [ATP], [ADP], and [P_i] are *not* 1 M under physiological conditions. Calculate [Y]/[X] for the ATP-coupled reaction when the values of [ATP], [ADP], and [P_i] are those found in rat myocytes (Table 14–5).

Answer

(a) The ratio $[Y]_{eq}/[X]_{eq}$ is equal to the equilibrium constant, K'_{eq}.

$$\ln K'_{eq} = -\Delta G'^{\circ}/RT$$
$$= -(20 \text{ kJ/mol})/(2.479 \text{ kJ/mol})$$
$$= -8.07$$
$$K'_{eq} = e^{-8.07} = 3.02 \times 10^{-4} = [Y]_{eq}/[X]_{eq}$$

This is a very small value of K'_{eq}; consequently, $\Delta G'^{\circ}$ is large and positive, making the reaction energetically unfavorable as written.

(b) First we need to calculate $\Delta G'^{\circ}$ for the overall reaction.

(1)	$X \longrightarrow Y$	$\Delta G_1'^{\circ} = 20$ kJ/mol
(2)	$ATP + H_2O \longrightarrow ADP + P_i$	$\Delta G_2'^{\circ} = -30.5$ kJ/mol
Sum:	$X + ATP \longrightarrow ADP + P_i + Y$	$\Delta G_{sum}'^{\circ} = -10.5$ kJ/mol

$$K'_{eq} = \frac{[Y]_{eq}[P_i]_{eq}[ADP]_{eq}}{[X]_{eq}[ATP]_{eq}}; \text{ note: water is omitted.}$$

Since [ADP], [ATP], and [P_i] are 1 M, this simplifies to K'_{eq} = [Y]/[X] in units of M.

$$\ln K'_{eq} = -\Delta G'^{\circ}/RT$$
$$= -(-10.5 \text{ kJ/mol})/(2.479 \text{ kJ/mol}) = 4.236$$
$$K'_{eq} = e^{4.236} = 69.1 = [Y]/[X]$$

$\Delta G'^{\circ}$ is fairly large and negative; the coupled reaction is favorable as written.

(c) Here we are dealing with the nonstandard conditions of the cell. Under physiological conditions, a favorable reaction (under standard conditions) becomes even more favorable.

$$K'_{eq} = \frac{[Y]_{eq}[P_i]_{eq}[ADP]_{eq}}{[X]_{eq}[ATP]_{eq}}$$
$$[Y]/[X] = \frac{K'_{eq}[ATP]}{[P_i][ADP]}$$
$$= \frac{(69.1 \text{ M})(8.05 \times 10^{-3} \text{ M})}{(8.05 \times 10^{-3} \text{ M})(0.93 \times 10^{-3} \text{ M})}$$
$$= 7.4 \times 10^4$$

12. Calculations of ΔG at Physiological Concentrations Calculate the physiological ΔG (not $\Delta G'^{\circ}$) for the reaction

$$\text{Phosphocreatine} + \text{ADP} \longrightarrow \text{creatine} + \text{ATP}$$

at 25 °C as it occurs in the cytosol of neurons, in which phosphocreatine is present at 4.7 mM, creatine at 1.0 mM, ADP at 0.73 mM, and ATP at 2.6 mM.

Answer

Using $\Delta G'^{\circ}$ values from Table 14–6:

(1)	Phosphocreatine + $H_2O \longrightarrow$ creatine + P_i	$\Delta G_1'^{\circ} = -43.0$ kJ/mol
(2)	ADP + $P_i \longrightarrow$ ATP + H_2O	$\Delta G_2'^{\circ} = 30.5$ kJ/mol
Sum:	Phosphocreatine + ADP $\longrightarrow$ creatine + ATP	$\Delta G_{sum}'^{\circ} = -12.5$ kJ/mol

$$\text{Mass action ratio, } Q = \frac{[\text{products}]}{[\text{reactants}]} = \frac{[\text{creatine}][\text{ATP}]}{[\text{phosphocreatine}][\text{ADP}]}$$
$$= \frac{(1 \times 10^{-3}\ \text{M})(2.6 \times 10^{-3}\ \text{M})}{(4.7 \times 10^{-3}\ \text{M})(7.3 \times 10^{-4}\ \text{M})}$$
$$= 0.75$$
$$\Delta G = \Delta G'^{\circ} + RT \ln Q$$
$$= -12.5 \text{ kJ/mol} + (2.479 \text{ kJ/mol}) \ln 0.75$$
$$= -13.2 \text{ kJ/mol} \approx -13 \text{ kJ/mol}$$

13. Free Energy Required for ATP Synthesis under Physiological Conditions In the cytosol of rat hepatocytes, the mass action ratio, Q, is

$$\frac{[\text{ATP}]}{[\text{ADP}][\text{P}_i]} = 5.33 \times 10^2\ \text{M}^{-1}$$

Calculate the free energy required to synthesize ATP in the rat hepatocyte.

Answer The reaction for the synthesis of ATP is:

$$\text{ADP} + \text{P}_i \longrightarrow \text{ATP} + \text{H}_2\text{O} \qquad \Delta G'^{\circ} = 30.5 \text{ kJ/mol}$$

The mass action ratio, Q, is

$$\frac{[\text{products}]}{[\text{reactants}]} = \frac{[\text{ATP}]}{[\text{P}_i][\text{ADP}]} = 5.33 \times 10^2\ \text{M}^{-1}$$

$$\text{Since } \Delta G = \Delta G'^{\circ} + RT \ln Q$$
$$= 30.5 \text{ kJ/mol} + (2.479 \text{ kJ/mol}) \ln 5.33 \times 10^2$$
$$= 46 \text{ kJ/mol}$$

14. Daily ATP Utilization by Human Adults

(a) A total of 30.5 kJ/mol of free energy is needed to synthesize ATP from ADP and P_i when the reactants and products are at 1 M concentration (standard state). Because the actual physiological concentrations of ATP, ADP, and P_i are not 1 M, the free energy required to synthesize ATP under physiological conditions is different from $\Delta G'^{\circ}$. Calculate the free energy required to synthesize ATP in the human hepatocyte when the physiological concentrations of ATP, ADP, and P_i are 3.5, 1.50, and 5.0 mM, respectively.

(b) A 68-kg (150-lb) adult requires a caloric intake of 2,000 kcal (8,360 kJ) of food per day (24 h). The food is metabolized and the free energy is used to synthesize ATP, which then provides energy for the body's daily chemical and mechanical work. Assuming that the efficiency of converting food energy into ATP is 50%, calculate the weight of ATP used by a human adult in 24 h. What percentage of the body weight does this represent?

(c) Although adults synthesize large amounts of ATP daily, their body weight, structure, and composition do not change significantly during this period. Explain this apparent contradiction.

Answer

(a) $\text{ADP} + \text{P}_i \longrightarrow \text{ATP} + \text{H}_2\text{O} \qquad \Delta G'^{\circ} = 30.5 \text{ kJ/mol}$

$$\text{Mass action ratio; } Q, \text{ is } = \frac{[\text{ATP}]}{[\text{P}_i][\text{ADP}]} = \frac{[3.5 \times 10^{-3}\ \text{M}]}{[1.5 \times 10^{-3}\ \text{M}][5.0 \times 10^{-3}\ \text{M}]} = 4.7 \times 10^2\ \text{M}^{-1}$$
$$\Delta G = \Delta G'^{\circ} + \text{RT} \ln Q$$
$$= 30.5 \text{ kJ/mol} + (2.479 \text{ kJ/mol}) \ln 4.7 \times 10^2\ \text{M}^{-1} = 46 \text{ kJ/mol}$$

(b) The energy going into ATP synthesis is 8,360 kJ × 50% = 4,180 kJ. Using the value of ΔG from (a), the amount of ATP synthesized is
(4,180 kJ)/(46 kJ/mol) = 91 mol

The molecular weight of ATP is 503 (calculated by summing atomic weights). Thus the weight of ATP synthesized is
(91 mol ATP)(503 g/mol) = 46 kg ATP

As a percentage of body weight:
100(46 kg ATP)/(68 kg body weight) = 68%

(c) The concentration of ATP in a healthy body is maintained in a steady state; this is an example of homeostasis, a condition in which the body synthesizes and breaks down ATP as needed.

15. Rates of Turnover of γ and β Phosphates of ATP If a small amount of ATP labeled with radioactive phosphorus in the terminal position, [γ-^{32}P]ATP, is added to a yeast extract, about half of the ^{32}P activity is found in P_i within a few minutes, but the concentration of ATP remains unchanged. Explain. If the same experiment is carried out using ATP labeled with ^{32}P in the central position, [β-^{32}P]ATP, the ^{32}P does not appear in P_i within such a short time. Why?

Answer To solve this problem, it is useful to represent ATP as A-P-P-P. A radiolabeled phosphate group can be denoted *P. One possible reaction for γ-labeled ATP would be

$$\text{A-P-P-*P} + \text{Glc} \longrightarrow \text{A-P-P} + \text{G6*P} \rightarrow\ \rightarrow\ \rightarrow \text{*P}_i$$

or more generally

$$\text{A-P-P-*P} + \text{H}_2\text{O} \longrightarrow \text{A-P-P} + \text{*P}_i$$

These reactions occur relatively quickly. As we saw in Problem 11(c), the concentration of ATP in the body remains in a steady state. The reaction below occurs more slowly:

$$\text{A-P-*P-P} + \text{H}_2\text{O} \longrightarrow \text{A-P} + \text{*P}_i + \text{P}_i$$

As a result, one would expect a faster turnover of *P_i when it is in the γ position than when it is in the β position.

16. Cleavage of ATP to AMP and PP_i during Metabolism The synthesis of the activated form of acetate (acetyl-CoA) is carried out in an ATP-dependent process:

$$\text{Acetate} + \text{CoA} + \text{ATP} \longrightarrow \text{acetyl-CoA} + \text{AMP} + \text{PP}_i$$

(a) The $\Delta G'^\circ$ for the hydrolysis of acetyl-CoA to acetate and CoA is −32.2 kJ/mol and that for hydrolysis of ATP to AMP and PP_i is −30.5 kJ/mol. Calculate $\Delta G'^\circ$ for the ATP-dependent synthesis of acetyl-CoA.

(b) Almost all cells contain the enzyme inorganic pyrophosphatase, which catalyzes the hydrolysis of PP_i to P_i. What effect does the presence of this enzyme have on the synthesis of acetyl-CoA? Explain.

Answer

(a) The $\Delta G'^\circ$ can be determined for the coupled reactions:

(1)	Acetate + CoA ⟶ acetyl-CoA + H_2O	$\Delta G_1'^\circ = 32.2$ kJ/mol
(2)	ATP + H_2O ⟶ AMP + PP_i	$\Delta G_2'^\circ = -30.5$ kJ/mol
Sum:	Acetate + CoA + ATP ⟶ acetyl-CoA + AMP + PP_i	$\Delta G_{sum}'^\circ = 1.7$ kJ/mol

(b) Hydrolysis of PP_i would drive the reaction forward, favoring the synthesis of acetyl-CoA.

17. Energy for H^+ Pumping The parietal cells of the stomach lining contain membrane "pumps" that transport hydrogen ions from the cytosol of these cells (pH 7.0) into the stomach, contributing to the acidity of gastric juice (pH 1.0). Calculate the free energy required to transport 1 mol of hydrogen ions through these pumps. (Hint: See Chapter 13.) Assume a temperature of 25 °C.

Answer The free energy required to transport 1 mol of H^+ from the interior of the cell, where $[H^+]$ is 10^{-7} M, across the membrane to where $[H^+]$ is 10^{-1} M is

$$\Delta G_t = RT \ln (C_2/C_1)$$

$$\begin{aligned}\Delta G_t &= RT \ln (10^{-1}/10^{-7}) \\ &= (8.315 \text{ J/mol·K})(298 \text{ K}) \ln (10^{-1}/10^{-7}) \\ &\approx 34.2 \text{ kJ/mol}\end{aligned}$$

18. Standard Reduction Potentials The standard reduction potential, E'°, of any redox pair is defined for the half-cell reaction:

$$\text{Oxidizing agent} + n \text{ electrons} \longrightarrow \text{reducing agent}$$

The E'° values for the NAD^+/NADH and pyruvate/lactate conjugate redox pairs are −0.32 and −0.19 V, respectively.

(a) Which conjugate pair has the greater tendency to lose electrons? Explain.

(b) Which is the stronger oxidizing agent? Explain.

(c) Beginning with 1 M concentrations of each reactant and product at pH 7, in which direction will the following reaction proceed?

$$\text{Pyruvate} + \text{NADH} + H^+ \rightleftharpoons \text{lactate} + NAD^+$$

(d) What is the standard free-energy change ($\Delta G'^\circ$) at 25 °C for the conversion of pyruvate to lactate?

(e) What is the equilibrium constant (K'_{eq}) for this reaction?

Answer

(a) The NAD^+/NADH pair is more likely to lose electrons. The equations in Table 14–7 are written in the direction of reduction (gain of electrons). E'° is positive if the oxidized member of a conjugate pair has a tendency to accept electrons. E'° is negative if the oxidized member of a conjugate pair does *not* have a tendency to accept electrons. Both NAD^+/NADH and pyruvate/lactate have negative E'° values. The E'° of NAD^+/NADH is more negative than that for pyruvate/lactate, so this pair has the greater tendency to accept electrons and is thus the stronger oxidizing system.

(b) The pyruvate/lactate pair is the more likely way to accept electrons and thus is the stronger oxidizing agent. For the same reason that NADH tends to donate electrons to pyruvate, pyruvate tends to accept electrons from NADH. Pyruvate will be reduced to form lactate; NADH will be oxidized to form NAD^+. Pyruvate is the oxidizing agent; NADH is the reducing agent.

(c) From the answers to (a) and (b), it is evident that the reaction will tend to go in the direction of lactate formation.

(d) The first step is to calculate $\Delta E'^\circ$ for the reaction, using the E'° values from Table 14–7 (Notice that the sign for the NAD^+/NADH pair changes.)

$$\begin{array}{ll} \text{NADH} + H^+ \longrightarrow NAD^+ + 2H^+ + 2e^- & E'^\circ = 0.320 \text{ V} \\ \text{Pyruvate} + 2H^+ + 2e^- \longrightarrow \text{lactate} & E'^\circ = -0.185 \text{ V} \\ \hline \text{NADH} + \text{pyruvate} \longrightarrow NAD^+ + \text{lactate} & \Delta E'^\circ = 0.135 \text{ V} \end{array}$$

$$\begin{aligned}\Delta G'^\circ &= -n\mathcal{F}\,\Delta E'^\circ \\ &= -2(96.5 \text{ kJ/mol·V})(0.135 \text{ V}) \\ &= -26 \text{ kJ/mol}\end{aligned}$$

(e) $\ln K'_{eq} = -\Delta G'^{\circ}/RT$
$= -(-26 \text{ kJ/mol})/(2.479 \text{ kJ/mol})$
$= -10.49$
$K'_{eq} = e^{10.49} = 3.59 \times 10^4$

19. Energy Span of the Respiratory Chain Electron transfer in the mitochondrial respiratory chain may be represented by the net reaction equation

$$\text{NADH} + \text{H}^+ + \tfrac{1}{2}\,\text{O}_2 \rightleftharpoons \text{H}_2\text{O} + \text{NAD}^+$$

(a) Calculate the value of $\Delta E'^{\circ}$ for the net reaction of mitochondrial electron transfer.
(b) Calculate $\Delta G'^{\circ}$ for this reaction.
(c) How many ATP molecules can *theoretically* be generated by this reaction if the free energy of ATP synthesis under cellular conditions is 52 kJ/mol?

Answer

(a) Using E'° values from Table 14–7:

$$\begin{array}{ll} \text{NADH} + \text{H}^+ \longrightarrow \text{NAD}^+ + 2\text{H}^+ + 2\text{e}^- & E'^{\circ} = 0.320 \text{ V} \\ \tfrac{1}{2}\text{O}_2 + 2\text{H}^+ + 2\text{e}^- \longrightarrow \text{H}_2\text{O} & E'^{\circ} = 0.816 \text{ V} \\ \hline \text{NADH} + \text{H}^+ + \tfrac{1}{2}\text{O}_2 \longrightarrow \text{H}_2\text{O} + \text{NAD}^+ & \Delta E'^{\circ} = 1.136 \text{ V} \\ & \approx 1.14 \text{ V} \end{array}$$

(b) $\Delta G'^{\circ} = -n\mathcal{F}\,\Delta E'^{\circ}$
$= -2(96.5 \text{ kJ/mol·V})(1.14 \text{ V})$
$= -220 \text{ kJ/mol}$

(c) For ATP synthesis, the reaction is

$\text{ADP} + \text{P}_i \longrightarrow \text{ATP} \qquad \Delta G'^{\circ} = 52 \text{ kJ/mol}$

Thus the number of ATP molecules that can, in theory, be generated is

$$\frac{220 \text{ kJ/mol}}{52 \text{ kJ/mol}} = 4.2 \approx 4$$

20. Dependence of Electromotive Force on Concentrations Calculate the electromotive force (in volts) registered by an electrode immersed in a solution containing the following mixtures of NAD^+ and NADH at pH 7.0 and 25 °C, with reference to a half-cell of E'° 0.00 V.

(a) 1.0 mM NAD^+ and 10 mM NADH
(b) 1.0 mM NAD^+ and 1.0 mM NADH
(c) 10 mM NAD^+ and 1.0 mM NADH

Answer The relevant equation for calculating E for this system is

$$E = E'^{\circ} + \frac{RT}{n\mathcal{F}} \ln \frac{[\text{NAD}^+]}{[\text{NADH}]}$$

At 25 °C, the $RT/n\mathcal{F}$ term simplifies to 0.026 V/n.

(a) From Table 14–7, E'° for the NAD^+/NADH redox pair is -0.320 V. Since two electrons are transferred, $n = 2$. Thus,

$E = (-0.320 \text{ V}) + (0.026 \text{ V}/2) \ln (1 \times 10^{-3} \text{ M})/(10 \times 10^{-3} \text{ M})$
$= -0.35 \text{ V}$

(b) The conditions specified here are "standard conditions," so we expect that $E' = E'^{\circ}$. As proof, we know that the value for $\ln 1 = 0$, so under standard conditions the term $(RT/n\mathcal{F}) \ln 1 = 0$, which makes $E = E'^{\circ} = -0.320$ V.

(c) Here the concentration of NAD^+ (the electron acceptor) is 10 times that of NADH (the electron donor). This affects the value of E'.

$$E = (-0.320 \text{ V}) + (0.026/2 \text{ V}) \ln (10 \times 10^{-3} \text{ M})/(1 \times 10^{-3})$$
$$= -0.29 \text{ V}$$

21. Electron Affinity of Compounds List the following substances in order of increasing tendency to accept electrons:
(a) α-ketoglutarate + CO_2 (yielding isocitrate);
(b) oxaloacetate;
(c) O_2;
(d) $NADP^+$.

Answer To solve this problem, first write the half-reactions as in Table 14–7, and then find the value for $E'°$ for each. Pay attention to the sign!

Half-reaction	$E'°$ (V)
(a) α-Ketoglutarate + CO_2 + $2H^+$ + $2e^-$ ⟶ isocitrate	−0.38
(b) Oxaloacetate + $2H^+$ + $2e^-$ ⟶ malate	−0.166
(c) $\frac{1}{2}O_2 + 2H^+ + 2e^- \longrightarrow H_2O$	0.816
(d) $NADP^+$ + $2H^+$ + $2e^-$ ⟶ NADPH + H^+	−0.324

The more positive the $E'°$, the more likely is the substance to accept electrons; thus we can list the substances in order of increasing tendency to accept electrons: (a), (d), (b), (c).

22. Direction of Oxidation-Reduction Reactions Which of the following reactions would you expect to proceed in the direction shown under standard conditions, assuming that the appropriate enzymes are present to catalyze them?
(a) Malate + NAD^+ ⟶ oxaloacetate + NADH + H^+
(b) Acetoacetate + NADH + H^+ ⟶ β-hydroxybutyrate + NAD^+
(c) Pyruvate + NADH + H^+ ⟶ lactate + NAD^+
(d) Pyruvate + β-hydroxybutyrate ⟶ lactate + acetoacetate
(e) Malate + pyruvate ⟶ oxaloacetate + lactate
(f) Acetaldehyde + succinate ⟶ ethanol + fumarate

Answer It is important to note that standard conditions do not exist in the cell. The value of $\Delta E'°$, as calculated in this problem, gives an indication of whether a reaction will or will not occur in a cell without additional energy being added (usually from ATP); but $\Delta E'°$ does not tell the entire story. The actual cellular concentrations of the electron donors and electron acceptors contribute significantly to the value of $E'°$ (see Problem 20, for example). The potential under nonstandard conditions can either add to an already favorable $\Delta E'°$ or be such a large positive number as to "overwhelm" an unfavorable $\Delta E'°$ and make ΔE favorable.

To solve this problem, write the two half-reactions with the appropriate E'° values (see Table 14–7). Pay attention to the sign! Then judge whether the reaction is likely to proceed favorably as written.

(a) Not favorable.

Malate $\longrightarrow$ oxaloacetate + $2H^+$ + $2e^-$	$E'^\circ = 0.166$ V
NAD^+ + $2H^+$ + $2e^-$ $\longrightarrow$ NADH + H^+	$E'^\circ = -0.320$ V
Malate + NAD^+ $\longrightarrow$ oxaloacetate + NADH	$\Delta E'^\circ = -0.154$ V

(b) Not favorable.

Acetoacetate + $2H^+$ + $2e^-$ $\longrightarrow$ β-hydroxybutyrate	$E'^\circ = -0.346$ V
NADH + $2H^+$ $\longrightarrow$ NAD^+ + $2H^+$ + $2e^-$	$E'^\circ = 0.320$ V
Acetoacetate + NADH $\longrightarrow$ β-hydroxybutyrate + NAD^+	$\Delta E'^\circ = -0.026$ V

(c) Favorable.

Pyruvate + $2H^+$ + $2e^-$ $\longrightarrow$ lactate	$E'^\circ = -0.185$ V
NADH + H^+ $\longrightarrow$ NAD^+ + $2H^+$ + $2e^-$	$E'^\circ = 0.320$ V
Pyruvate + NADH $\longrightarrow$ lactate + NAD^+	$\Delta E'^\circ = 0.135$ V

(d) Favorable.

Pyruvate + $2H^+$ + $2e^-$ $\longrightarrow$ lactate	$E'^\circ = -0.185$ V
β-hydroxybutyrate $\longrightarrow$ acetoacetate + $2H^+$ + $2e^-$	$E'^\circ = 0.346$ V
Pyruvate + β-hydroxybutyrate $\longrightarrow$ acetoacetate + lactate	$\Delta E'^\circ = 0.161$ V

(e) Not favorable.

Malate $\longrightarrow$ oxaloacetate + $2H^+$ + $2e^-$	$E'^\circ = 0.166$ V
Pyruvate + $2H^+$ + $2e^-$ $\longrightarrow$ lactate	$E'^\circ = -0.185$ V
Pyruvate + malate $\longrightarrow$ lactate + oxaloacetate	$\Delta E'^\circ = -0.019$ V

(f) Not favorable.

Acetaldehyde + $2H^+$ + $2e^-$ $\longrightarrow$ ethanol	$E'^\circ = -0.197$ V
Succinate $\longrightarrow$ fumarate + $2H^+$ + $2e^-$	$E'^\circ = -0.031$ V
Acetaldehyde + succinate $\longrightarrow$ ethanol + fumarate	$\Delta E'^\circ = -0.228$ V

chapter

15 Glycolysis and the Catabolism of Hexoses

1. **Equation for the Preparatory Phase of Glycolysis** Write balanced biochemical equations for all the reactions in the catabolism of D-glucose to two molecules of D-glyceraldehyde 3-phosphate (the preparatory phase of glycolysis), including the standard free-energy change for each reaction. Then write the overall or net equation for the preparatory phase of glycolysis, with the net standard free-energy change.

Answer The initial phase of glycolysis requires ATP; it is endergonic. There are five reactions in this phase (see pp. 532–534):

1. Glucose + ATP ⟶ glucose 6-phosphate + ADP $\Delta G'^{\circ} = -16.7$ kJ/mol
2. Glucose 6-phosphate ⟶ fructose 6-phosphate $\Delta G'^{\circ} = 1.7$ kJ/mol
3. Fructose 6-phosphate + ATP ⟶ fructose 1,6-bisphosphate $\Delta G'^{\circ} = -14.2$ kJ/mol
4. Fructose 1,6-bisphosphate ⟶
dihydroxyacetone phosphate + glyceraldehyde 3-phosphate $\Delta G'^{\circ} = 23.8$ kJ/mol
5. Dihydroxyacetone phosphate ⟶ glyceraldehyde 3-phosphate $\Delta G'^{\circ} = 7.5$ kJ/mol

The net equation for this phase is

Glucose + 2ATP ⟶ 2 glyceraldehyde 3-phosphate + 2ADP + $2H^+$

The overall standard free-energy change can be calculated by summing the individual reactions: $\Delta G'^{\circ} = 2.1$ kJ/mol (endergonic).

2. **The Payoff Phase of Glycolysis in Skeletal Muscle** In working skeletal muscle under anaerobic conditions, glyceraldehyde 3-phosphate is converted to pyruvate (the payoff phase of glycolysis), and the pyruvate is reduced to lactate. Write balanced biochemical equations for all the reactions in this process, with the standard free-energy change for each reaction. Then write the overall or net equation for the payoff phase of glycolysis (with lactate as the end product), including the net standard free-energy change.

Answer The payoff phase of glycolysis produces ATP, making it is exergonic. This phase consists of five reactions, designated 6–10 in the text (see pp. 535–539):

6. Glyceraldehyde 3-phosphate + P_i + NAD^+ ⟶
1,3-bisphosphoglycerate + NADH + H^+ $\Delta G'^{\circ} = 6.3$ kJ/mol
7. 1,3-Bisphosphoglycerate + ADP ⟶
3-phosphoglycerate + ATP $\Delta G'^{\circ} = -18.5$ kJ/mol
8. 3-Phosphoglycerate ⟶ 2-phosphoglycerate $\Delta G'^{\circ} = 4.4$ kJ/mol
9. 2-Phosphoglycerate ⟶ phosphoenolpyruvate $\Delta G'^{\circ} = 7.5$ kJ/mol
10. Phosphoenolpyruvate + ADP ⟶ pyruvate + ATP $\Delta G'^{\circ} = -31.4$ kJ/mol

The pyruvate is then converted to lactate:

Pyruvate + NADH + H^+ ⟶ lactate + NAD^+ $\quad \Delta G'^\circ = -25.1$ kJ/mol

The net equation is:

Glyceraldehyde 3-phosphate + 2ADP + P_i ⟶ lactate + NAD^+ $\quad \Delta G'^\circ = -56.8$ kJ/mol

Since the payoff phase utilizes two glyceraldehyde 3-phosphate molecules from each glucose entering glycolysis, the energetic payoff for the net reaction should be doubled, which makes the net $\Delta G'^\circ = -113.6$ kJ/mol.

3. Pathway of Atoms in Fermentation A "pulse-chase" experiment using ^{14}C-labeled carbon sources is carried out on a yeast extract maintained under strictly anaerobic conditions to produce ethanol. The experiment consists of incubating a small amount of ^{14}C-labeled substrate (the pulse) with the yeast extract just long enough for each intermediate in the fermentation pathway to become labeled. The label is then "chased" through the pathway by the addition of excess unlabeled glucose. The chase effectively prevents any further entry of labeled glucose into the pathway.

(a) If [1-^{14}C]glucose (glucose labeled at C-1 with ^{14}C) is used as a substrate, what is the location of ^{14}C in the product ethanol? Explain.

(b) Where would ^{14}C have to be located in the starting glucose to ensure that all the ^{14}C activity is liberated as $^{14}CO_2$ during fermentation to ethanol? Explain.

Answer Anaerobiosis requires the regeneration of NAD^+ from NADH in order to allow glycolysis to continue.

(a) Figure 15–4 illustrates the fate of the carbon atoms of glucose. C-1 (or C-6) becomes C-3 of glyceraldehyde 3-phosphate and subsequently pyruvate. When pyruvate is decarboxylated and reduced to ethanol, C-3 of pyruvate becomes the C-2 of ethanol ($^*CH_3—CH_2—OH$).

(b) In order for all of the labeled carbon from glucose to be converted to $^{14}CO_2$ during ethanol fermentation, the original label would have to be on C-3 and/or C-4 of glucose, since these are converted to the carboxyl group of pyruvate.

4. Fermentation to Produce Soy Sauce Soy sauce is prepared by fermenting a salted mixture of soybeans and wheat with several microorganisms, including yeast, over a period of 8 to 12 months. The resulting sauce (after the solids are removed) is rich in lactate and ethanol. How are these two compounds produced? To prevent the soy sauce from having a strong vinegar taste (vinegar is dilute acetic acid), oxygen must be kept out of the fermentation tank. Why?

Answer Soybeans and wheat contain starch, a polymer of glucose, which is broken down to glucose by the microorganisms. The glucose is then broken down to pyruvate via glycolysis. Because the process is carried out in the absence of oxygen (i.e., it is a fermentation), pyruvate is reduced to lactic acid and ethanol by the microorganisms. If oxygen were present, pyruvate would be oxidized to acetyl-CoA and then to CO_2 and H_2O. Some of the acetyl-CoA, however, would also be hydrolyzed to acetic acid (vinegar) in the presence of oxygen.

5. Equivalence of Triose Phosphates ^{14}C-Labeled glyceraldehyde 3-phosphate was added to a yeast extract. After a short time, fructose 1,6-bisphosphate labeled with ^{14}C at C-3 and C-4 was isolated. What was the location of the ^{14}C label in the starting glyceraldehyde 3-phosphate? Where did the second ^{14}C label in fructose 1,6-bisphosphate come from? Explain.

Answer Problem 1 outlines the steps in glycolysis involving fructose 1,6-bisphosphate, glyceraldehyde 3-phosphate, and dihydroxyacetone phosphate. Keep in mind here that the aldolase reaction is readily reversible and the triose phosphate isomerase reaction catalyzes extremely rapid interconversion of its substrates. Thus, the label in the C-1 position of glyceraldehyde 3-phosphate would equilibrate with C-1 of dihydroxyacetone phosphate ($\Delta G'^{\circ} =$ 7.5 kJ/mol). Since the aldolase reaction has a $\Delta G'^{\circ} = -23.8$ kJ/mol, favorable in the direction of hexose formation, fructose 1,6-bisphosphate would be readily formed and would be labeled in C-3 and C-4 (see Fig. 15–4).

6. Glycolysis Shortcut Suppose you discovered a mutant yeast whose glycolytic pathway was shorter because of the presence of a new enzyme catalyzing the reaction:

$$\text{Glyceraldehyde 3-phosphate} + H_2 \xrightarrow{NAD^+ \;\rightarrow\; NADH + H^+} \text{3-phosphoglycerate}$$

Would shortening the glycolytic pathway in this way benefit the cell? Explain.

Answer Under anaerobic conditions, the phosphoglycerate kinase and pyruvate kinase reactions are essential. The shortcut in the mutant yeast would bypass the formation of an acylphosphate by glyceraldehyde 3-phosphate dehydrogenase and therefore would not allow the formation of 1,3-bisphosphoglycerate. Without the formation of a substrate for 3-phosphoglycerate kinase, no ATP would be formed. Under anaerobic conditions, the net reaction for glycolysis normally produces 2 ATPs per glucose. In the mutant yeast, net production of ATP would be zero and growth could not occur. Under aerobic conditions, however, since the majority of ATP formation occurs via oxidative phosphorylation, the mutation would have no observable effect.

7. Role of Lactate Dehydrogenase During strenuous activity, the demand for ATP in muscle tissue is vastly increased. In rabbit leg muscle or turkey flight muscle, the ATP is produced almost exclusively by lactic acid fermentation. ATP is formed in the payoff phase of glycolysis by two reactions, promoted by phosphoglycerate kinase and pyruvate kinase. Suppose skeletal muscle were devoid of lactate dehydrogenase. Could it carry out strenuous physical activity; that is, could it generate ATP at a high rate by glycolysis? Explain.

Answer The key point here is that NAD^+ must be regenerated from NADH in order for the glycolytic pathway to continue to function. Some tissues, such as skeletal muscle, obtain almost all of their ATP through the glycolytic pathway and are capable of short-term exercise only (see Box 15–1). In order to generate ATP at a high rate, the NADH formed during glycolysis must be oxidized. In the absence of significant amounts of O_2 in the tissues, pyruvate and NADH are converted to lactate and NAD^+ by lactate dehydrogenase. In the absence of this enzyme, NAD^+ could not be regenerated and glycolytic generation of ATP would stop, and as a consequence, muscle activity could not be maintained.

8. Efficiency of ATP Production in Muscle The transformation of glucose to lactate in myocytes releases only about 7% of the free energy released when glucose is completely oxidized to CO_2 and H_2O. Does this mean that anaerobic glycolysis in muscle is a wasteful use of glucose? Explain.

Answer The transformation of glucose to lactate occurs when muscle cells are low in oxygen and provides a means of generating ATP under oxygen-deficient conditions. Since lactate can be transformed back to pyruvate, glucose is not wasted because pyruvate can be oxidized by aerobic reactions when oxygen becomes plentiful. This metabolic flexibility gives the organism a greater capacity to adapt to its environment.

9. Free-Energy Change for Triose Phosphate Oxidation The oxidation of glyceraldehyde 3-phosphate to 1,3-bisphosphoglycerate, catalyzed by glyceraldehyde 3-phosphate dehydrogenase, proceeds with an unfavorable equilibrium constant ($K'_{eq} = 0.08$; $\Delta G'^{\circ} = 6.3$ kJ/mol), yet the flow through this point in the glycolytic pathway proceeds smoothly. How does the cell overcome the unfavorable equilibrium?

Answer In living organisms, where directional flow in a pathway is required, exergonic reactions are coupled to endergonic reactions to overcome unfavorable free-energy changes. The endergonic glyceraldehyde 3-phosphate dehydrogenase reaction is followed by the phosphoglycerate kinase reaction, which rapidly removes the product of the former reaction. Consequently, the dehydrogenase reaction does not reach equilibrium and its unfavorable free-energy change is thus circumvented. The net $\Delta G'^{\circ}$ of the two reactions, when coupled, is $(-18.5 + 6.3)$ kJ/mol $= -12.2$ kJ/mol.

10. Are All Metabolic Reactions at Equilibrium?

(a) Phosphoenolpyruvate (PEP) is one of the two phosphoryl donors in the synthesis of ATP during glycolysis. In human erythrocytes, the steady-state concentration of ATP is 2.24 mM, that of ADP is 0.25 mM, and that of pyruvate is 0.051 mM. Calculate the concentration of PEP at 25 °C, assuming that the pyruvate kinase reaction (see Fig. 15–2) is at equilibrium in the cell.

(b) The physiological concentration of PEP in human erythrocytes is 0.023 mM. Compare this with the value obtained in (a). Explain the significance of this difference.

Answer

(a) First we must calculate the overall $\Delta G'^{\circ}$ for the pyruvate kinase reaction, breaking this process into two opposing hydrolyses:

$$\begin{array}{ll} (1)\quad \text{PEP} + H_2O \longrightarrow \text{pyruvate} + P_i & \Delta G_1'^{\circ} = -61.9 \text{ kJ/mol} \\ (2)\quad \text{ADP} + P_i \longrightarrow \text{ATP} + H_2O & \Delta G_2'^{\circ} = 30.5 \text{ kJ/mol} \\ \hline \text{Sum: PEP} + \text{ADP} \longrightarrow \text{pyruvate} + \text{ATP} & \Delta G_{sum}'^{\circ} = -31.4 \text{ kJ/mol} \end{array}$$

$$\begin{aligned} \ln K'_{eq} &= -\Delta G'^{\circ}/RT \\ &= -(-31.4 \text{ kJ/mol})/(2.479 \text{ kJ/mol}) \\ &= 12.67 \\ K'_{eq} &= 3.17 \times 10^5 \end{aligned}$$

$$\text{Since } K'_{eq} = \frac{[\text{pyruvate}][\text{ATP}]}{[\text{ADP}][\text{PEP}]}$$

$$\begin{aligned} [\text{PEP}] &= \frac{[\text{pyruvate}][\text{ATP}]}{[\text{ADP}]\,K'_{eq}} \\ &= \frac{(5.1 \times 10^{-5} \text{ M})(2.24 \times 10^{-3} \text{ M})}{(2.5 \times 10^{-4} \text{ M})(3.17 \times 10^{5} \text{ M})} \\ &= 1.44 \times 10^{-9} \text{ M} \end{aligned}$$

(b) The physiological [PEP] of 0.023 mM is

$$\frac{0.023 \times 10^{-3} \text{ M}}{1.44 \times 10^{-9} \text{ M}} = 16{,}000 \text{ times the equilibrium concentration}$$

This reaction, like many others in the cell, does not reach equilibrium.

11. Arsenate Poisoning Arsenate is structurally and chemically similar to inorganic phosphate (P_i), and many enzymes that require phosphate will also use arsenate. Organic compounds of arsenate are less stable than analogous phosphate compounds, however. For example, acyl *arsenates* decompose rapidly by hydrolysis:

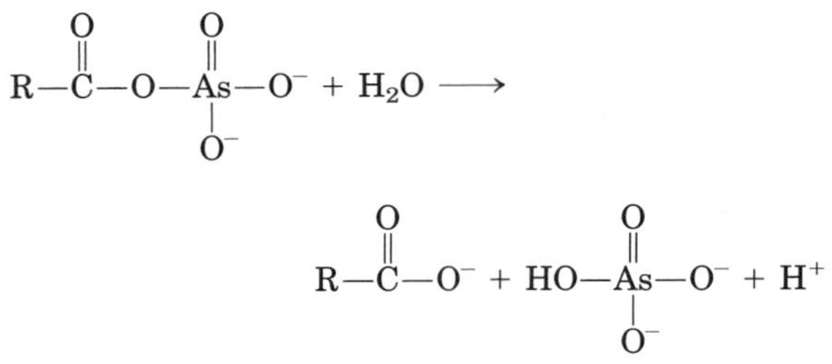

$$R{-}\overset{\overset{\large O}{\|}}{C}{-}O{-}\underset{\underset{\large O^-}{|}}{\overset{\overset{\large O}{\|}}{As}}{-}O^- + H_2O \longrightarrow$$

$$R{-}\overset{\overset{\large O}{\|}}{C}{-}O^- + HO{-}\underset{\underset{\large O^-}{|}}{\overset{\overset{\large O}{\|}}{As}}{-}O^- + H^+$$

On the other hand, acyl *phosphates,* such as 1,3-bisphosphoglycerate, are more stable and undergo further enzyme-catalyzed transformation in cells.

(a) Predict the effect on the net reaction catalyzed by glyceraldehyde 3-phosphate dehydrogenase if phosphate were replaced by arsenate.

(b) What would be the consequence to an organism if arsenate were substituted for phosphate? Arsenate is very toxic to most organisms. Explain why.

Answer

(a) In the presence of arsenate, the product of the glyceraldehyde 3-phosphate dehydrogenase reaction will be 1-arseno-3-phosphoglycerate, which nonenzymatically decomposes to 3-phosphoglycerate and arsenate, and the substrate for the phosphoglycerate kinase will be bypassed.

(b) No ATP can be formed in the presence of arsenate because 1,3-bisphosphoglycerate is not formed. Under anaerobic conditions, this would result in no net glycolytic synthesis of ATP. Arsenate poisoning can be used as a test for the presence of an acyl phosphate intermediate in a reaction pathway.

12. Requirement for Phosphate in Alcohol Fermentation In 1906, Harden and Young, in a series of classic studies on the fermentation of glucose to ethanol and CO_2 by extracts of brewer's yeast, made the following observations. (1) Inorganic phosphate was essential to fermentation; when the supply of phosphate was exhausted, fermentation ceased before all the glucose was used. (2) During fermentation under these conditions, ethanol, CO_2, and a bisphosphohexose accumulated. (3) When arsenate was substituted for phosphate, no bisphosphohexose accumulated, but the fermentation proceeded until all the glucose was converted to ethanol and CO_2.

(a) Why did fermentation cease when the supply of phosphate was exhausted?

(b) Why did ethanol and CO_2 accumulate? Was the conversion of pyruvate to ethanol and CO_2 essential? Why? Identify the bisphosphohexose that accumulated. Why did it accumulate?

(c) Why did the substitution of arsenate for phosphate prevent the accumulation of the bisphosphohexose yet allow fermentation to ethanol and CO_2 to go to completion? (See Problem 11.)

Answer Alcohol fermentation in yeast has the following overall equation

$$\text{Glucose} + 2\text{ADP} + 2\text{P}_i \longrightarrow 2 \text{ ethanol} + 2\text{CO}_2 + 2\text{ATP} + 2\text{H}_2\text{O}$$

It is clear that phosphate is required for the continued operation of glycolysis and ethanol formation. In extracts to which glucose is added, fermentation will proceed until ADP and P_i (present in the extracts) are exhausted.

(a) Phosphate is required in the glyceraldehyde 3-phosphate dehydrogenase reaction, and glycolysis will stop at this step when P_i is exhausted. Since glucose remains it will be phosphorylated by ATP, but P_i will not be released.

(b) Fermentation in yeast cells produces ethanol and CO_2 rather than lactate—see Box 15–2. Without these reactions (in the absence of oxygen), NADH would accumulate and no new NAD^+ would be available for further glycolysis—see Problem 7. The biphosphorylated intermediate (of glycolysis) that accumulates is fructose 1,6-bisphosphate. FBP accumulates because in terms of energetics, it lies at a "low point" or valley in the pathway, between the energy-input reactions that precede it and the energy-payoff reactions that follow.

(c) Arsenate replaces P_i in the glyceraldehyde 3-phosphate dehydrogenase reaction to yield an acylarsenate, which spontaneously hydrolyzes. This prevents the formation of FBP and ATP but allows the formation of 3-phosphoglycerate, which continues through the pathway.

13. Role of the Vitamin Niacin Adults engaged in strenuous physical activity require an intake of about 160 g of carbohydrate daily, but only about 20 mg of niacin for optimal nutrition. Given the role of niacin in glycolysis, how do you explain the observation?

Answer Dietary niacin is used to synthesize NAD^+. Oxidations carried out by NAD^+ are part of cyclic processes, with NAD^+ as an electron carrier (reducing agent). Because of this cycling, one molecule can oxidize many thousands of molecules of glucose, and thus the dietary requirement for the precursor vitamin (niacin) is relatively small.

14. Cellular Glucose Concentration The concentration of glucose in human blood plasma is maintained at about 5 mM. The concentration of free glucose inside a myocyte is much lower. Why is the concentration so low in the cell? What happens to glucose after entry into the cell? Glucose is administered intravenously as a food source in certain clinical situations. Given that the transformation of glucose to glucose 6-phosphate consumes ATP, why not administer intravenous glucose 6-phosphate instead?

Answer Glucose enters cells and is immediately exposed to hexokinase, which converts it to glucose 6-phosphate using the energy of ATP. This reaction is highly exergonic ($\Delta G'^\circ = -16.7$ kJ/mol), and formation of glucose 6-phosphate is strongly favored. Because the glucose transporter is specific for glucose, glucose 6-phosphate cannot leave the cell and must be stored (as glycogen) or metabolized via glycolysis. For this same reason, glucose 6-phosphate added intravenously cannot enter cells.

15. Metabolism of Glycerol Glycerol obtained from the breakdown of fat is metabolized by conversion to dihydroxyacetone phosphate, a glycolytic intermediate, in two enzyme-catalyzed reactions. Propose a reaction sequence for glycerol metabolism. On which known enzyme-catalyzed reactions is your proposal based? Write the net equation for the conversion of glycerol to pyruvate according to your scheme.

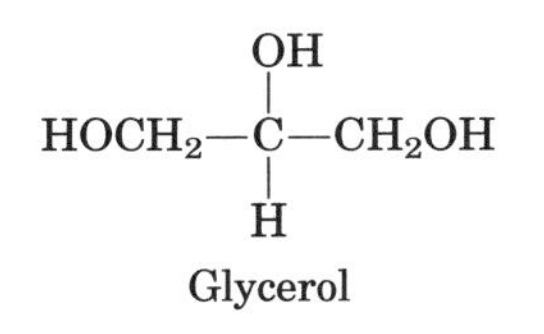

Glycerol

Answer Glycerol enters metabolism at the center of the glycolytic sequence and is converted to dihydroxyacetone phosphate in two steps, catalyzed by kinase and a dehydrogenase, respectively:

$$\text{Glycerol} + \text{ATP} \longrightarrow \text{glycerol 3-phosphate} + \text{ADP}$$
$$\text{Glycerol 3-phosphate} + \text{NAD}^+ \longrightarrow \text{dihydroxyacetone phosphate} + \text{NADH} + \text{H}^+$$

Dihydroxyacetone phosphate is converted to pyruvate in the payoff portion of glycolysis. The overall equation for conversion of glycerol to pyruvate is

$$\text{Glycerol} + 2\text{NAD}^+ + \text{ADP} + \text{P}_\text{i} \longrightarrow \text{pyruvate} + 2\text{NADH} + 2\text{H}^+ + \text{ATP}$$

16. Measurement of Intracellular Metabolite Concentrations Measuring the concentrations of metabolic intermediates in a living cell presents great experimental difficulties. Enzymes catalyze metabolic interconversions very rapidly, so a common problem associated with these measurements is that they reflect not the physiological concentrations of metabolites but the equilibrium concentrations in the cell extracts. A reliable experimental technique requires all enzyme-catalyzed reactions to be instantaneously stopped in the *intact* tissue so that intermediates do not undergo further change. This objective is accomplished by rapidly compressing the tissue between large aluminum plates cooled with liquid nitrogen (−190 °C), a process called **freeze-clamping.** After freezing, the tissue is powdered and the enzymes are inactivated by precipitation with perchloric acid. The precipitate is removed by centrifugation, and the clear supernatant extract is analyzed for metabolites. To calculate intracellular concentrations, the intracellular volume is determined from the total water content of the tissue and a measurement of the extracellular volume.

The intracellular concentrations of the substrates and products of the phosphofructokinase-1 reaction in rat heart tissue are given in the table below.

Metabolite	Concentration (mM)*
Fructose 6-phosphate	0.087
Fructose 1,6-bisphosphate	0.022
ATP	11.4
ADP	1.32

Source: From Williamson, J.R. (1965) Glycolytic control mechanisms I. Inhibition of glycolysis by acetate and pyruvate in the isolated, perfused rat heart. *J. Biol. Chem.* **240,** 2308–2321.

*Calculated as μmol/mL of intracellular water.

(a) Calculate the mass-action ratio, [fructose 1,6-bisphosphate][ADP]/[fructose 6-phosphate][ATP], for the PFK-1 reaction under physiological conditions.

(b) Given that $\Delta G'^\circ$ for the PFK-1 reaction is −14.2 kJ/mol, calculate the equilibrium constant for the reaction.

(c) Compare the values of the mass-action ratio and K'_{eq}. Is the physiological reaction at equilibrium? Explain. What does this experiment tell you about the role of PFK-1 as a regulatory enzyme?

Answer

(a) The mass-action ratio is $\frac{(22\ \mu\text{M})(1{,}320\ \mu\text{M})}{(87\ \mu\text{M})(11{,}400\ \mu\text{M})} = 0.029$

(b) $\Delta G'^\circ = -RT \ln K'_{eq}$

$\ln K'_{eq} = -\Delta G'^\circ/RT$

$= -(-14.2\ \text{kJ/mol})/(2.479\ \text{kJ/mol})$

$= 5.728$

$K'_{eq} = 307$

(c) It is clear that the PFK-1 reaction does not approach equilibrium in vivo. This indicates that the product, FBP, does not approximate equilibrium concentrations in vivo, because the pathways are "open systems," operating under near-steady state conditions, with substrates flowing in and products flowing out at all times. Thus, all FBP formed is utilized or turned over rapidly. The PFK-catalyzed reaction, being the rate-limiting step in glycolysis, is thus an excellent candidate for the critical regulatory point of the pathway.

17. Effect of O_2 Supply on Glycolytic Rates The regulated steps of glycolysis in intact cells can be identified by studying the catabolism of glucose in whole tissues or organs. For example, glucose consumption by heart muscle can be measured by artificially circulating blood through an isolated intact heart and measuring the blood glucose before and after the blood passes through the heart. If the circulating blood is deoxygenated, heart muscle consumes glucose at a steady rate. When oxygen is added to the blood, the rate of glucose consumption drops dramatically, then is maintained at the new, lower rate. Why?

Answer In the absence of O_2, the ATP needs of the cell are met by anaerobic glucose metabolism (fermentation) to form lactate. This produces a maximum of 2 ATPs per glucose. Because the aerobic metabolism of glucose produces far more ATP per glucose (by oxidative phosphorylation), far less glucose is needed to produce the same amount of ATP. The Pasteur effect was the first demonstration of the primacy of energy production—that is, of ATP levels—in controlling the rate of glycolysis.

18. Regulation of Phosphofructokinase-1 The effect of ATP on the allosteric enzyme PFK-1 is shown below. For a given concentration of fructose 6-phosphate, PFK-1 activity increases with increasing [ATP], but a point is reached beyond which increasing [ATP] inhibits the enzyme.

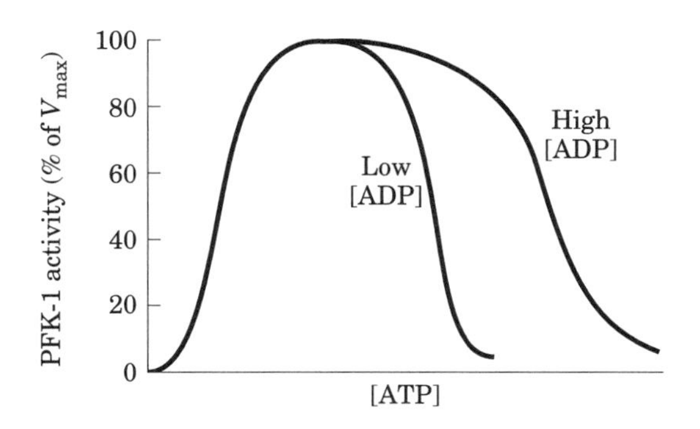

(a) Explain how ATP can be both a substrate and an inhibitor of PFK-1. How is the enzyme regulated by ATP?

(b) How does [ATP] regulate glycolysis?

(c) Explain why the inhibition of PFK-1 by ATP is diminished when [ADP] is high (see graph).

Answer

(a) In addition to binding sites for substrate(s), allosteric enzymes have binding sites for regulatory metabolites. Binding of effectors to these regulatory sites leads to a modification of enzyme activity by altering its V_{max} or K_m value. ATP is both a substrate and an allosteric inhibitor of PFK-1. Binding of ATP to the catalytic site increases activity, whereas binding to allosteric site inhibits activity.

(b) Because ATP is a negative regulator of PFK-1, elevation of ATP when energy is abundant inhibits the enzyme and thus the flux of metabolites through the glycolytic pathway.

(c) The graph indicates that the addition of ADP suppresses the inhibition of PFK-1 by ATP. Since the total adenine nucleotide pool is fairly constant in all cells, utilization of ATP leads to an increase in ADP. The data indicate that the activity of the enzyme may be regulated in vivo by the ratio of [ATP]/[ADP].

19. Enzyme Activity and Physiological Function The V_{max} of glycogen phosphorylase from skeletal muscle is much larger than the V_{max} of the same enzyme from liver.

(a) What is the physiological function of glycogen phosphorylase in skeletal muscle? In liver?

(b) Why does the V_{max} of the muscle enzyme need to be larger than that of the liver enzyme?

Answer

(a) The role of glycogen and its metabolism differs in muscle and liver. *In muscle*, glycogen is broken down to supply energy (ATP), via glycolysis and lactic acid fermentation. Glycogen phosphorylase catalyzes the conversion of stored glycogen to glucose 1-phosphate, which is converted to glucose 6-phosphate and thus enters glycolysis. During strenuous activity, muscle becomes anaerobic and large quantities of glucose 6-phosphate undergo lactic acid fermentation to form the necessary ATP.

In the liver, glycogen is used to maintain the level of glucose in the blood (primarily between meals). In this case, the glucose 6-phosphate is converted to glucose and transported into the bloodstream.

(b) Strenuous muscular activity requires large amounts of ATP, which must be formed rapidly and efficiently. This requires that glycogen phosphorylase have a high ratio of V_{max}/K_m in muscle. This is not a critical requirement in liver tissue.

20. Glycogen Phosphorylase Equilibrium Glycogen phosphorylase catalyzes the removal of glucose from glycogen. The $\Delta G'^\circ$ for this reaction is 3.1 kJ/mol.

(a) Calculate the ratio of [P_i] to [glucose 1-phosphate] when the reaction is at equilibrium. (Hint: The removal of glucose units from glycogen does not change the glycogen concentration.)

(b) The measured ratio of [P_i] to [glucose 1-phosphate] in myocytes under physiological conditions is more than 100 to 1. Why are the equilibrium and physiological ratios different? What is the possible significance of this difference?

Answer

(a) First we need to calculate the equilibrium constant from the equation:

$$\begin{aligned}\Delta G'^\circ &= -RT \ln K'_{eq}\\ \ln K'_{eq} &= -\Delta G'^\circ/RT\\ &= -(3.1\ \text{kJ/mol})/(2.479\ \text{kJ/mol})\\ &= -1.251\\ K'_{eq} &= 0.29\end{aligned}$$

For the glycogen phosphorylase reaction:

$$\text{Glycogen}_n + P_i = \text{glycogen}_{n-1} + \text{glucose 1-phosphate}$$

$$\text{so, } K'_{eq} = \frac{(\text{glycogen}_{n-1})(\text{glucose 1-phosphate})}{(\text{glycogen}_n)(P_i)}$$

but since the concentration of glycogen remains constant, these terms cancel and the expression becomes

$$K'_{eq} = \frac{\text{glucose 1-phosphate}}{P_i}$$

This may be arranged to $\dfrac{P_i}{\text{glucose 1-phosphate}} = \dfrac{1}{K'_{eq}} = \dfrac{1}{0.29} \approx 3.5$

(b) Given that the ration of [P_i] to [glucose 1-phosphate] in muscle cells is greater than 100/1, the rate at which glucose 1-phosphate can be removed by phosphoglucomutase is greater than the rate at which it can be produced by glycogen phosphorylase. This indicates that the glycogen phosphorylase-catalyzed reaction is the rate-limiting step in glycogen breakdown.

21. Regulation of Glycogen Phosphorylase In muscle tissue, the rate of conversion of glycogen to glucose 6-phosphate is determined by the ratio of phosphorylase *a* (active) to phosphorylase *b* (less active). What happens to the rate of glycogen breakdown if a muscle preparation containing glycogen phosphorylase is treated with (a) phosphorylase kinase and ATP; (b) phosphorylase phosphatase; (c) epinephrine?

Answer

(a) Treatment with the kinase and ATP converts glycogen phosphorylase to the more active, phosphorylated form; glycogen breakdown will accelerate.

(b) Treatment with the phosphatase converts the active phosphorylase *a* to the less active phosphorylase *b*; glycogen breakdown will slow down.

(c) Addition of epinephrine to muscle tissue causes the synthesis of cyclic AMP, which in turn activates the phosphorylase *b* kinase (see Fig. 13–15). The kinase converts phosphorylase *b* (less active) into phosphorylase *a* (more active); glycogen breakdown will accelerate.

22. Enzyme Defects in Carbohydrate Metabolism Summaries of three clinical case studies follow. For each case determine which enzyme is defective and designate the appropriate treatment, from the lists provided. Justify your choices. Answer the questions contained in each case study.

Case A The patient develops vomiting and diarrhea shortly after milk ingestion. A lactose tolerance test is administered. The patient ingests a standard amount of lactose, and the glucose and galactose concentrations in blood plasma are measured at intervals. In lactose-tolerant individuals the levels increase to a maximum in about 1 h, then decline. Explain why. The patient's blood glucose and galactose levels do not increase during the test. Explain why.

Case B The patient develops vomiting and diarrhea after ingestion of milk. His blood is found to have a low concentration of glucose but a much higher than normal concentration of reducing sugars. A urine test reveals galactose in the urine. Why is the level of reducing sugar in the blood high? Why does galactose appear in the urine?

Case C The patient is lethargic and her liver is enlarged. A liver biopsy shows large amounts of excess glycogen. She also has lower than normal blood glucose. Why does this patient have low blood glucose?

Defective Enzyme

(a) Muscle phosphofructokinase-1
(b) Phosphomannose isomerase
(c) Galactose 1-phosphate uridylyltransferase
(d) Liver glycogen phosphorylase
(e) Triose kinase
(f) Lactase of intestinal epithelial cells
(g) Maltase of intestinal epithelial cells

Treatment

1. Frequent regular feedings
2. Fat-free diet
3. Low-lactose diet
4. Large doses of niacin (the precursor of NAD^+)

Answer

Case A: (f). Lactase in the intestinal mucosa hydrolyzes milk lactose to glucose and galactose, causing the levels of these sugars to increase transiently after milk ingestion. A patient lacking the enzyme would not exhibit an increase in these sugars but would demonstrate symptoms of lactose toxicity. Such a patient should exclude lactose (milk) from the diet (treatment 3).

Case B: (c). Galactose 1-phosphate uridylyltransferase is an enzyme involved in conversion of galactose to glucose so that the former can enter glycolysis. Absence of this enzyme leads to accumulation of galactose in the blood and excretion in urine. Patients with this deficiency should be on a low-lactose diet (treatment 3).

Case C: (d). Liver glycogen functions as a source of blood glucose. The liver glycogen phosphorylase must be low or defective if glycogen accumulates and blood glucose is low. The patient should eat light meals regularly and frequently (treatment 1).

23. Severity of Clinical Symptoms Due to Enzyme Deficiency The clinical symptoms of the two forms of galactosemia—deficiency of either galactokinase or UDP glucose → galactose 1-phosphate uridylyltransferase—show radically different severity. Although both types produce gastric discomfort after milk ingestion, deficiency of the latter enzyme also leads to liver, kidney, spleen, and brain dysfunction and eventual death. What products accumulate in the blood and tissues with each type of enzyme deficiency? Estimate the relative toxicities of these products from the above information.

Answer In galactokinase deficiency, galactose accumulates; in galactose 1-phosphate uridylyltransferase deficiency, galactose 1-phosphate accumulates (see Fig. 15–14). The latter is clearly more toxic.

chapter

16

The Citric Acid Cycle

1. **Balance Sheet for the Citric Acid Cycle** The citric acid cycle has eight enzymes: citrate synthase, aconitase, isocitrate dehydrogenase, α-ketoglutarate dehydrogenase, succinyl-CoA synthetase, succinate dehydrogenase, fumarase, and malate dehydrogenase.
   (a) Write a balanced equation for the reaction catalyzed by each enzyme.
   (b) Name the cofactor(s) required by each enzyme reaction.
   (c) For each enzyme determine which of the following describes the type of reaction(s) catalyzed: condensation (carbon–carbon bond formation); dehydration (loss of water); hydration (addition of water); decarboxylation (loss of CO_2); oxidation-reduction; substrate-level phosphorylation; isomerization.
   (d) Write a balanced net equation for the catabolism of acetyl-CoA to CO_2.

Answer

Citrate synthase
(a) Acetyl-CoA + oxaloacetate + $H_2O \longrightarrow$ citrate + CoA + H^+
(b) CoA
(c) Condensation

Aconitase
(a) Citrate $\longrightarrow$ isocitrate
(b) No cofactors
(c) Isomerization

Isocitrate dehydrogenase
(a) Isocitrate + $NAD^+ \longrightarrow$ α-ketoglutarate + CO_2 + NADH + H^+
(b) NAD^+
(c) Oxidative decarboxylation

α-Ketoglutarate dehydrogenase
(a) α-ketoglutarate + NAD^+ + CoA $\longrightarrow$ succinyl-CoA + CO_2 + NADH + H^+
(b) NAD^+, CoA, thiamine pyrophosphate
(c) Oxidative decarboxylation

Succinyl-CoA synthetase
(a) Succinyl-CoA + P_i + GDP $\longrightarrow$ succinate + CoA + GTP
(b) CoA
(c) Phosphorylation and acyl transfer

Succinate dehydrogenase

(a) Succinate + FAD $\longrightarrow$ fumarate + $FADH_2$

(b) FAD

(c) Oxidation

Fumarase

(a) Fumarate + $H_2O \longrightarrow$ malate

(b) None

(c) Hydration

Malate dehydrogenase

(a) Malate + $NAD^+ \longrightarrow$ oxaloacetate + NADH + H^+

(b) NAD^+

(c) Oxidation

(d) The net equation for the catabolism of acetyl-CoA is

$$\text{Acetyl-CoA} + 3NAD^+ + FAD + GDP + P_i + 2H_2O \longrightarrow 2CO_2 + CoA + 3NADH + FADH_2 + GTP + 2H^+$$

2. Recognizing Oxidation and Reduction Reactions One biochemical strategy of many living organisms is the stepwise oxidation of organic compounds to CO_2 and H_2O and the conservation of a major part of the energy thus produced in the form of ATP. It is important to be able to recognize oxidation-reduction processes in metabolism. Reduction of an organic molecule results from the hydrogenation of a double bond (Eqn 1 below) or of a single bond with accompanying cleavage (Eqn 2). Conversely, oxidation results from dehydrogenation. In biochemical redox reactions, the coenzymes NAD and FAD dehydrogenate/hydrogenate organic molecules in the presence of the proper enzymes.

For each of the metabolic transformations in (a) through (h), determine whether oxidation or reduction has occurred. Balance each transformation by inserting H—H and, where necessary, H_2O.

$$\underset{\text{Acetaldehyde}}{CH_3\text{—}C(=O)\text{—}H} + H\text{—}H \underset{\text{oxidation}}{\overset{\text{reduction}}{\rightleftharpoons}} \left[CH_3\text{—}C(=O)(H)\cdots H\text{—}H \right] \underset{\text{oxidation}}{\overset{\text{reduction}}{\rightleftharpoons}} \underset{\text{Ethanol}}{CH_3\text{—}CH(OH)\text{—}H} \quad (1)$$

$$\underset{\text{Acetate}}{CH_3\text{—}C(=O)\text{—}O^-} + H^+ + H\text{—}H \underset{\text{oxidation}}{\overset{\text{reduction}}{\rightleftharpoons}} \left[CH_3\text{—}C(=O)\text{—}O^- \cdots H\text{—}H \;\; H^+ \right] \underset{\text{oxidation}}{\overset{\text{reduction}}{\rightleftharpoons}} \underset{\text{Acetaldehyde}}{CH_3\text{—}C(=O)\text{—}H} + H\text{—}O\text{—}H \quad (2)$$

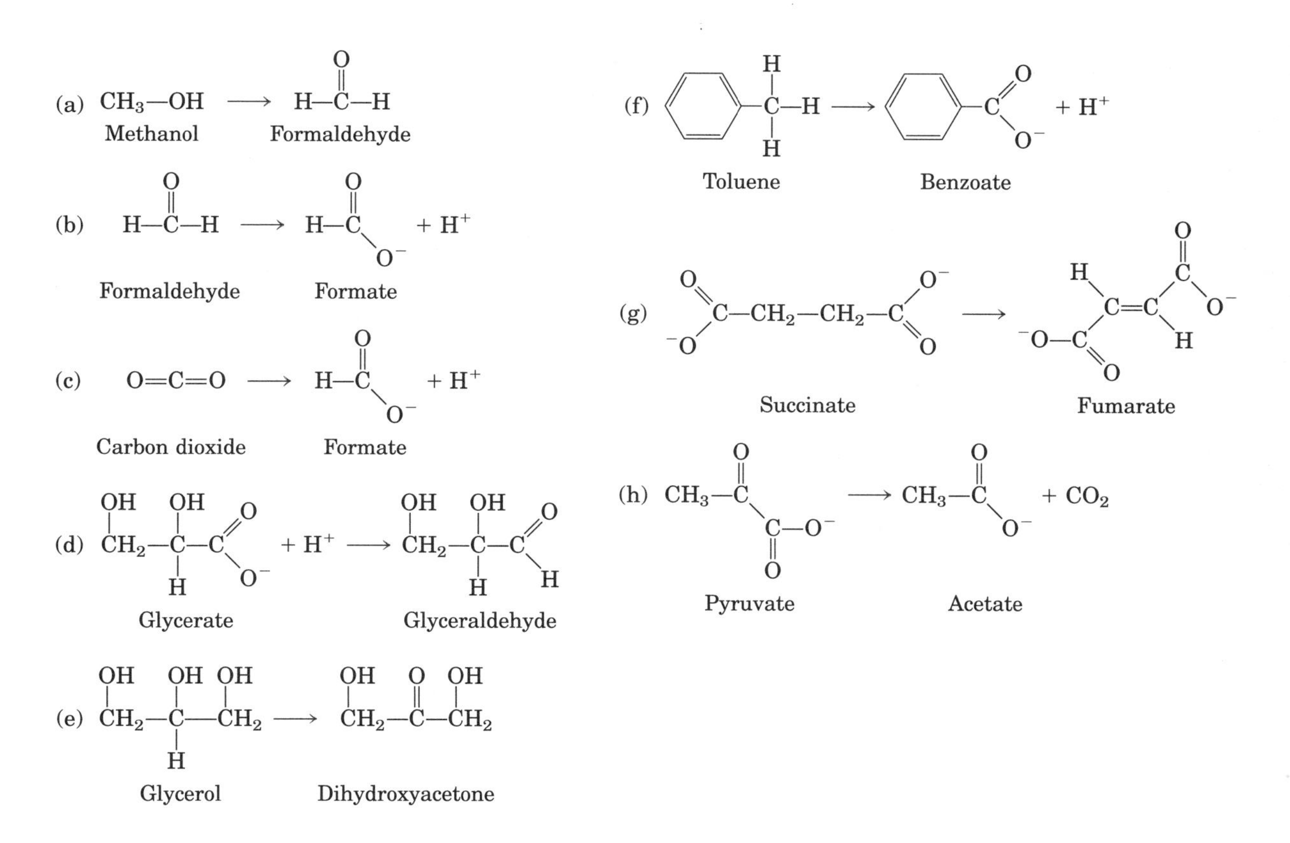

Answer Keep in mind that oxidation is the loss of electrons and accompanying H^+, whereas reduction is the gain of electrons (or H—H).

(a) Oxidation: Methanol ⟶ formaldehyde + H—H

(b) Oxidation: Formaldehyde ⟶ formate + H—H

(c) Reduction: CO_2 + H—H ⟶ formate + H^+

(d) Reduction: Glycerate + H—H + H^+ ⟶ glyceraldehyde + H_2O

(e) Oxidation: Glycerol ⟶ dihydroxyacetone + H—H

(f) Oxidation: Toluene + $2H_2O$ ⟶ benzoate + H^+ + 3H—H

(g) Oxidation: Succinate ⟶ fumarate + H—H

(h) Oxidation: Pyruvate + H_2O ⟶ acetate + CO_2 + H—H

3. Relationship between Energy Release and the Oxidation State of Carbon A eukaryotic cell can use glucose ($C_6H_{12}O_6$) and hexanoic acid ($C_6H_{14}O_2$) as fuels for cellular respiration. On the basis of their structural formulas, which substance releases more energy on complete combustion to CO_2 and H_2O?

Answer From the structural formulas, we see that the carbon-bound H/C ratio of hexanoic acid (11/6) is higher than that of glucose (7/6). Hexanoic acid is more reduced and yields more energy upon complete combustion to CO_2 and H_2O.

4. Nicotinamide Coenzymes as Reversible Redox Carriers The nicotinamide coenzymes (see Fig. 14–15) can undergo reversible oxidation-reduction reactions with specific substrates in the presence of the appropriate dehydrogenase. In these reactions, NADH + H^+ serves as the hydrogen source, as described in Problem 2. Whenever the coenzyme is oxidized, a substrate must be simultaneously reduced:

$$\underset{\text{Oxidized}}{\text{Substrate}} + \underset{\text{Reduced}}{\text{NADH}} + H^+ \rightleftharpoons \underset{\text{Reduced}}{\text{product}} + \underset{\text{Oxidized}}{\text{NAD}^+}$$

For each of the reactions in (a) through (f), determine whether the substrate has been oxidized or reduced or is unchanged in oxidation state (see Problem 2). If a redox change has occurred, balance the reaction with the necessary amount of NAD^+, NADH, H^+, and H_2O. The objective is to recognize when a redox coenzyme is necessary in a metabolic reaction.

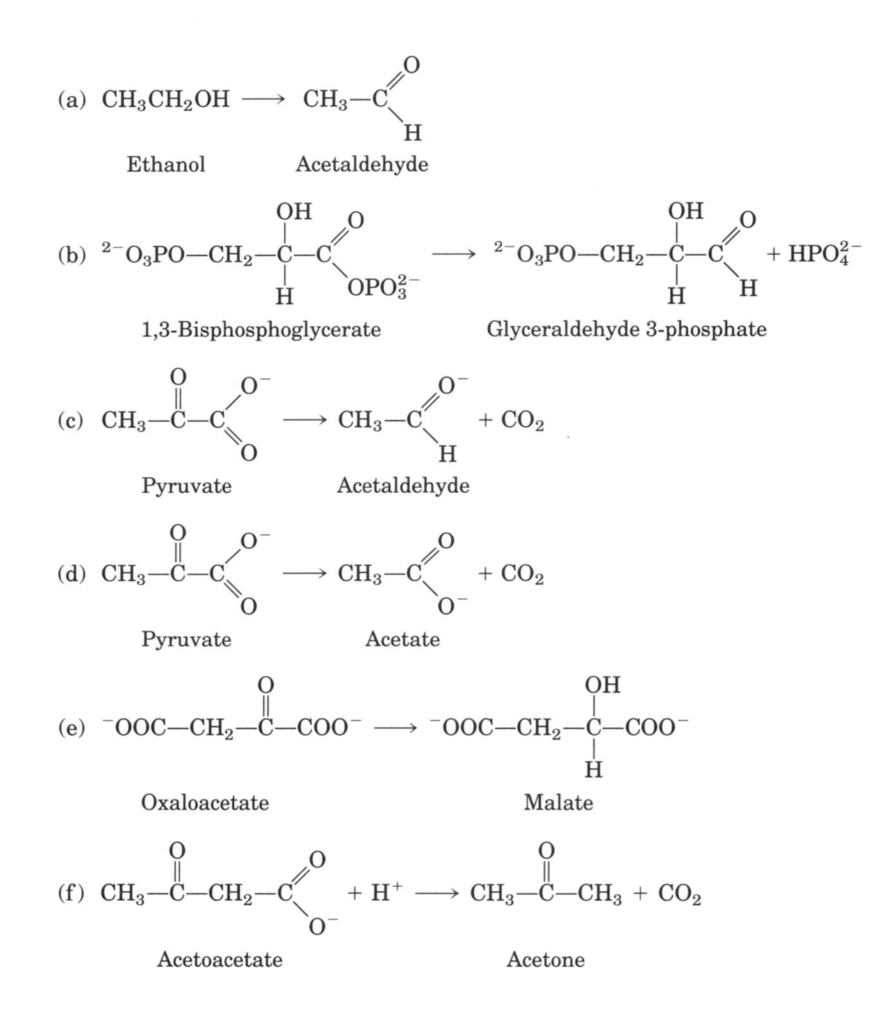

Answer

(a) Oxidized: Ethanol + $NAD^+ \longrightarrow$ acetaldehyde + NADH + H^+

(b) Reduced: 1,3-Bisphosphoglycerate + NADH + $H^+ \longrightarrow$ glyceraldehyde 3-phosphate + NAD^+ + P_i

(c) Unchanged: Pyruvate + $H^+ \longrightarrow$ acetaldehyde + CO_2

(d) Oxidized: Pyruvate + $NAD^+ \longrightarrow$ acetate + CO_2 + NADH + H^+

(e) Reduced: Oxaloacetate + NADH + $H^+ \longrightarrow$ malate + NAD^+

(f) Unchanged: Acetoacetate + $H^+ \longrightarrow$ acetone + CO_2

5. Stimulation of Oxygen Consumption by Oxaloacetate and Malate In the early 1930s, Albert Szent-Györgyi reported the interesting observation that the addition of small amounts of oxaloacetate or malate to suspensions of minced pigeon-breast muscle stimulated the oxygen consumption of the preparation. Surprisingly, the amount of oxygen consumed was about seven times more than the amount necessary for complete oxidation (to CO_2 and H_2O) of the added oxaloacetate or malate. Why did the addition of oxaloacetate or malate stimulate oxygen consumption? Why was the amount of oxygen consumed so much greater than the amount necessary to completely oxidize the added oxaloacetate or malate?

Answer

Oxygen consumption is a measure of the activity of the first two stages of cellular respiration: glycolysis and the citric acid cycle. Initial nutrients being oxidized are carbohydrates and lipids. Since several intermediates of the citric acid cycle can be siphoned off into biosynthetic pathways, the cycle may slow down for lack of oxaloacetate in the citrate synthase reaction, and acetyl-CoA will accumulate. Addition of oxaloacetate or malate (converted to oxaloacetate by malate dehydrogenase) will stimulate the cycle and allow it to use the accumulated acetyl-CoA. This stimulates respiration. Oxaloacetate is regenerated in the cycle, so addition of oxaloacetate (or malate) stimulates the oxidation of a much larger amount of acetyl-CoA.

6. Formation of Oxaloacetate in a Mitochondrion In the last reaction of the citric acid cycle, malate is dehydrogenated to regenerate the oxaloacetate necessary for the entry of acetyl-CoA into the cycle:

$$\text{L-Malate} + \text{NAD}^+ \longrightarrow \text{oxaloacetate} + \text{NADH} + \text{H}^+ \qquad \Delta G'^\circ = 30 \text{ kJ/mol}$$

(a) Calculate the equilibrium constant for this reaction at 25 °C.

(b) Because $\Delta G'^\circ$ assumes a standard pH of 7, the equilibrium constant calculated in (a) corresponds to

$$K'_{eq} = \frac{[\text{oxaloacetate}][\text{NADH}]}{[\text{L-malate}][\text{NAD}^+]}$$

The measured concentration of L-malate in rat liver mitochondria is about 0.20 mM when $[\text{NAD}^+]/[\text{NADH}]$ is 10. Calculate the concentration of oxaloacetate at pH 7 in these mitochondria.

(c) Rat liver mitochondria are roughly spherical, with a diameter of about 2 μm. To appreciate the magnitude of the mitochondrial oxaloacetate concentration, calculate the number of oxaloacetate molecules in a single rat liver mitochondrion.

Answer

(a)
$$\begin{aligned}\Delta G'^\circ &= -RT \ln K'_{eq} \\ \ln K'_{eq} &= -\Delta G'^\circ/RT \\ &= -(30 \text{ kJ/mol})/(2.479 \text{ kJ/mol}) \\ &= -12.10 \\ K'_{eq} &= 5.6 \times 10^{-6}\end{aligned}$$

(b)
$$\begin{aligned}\text{Since } K'_{eq} &= ([\text{OAA}][\text{NADH}])/([\text{malate}][\text{NAD}^+]) \\ [\text{oxaloacetate}] &= (K'_{eq})\,[\text{malate}][\text{NAD}^+]/[\text{NADH}] \\ &= (5.6 \times 10^{-6})(2 \times 10^{-4})(10) \\ &= 1.12 \times 10^{-8}\ \text{M} \approx 1.1 \times 10^{-8}\ \text{M}\end{aligned}$$

(c) The volume of a sphere is 4/3 πr^3, thus the volume of a mitochondrion is

$$\begin{aligned}1.33(3.14)(1 \times 10^{-3} \text{ mm})^3 &= 4.19 \times 10^{-9} \text{ mm}^3 \\ &= 4.19 \times 10^{-15} \text{ L}\end{aligned}$$

Given the concentration of oxaloacetate and Avogadro's number, we can calculate the number of molecules in a mitochondrion:

$$(1.1 \times 10^{-8} \text{ mol/L})(6.023 \times 10^{23} \text{ molecules/mol})(4.19 \times 10^{-15} \text{ L}) = 28.3 \approx 28 \text{ molecules}$$

7. Energy Yield from the Citric Acid Cycle The reaction catalyzed by succinyl-CoA synthetase produces the high-energy compound GTP. How is the free energy contained in GTP incorporated into the cellular ATP pool?

Answer The terminal phosphoryl group in GTP can be transferred to ADP in a reaction that is catalyzed by nucleoside diphosphate kinase and has an equilibrium constant of unity: $\text{GTP} + \text{ADP} \rightarrow \text{GDP} + \text{ATP}$

8. Respiration Studies in Isolated Mitochondria Cellular respiration can be studied in isolated mitochondria by measuring oxygen consumption under different conditions. If 0.01 M sodium malonate is added to actively respiring mitochondria that are using pyruvate as fuel source, respiration soon stops and a metabolic intermediate accumulates.

(a) What is the structure of this intermediate?

(b) Explain why it accumulates.

(c) Explain why oxygen consumption stops.

(d) Aside from removing the malonate, how can this inhibition of respiration be overcome? Explain.

Answer Malonate is a structural analog of succinate and a competitive inhibitor of succinate dehydrogenase.

(a) Succinate: $^{-}OOC{-}CH_2{-}CH_2{-}COO^{-}$

(b) When succinate dehydrogenase is inhibited, succinate accumulates.

(c) Inhibition of any reaction in a pathway causes the substrate of that reaction to accumulate. Because this substrate is also the product of the preceding reaction, its accumulation changes the effective ΔG of that reaction, and so on for all the steps in the pathway. The net rate of the pathway (or cycle) slows down and eventually becomes almost negligible. In the case of the citric acid cycle, ceasing to produce the primary product, NADH, has the effect of stopping electron transport and consumption of oxygen, the end acceptor of electrons derived from NADH.

(d) Since malonate is a competitive inhibitor, the addition of large amounts of succinate will overcome the inhibition.

9. Labeling Studies in Isolated Mitochondria The metabolic pathways of organic compounds have often been delineated by using a radioactively labeled substrate and following the fate of the label.

(a) How can you determine whether glucose added to a suspension of isolated mitochondria is metabolized to CO_2 and H_2O?

(b) Suppose you add a brief pulse of [3-^{14}C]pyruvate (labeled in the methyl position) to the mitochondria. After one turn of the citric acid cycle, what is the location of the ^{14}C in the oxaloacetate? Explain by tracing the ^{14}C label through the pathway. How many turns of the cycle are required to release all of the [3-^{14}C]pyruvate as CO_2?

Answer

(a) The use of uniformly labeled glucose (^{14}C in all carbon atoms) allows one to observe its metabolism. Since glucose is converted to pyruvate, which produces CO_2 in the citric acid cycle, the evolution of $^{14}CO_2$ by mitochondria would confirm the metabolism of glucose to CO_2 and H_2O.

(b) One turn of the cycle would produce oxaloacetate with label equally distributed between C-2 and C-3 (see Box 16–2). Since the second turn of the cycle would release half of the label and every subsequent turn would release another half, it would take an infinite number of turns to release *all* of the labeled carbon.

10. [1-^{14}C]Glucose Catabolism An actively respiring bacterial culture is briefly incubated with [1-^{14}C]glucose, and the glycolytic and citric acid cycle intermediates are isolated. Where is the ^{14}C in each of the intermediates listed below? Consider only the initial incorporation of ^{14}C, in the first pass of labeled glucose through the pathways.

(a) Fructose 1,6-bisphosphate
(b) Glyceraldehyde 3-phosphate
(c) Phosphoenolpyruvate
(d) Acetyl-CoA
(e) Citrate
(f) α-Ketoglutarate
(g) Oxaloacetate

Answer Figures 15–2 and 15–4 and Box 16–2 outline the fate of all the carbon atoms of glucose. If [1-^{14}C] glucose is fed to bacteria for a very short time, the label can be found as follows:

(a) C-1
(b) C-3
(c) C-3
(d) C-2 (methyl group)
(e) Equally distributed in the methylene ($—CH_2—$) carbons
(f) C-4
(g) Equally distributed in C-2 and C-3, the methylene ($—CH_2—$) carbons

11. Role of the Vitamin Thiamin People with beriberi, a disease caused by thiamin deficiency, have elevated levels of blood pyruvate and α-ketoglutarate, especially after consuming a meal rich in glucose. How are these effects related to thiamin deficiency?

Answer Thiamin is required for the synthesis of thiamin pyrophosphate (TPP), a prosthetic group in the pyruvate dehydrogenase and α-ketoglutarate dehydrogenase enzyme complexes. A thiamin deficiency reduces the activity of these enzyme complexes and causes the observed accumulation of precursors.

12. Synthesis of Oxaloacetate by the Citric Acid Cycle Oxaloacetate is formed in the last step of the citric acid cycle by the NAD^+-dependent oxidation of L-malate. Can a net synthesis of oxaloacetate from acetyl-CoA occur using only the enzymes and cofactors of the citric acid cycle, without depleting the intermediates of the cycle? Explain. How is oxaloacetate that is lost from the cycle (to biosynthetic reactions) replenished?

Answer In the citric acid cycle, the entering acetyl-CoA combines with oxaloacetate to form citrate. One turn of the cycle regenerates oxaloacetate and produces two CO_2 molecules. There is *no* net synthesis of oxaloacetate in the cycle. If any cycle intermediates are channeled into biosynthetic reactions, a formation of oxaloacetate is essential. Four enzymes can produce oxaloacetate (or malate) from pyruvate or phosphoenolpyruvate. Pyruvate carboxylase (liver) and PEP carboxykinase (muscle) are the most important in animals, and PEP carboxylase is most important in plants. Malic enzyme produces malate from pyruvate in many organisms (see Table 16–2).

13. Mode of Action of the Rodenticide Fluoroacetate Fluoroacetate, prepared commercially for rodent control, is also produced by a South African plant. After entering a cell, fluoroacetate is converted to fluoroacetyl-CoA in a reaction catalyzed by the enzyme acetate thiokinase:

$$F—CH_2COO^- + CoA\text{-}SH + ATP \longrightarrow F—CH_2(CO)—S\text{-}CoA + AMP + PP_i$$

The toxic effect of fluoroacetate was studied in an experiment using intact isolated rat heart. After the heart was perfused with 0.22 mM fluoroacetate, the measured rate of glucose uptake and glycolysis decreased, and glucose 6-phosphate and fructose 6-phosphate accumulated. Examination of the citric acid cycle intermediates revealed that their concentrations were below normal, except for citrate, with a concentration 10 times higher than normal.

(a) Where did the block in the citric acid cycle occur? What caused citrate to accumulate and the other cycle intermediates to be depleted?

(b) Fluoroacetyl-CoA is enzymatically transformed in the citric acid cycle. What is the structure of the end product of fluoroacetate metabolism? Why does it block the citric acid cycle? How might the inhibition be overcome?

(c) In the heart perfusion experiments, why did glucose uptake and glycolysis decrease? Why did hexose monophosphates accumulate?

(d) Why is fluoroacetate poisoning fatal?

Answer

(a) Fluoroacetate, an analog of acetate, can be activated to fluoroacetyl-CoA and can condense with oxaloacetate to form fluorocitrate. However, it is a strong competitive inhibitor of aconitase.

(b) Fluorocitrate is a structural analog of citrate and a strong competitive inhibitor of aconitase. The inhibition can be overcome by addition of large amounts of citrate.

(c) Both citrate and fluorocitrate are allosteric inhibitors of phosphofructokinase-1. Their accumulation thus inhibits glycolysis and glucose uptake, and causes the accumulation of hexose monophosphates.

(d) The net effect of fluoroacetate is to shut down ATP synthesis, both aerobic (oxidative) and anaerobic (fermentative).

14. Synthesis of L-Malate in Wine Making The tartness of some wines is due to high concentrations of L-malate. Write a sequence of reactions showing how yeast cells synthesize L-malate from glucose under anaerobic conditions in the presence of dissolved CO_2 (HCO_3^-). Note that the overall reaction for this fermentation cannot involve the consumption of nicotinamide coenzymes or citric acid cycle intermediates.

Answer In glycolysis,

$$\text{Glucose} + 2P_i + 2ADP + 2NAD^+ \longrightarrow 2\text{ pyruvate} + 2ATP + 2NADH + 2H^+ + 2H_2O$$

Then pyruvate carboxylase catalyzes the reaction

$$2\text{ Pyruvate} + 2CO_2 + 2ATP + 2H_2O \longrightarrow 2\text{ oxaloacetate} + 2ADP + 2P_i + 4H^+$$

In the citric acid cycle, malate dehydrogenase then catalyzes the reaction

$$2\text{ Oxaloacetate} + 2NADH + 2H^+ \longrightarrow 2\text{ L-malate} + 2NAD^+$$

which recycles nicotinamide coenzymes under anaerobic conditions. The overall reaction is

$$\text{Glucose} + 2CO_2 \longrightarrow 2\text{ L-malate} + 4H^+$$

The overall reaction produces four H^+, which increases the acidity and thus the tartness of the wine.

15. Net Synthesis of α-Ketoglutarate α-Ketoglutarate plays a central role in the biosynthesis of several amino acids. Write a sequence of enzymatic reactions that could result in the net synthesis of α-ketoglutarate from pyruvate. Your proposed sequence must not involve the net consumption of other citric acid cycle intermediates. Write an equation for the overall reaction and identify the source of each reactant.

Answer

Anaplerotic reactions replenish lost intermediates in the citric acid cycle. Synthesis of α-ketoglutarate from pyruvate occurs by the sequential action of pyruvate carboxylase (which makes extra molecules of oxaloacetate); plus pyruvate dehydrogenase and the citric acid cycle enzymes citrate synthase, aconitase, and isocitrate dehydrogenase. These make up a series of reactions.

$$\text{Pyruvate} + \text{ATP} + CO_2 \longrightarrow \text{oxaloacetate} + \text{ADP} + P_i$$

$$\text{Pyruvate} + NAD^+ + \text{CoA} \longrightarrow \text{Ac-CoA} + CO_2 + \text{NADH} + H^+$$

$$\text{Oxaloacetate} + \text{Ac-CoA} \longrightarrow \text{citrate} + \text{CoA}$$

$$\text{Citrate} \longrightarrow \text{Isocitrate}$$

$$\text{Isocitrate} + NAD^+ \longrightarrow \alpha\text{-ketoglutarate} + \text{NADH} + H^+ + CO_2$$

Net reaction:

$$2\ \text{Pyruvate} + \text{ATP} + 2NAD^+ + H_2O \longrightarrow \alpha\text{-ketoglutarate} + CO_2 + 2\text{NADH} + \text{ADP} + P_i + 3H^+$$

16. Regulation of Pyruvate Dehydrogenase In animal tissue, the rate of conversion of pyruvate to acetyl-CoA is regulated by the ratio of active, phosphorylated pyruvate dehydrogenase to inactive, unphosphorylated pyruvate dehydrogenase phosphate. Determine what happens to the rate of this reaction when a preparation of rabbit muscle mitochondria containing pyruvate dehydrogenase is treated with (a) pyruvate dehydrogenase kinase, ATP, and NADH; (b) pyruvate dehydrogenase phosphatase and Ca^{2+}; (c) malonate.

Answer Pyruvate dehydrogenase is regulated by covalent modification and by allosteric inhibitors. The mitochondrial preparation responds as follows: (a) Active pyruvate dehydrogenase (dephosphorylated) is converted to inactive pyruvate dehydrogenase (phosphorylated) and the rate of conversion of pyruvate to acetyl-CoA decreases. (b) The phosphoryl group on pyruvate dehydrogenase phosphate is removed enzymatically to yield active pyruvate dehydrogenase, which increases the rate of conversion of pyruvate to acetyl-CoA. (c) Malonate inhibits succinate dehydrogenase and citrate accumulates. The accumulated citrate inhibits citrate synthase, and acetyl-CoA accumulates. High levels of acetyl-CoA inhibit pyruvate dehydrogenase, and the rate of conversion of pyruvate to acetyl-CoA is reduced.

17. Commercial Synthesis of Citric Acid Citric acid is used as a flavoring agent in soft drinks, fruit juices, and numerous other foods. Worldwide, the market for citric acid is hundreds of millions of dollars per year. Commercial production uses the mold *Aspergillus niger*, acting on sucrose under carefully controlled conditions.

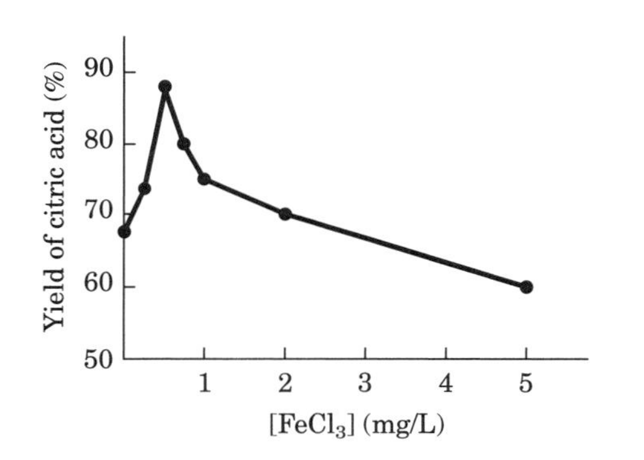

(a) The yield of citric acid is strongly dependent on the concentration of $FeCl_3$ in the culture medium, as indicated in the graph. Why does the yield decrease when the concentration of Fe^{3+} is above or below the optimal value of 0.5 mg/L?

(b) Write the sequence of reactions by which *A. niger* synthesizes citric acid from sucrose. Write an equation for the overall reaction.

(c) Does the commercial process require the culture medium to be aerated—that is, is this a fermentation or an aerobic process? Explain.

Answer

(a) Citrate is produced through the action of citrate synthase on oxaloacetate and acetyl-CoA. Although the citric acid cycle does not normally lead to the accumulation of intermediates, citrate synthase can be used for net synthesis of citrate when (1) there is a continuous influx of new oxaloacetate and acetyl-CoA, and (2) the transformation of citrate to isocitrate is blocked or at least restricted. *A. niger,* grown in a medium rich in sucrose but low in Fe^{3+}, meets those requirements. Citrate is normally transformed to isocitrate by aconitase, an Fe^{3+}-containing enzyme. An Fe^{3+} deficient medium restricts the synthesis of aconitase, and thus the breakdown of citrate is partially blocked; this intermediate accumulates and can be isolated in commercial quantities. Note that some aconitase activity is necessary—iron cannot be withheld entirely—or the mold will not thrive, as shown by the reduced yield at low $[Fe^{3+}]$. At high $[Fe^{3+}]$, aconitase is synthesized in increasing amounts. This leads to a decrease in the yield of citrate because citrate cycles through the citric acid cycle.

(b) Sucrose + H_2O $\longrightarrow$ glucose + fructose

$$\text{Glucose} + 2P_i + 2ADP + 2NAD^+ \longrightarrow 2\ \text{pyruvate} + 2ATP + 2NADH + 2H^+ + 2H_2O$$

$$\text{Fructose} + 2P_i + 2ADP + 2NAD^+ \longrightarrow 2\ \text{pyruvate} + 2ATP + 2NADH + 2H^+ + 2H_2O$$

$$2\ \text{Pyruvate} + 2NAD^+ + 2\text{CoA-SH} \longrightarrow 2\ \text{acetyl-SCoA} + 2NADH + 2CO_2$$

$$2\ \text{Pyruvate} + 2CO_2 + 2ATP + 2H_2O \longrightarrow 2\ \text{oxaloacetate} + 2ADP + 2P_i + 4H^+$$

$$2\ \text{Acetyl-SCoA} + 2\ \text{oxaloacetate} + 2H_2O \longrightarrow 2\ \text{citrate} + 2\text{CoA-SH} + 2H^+$$

The overall reaction is

$$\text{Sucrose} + H_2O + 2P_i + 2ADP + 6NAD^+ \longrightarrow 2\ \text{citrate} + 2ATP + 6NADH + 10H^+$$

(c) Note that the overall reaction consumes NAD^+. Since the cellular pool of this oxidized coenzyme is limited, it must be recycled by the electron-transfer chain with consumption of oxygen. Consequently, the overall conversion of sucrose to citric acid is an aerobic process and requires molecular oxygen.

18. Regulation of Citrate Synthase In the presence of saturating amounts of oxaloacetate, the activity of citrate synthase from pig heart tissue shows a sigmoid dependence on the concentration of acetyl-CoA, as shown in the graph. When succinyl-CoA is added, the curve shifts to the right and the sigmoid dependence is more pronounced.

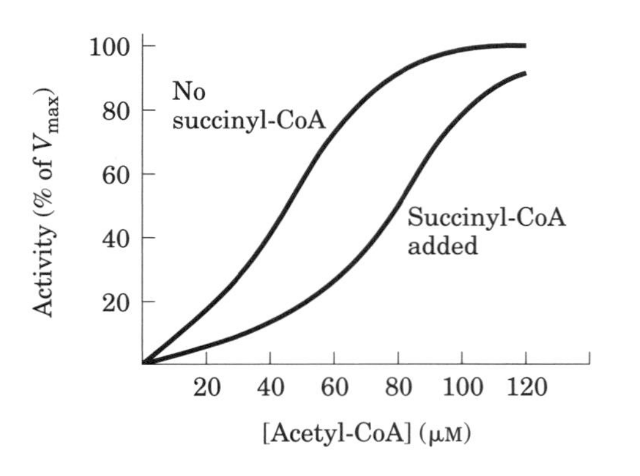

On the basis of these observations, suggest how succinyl-CoA regulates the activity of citrate synthase. (Hint: See Figure 8–26.) Why is succinyl-CoA an appropriate signal for regulation of the citric acid cycle? How does the regulation of citrate synthase control the rate of cellular respiration in pig heart tissue?

Answer Succinyl-CoA is the first 4-C intermediate of the citric acid cycle, occurring just prior to steps that form $FADH_2$ and cause overall transformation to oxaloacetate. Accumulation of ATP (a negative feedback signal) results in decreased rates of $FADH_2$ oxidation by the electron transport chain, which would in turn cause accumulation of succinyl-CoA, making this intermediate an appropriate signal from within the citric acid cycle for events occurring "downstream."

The curves in the graph show that succinyl-CoA shifts the half-saturation point ($S_{0.5}$) of the sigmoidal curve for Ac-CoA to the right, but does not alter V_{max}. This indicates that it may act directly as a competitive inhibitor vs. Ac-CoA, or that it may act by binding to a site separate from the active site.

Citrate synthase catalyzes the "first" step in the citric acid cycle, the entry point for input of Ac-CoA. In the absence of sufficient uptake of Ac-CoA by the cycle, it can instead be shunted toward fatty acid synthesis.

19. Regulation of Pyruvate Carboxylase The carboxylation of pyruvate by pyruvate carboxylase occurs at a very low rate unless acetyl-CoA, a positive allosteric modulator, is present. If you have just eaten a meal rich in fatty acids (triacylglycerols) but low in carbohydrates (glucose), how does this regulatory property shut down the oxidation of glucose to CO_2 and H_2O but increase the oxidation of acetyl-CoA derived from fatty acids?

Answer Fatty acid catabolism increases the level of acetyl-CoA, which stimulates pyruvate carboxylase. The resulting increase in oxaloacetate stimulates acetyl-CoA consumption through the citric acid cycle, causing the citrate and ATP concentrations to rise. These metabolites inhibit glycolysis at PFK-1 and inhibit pyruvate dehydrogenase, effectively slowing the utilization of sugars and pyruvate.

20. Relationship between Respiration and the Citric Acid Cycle Although oxygen does not participate directly in the citric acid cycle, the cycle operates only when O_2 is present. Why?

Answer Oxygen is the terminal electron acceptor in oxidative phosphorylation, and thus is needed to recycle NAD^+ from NADH. NADH is produced in greatest quantities by the oxidative reactions of the citric acid cycle. In the absence of O_2, NADH accumulates and allosterically inhibits pyruvate dehydrogenase and α-ketoglutarate dehydrogenase (see Fig. 16–15).

21. Regulation of Metabolism in Rabbit Muscle The intracellular use of glucose and glycogen is tightly regulated at four points. To compare the regulation of glycolysis when oxygen is plentiful and when it is depleted, consider the utilization of glucose and glycogen by rabbit leg muscle in two physiological settings: a resting rabbit, with low ATP demands, and a rabbit that sights its mortal enemy, the coyote, and dashes for shelter. For each setting, determine the relative levels (high, intermediate, or low) of AMP, ATP, citrate, and acetyl-CoA and how these levels affect the flow of metabolites through glycolysis by regulating specific enzymes. (In periods of stress, rabbit leg muscle produces much of its ATP by anaerobic glycolysis (lactic acid fermentation) and very little by oxidation of acetyl-CoA derived from fat breakdown.)

Answer A primary role of glycolysis is the production of ATP, and the pathway is regulated to ensure efficient ATP formation. The utilization of glycogen and glucose for energy purposes is regulated at the following steps: glycogen phosphorylase, phosphofructokinase-1, pyruvate kinase, and entry of actyl CoA into the citric aci cycle. In muscle, the primary regulatory metabolites are ATP, AMP, citrate, and acetyl-CoA. ATP is an inhibitor of glycogen phosphorylase and PFK-1; AMP stimulates both. Citrate inhibits PFK-1, and acetyl-CoA inhibits pyruvate kinase. Lack of O_2 leads to elevated levels of NADH, which inhibits pyruvate dehydrogenase and promotes fermentation of pyruvate to lactate.

Under resting conditions, ATP levels will be high and AMP levels low since the total adenine nucleotide pools are constant. Citrate and acetyl-CoA levels will be intermediate because O_2 is not limiting and the citric acid cycle will function. Under conditions of active exertion (running), O_2 becomes limiting and ATP synthesis decreases. Consequently, [ATP] becomes relatively low and [AMP] becomes relatively high, compared to conditions in the presence of O_2. Citrate and acetyl-CoA are low. These changes release the inhibition of glycolysis and stimulate lactic acid production.

22. Regulation of Metabolism in Migrating Birds Unlike the rabbit with its short dash, migratory birds require energy for extended periods of time. For example, ducks generally fly several thousand miles during their annual migration. The flight muscles of migratory birds have a high oxidative capacity and obtain the necessary ATP through the oxidation of acetyl-CoA (obtained from fats) via the citric acid cycle. Compare the regulation of muscle glycolysis during short-term intense activity, as in the fleeing rabbit, and during extended activity, as in the migrating duck. Why must the regulation in these two settings be different?

Answer Migratory birds have a very efficient respiratory system to ensure that O_2 is available to flight muscles under stress (see Box 15–1). Birds also rely on the aerobic oxidation of fat, since this produces the greatest amount of energy per gram of fuel. Consequently, the migratory bird must regulate glycolysis so that glycogen is used only for short bursts of energy, not for the long-term stress of prolonged flight. The sprinting rabbit relies on breakdown of stored (liver) glycogen plus subsequent anaerobic glycolysis for short-term production of ATP for muscle action.

The regulation of these two means of ATP production is very different. Under aerobic conditions (see answer to Problem 21) glycolysis is inhibited by ATP levels that remain relatively high, due to feeding acetyl-CoA units derived from fat into the TCA cycle, followed by oxidative phosphorylation. Under anaerobic conditions, glycolysis is stimulated and metabolism of fats does not occur appreciably, since citrate and acetyl CoA are low and oxygen (the final acceptor of electronis in oxidative phosphorylation) is absent.

23. Thermodynamics of Citrate Synthase Reaction in Cells Citrate is formed by the condensation of acetyl-CoA with oxaloacetate, catalyzed by citrate synthase:

$$\text{Oxaloacetate} + \text{acetyl-CoA} + H_2O \longrightarrow \text{Citrate} + \text{CoA} + H^+$$

In rat heart mitochondria at pH 7.0 and 25 °C, the concentrations of reactants and products are: oxaloacetate, 1 μM; acetyl-CoA, 1 μM; citrate, 220 μM; and CoA, 65 μM. Given that the standard free-energy change for the citrate synthase reaction is −32.2 kJ/mol, what is the direction of metabolite flow through the citrate synthase reaction in cells of the rat heart? Explain.

Answer The free-energy change of the citrate synthase reaction in the cell is

$$\begin{aligned}\Delta G &= \Delta G'^{\circ} + RT \ln \frac{[\text{citrate}][\text{CoA}]}{[\text{OAA}][\text{Acetyl-CoA}]} \\ &= -32.2 \text{ kJ/mol} + (2.479 \text{ kJ/mol}) \ln \frac{(220 \times 10^{-6} \text{ M})(65 \times 10^{-6} \text{ M})}{(1 \times 10^{-6} \text{ M})(1 \times 10^{-6} \text{ M})} \\ &= -8.48 \text{ kJ/mol}\end{aligned}$$

Thus the citrate synthase reaction is exergonic and proceeds in the direction of citrate formation.

24. Reactions of the Pyruvate Dehydrogenase Complex Two of the steps in the oxidative decarboxylation of pyruvate (steps ④ and ⑤ in Fig. 16–6) do not involve any of the three carbons of pyruvate yet are essential to the operation of the pyruvate dehydrogenase complex. Explain.

Answer The pyruvate dehydrogenase complex can be thought of as performing five enzymatic reactions. The first three (see Fig. 16–6) catalyze the oxidation of pyruvate to acetyl-CoA and reduction of the enzyme. The last two reactions convert reduced enzyme and NAD^+ to oxidized enzyme and NADH + H^+. The moiety on the enzyme that is oxidized/reduced is the lipoamide cofactor.

chapter

17 Oxidation of Fatty Acids

1. **Energy in Triacylglycerols** On a per-carbon basis, where does the largest amount of biologically available energy in triacylglycerols reside: in the fatty acid portions or the glycerol portion? Indicate how knowledge of the chemical structure of triacylglycerols provides the answer.

   Answer The fatty acids of triacylglycerols are hydrocarbon in nature (except for the carboxyl group) and require a great deal of oxidation for complete catabolism. Glycerol on the other hand is partially oxidized, having an —OH group on each carbon. Thus, the fatty acids produce far more energy per carbon during oxidation than does glycerol. Triacylglycerols have an energy of oxidation more than twice that for the same weight of carbohydrates or proteins.

2. **Fuel Reserves in Adipose Tissue** Triacylglycerols, with their hydrocarbon-like fatty acids, have the highest energy content of the major nutrients.
   (a) If 15% of the body mass of a 70 kg adult consists of triacylglycerols, calculate the total available fuel reserve, in both kilojoules and kilocalories, in the form of triacylglycerols. Recall that 1.00 kcal = 4.18 kJ.
   (b) If the basal energy requirement is approximately 8,400 kJ/day (2,000 kcal/day), how long could this person survive if the oxidation of fatty acids stored as triacylglycerols were the only source of energy?
   (c) What would be the weight loss in pounds per day under such starvation conditions (1 lb = 0.454 kg)?

   Answer
   (a) Knowing that the energy value of stored triacylglycerol is 38 kJ/g, the available fuel reserve is

$$\begin{aligned}(0.15)(70 \times 10^3 \text{ g})(38 \text{ kJ/g}) &= 4 \times 10^5 \text{ kJ} \\ &= 9.5 \times 10^4 \text{ kcal}\end{aligned}$$

   (b) At a rate of 8.4×10^3 kJ/day, the fuel supply would last

$$(4 \times 10^5 \text{ kJ})/(8.4 \times 10^3 \text{ kJ/day}) = 48 \text{ days}$$

   (c) If the total triacylglycerol is used over a 48 day period, this represents a rate of weight loss of

$$\frac{(0.15)(70 \times 10^3 \text{ g})}{48 \text{ days}} = 218 \text{ g/day} \approx 0.5 \text{ lb/day}$$

3. **Common Reaction Steps in the Fatty Acid Oxidation Cycle and Citric Acid Cycle** Cells often use the same enzyme reaction pattern for analogous metabolic conversions. For example, the steps in the oxidation of pyruvate to acetyl-CoA and α-ketoglutarate to succinyl-CoA, although catalyzed by

different enzymes, are very similar. The first stage of fatty acid oxidation follows a reaction sequence closely resembling a sequence in the citric acid cycle. Use equations to show the analogous reaction sequences in the two pathways.

Answer The first three reactions in the β oxidation of fatty acyl–CoA molecules are analagous to three reactions of the citric acid cycle.

The fatty acyl–CoA dehydrogenase reaction is analogous to the succinate dehydrogenase reaction, as both are FAD-requiring oxidations:

$$\text{Succinate} + \text{FAD} \longrightarrow \text{fumarate} + \text{FADH}_2$$

$$\text{Fatty acyl–CoA} + \text{FAD} \longrightarrow \text{fatty acyl-trans-}\Delta^2\text{-enoyl-CoA} + \text{FADH}_2$$

The enoyl-CoA hydratase reaction is analogous to the fumarase reaction (both add water to an olefinic bond):

$$\text{Fumarate} + H_2O \longrightarrow \text{malate}$$

$$\text{Fatty acyl-trans-}\Delta^2\text{-enoyl-CoA} + H_2O \longrightarrow \text{L-}\beta\text{-hydroxyacyl-CoA}$$

The β-hydroxyacyl-CoA dehydrogenase reaction is analogous to the malate dehydrogenase reaction (both are NAD-requiring and act upon β-hydroxy acyl compounds):

$$\text{Malate} + \text{NAD}^+ \longrightarrow \text{oxalacetate} + \text{NADH}$$

$$\text{L-}\beta\text{-Hydroxyacyl-CoA} + \text{NAD}^+ \longrightarrow \beta\text{-ketoacyl-CoA} + \text{NADH}$$

4. Chemistry of the Acyl-CoA Synthetase Reaction Fatty acids are converted to their coenzyme A esters in a reversible reaction catalyzed by acyl-CoA synthetase:

$$\text{R—COO}^- + \text{ATP} + \text{CoA} \rightleftharpoons \text{R—}\overset{\overset{\displaystyle O}{\|}}{C}\text{—CoA} + \text{AMP} + \text{PP}_i$$

(a) The enzyme-bound intermediate in this reaction has been identified as the mixed anhydride of the fatty acid and adenosine monophosphate (AMP), acyl-AMP:

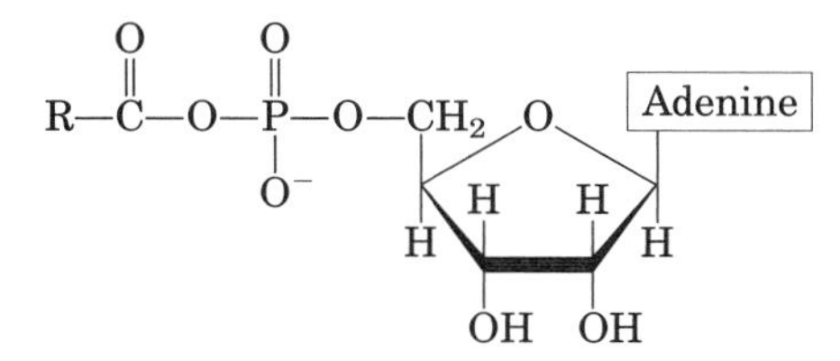

Write two equations corresponding to the two steps of the reaction catalyzed by acyl-CoA synthetase.

(b) The reaction in the above equation is readily reversible, with an equilibrium constant near 1. How can this reaction be made to favor formation of fatty acyl-CoA?

Answer The activation of carboxyl groups by ATP could in theory be accomplished by three types of reactions: the formation of acyl-phosphate + ADP, or acyl-ADP + P_i, or acyl-AMP + PP_i. All of these reactions are readily reversible. In order to create an activation reaction with a highly negative $\Delta G'^{\circ}$ (effectively irreversible), the third type of reaction (only) can be coupled to a pyrophosphatase, as in the synthesis of fatty acyl–CoA molecules.

(a) $\text{R—COO}^- + \text{ATP} \longrightarrow \text{acyl-AMP} + \text{PP}_i$

$\text{Acyl-AMP} + \text{CoA} \longrightarrow \text{acyl-CoA} + \text{AMP}$

(b) The hydrolysis of PP_i by an inorganic pyrophosphatase

5. **Oxidation of Tritiated Palmitate** Palmitate uniformly labeled with tritium (^3H) to a specific activity of 2.48×10^8 counts per minute (cpm) per micromole of palmitate is added to a mitochondrial preparation that oxidizes it to acetyl-CoA. The acetyl-CoA is isolated and hydrolyzed to acetate. The specific activity of the isolated acetate is 1.00×10^7 cpm/μmol. Is this result consistent with the β-oxidation pathway? Explain. What is the final fate of the removed tritium?

   Answer The β-oxidation pathway includes two dehydrogenase enzymes that remove hydrogen (H—H) from the fatty acyl-CoA chain at both $—CH_2—CH_2—$ and CH—OH structures (see answer to Problem 2). The net result of these two reactions is removal of one of the two hydrogens present at the point of formation of the enoyl-CoA intermediate. The two other H's that appear in the methyl group of acetyl-CoA come from water. If palmitate contains 16 carbons, each containing $(14 \times 2) + 3 = 31$ H's, then each 2-carbon unit contains roughly 4 or 1/8 of the total ^3H present. Thus, the amount expected to be present per acetyl-CoA is $0.5(2.48 \times 10^8 \text{ cpm})/8 = 1.55 \times 10^7$ cpm, somewhat higher than observed. Exchange with solvent water from β-ketoacyl-CoA could cause additional loss of ^3H.

6. **Compartmentation in β Oxidation** Free palmitate is activated to its coenzyme A derivative (palmitoyl-CoA) in the cytosol before it can be oxidized in the mitochondrion. If palmitate and [^{14}C]coenzyme A are added to a liver homogenate, palmitoyl-CoA isolated from the cytosolic fraction is radioactive, but that isolated from the mitochondrial fraction is not. Explain.

   Answer The transport of fatty acid molecules into mitochondria requires a shuttle system involving a fatty acyl–carnitine intermediate. Fatty acids are first converted to fatty acyl-CoA molecules by the action of acyl-CoA synthetases in the outer mitochondrial membrane, followed by (at the inner membrane of the mitochondrion) transfer to carnitine and release of CoA. After transport of fatty acyl–carnitine through the membrane, the fatty acyl group is transferred to mitochondrial CoA and is oxidized. The cytosolic and mitochondrial pools of CoA are kept separate, and no labeled CoA from the cytosolic pool will enter the mitochondrion.

7. **Comparative Biochemistry: Energy-Generating Pathways in Birds** One indication of the relative importance of various ATP-producing pathways is the V_{max} of certain enzymes of these pathways. The V_{max} of several enzymes from the pectoral muscles (chest muscles used for flying) of pigeon and pheasant are listed below.

Enzyme	V_{max} (μmol substrate/min/g tissue)	
	Pigeon	Pheasant
Hexokinase	3.0	2.3
Glycogen phosphorylase	18.0	120.0
Phosphofructokinase-1	24.0	143.0
Citrate synthase	100.0	15.0
Triacylglycerol lipase	0.07	0.01

   (a) Discuss the relative importance of glycogen metabolism and fat metabolism in generating ATP in the pectoral muscles of these birds.
   (b) Compare oxygen consumption in the two birds.
   (c) Judging from the data in the table, which bird is the long-distance flyer? Justify your answer.
   (d) Why were these particular enzymes selected for comparison? Would the activities of triose phosphate isomerase and malate dehydrogenase be equally good bases for comparison? Explain.

Answer

(a) In the pigeon, β oxidation predominates; in the pheasant, anaerobic glycolysis of glycogen predominates.

(b) Pigeon muscle would consume more oxygen.

(c) Fat contains more energy per gram than glycogen does. In addition, the anaerobic breakdown of glycogen is limited by the tissue's tolerance to lactate buildup. Thus, the pigeon, operating on the oxidative catabolism of fats, is the long-distance flyer.

(d) These enzymes are the regulatory enzymes of their respective pathways and thus limit ATP production rates.

8. Effect of Carnitine Deficiency An individual developed a condition characterized by progressive muscular weakness and aching muscle cramps. The symptoms were aggravated by fasting, exercise, and a high-fat diet. The homogenate of a skeletal muscle specimen from the patient oxidized added oleate more slowly than did control homogenates of muscle specimens from healthy individuals. When carnitine was added to the patient's muscle homogenate, the rate of oleate oxidation equaled that in the control homogenates. The patient was diagnosed as having a carnitine deficiency.

(a) Why did added carnitine increase the rate of oleate oxidation in the patient's muscle homogenate?

(b) Why were the patient's symptoms aggravated by fasting, exercise, and a high-fat diet?

(c) Suggest two possible reasons for the deficiency of muscle carnitine in this individual.

Answer

(a) The carnitine-mediated transport of fatty acids into mitochondria is the rate-limiting step in β oxidation (see Fig. 17–6). Carnitine deficiency decreases the rate of transport of fatty acids into mitochondria and thus the rate of β oxidation, so addition of carnitine would increase the rate of oxidation.

(b) Fasting, exercise, and a high fat diet cause an increased need for β oxidation of fatty acids. These conditions increase the need for the carnitine shuttle system and thus the severity of the symptoms of carnitine deficiency.

(c) The deficiency of carnitine may result from a deficiency of its precursor, lysine, or from a defect in one of the enzymes that synthesizes carnitine from this precursor.

9. Fatty Acids as a Source of Water Contrary to legend, camels do not store water in their humps, which actually consist of large fat deposits. How can these fat deposits serve as a source of water? Calculate the amount of water (in liters) that a camel can produce from 1 kg of fat. Assume for simplicity that the fat consists entirely of tripalmitoylglycerol.

Answer Oxidation of fatty acids produces water in significant amounts. From Equation 17–6:

$$\text{Palmitoyl-CoA} + 23O_2 + 108P_i + 108ADP \longrightarrow \text{CoA} + 16CO_2 + 108ATP + 23H_2O$$

we know that the oxidation of 1 mol of palmitoyl-CoA produces 23 mol of water. Tripalmitoin (glycerol plus three palmitates in ester linkage) has a molecular weight of 885, so 1 kg of tripalmitoin is (1000 g/kg)/(885 g/mol) or 1.13 mol. Complete oxidation of its substituent palmitates will produce:

$$(1.13 \text{ mol tripalmitoin})(3 \text{ mol palmitate/mol tripalmitoin})(23 \text{ mol } H_2O\text{/mol palmitate}) = 78 \text{ mol } H_2O$$

$$(78 \text{ mol})(18 \text{ g/mol}) = 1400 \text{ g} = 1.4 \text{ kg} = 1.4 \text{ L.}$$

Note: This may be an overestimate, since the fatty acyl groups on the triacylglycerol may be less highly reduced than palmitate and because of the weight contribution of the tripalmitoin (tripalmitoylglycerol).

10. Petroleum as a Microbial Food Source Some microorganisms of the genera *Nocardia* and *Pseudomonas* can grow in an environment where hydrocarbons are the only food source. These bacteria oxidize straight-chain aliphatic hydrocarbons, such as octane, to their corresponding carboxylic acids:

$$CH_3(CH_2)_6CH_3 + NAD^+ + O_2 \rightleftharpoons CH_3(CH_2)_6COOH + NADH + H^+$$

How can these bacteria be used to clean up oil spills? What are some of the limiting factors to the efficiency of this process?

Answer

The oxidation of hydrocarbons to the corresponding fatty acids allows these microbes to obtain all of the energy needed for growth from β oxidation and oxidative phosphorylation. Thus the hydrocarbons can be converted to CO_2 and H_2O. Theoretically, the oil in spills could be broken down by treatment with these microbes.

Because of the extreme hydrophobicity of hydrocarbons, close contact between the substrate (octane) and the enzymes of the bacteria may be difficult to achieve. Under field conditions (e.g., an oil spill), detergents are often added to improve this contact. In addition, other nutrients such as nitrogen or phosphorus may be limiting to the bacterial populations and are often added to increase the rate of growth of hydrocarbon-oxidizers.

11. Metabolism of a Straight-Chain Phenylated Fatty Acid A crystalline metabolite was isolated from the urine of a rabbit that had been fed a straight-chain fatty acid containing a terminal phenyl group:

$$C_6H_5{-}CH_2{-}(CH_2)_n{-}COO^-$$

A 302 mg sample of the metabolite in aqueous solution was completely neutralized by adding 22.2 mL of 0.1 M NaOH.

(a) What is the probable molecular weight and structure of the metabolite?

(b) Did the straight-chain fatty acid contain an even or an odd number of methylene ($-CH_2-$) groups (i.e., is n even or odd)? Explain.

Answer

(a) 22.2 mL of 0.1 M NaOH is equivalent to $(22.3 \times 10^{-3}\ L)(0.1\ mol/L) = 22.2 \times 10^{-4}$ mol of unknown metabolite (assuming that it contains only one carboxyl group). Thus the M_r of the metabolite is

$$\frac{302 \times 10^{-3}\ g}{22.2 \times 10^{-4}\ mol} = 136$$

This is the M_r of phenylacetic acid.

(b) Since β oxidation removes two-carbon units, and since the end product is a two-carbon unit, the original fatty acyl chain must have been even-numbered (with the phenyl group counted as equivalent to a terminal methyl group). An odd-numbered fatty acid would have produced phenylpropionate.

12. Fatty Acid Oxidation in Diabetes When the acetyl-CoA produced during β oxidation in the liver exceeds the capacity of the citric acid cycle, the excess acetyl-CoA forms ketone bodies—acetone, acetoacetate, and D-β-hydroxybutyrate. This occurs in severe diabetes: because the tissues cannot use glucose, they oxidize large amounts of fatty acids instead. Although acetyl-CoA is not toxic, the mitochondrion must divert the acetyl-CoA to ketone bodies. What problem would arise if acetyl-CoA were not converted to ketone bodies? How does this diversion solve the problem?

Answer Diabetics oxidize large quantities of fat because of their inability to use glucose efficiently. This leads to a decrease in the activity of the citric acid cycle (see Problem 13; see also Chapter 16, Problem 12) and an increase in the pool of acetyl-CoA. If acetyl-CoA were not converted to ketone bodies, the CoA pool would become depleted. Because the mitochondrial CoA pool is small, liver mitochondria recycle CoA by condensing two acetyl-CoA molecules to form acetoacetyl-CoA + CoA (see Fig. 17–16). The acetoacetyl-CoA is converted to other ketones, and the CoA is recycled for use in the β-oxidation pathway and energy production.

13. Consequences of a High-Fat Diet with No Carbohydrates Suppose you had to subsist on a diet of whale blubber and seal blubber with little or no carbohydrate.

(a) What would be the effect of carbohydrate deprivation on the utilization of fats for energy?

(b) If your diet were totally devoid of carbohydrate, would it be better to consume odd-or even-numbered fatty acids? Explain.

Answer

(a) Pyruvate, formed from glucose via glycolysis, is the main source of the oxaloacetate needed to replenish citric acid cycle intermediates (see Table 16–2). In the absence of carbohydrate in the diet, the oxaloacetate level drops and the citric acid cycle slows. This increases the rate of β oxidation of fatty acids and leads to ketosis.

(b) The last cycle of β oxidation produces 2 acetyl-CoA molecules from an even-numbered fatty acid, or propionyl-CoA + acetyl-CoA from an odd-numbered fatty acid. Propionyl-CoA can be converted to succinyl-CoA (see Fig. 17–11), which when converted to oxaloacetate stimulates the citric acid cycle and relieves the conditions leading to ketosis. Thus it would be better to consume odd-numbered fatty acids.

14. Metabolic Consequences of Ingesting ω-Fluorooleate The shrub *Dichapetalum toxicanium* which grows in Sierra Leone, produces ω-fluorooleate, which is highly toxic to warm-blooded animals.

$$F{-}CH_2{-}(CH_2)_7{-}\overset{\displaystyle H}{\overset{|}{C}}{=}\overset{\displaystyle H}{\overset{|}{C}}{-}(CH_2)_7{-}COO^-$$

ω-Fluorooleate

This substance has been used as an arrow poison, and powdered fruit from the plant is sometimes used as a rat poison (hence the plant's common name, ratsbane). Why is this substance so toxic? (Hint: Review Chapter 16, Problem 13).

Answer Oxidation of ω-fluorooleate in the β-oxidation pathway forms fluoroacetyl-CoA in the last pass through the sequence. Entry of fluoroacetyl-CoA into the citric acid cycle produces fluorocitrate, a powerful inhibitor of the citric acid cycle enzyme aconitase. As a result of this inhibition, the citric acid cycle shuts down and the flow of reducing equivalents to oxidative phosphorylation is fatally impaired.

15. Role of FAD as Electron Acceptor Acyl-CoA dehydrogenase uses enzyme-bound FAD as a prosthetic group to dehydrogenate the α and β carbons of fatty acyl-CoA. What is the advantage of using FAD as an electron acceptor rather than NAD^+? Explain in terms of the standard reduction potentials for the Enz-FAD/$FADH_2$ ($E'° = -0.219$ V) and NAD^+/NADH ($E'° = -0.320$ V) half-reactions.

Answer Enz-FAD, having a more positive standard reduction potential, is a better electron acceptor than NAD^+, and the reaction is driven in the direction of fatty acyl-CoA oxidation (a negative free-energy change; see Eqn 14–8). This more favorable free energy change is obtained at the expense of 1 ATP; only 1.5 ATP molecules are formed per $FADH_2$ oxidized in the respiratory chain (compared with 2.5 ATP per NADH).

16. β Oxidation of Arachidic Acid How many turns of the fatty acid oxidation cycle are required for complete oxidation of arachidic acid (see Table 11–1) to acetyl-CoA?

Answer Arachidic acid is a 20-carbon saturated fatty acid. Nine cycles of the β-oxidation pathway are required to produce 10 molecules of acetyl-CoA, the last two in the ninth turn.

17. Fate of Labeled Propionate If [3-^{14}C]propionate (^{14}C in the methyl group) is added to a liver homogenate, ^{14}C-labeled oxaloacetate is rapidly produced. Draw a flow chart for the pathway by which propionate is transformed to oxaloacetate, and indicate the location of the ^{14}C in oxaloacetate.

Answer Propionate is first converted to the CoA derivative. Fig. 17–11 shows the three-step pathway that converts (3-C) propionyl-CoA to (4-C) succinyl-CoA, which can be summarized as:

(a) Propionyl-CoA carboxylase uses CO_2 and ATP to form D-methylmalonyl-CoA by carboxylation at C-2 of the propionyl group.

(b) Methylmalonyl-CoA epimerase shifts the CoA thioester from C-1 (of the original propionyl group) to the newly added carboxylate, making the product L-methylmalonyl-CoA.

(c) Methylmalonyl-CoA mutase moves the carboxy-CoA group from C-2 to C-3 within the original propionyl unit, forming succinyl-CoA.
Once succinyl-CoA is formed, the citric acid cycle can convert it to oxaloacetate. Based on the above reactions and the stereochemistry involved, the [^{14}C]-label is equilibrated at C-2 and C-3 of oxaloacetate.

Note: You should use these descriptions to prepare your own flow diagram.

18. Sources of H_2O Produced in β Oxidation The complete oxidation of palmitoyl-CoA to carbon dioxide and water is represented by the overall equation

$$\text{Palmitoyl-CoA} + 23O_2 + 108P_i + 108ADP \longrightarrow CoA + 16CO_2 + 108ATP + 23H_2O$$

Water is also produced in the reaction

$$ADP + P_i \longrightarrow ATP + H_2O$$

Why is this water not included as a net product of the reaction?

Answer ATP hydrolysis in the energy requiring reactions of a cell takes up water in the reaction

$$ATP + H_2O \longrightarrow ADP + P_i$$

so that there is no net production of H_2O by this mechanism.

19. Biological Importance of Cobalt In cattle, deer, sheep, and other ruminant animals, large amounts of propionate are produced in the rumen through the bacterial fermentation of ingested plant matter. Propionate is the principal source of glucose for these animals via the route propionate → oxaloacetate → glucose. In some areas of the world, notably Australia, ruminant animals sometimes show symptoms of anemia with concomitant loss of appetite and retarded growth, resulting from an inability to transform propionate to oxaloacetate. This condition is due to a cobalt deficiency caused by very low cobalt levels in the soil and thus in plant matter. Explain.

Answer One of the enzymes necessary for the conversion of propionate to oxaloacetate is methylmalonyl-CoA mutase (see Fig. 17–11). This enzyme requires as an essential cofactor the cobalt-containing coenzyme B_{12}, which is synthesized from vitamin B_{12}. A cobalt deficiency in animals would result in vitamin B_{12} deficiency.

20. Fat Loss during Hibernation Bears expend about 25×10^6 J/day during periods of hibernation, which may last as long as seven months. The energy required to sustain life is obtained from fatty acid oxidation. How much weight loss (in kilograms) has occurred after seven months? How might ketosis be minimized during hibernation?

Answer The catabolism of fat yields about 9 kcal/g or 37.6 kJ/g or 3.8×10^4 kJ/kg. Each day the bear expends 2.5×10^4 kJ of energy.

$$(2.5 \times 10^4 \text{ kJ/day})/(3.8 \times 10^4 \text{ kJ/kg}) = 0.66 \text{ kg/day}$$

0.66 kg/day $\times$ 210 days $= 138$ kg, the mass lost over the 210 day hibernation period.

To minimize ketosis, a slow but steady degradation of nonessential proteins would provide three-, four-, and five-carbon products essential to the formation of glucose by gluconeogenesis. This would avoid the inhibition of the citric acid cycle that occurs when oxaloacetate is withdrawn from the cycle to be used for gluconeogenesis. The citric acid cycle could continue to degrade acetyl-CoA to CO_2, rather than shunting it off toward ketone body formation.

Amino Acid Oxidation and the Production of Urea

1. **Products of Amino Acid Transamination** Name and draw the structure of the α-keto acid resulting when the following amino acids undergo transamination with α-ketoglutarate:
   (a) aspartate
   (b) glutamate
   (c) alanine
   (d) phenylalanine

   Answer

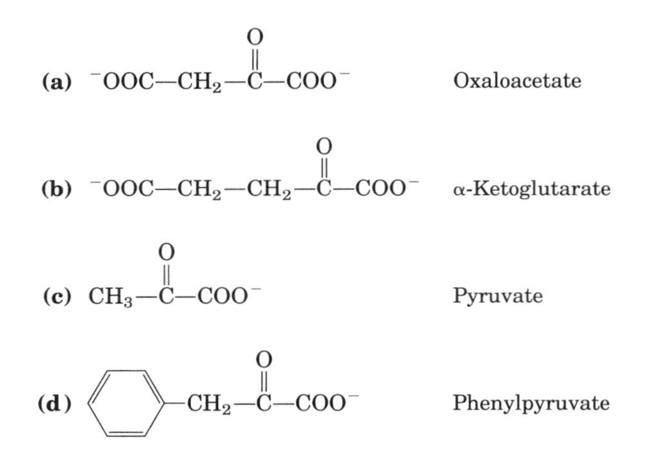

(a) Oxaloacetate
(b) α-Ketoglutarate
(c) Pyruvate
(d) Phenylpyruvate

2. **Measurement of Alanine Aminotransferase Activity** The activity (reaction rate) of alanine aminotransferase is usually measured by including an excess of pure lactate dehydrogenase and NADH in the reaction system. The rate of alanine disappearance is equal to the rate of NADH disappearance measured spectrophotometrically. Explain how this assay works.

   Answer The measurement of the activity of alanine aminotransferase by measurement of the reaction of its product with lactate dehydrogenase is an example of a "coupled" assay. The product of the transamination (pyruvate) is rapidly consumed in the subsequent "indicator reaction," catalyzed by an excess of lactate dehydrogenase. The dehydrogenase uses the cofactor NADH, the disappearance of which is conveniently measured by observing the rate of decrease in NADH absorption at 340 nm. Thus, the rate of disappearance of NADH is a measure of the rate of the aminotransferase reaction, *if NADH and lactate dehydrogenase are added in excess.*

3. Distribution of Amino Nitrogen If your diet is rich in alanine but deficient in aspartate, will you show signs of aspartate deficiency? Explain.

Answer No; aspartate is readily formed by the transfer of the amino group of alanine to oxaloacetate. Cellular levels of aminotransferases are sufficient to provide all of the amino acids in this fashion if the keto acids are available.

4. A Genetic Defect in Amino Acid Metabolism: A Case History A two-year-old child was taken to the hospital. His mother said that he vomited frequently, especially after feedings. The child's weight and physical development were below normal. His hair, although dark, contained patches of white. A urine sample treated with ferric chloride ($FeCl_3$) gave a green color characteristic of the presence of phenylpyruvate. Quantitative analysis of urine samples gave the results shown in the table.

	Concentration (mM)	
Substance	**Patient's urine**	**Normal urine**
Phenylalanine	7.0	0.01
Phenylpyruvate	4.8	0
Phenyllactate	10.3	0

(a) Suggest which enzyme might be deficient in this child. Propose a treatment.
(b) Why does phenylalanine appear in the urine in large amounts?
(c) What is the source of phenylpyruvate and phenyllactate? Why does this pathway (normally not functional) come into play when the concentration of phenylalanine rises?
(d) Why does the boy's hair contain patches of white?

Answer

(a) Since phenylalanine (and its related phenylketones) accumulate in this patient, it is likely that the first enzyme in phenylalanine catabolism, phenylalanine hydroxylase (also called phenylalanine-4-monooxygenase), is defective or missing (see Fig. 18–22). The most appropriate treatment for patients with this disease, known as phenylketonuria (PKU), is to establish a low-phenylalanine diet that provides just enough of the amino acid to meet the needs for protein synthesis.

(b) Phenylalanine appears in the urine because high levels of this amino acid accumulate in the bloodstream and the body attempts to dispose of it.

(c) Phenylalanine is converted to phenylpyruvate by transamination, a reaction that has an equilibrium constant of about 1.0. Phenyllactate is formed from phenylpyruvate by reduction (see Fig. 18–24). This pathway is of importance only when phenylalanine hydroxylase is defective.

(d) The normal catabolic pathway of phenylalanine is through tyrosine, a precursor of melanin, the dark pigment normally present in hair. Decreased tyrosine levels in such patients result in varying degrees of pigment loss.

5. Role of Cobalamin in Amino Acid Catabolism Pernicious anemia is caused by impaired absorption of vitamin B_{12}. What is the effect of this impairment on the catabolism of amino acids? Are all amino acids equally affected? (Hint: See Box 17–2.)

Answer The catabolism of the carbon skeletons of valine, isoleucine, and methionine is impaired because of the absence of a functional methylmalonyl-CoA mutase. This enzyme requires coenzyme B_{12} as a cofactor, and a deficiency of this vitamin leads to elevated methylmalonic acid levels (methylmalonic acidemia). The symptoms and effects of this deficiency are severe (see Table 18–2 and Box 18–2).

6. Lactate versus Alanine as Metabolic Fuel: The Cost of Nitrogen Removal The three carbons in lactate and alanine have identical oxidation states, and animals can use either carbon source as a metabolic fuel. Compare the net ATP yield (moles of ATP per mole of substrate) for the complete oxidation (to CO_2 and H_2O) of lactate versus alanine when the cost of nitrogen excretion as urea is included.

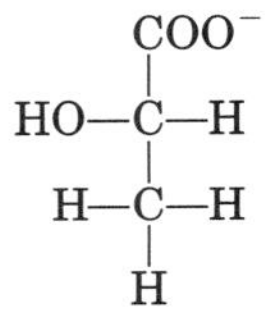

```
    COO⁻              COO⁻
    |           +     |
HO—C—H        H3N—C—H
    |                 |
 H—C—H             H—C—H
    |                 |
    H                 H
 Lactate           Alanine
```

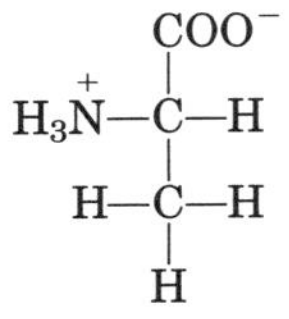

Answer Both lactate and alanine are converted to pyruvate by their respective dehydrogenases, lactate dehydrogenase and alanine dehydrogenase, producing pyruvate and NADH + H^+ and in the case of alanine, NH_4^+. Complete oxidation of pyruvate to CO_2 and H_2O produces 12.5 ATP via the citric acid cycle and oxidative phosphorylation (see Table 16–1). In addition, the NADH from each dehydrogenase reaction produces 2.5 more ATP. Thus oxidation produces 15 moles of ATP per mole of lactate. Urea formation uses the equivalent of 4 ATP per urea molecule formed (Fig. 18–9), or 2 ATP per NH_4^+. Subtracting this value from the energy yield of alanine results in 13 ATP for the oxidation of alanine.

7. Pathway of Carbon and Nitrogen in Glutamate Metabolism When [2-^{14}C,^{15}N] glutamate undergoes oxidative degradation in the liver of a rat, in which atoms of the following metabolites will each isotope be found?

(a) urea
(b) succinate
(c) arginine
(d) citrulline
(e) ornithine
(f) aspartate

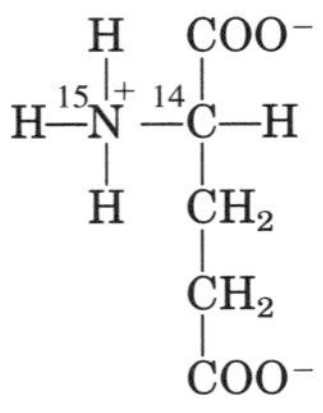

```
     H   COO⁻
   15|+ 14|
  H—N —C—H
     |    |
     H   CH2
          |
         CH2
          |
         COO⁻
   Glutamate
```

Answer

(a) $^{15}NH_2{-}CO{-}^{15}NH_2$

(b) $^{-}OO^{14}C{-}CH_2{-}CH_2{-}^{14}COO^{-}$

(c) $R{-}NH{-}C({=}^{15}NH){-}^{15}NH_2$

(d) $R{-}NH{-}C({=}O){-}^{15}NH_2$

(e) No label

(f) $^{-}OO^{14}C{-}CH(^{15}NH_2){-}CH_2{-}^{14}COO^{-}$

(a) The amino groups of urea will contain ^{15}N, a result of glutamate dehydrogenase producing $^{15}NH_4^+$ or of a transaminase producing ^{15}N-labeled aspartate.

(b) After loss of the amino group, the [2-^{14}C] α-ketoglutarate will be metabolized in the citric acid cycle. Succinate thus formed will be labeled in the carboxyl groups.

(c) The arginine formed in the urea cycle will contain ^{15}N in both guanidino nitrogens.

(d) Citrulline formed in the urea cycle will contain ^{15}N in the carboxamide group.

(e) No labeled N will be found in ornithine.

(f) Aspartate will contain ^{15}N in its amino group as a result of transamination from glutamate. It will also contain ^{14}C in its carboxyl groups as a result of succinate (see **b**) conversion to oxaloacetate. Note: In **(c)**, **(d)**, and **(e)**, these intermediates in the urea cycle will contain low levels of ^{14}C as a result of a very weak synthesis of ornithine from glutamate.

8. Chemical Strategy of Isoleucine Catabolism Isoleucine is degraded in six steps to propionyl-CoA and acetyl-CoA:

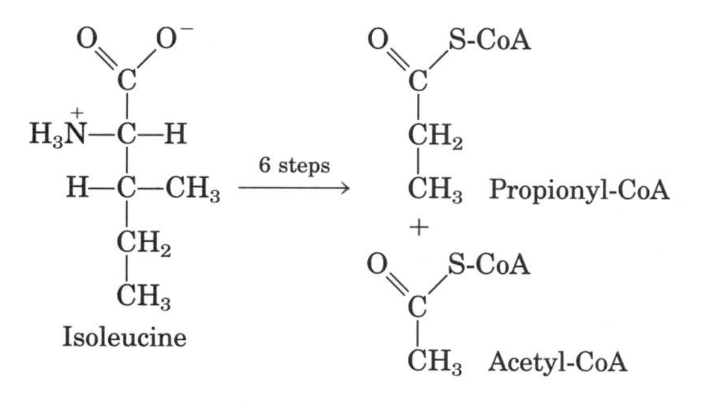

(a) The chemical process of isoleucine degradation includes strategies analogous to those used in the citric acid cycle and the β oxidation of fatty acids. The intermediates of isoleucine degradation (I to V) shown below are not in the proper order. Use your knowledge and understanding of the citric acid cycle and β-oxidation pathway to arrange the intermediates into the proper metabolic sequence for isoleucine degradation.

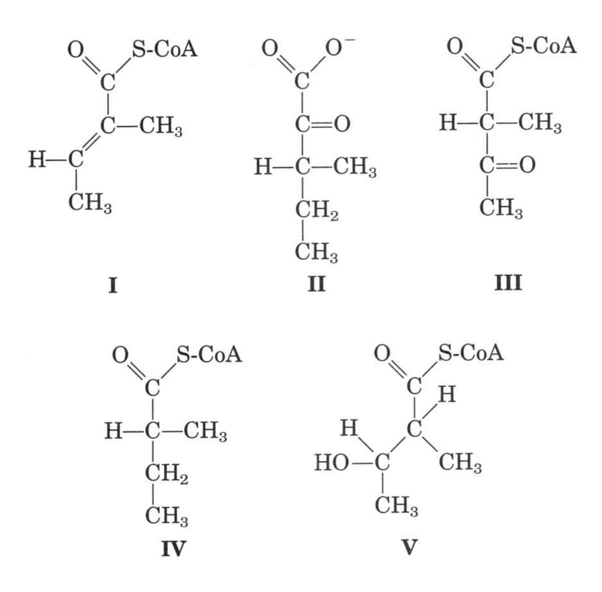

(b) For each step you propose, describe the chemical process, provide an analogous example from the citric acid cycle or β-oxidation pathway (where possible), and indicate any necessary cofactors.

Answer

(a) Isoleucine $\xrightarrow{1}$ II $\xrightarrow{2}$ IV $\xrightarrow{3}$ I $\xrightarrow{4}$ V $\xrightarrow{5}$ III $\xrightarrow{6}$ acetyl-CoA + propionyl-CoA

(b) Step 1 is a transamination that has no analogous reaction and requires PLP. Step 2 is an oxidative decarboxylation similar to the pyruvate dehydrogenase reaction and requires NAD^+. Step 3 is an oxidation similar to the succinate dehydrogenase reaction and requires FAD. Step 4 is a hydration analogous to the fumarase reaction; no cofactor is required. Step 5 is an oxidation analogous to the malate dehyrogenase reaction of the citric acid cycle; it requires NAD^+. Step 6 is a thiolysis analogous to the final cleavage step of β oxidation catalyzed by thiolase; it requires CoA.

9. Ammonia Intoxication Resulting from an Arginine-Deficient Diet In a study conducted some years ago, cats were fasted overnight then given a single meal complete in all amino acids except arginine. Within 2 hours, blood ammonia levels increased from a normal level of 18 μg/L to 140 μg/L, and the cats showed the clinical symptoms of ammonia toxicity. A control group fed a complete amino acid diet or an amino acid diet in which arginine was replaced by ornithine showed no unusual clinical symptoms.

(a) What was the role of fasting in the experiment?

(b) What caused the ammonia levels to rise in the experimental group? Why did the absence of arginine lead to ammonia toxicity? Is arginine an essential amino acid in cats? Why or why not?

(c) Why can ornithine be substituted for arginine?

Answer

(a) The fasting resulted in lowering of blood glucose levels. Subsequent feeding of an arginine-free diet led to a rapid catabolism of all of the amino acids, especially the glucogenic ones. This catabolism was exacerbated by the lack of an essential amino acid, which prevented protein synthesis.

(b) Oxidative deamination of amino acids caused the elevation of ammonia levels. In addition, the lack of arginine (an intermediate in the urea cycle) slowed the conversion of ammonia to urea. Arginine (or ornithine) synthesis in the cat is not sufficient to meet the needs imposed by the stress of this experiment, suggesting that arginine is an essential amino acid.

(c) Ornithine (or citrulline) can be substituted for arginine because it also is an intermediate in the urea cycle.

10. Oxidation of Glutamate Write a series of balanced equations, and an overall equation for the net reaction, describing the oxidation of 2 moles of glutamate to 2 moles of α-ketoglutarate and 1 mole of urea.

Answer

H_2O + glutamate + $NAD^+ \longrightarrow$ α-ketoglutarate + NH_4^+ + NADH + H^+

NH_4^+ + 2ATP + H_2O + $CO_2 \longrightarrow$ carbamoyl phosphate + 2ADP + P_i + $3H^+$

Carbamoyl phosphate + ornithine $\longrightarrow$ citrulline + P_i + H^+

Citrulline + aspartate + ATP $\longrightarrow$ argininosuccinate + AMP + PP_i + H^+

Argininosuccinate $\longrightarrow$ arginine + fumarate

Fumarate + $H_2O \longrightarrow$ malate

Malate + $NAD^+ \longrightarrow$ oxaloacetate + NADH + H^+

Oxaloacetate + glutamate $\longrightarrow$ aspartate + α-ketoglutarate

Arginine + $H_2O \longrightarrow$ urea + ornithine

The sum of these reactions is

2 Glutamate + CO_2 + $4H_2O$ + $2NAD^+$ + 3ATP $\longrightarrow$

2 α-ketoglutarate + 2NADH + $7H^+$ + urea + 2ADP + AMP + PP_i + $2P_i$ **(1)**

Three additional reactions need to be considered:

AMP + ATP $\longrightarrow$ 2ADP **(2)**

O_2 + $8H^+$ + 2NADH + 6ADP + $6P_i \longrightarrow 2NAD^+$ + 6ATP + $8H_2O$ **(3)**

H_2O + $PP_i \longrightarrow 2P_i$ + H^+ **(4)**

Summing the last four equations:

2 Glutamate + CO_2 + O_2 + 2ADP + $2P_i \longrightarrow$ 2 α-ketoglutarate + urea + $3H_2O$ + 2ATP

11. Role of Pyridoxal Phosphate in Glycine Metabolism The enzyme serine hydroxymethyl transferase (Fig. 18–19) requires pyridoxal phosphate as cofactor. Propose a mechanism for the reaction that explains this requirement. (Hint: See Figure 18–6.)

Answer See the mechanism on the following page. The formaldehyde produced in the second step reacts rapidly with tetrahydrofolate at the enzyme active site to produce N^5,N^{10}-methylene tetrahydrofolate (see Fig. 18–16).

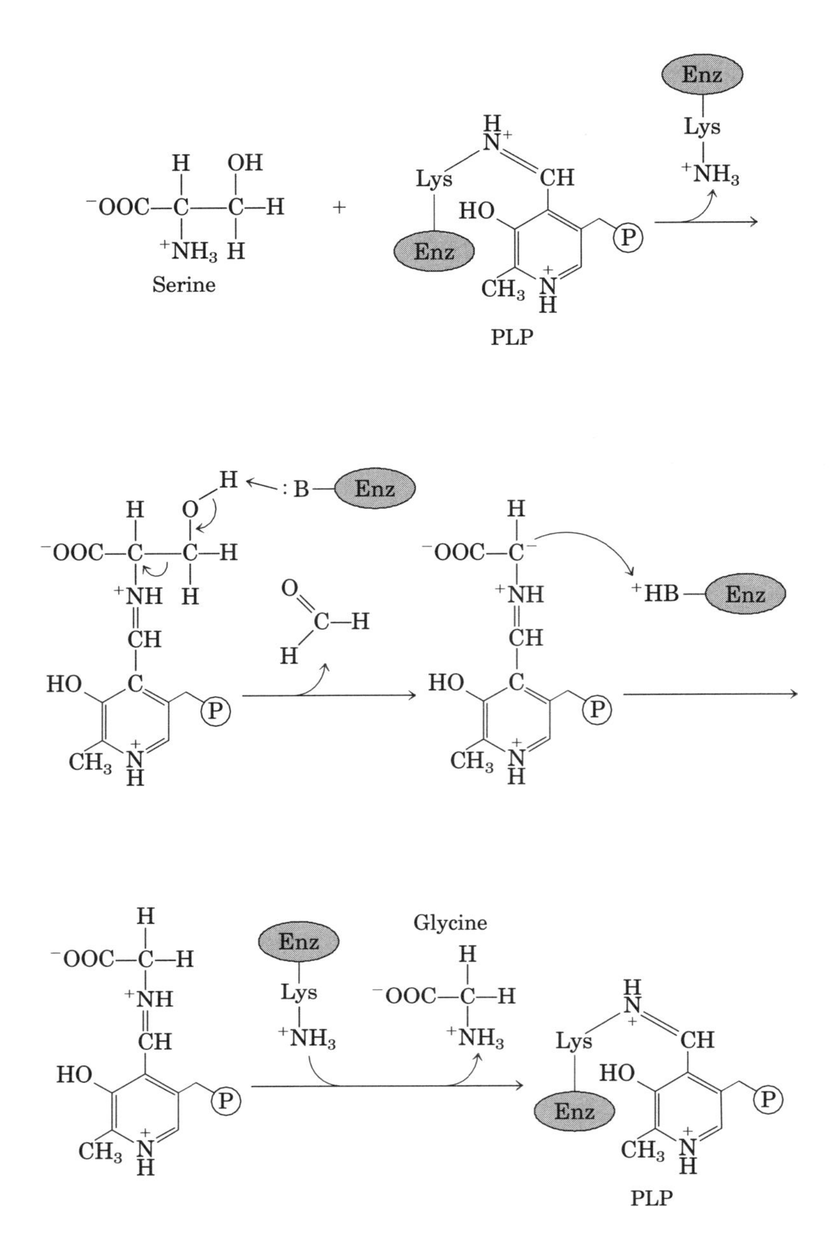

12. Parallel Pathways for Amino Acid and Fatty Acid Degradation The carbon skeleton of leucine is degraded by a series of reactions closely analogous to those of the citric acid cycle and β oxidation. For each reaction, **(a)** through **(f),** indicate its type, provide an analogous example from the citric acid cycle or β-oxidation pathway (where possible), and note any necessary cofactors.

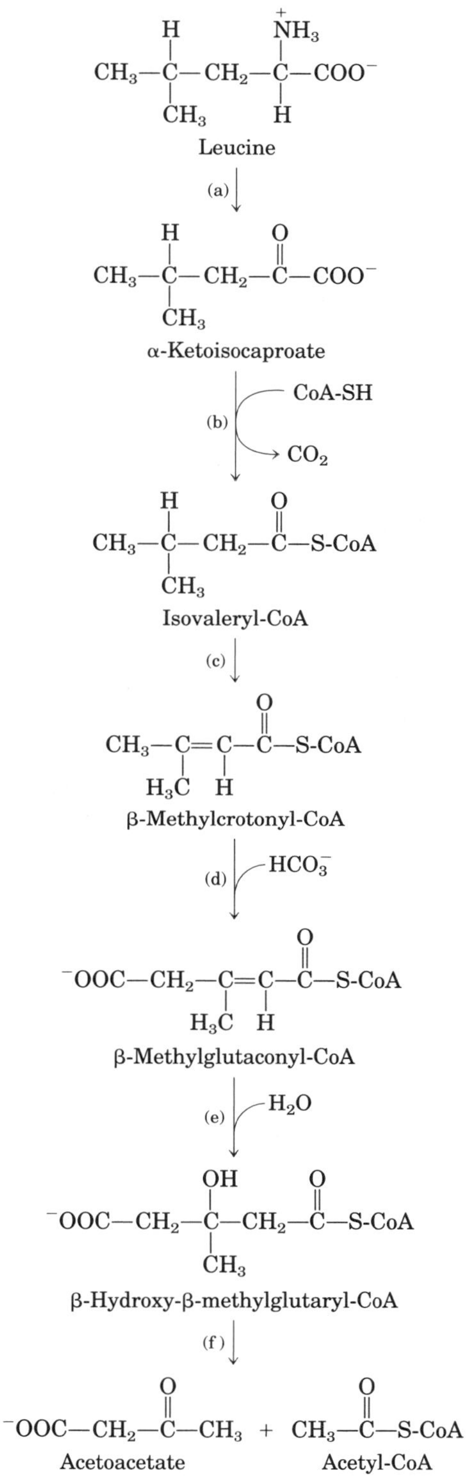

$CH_3{-}CH(CH_3){-}CH_2{-}CH(\overset{+}{N}H_3){-}COO^-$

Leucine

↓ (a)

$CH_3{-}CH(CH_3){-}CH_2{-}C({=}O){-}COO^-$

α-Ketoisocaproate

↓ (b) CoA-SH → CO_2

$CH_3{-}CH(CH_3){-}CH_2{-}C({=}O){-}S\text{-}CoA$

Isovaleryl-CoA

↓ (c)

$CH_3{-}C(H_3C){=}CH{-}C({=}O){-}S\text{-}CoA$

β-Methylcrotonyl-CoA

↓ (d) HCO_3^-

$^-OOC{-}CH_2{-}C(H_3C){=}CH{-}C({=}O){-}S\text{-}CoA$

β-Methylglutaconyl-CoA

↓ (e) H_2O

$^-OOC{-}CH_2{-}C(OH)(CH_3){-}CH_2{-}C({=}O){-}S\text{-}CoA$

β-Hydroxy-β-methylglutaryl-CoA

↓ (f)

$^-OOC{-}CH_2{-}C({=}O){-}CH_3 \;+\; CH_3{-}C({=}O){-}S\text{-}CoA$

Acetoacetate + Acetyl-CoA

Answer

(a) Transamination; no analogies in either pathway; requires PLP.

(b) Oxidative decarboxylation; analogous to oxidative decarboxylation of pyruvate to acetyl-CoA prior to entry into the citric acid cycle, and of α-ketoglutarate to succinyl-CoA in the citric acid cycle; requires NAD^+, FAD, lipoate, thiamine pyrophosphate.

(c) Dehydrogenation (oxidation); analogous to dehydrogenation of succinate to fumarate in the citric acid cycle and of fatty acyl-CoA to enoyl-CoA in β oxidation; requires FAD.

(d) Carboxylation; analogous to carboxylation of pyruvate to oxaloacetate in the citric acid cycle; requires ATP and biotin.

(e) Hydration; analogous to hydration of fumarate to malate in the citric acid cycle and of enoyl-CoA to 3-hydroxyacyl-CoA in β oxidation; no cofactors.

(f) Reverse aldol reaction; analogous to reverse of citrate synthase reaction in the citric acid cycle and identical to cleavage of β-hydroxy-β-methylglutaryl-CoA in formation of ketone bodies; no cofactors.

13. Transamination and the Urea Cycle Aspartate aminotransferase has the highest activity of all the mammalian liver aminotransferases. Why?

Answer The second amino group introduced into urea is transferred from aspartate. This amino acid is generated in large quantities by transamination between oxaloacetate and glutamate (and many other amino acids), catalyzed by aspartate aminotransferase. Approximately one half of all the amino groups that are excreted as urea must pass through the aspartate aminotransferase reaction, and liver contains higher levels of this aminotransferase than any other.

14. The Case against the Liquid Protein Diet A weight-reducing diet heavily promoted some years ago required the daily intake of "liquid protein" (soup of hydrolyzed gelatin), water, and an assortment of vitamins. All other food and drink were to be avoided. People on this diet typically lost 10 to 14 lb in the first week.

(a) Opponents argued that the weight loss was almost entirely due to water loss and would be regained almost immediately when a normal diet was resumed. What is the biochemical basis for this argument?

(b) A number of people on this diet died. What are some of the dangers inherent in the diet and how can they lead to death?

Answer

(a) A person on a diet consisting only of protein must use amino acids as the principal source of metabolic fuel. Because the catabolism of amino acids requires the removal of nitrogen as urea, the process consumes large quantities of water to dilute and excrete the urea in the urine. Furthermore, electrolytes in the "liquid protein" must be diluted with water and excreted. If this abnormally large daily water loss through the kidney is not balanced by a sufficient water intake, a net loss of body water results.

(b) When considering the nutritional benefits of protein, keep in mind the total amount of amino acids needed for protein synthesis and the distribution of amino acids in the dietary protein. Gelatin contains a nutritionally unbalanced distribution of amino acids. As large amounts of gelatin are ingested and the excess amino acids are catabolized, the capacity of the urea cycle may be exceeded, leading to ammonia toxicity. This is further complicated by the dehydration that may result from excretion of large quantities of urea. A combination of these two factors could produce coma and death.

15. Alanine and Glutamine in the Blood Normal human blood plasma contains all the amino acids required for the synthesis of body proteins, but not in equal concentrations. Alanine and glutamine are present in much higher concentrations than any other amino acids. Suggest why.

Answer Muscle tissue is capable of converting amino acids to their keto acids plus ammonia, and of oxidizing the keto acids to produce ATP for muscle contraction. However, urea cannot be formed in muscle. Alanine and glutamine transport amino groups to the liver (see Fig. 18–2) from muscle and other nonhepatic tissues. In muscle, amino groups from all other amino acids are transferred to pyruvate or glutamate to form alanine or glutamine, and these later amino acids are transported to the liver in the blood.

Oxidative Phosphorylation and Photophosphorylation

1. **Oxidation-Reduction Reactions** The NADH dehydrogenase complex of the mitochondrial respiratory chain promotes the following series of oxidation-reduction reactions, in which Fe^{3+} and Fe^{2+} represent the iron in iron-sulfur centers, Q is ubiquinone, QH_2 is ubiquinol, and E is the enzyme:

   (1) $\text{NADH} + \text{H}^+ + \text{E-FMN} \longrightarrow \text{NAD}^+ + \text{E-FMNH}_2$

   (2) $\text{E-FMNH}_2 + 2\text{Fe}^{3+} \longrightarrow \text{E-FMN} + 2\text{Fe}^{2+} + 2\text{H}^+$

   (3) $2\text{Fe}^{2+} + 2\text{H}^+ + \text{Q} \longrightarrow 2\text{Fe}^{3+} + \text{QH}_2$

   Sum: $\text{NADH} + \text{H}^+ + \text{Q} \longrightarrow \text{NAD}^+ + \text{QH}_2$

   For each of the three reactions catalyzed by the NADH dehydrogenase complex, identify **(a)** the electron donor, **(b)** the electron acceptor, **(c)** the conjugate redox pair, **(d)** the reducing agent, and **(e)** the oxidizing agent.

   Answer Oxidation-reduction reactions require an electron donor and an electron acceptor. Recall that electron donors are reducing agents; electron acceptors are oxidizing agents.

   Reaction (1): NADH is the electron donor **(a)** and the reducing agent **(d);** E-FMN is the electron acceptor **(b)** and the oxidizing agent **(e);** NAD^+/NADH and E-FMN/E-$FMNH_2$ are conjugate redox pairs **(c).**

   Reaction (2): E-$FMNH_2$ is the electron donor **(a)** and reducing agent **(d);** Fe^{3+} is the electron acceptor **(b)** and oxidizing agent **(e);** E-FMN/E-$FMNH_2$ and Fe^{3+}/Fe^{2+} are redox pairs **(c).**

   Reaction (3): Fe^{2+} is the electron donor **(a)** and reducing agent **(d);** Q is the electron acceptor **(b)** and oxidizing agent **(d);** and Fe^{3+}/Fe^{2+} and Q/QH_2 are redox pairs **(c).**

2. **All Parts of Ubiquinone Have a Function** In electron transfer, only the quinone portion of ubiquinone undergoes oxidation-reduction; the isoprenoid side chain remains unchanged. What is the function of this chain?

   Answer The long isoprenoid side chain makes ubiquinone very soluble in lipids and allows it to diffuse in the semifluid membrane. This is important because ubiquinone transfers electrons from Complexes I and II to Complex III, all three of which are embedded in the inner membrane of the mitochondrion.

3. **Use of FAD Rather Than NAD^+ in Succinate Oxidation** All the dehydrogenases of glycolysis and the citric acid cycle use NAD^+ ($E'°$ for NAD^+/NADH is −0.32 V) as electron acceptor except succinate dehydrogenase, which uses covalently bound FAD ($E'°$ for FAD/$FADH_2$ in this enzyme is 0.05 V). Suggest why FAD is a more appropriate electron acceptor than NAD^+ in the dehydrogenation of succinate, based on the $E'°$ values of fumarate/succinate ($E'° = 0.03$), NAD^+/NADH, and the succinate dehydrogenase FAD/$FADH_2$.

Answer From the difference in standard reduction potential ($\Delta E'°$) for each pair of half-reactions, we can calculate the $\Delta G'°$ values for the oxidation of succinate using NAD^+ and the oxidation using E-FAD.

For NAD^+:

$$\begin{aligned}\Delta G'° &= -n\mathcal{F}\Delta E'° \\ &= -2(96.5 \text{ kJ/V·mol})(-0.32 \text{ V} - 0.03 \text{ V}) \\ &= 67.6 \text{ kJ/mol}\end{aligned}$$

For E-FAD:

$$\begin{aligned}\Delta G'° &= -2(96.5 \text{ kJ/V·mol})(0.05 \text{ V} - 0.03 \text{ V}) \\ &= -3.86 \text{ kJ/mol}\end{aligned}$$

The oxidation of succinate by E-FAD is favored by the negative standard free-energy change, which is consistent with a K'_{eq} of greater than 1. Oxidation by NAD^+ would require a large, positive, standard free-energy change and have a K'_{eq} favoring the synthesis of succinate.

4. **Degree of Reduction of Electron Carriers in the Respiratory Chain** The degree of reduction of each carrier in the respiratory chain is determined by conditions in the mitochondrion. For example, when NADH and O_2 are abundant, the steady-state degree of reduction of the carriers decreases as electrons pass from the substrate to O_2. When electron transfer is blocked, the carriers before the block become more reduced and those beyond the block become more oxidized (see Fig. 19–6). For each of the conditions below, predict the state of oxidation of ubiquinone and cytochromes b, c_1, c, and $a + a_3$.

(a) Abundant NADH and O_2, but cyanide added
(b) Abundant NADH, but O_2 exhausted
(c) Abundant O_2, but NADH exhausted
(d) Abundant NADH and O_2

Answer As shown in Figure 19–6, the oxidation-reduction state of the carriers in the electron transport system will vary with the conditions.

(a) Cyanide inhibits cytochrome oxidase ($a + a_3$); all carriers become reduced.
(b) In the absence of O_2, no terminal electron acceptor is present; all carriers become reduced.
(c) In the absence of NADH, no carrier can be reduced; all carriers become oxidized.
(d) These are normal circumstances (abundant NADH and O_2); the early carriers (Q, for example) are somewhat reduced while the late ones (cytochrome c, for example) are oxidized.

5. **Effect of Rotenone and Antimycin A on Electron Transfer** Rotenone, a toxic natural product from plants, strongly inhibits NADH dehydrogenase of insect and fish mitochondria. Antimycin A, a toxic antibiotic, strongly inhibits the oxidation of ubiquinol.

(a) Explain why rotenone ingestion is lethal to some insect and fish species.
(b) Explain why antimycin A is a poison.

(c) Assuming that rotenone and antimycin A are equally effective in blocking their respective sites in the electron-transfer chain, which would be a more potent poison? Explain.

Answer

(a) The inhibition of NADH dehydrogenase by rotenone decreases the rate of electron flow through the respiratory chain, which in turn decreases the rate of ATP production. If this reduced rate is unable to meet its ATP requirements, the organism dies.

(b) Antimycin A strongly inhibits the oxidation of Q in the respiratory chain, severely limiting the rate of electron transfer and ATP production.

(c) Electrons flow into the system at Complex I (see Fig. 19–14) from the NAD^+-linked reactions and at Complex II from succinate and fatty acyl–CoA (see Fig. 19–8). Antimycin A inhibits electron flow (through Q) from all of these sources whereas rotenone inhibits flow only through Complex I. Thus antimycin A is a more potent poison.

6. Uncouplers of Oxidative Phosphorylation In normal mitochondria the rate of electron transfer is tightly coupled to the demand for ATP. When the rate of use of ATP is relatively low, the rate of electron transfer is low. When demand for ATP increases, electron-transfer rate increases. Under these conditions of tight coupling, the number of ATP molecules produced per atom of oxygen consumed when NADH is the electron donor—the P/O ratio—is about 2.5.

(a) Predict the effect of a relatively low and a relatively high concentration of uncoupling agent on the rate of electron transfer and the P/O ratio.

(b) Ingestion of uncouplers causes profuse sweating and an increase in body temperature. Explain this phenomenon in molecular terms. What happens to the P/O ratio in the presence of uncouplers?

(c) The uncoupler 2,4-dinitrophenol was once prescribed as a weight-reducing drug. How could this agent, in principle, serve as a weight-reducing aid? Uncoupling agents are no longer prescribed because some deaths occurred following their use. Why might the ingestion of uncouplers lead to death?

Answer Uncouplers of oxidative phosphorylation stimulate the rate of electron flow but not ATP synthesis.

(a) At relatively low levels of an uncoupling agent, P/O ratios will drop somewhat, but the cell will be able to compensate for this by increasing the rate of electron flow; ATP levels can be kept relatively normal. At high levels of uncoupler, P/O ratios approach zero and the cell cannot maintain ATP levels.

(b) As amounts of an uncoupler increase, the P/O ratio will decrease and the body struggles to make sufficient ATP by oxidizing more fuel. The heat produced by this increased rate of oxidation will increase the body temperature. The P/O ratio is affected as noted in **(a).**

(c) Increased activity of the respiratory chain in the presence of an uncoupler requires the degradation of additional energy stores (glycogen and fat). By oxidizing more fuel in an attempt to produce the same amount of ATP, the body loses weight. If the P/O ratio nears zero, however, the lack of ATP will be lethal.

7. Effects of Valinomycin on Oxidative Phosphorylation When the antibiotic valinomycin is added to actively respiring mitochondria, several things happen: the yield of ATP decreases, the rate of O_2 consumption increases, heat is released, and the pH gradient across the inner mitochondrial membrane increases. Does valinomycin act as an uncoupler or an inhibitor of oxidative phosphorylation? Explain the experimental observations in terms of the antibiotic's ability to transfer K^+ ions across the inner mitochondrial membrane.

Answer The observed effects are consistent with the action of an uncoupler—that is, an agent that causes the free energy released in electron transfer to appear as heat rather than in ATP. In respiring mitochondria, H^+ ions are translocated from the matrix to the outside during electron transfer, creating a proton gradient and an electrical potential across the membrane. A significant portion of the free energy used to synthesize ATP originates from this electric potential. Valinomycin combines with K^+ ions to form a complex that passes through the inner mitochondrial membrane. As a proton is translocated by electron transfer, a K^+ ion moves in the opposite direction. As a result, the positive charges on both sides of the membrane are always balanced and the potential across the membrane is lost. This reduces the yield of ATP per mole of protons flowing through F_oF_1ATPase. In other words, electron transfer and phosphorylation become uncoupled. In response to the decreased efficiency of ATP synthesis, the rate of electron transfer increases markedly. This results in an increase in the H^+ gradient, in oxygen consumption, and in the amount of heat released.

8. Mode of Action of Dicyclohexylcarbodiimide (DCCD) When DCCD is added to a suspension of tightly coupled, actively respiring mitochondria, the rate of electron transfer (measured by O_2 consumption) and the rate of ATP production dramatically decrease. If a solution of 2,4-dinitrophenol is now added to the preparation, O_2 consumption returns to normal but ATP production remains inhibited.

(a) What process in electron transfer or oxidative phosphorylation is affected by DCCD?

(b) Why does DCCD affect the O_2 consumption of mitochondria? Explain the effect of 2,4-dinitrophenol on the inhibited mitochondrial preparation.

(c) Which of the following inhibitors does DCCD most resemble in its action: antimycin A, rotenone, or oligomycin?

Answer

(a) DCCD inhibits ATP synthesis. In tightly coupled mitochondria, this inhibition will lead to an inhibition of electron transfer also.

(b) A decrease in electron transfer causes a decrease in O_2 consumption. 2,4-Dinitrophenol uncouples electron transfer from ATP synthesis, allowing respiration to increase. No ATP is synthesized and the P/O ratio will decrease.

(c) Both DCCD and oligomycin inhibit ATP synthesis (see Table 19–4).

9. Compartmentalization of Citric Acid Cycle Components Isocitrate dehydrogenase is found only in the mitochondria, but malate dehydrogenase is found in both the cytosol and mitochondria. What is the role of cytosolic malate dehydrogenase?

Answer Malate dehydrogenase catalyzes the transformation of malate to oxaloacetate in the citric acid cycle, which takes place in the mitochondrion, but also plays a key role in the transport of reducing equivalents across the inner mitochondrial membrane via the malate-aspartate shuttle. This shuttle requires the presence of malate dehydrogenase in both the cytosol and the mitochondria.

10. The Malate–α-Ketoglutarate Transport System The transport system that conveys malate and α-ketoglutarate across the inner mitochondrial membrane (Fig. 19–26) is inhibited by *n*-butylmalonate. Suppose *n*-butylmalonate is added to an aerobic suspension of kidney cells using glucose exclusively as fuel. Predict the effect of this inhibitor on **(a)** glycolysis, **(b)** oxygen consumption, **(c)** lactate formation, and **(d)** ATP synthesis.

Answer NADH produced in the cytosol cannot cross the mitochondrial membrane, but must be oxidized if glycolysis is to continue. Reducing equivalents from NADH enter the mitochondrion by way of the malate-aspartate shuttle. NADH reduces oxaloacetate to form malate and NAD^+, and the malate is transported into the mitochondrion. Cytosolic oxidation of glucose can continue, and the malate is converted back to oxaloacetate and NADH in the mitochondrion (see Fig. 19–26).

(a) If *n*-butylmalonate, an inhibitor of the shuttle, is added to cells, NADH accumulates in the cytosol. This forces glycolysis to operate anaerobically, with reoxidation of NADH in the lactate dehydrogenase reaction.

(b) Since reducing equivalents from the oxidation reactions of glycolysis do not enter the mitochondrion, oxygen consumption ceases.

(c) The end product of anaerobic glycolysis, lactate, will accumulate.

(d) ATP will not be formed aerobically because the cells have converted to anaerobic glycolysis. Overall ATP synthesis will decrease drastically, to 2 ATP per glucose molecule.

11. Cellular ADP Concentration Controls ATP Formation Although both ADP and P_i are required for the synthesis of ATP, the rate of synthesis depends mainly on the concentration of ADP, not P_i. Why?

Answer The steady-state concentration of P_i in the cell is much higher than that of ADP. As the ADP concentration rises as a result of ATP consumption, there is little change in the P_i concentration and so P_i cannot serve as a regulator.

12. The Pasteur Effect When O_2 is added to an anaerobic suspension of cells consuming glucose at a high rate, the rate of glucose consumption declines dramatically as the O_2 is used up, and accumulation of lactate ceases. This effect, first observed by Louis Pasteur in the 1860s, is characteristic of most cells capable of both aerobic and anaerobic glucose catabolism.

(a) Why does the accumulation of lactate cease after O_2 is added?

(b) Why does the presence of O_2 decrease the rate of glucose consumption?

(c) How does the onset of O_2 consumption slow down the rate of glucose consumption? Explain in terms of specific enzymes.

Answer The addition of oxygen to an anaerobic suspension allows cells to convert from fermentation to oxidative phosphorylation as a mechanism for reoxidizing NADH and making ATP. Since ATP synthesis is much more efficient under aerobic conditions, the amount of glucose needed will decrease (the Pasteur effect). This decreased utilization of glucose in the presence of oxygen can be demonstrated in any tissue that is capable of aerobic and anaerobic glycolysis.

(a) Oxygen allows the tissue to convert to oxidative phosphorylation rather than lactic acid fermentation as the mechanism for NADH oxidation.

(b) Cells produce much more ATP per glucose molecule oxidized aerobically, so less glucose is needed.

(c) As [ATP] rises in the cells, phosphofructokinase-1 is inhibited, thus slowing the rate of glucose entry into the glycolytic pathway.

13. Respiration-Deficient Yeast Mutants and Ethanol Production Respiration-deficient yeast mutants (p^-; "petites") can be produced from wild-type parents by treatment with mutagenic agents. The mutants lack cytochrome oxidase, a deficit that markedly affects their metabolic behavior. One striking effect is that fermentation is not suppressed by O_2—that is, the mutants lack the Pasteur effect (see Problem 12). Some companies are very interested in using these mutants to ferment wood chips to ethanol for energy use. Explain the advantages of using these mutants rather than wild-type yeast for large-scale ethanol production. Why does the absence of cytochrome oxidase eliminate the Pasteur effect?

Answer The absence of cytochrome oxidase prevents these mutants from oxidizing the products of fermentation (ethanol, acetate, lactate, or glycerol) via the normal respiratory route. These mutants do not have a working citric acid cycle because they cannot recycle NADH through the O_2-dependent electron-transfer chain. Thus, catabolism of glucose stops at the ethanol stage, even in the presence of oxygen. The ability to carry out these fermentations in the presence of oxygen is a major practical advantage since completely anaerobic conditions are difficult to maintain. Also the Pasteur effect—the decrease in glucose consumption that occurs when oxygen is introduced—is not observed in the absence of an active citric acid cycle and electron-transfer chain.

14. How Many Protons in a Mitochondrion? Electron transfer translocates protons from the mitochondrial matrix to the external medium, establishing a pH gradient across the inner membrane (outside more acidic than inside). The tendency of protons to diffuse back into the matrix is the driving force for ATP synthesis by ATP synthase. During oxidative phosphorylation by a suspension of mitochondria in a medium of pH 7.4, the pH of the matrix has been measured as 7.7.

(a) Calculate $[H^+]$ in the external medium and in the matrix under these conditions.

(b) What is the outside:inside ratio of $[H^+]$? Comment on the energy inherent in this concentration. (Hint: See p. 416, Eqn 12–3.)

(c) Calculate the number of protons in a respiring liver mitochondrion, assuming its inner matrix compartment is a sphere of diameter 1.5 μm.

(d) From these, is the pH gradient alone sufficiently great to generate ATP?

(e) If not, suggest how the necessary energy for synthesis of ATP arises.

Answer

(a) Using the equation pH = $-\log [H^+]$, we can calculate the external $[H^+] = 4.0 \times 10^{-8}$ M (at pH 7.4), and the internal $[H^+] = 2.0 \times 10^{-8}$ M (at pH 7.7).

(b) From **(a),** the ratio is 2:1. We can calculate the free energy inherent in this concentration difference across the membrane:

$$\begin{aligned}\Delta G'^{\circ} &= RT \ln (C_2/C_1)\\ &= (2.479 \text{ kJ/mol}) \ln 2\\ &= -1.72 \text{ kJ/mol}\end{aligned}$$

(c) Given that the volume of the mitochondrion = $4/3\ \pi(0.75 \times 10^{-3}\text{ mm})^3$, $r = 7.5 \times 10^{-4}$ mm^3, and $[H^+] = 2.0 \times 10^{-8}$ M, the number of protons is

$$\frac{(4.19)(4.2 \times 10^{-10}\text{ mm}^3)(2.0 \times 10^{-8}\text{ mol/L})(6.023 \times 10^{23}\text{ protons/mol})}{(10^6\text{ mm}^3\text{/L})} = 21 \text{ protons}$$

(d) No; the total energy inherent in the pH gradient (1.72 kJ/mol—p. 673) is insufficient to synthesize 1 mol of ATP.

(e) The overall transmembrane potential is the main factor in producing a sufficiently large ΔG (see Eqns 19–8 and 19–9).

15. Rate of ATP Turnover in Rat Heart Muscle Rat heart muscle operating aerobically fills more than 90% of its ATP needs by oxidative phosphorylation. Each gram of tissue consumes O_2 at the rate of 10 μmol/min, with glucose as the fuel source.

(a) Calculate the rate at which the heart muscle consumes glucose and produces ATP.

(b) For a steady-state concentration of ATP of 5 μmol/g of heart muscle tissue, calculate the time required (in seconds) to completely turn over the cellular pool of ATP. What does this result indicate about the need for tight regulation of ATP production? (Note: Concentrations are expressed as micromoles per gram of muscle tissue because the tissue is mostly water.)

Answer ATP turns over very rapidly in all types of tissues and cells.

(a) Glucose oxidation requires 6 mol of O_2 per mol of glucose. Therefore, glucose is consumed at the rate of $10/6 = 1.7$ μmol/min·g of tissue. If each glucose produces 32 ATP (see Table 19–5), the muscle produces ATP at the rate of about $1.7 \times 32 = 54.4 =$ 54.4 μmol/min·g, or about 0.91 μmol/sec/g.

(b) It will take about 5.5 s to produce ATP at a level of 5 μmol/g. This indicates that the entire pool of ATP in muscle must be regenerated (turned over) every 5.5 s. In order to do this, the cell must regulate ATP synthesis very precisely.

16. Rate of ATP Breakdown in Flight Muscle ATP production in the flight muscle of the fly *Lucilia sericata* results almost exclusively from oxidative phosphorylation. During flight, 187 ml of O_2/h·g of body weight is needed to maintain an ATP concentration of 7 μmol/g of flight muscle. Assuming that flight muscle makes up 20% of the weight of the fly, calculate the rate at which the flight-muscle ATP pool turns over. How long would the reservoir of ATP last in the absence of oxidative phosphorylation? Assume that reducing equivalents are transferred by the glycerol 3-phosphate shuttle and that O_2 is at 25 °C and 101.3 kPa (1 atm). (Note: Concentrations are expressed in micromoles per gram of flight muscle.)

Answer Using the gas laws ($PV = nRT$) we can calculate that 187 mL of O_2 contains

$$n = PV/RT = (1\text{ atm})(0.187\text{ L})/(0.08205\text{ L·atm/mol·K})(298\text{ K}) \approx 7650\ \mu\text{mol of } O_2.$$

Thus the rate of oxygen consumption by flight muscle

$$= (7650\ \mu\text{mol/h})/(0.2\text{ g})(3600\text{ s/h}) = 10.6\ \mu\text{mol/s·g}$$

Assuming that ATP is formed in the ratio of 30 ATP per 6 O_2 per glucose (see Table 19–5; the glycerol 3-P shuttle produces 3 ATP from glycolysis), the amount of ATP formed is
(30/6)(10.6 μmol/s·g) = 53 μmol/s·g
A reservoir of 7 μmol/g would last 0.13 s.

17. Transmembrane Movement of Reducing Equivalents Under aerobic conditions, extramitochondrial NADH must be oxidized by the mitochondrial electron-transfer chain. Consider a preparation of rat hepatocytes containing mitochondria and all the cytosolic enzymes. If [4-^{3}H]NADH is introduced, radioactivity soon appears in the mitochondrial matrix. However, if [7-^{14}C]NADH is introduced, no radioactivity appears in the matrix. What do these observations reveal about the oxidation of extramitochondrial NADH by the electron transfer chain?

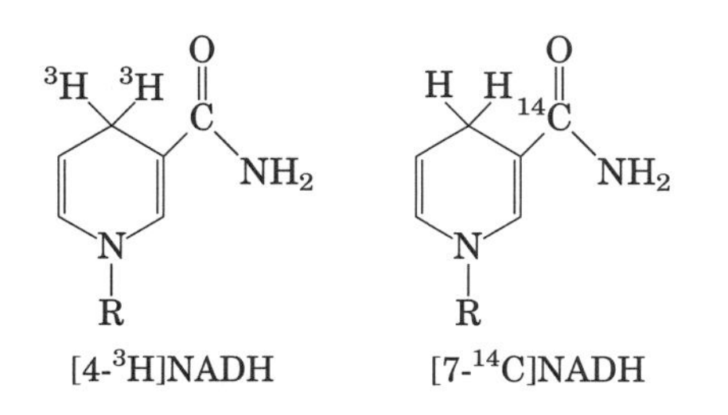

Answer The malate-aspartate shuttle transfers electrons and protons from the cytoplasm into the mitochondrion. Neither NAD^+ nor NADH passes through the inner membrane, thus the labeled NAD moiety of [7-^{14}C]NADH remains in the cytosol. The ^{3}H on [4-^{3}H]NADH enters the mitochondrion via the malate-aspartate shuttle (see Fig 19–26). In the cytosol, [4-^{3}H]NADH transfers its ^{3}H to oxaloacetate to form [^{3}H]malate, which is transported into the mitochondrion. The unlabeled NAD^+ that is produced remains in the cytosol. In the mitochondrion, [^{3}H]malate donates the ^{3}H to NAD^+ to form [4-^{3}H]NADH.

18. NAD Pools and Dehydrogenase Activities Although both pyruvate dehydrogenase and glyceraldehyde 3-phosphate dehydrogenase use NAD^+ as their electron acceptor, the two enzymes do not compete for the same cellular NAD pool. Why?

Answer Pyruvate dehydrogenase is located in the mitochondrion, and glyceraldehyde phosphate dehydrogenase is located in the cytosol. Because the mitochondrial and cytosolic pools of NAD are separated by the inner mitochondrial membrane, the enzymes do not compete for the same NAD pool. However, reducing equivalents are transferred from one nicotinamide coenzyme pool to the other via shuttle mechanisms.

19. Photochemical Efficiency of Light at Different Wavelengths The rate of photosynthesis, measured by O_2 production, is higher when a green plant is illuminated with light of wavelength 680 nm than with light of 700 nm. However, illumination by a combination of light of 680 nm and 700 nm gives a higher rate of photosynthesis than light of either wavelength alone. Explain.

Answer There are two photosystems that drive photosynthesis and photophosphorylation in plants. Photosystem I absorbs light maximally at 700 nm and catalyzes cyclic photophosphorylation and $NADP^+$ reduction (see Fig. 19–46). Photosystem II absorbs light maximally at 680 nm, splits H_2O to O_2 and H^+, and donates electrons and H^+ to Photosystem I. Therefore, light of 680 nm is better in promoting O_2 production, but maximum photosynthetic rates are observed only when plants are illuminated with light of both wavelengths.

20. Balance Sheet for Photosynthesis In 1804 Theodore de Saussure observed that the total weight of oxygen and dry organic matter produced by plants is greater than the weight of carbon dioxide consumed during photosynthesis. Where does the extra weight come from?

Answer Since the general reaction for plant photosynthesis is

$$CO_2 + H_2O \longrightarrow O_2 + \text{organic matter},$$

the extra weight must come from the water consumed in the overall reaction.

21. Role of H_2S in Some Photosynthetic Bacteria Illuminated purple sulfur bacteria carry out photosynthesis in the presence of H_2O and $^{14}CO_2$, but only if H_2S is added and O_2 is absent. During the course of photosynthesis, measured by formation of [^{14}C]carbohydrate, H_2S is converted to elemental sulfur, but no O_2 is evolved. What is the role of the conversion of H_2S to sulfur? Why is no O_2 evolved?

Answer Purple sulfur bacteria use H_2S as a source of electrons and protons

$$H_2S \rightleftharpoons S + 2H^+ + 2e^-$$

The electrons are "activated" by a light energy-capturing photosystem. These cells produce their ATP by photophosphorylation and their NADPH from H_2S oxidation. Since H_2O is not split, O_2 is not evolved. Photosystem II is absent in these bacteria, further explaining why O_2 is not evolved.

22. Boosting the Reducing Power of Photosystem I by Light Absorption When photosystem I absorbs red light at 700 nm, the standard reduction potential of P700 changes from 0.4 V to about −1.2 V. What fraction of the absorbed light is trapped in the form of reducing power?

Answer For a change in standard reduction potential of 0.4 V to −1.2 V, the free-energy change per electron is

$$\begin{aligned}\Delta G'^{\circ} &= n\mathcal{J}\Delta E \\ &= -(96.5\ \text{kJ/V·mol})(-1.6\ \text{V}) \\ &= 154\ \text{kJ/mol}\end{aligned}$$

Two photons are absorbed per electron elevated to a higher energy level, which for 700 nm light is equivalent to 2(170 kJ/mol) = 340 kJ/mol (see Fig. 19–36). Thus the fraction of light energy trapped as reducing power is

$$(154\ \text{kJ/mol})/(340\ \text{kJ/mol}) = 0.45$$

23. Limited ATP Synthesis in the Dark Spinach chloroplasts are illuminated in the absence of ADP and P_i, then the light is turned off and ADP and P_i are added. ATP is then synthesized for a short time in the dark. Explain this finding.

Answer Illumination of chloroplasts in the absence of ADP and P_i sets up a proton gradient across the thylakoid membrane. When ADP and P_i are added, ATP synthesis is driven by the gradient. In the absence of continuous illumination, the gradient soon becomes exhausted and ATP synthesis stops.

24. Mode of Action of the Herbicide DCMU When chloroplasts are treated with 3-(3,4-dichlorophenyl)-1,1-dimethylurea (DCMU, or Diuron), a potent herbicide, O_2 evolution and photophosphorylation cease. Oxygen evolution, but not photophosphorylation, can be restored by addition of an external electron acceptor, or Hill reagent. How does DCMU act as a weed killer? Suggest a location for the inhibitory action of this herbicide in the scheme shown in Figure 19–46. Explain.

Answer DCMU must inhibit the electron transfer system linking Photosystem II and Photosystem I at a position ahead of the first site of ATP production. DCMU competes with pQ_B for electrons from pQ_A (Table 19–4). Thus, addition of a Hill reagent allows H_2O to be split and O_2 to be evolved, but electrons are pulled out of the system before the point of ATP synthesis and before the production of NADPH. DCMU kills plants by inhibiting ATP production.

25. Bioenergetics of Photophosphorylation The steady-state concentrations of ATP, ADP, and P_i in isolated spinach chloroplasts under full illumination at pH 7.0 are 120, 6, and 700 μM, respectively.

(a) What is the free-energy requirement for the synthesis of 1 mol of ATP under these conditions?

(b) The energy for ATP synthesis is furnished by light-induced electron transfer in the chloroplasts. What is the minimum voltage drop necessary (during transfer of a pair of electrons) to synthesize ATP under these conditions? (You may need to refer to p. 517, Eqn 14–6.)

Answer

(a) $$\begin{aligned}\Delta G &= \Delta G'^{\circ} + RT \ln \frac{[\text{ATP}]}{[\text{ADP}][\text{P}_i]} \\ &= 30.5\ \text{kJ/mol} + (2.479\ \text{kJ/mol}) \ln \frac{1.2 \times 10^{-4}\ \text{M}}{(6 \times 10^{-6}\ \text{M})(7 \times 10^{-4}\ \text{M})} \\ &= 30.5\ \text{kJ/mol} + 25.4\ \text{kJ/mol} \\ &= 55.9\ \text{kJ/mol} \approx 56\ \text{kJ/mol}\end{aligned}$$

(b) $\Delta G = -n\mathcal{F}\Delta E$

$$\Delta E = -\Delta G/n\mathcal{F}$$
$$= \frac{-(55.9 \text{ kJ/mol})}{2(96.5 \text{ kJ/V·mol})}$$
$$= 0.29 \text{ V}$$

26. Light Energy for a Redox Reaction Suppose you have isolated a new photosynthetic microorganism that oxidizes H_2S and passes the electrons to NAD^+. What wavelength of light would provide enough energy for H_2S to reduce NAD^+ under standard conditions? Assume 100% efficiency in the photochemical event, and use E'° of −243 mV for H_2S and −320 mV for NAD^+. See Figure 19–36 for energy equivalents of wavelengths of light.

Answer First, calculate the standard free-energy change ($\Delta G'^\circ$) of the redox reaction:

$$NAD^+ + H_2S \longrightarrow NADH + S^- + H^+$$

$$\Delta G'^\circ = -n\mathcal{F}\Delta E'^\circ, \text{ where } \mathcal{F} \text{ is the Faraday constant, 96.5 kJ/V·mol}$$

$$\Delta G'^\circ = (-2)(96.5 \text{ kJ/V·mol})(-0.09 \text{ V}) = 17.4 \text{ kJ/mol},$$

which is the minimum energy needed to drive the reduction of NAD^+ by H_2S. Inspection of Figure 19–36 shows that the energy in a "mole" of photons (an einstein) throughout the visible part of the spectrum ranges from 170 to 300 kJ. Any visible light should have sufficient energy to drive the reduction of NADH by H_2S. In principle, and assuming 100% efficiency, even infrared light should have enough energy to drive this reaction.

27. Equilibrium Constant for Water-Splitting Reactions The coenzyme $NADP^+$ is the terminal electron acceptor in chloroplasts, according to the reaction

$$2H_2O + 2NADP^+ \longrightarrow 2NADPH + 2H^+ + O_2$$

Use the information in Table 19–2 to calculate the equilibrium constant for this reaction at 25 °C. (The relationship between K'_{eq} and $\Delta G'^\circ$ is discussed on p. 494.) How can the chloroplast overcome this unfavorable equilibrium?

Answer Using standard reduction potentials from Table 19–2, $\Delta E'^\circ$ for the reaction is −0.324 V − 0.816 V = −1.14 V.

$$\Delta G'^\circ = -n\mathcal{F}\Delta E'^\circ$$
$$= -4(96.5 \text{ kJ/V·mol})(-1.14 \text{ V})$$
$$= 440 \text{ kJ/mol}$$

(Note that $n = 4$ because 4 electrons are required to produce 1 mol of O_2.)

$$\Delta G'^\circ = -RT \ln K'_{eq}$$
$$\ln K'_{eq} = -\Delta G'^\circ/RT$$
$$= (-440 \text{ kJ/mol})/(2.479 \text{ kJ/mol})$$
$$= -177.5$$
$$K'_{eq} = 1.2 \times 10^{-77}$$

The equilibrium is clearly very unfavorable. In chloroplasts, the input of light energy overcomes this barrier.

28. Energetics of Phototransduction During photosynthesis, eight photons must be absorbed (four by each photosystem) for every O_2 molecule produced:

$$2H_2O + 2NADP^+ + 8 \text{ photons} \longrightarrow 2NADPH + 2H^+ + O_2$$

Assuming that these photons have a wavelength of 700 nm (red) and that the absorption and use of light energy are 100% efficient, calculate the free-energy change for the process.

Answer From Problem 27, $\Delta G'^\circ$ for the production of 1 mol of O_2 is 440 kJ/mol. A light input of 8 photons (700 nm) is equivalent to (8 photons)(170 kJ/einstein) = 1360 kJ/einstein (see Fig. 19–36). (Note that an einstein is the energy in 6.02×10^{23} photons—a "mole" of photons.) Thus coupling these two processes, the overall standard free-energy change is

$$\Delta G'^\circ = (440 - 1360) \text{ kJ/mol} = -920 \text{ kJ/mol}$$

29. Electron Transfer to a Hill Reagent Isolated spinach chloroplasts evolve O_2 when illuminated in the presence of potassium ferricyanide (the Hill reagent), according to the equation

$$2H_2O + 4Fe^{3+} \longrightarrow O_2 + 4H^+ + 4Fe^{2+}$$

where Fe^{3+} represents ferricyanide and Fe^{2+}, ferrocyanide. Is NADPH produced in this process? Explain.

Answer No NADPH is produced. Artificial electron acceptors can remove electrons from the photosynthetic apparatus and stimulate O_2 production. Ferricyanide competes with the cytochrome b_6f complex for electrons and removes them from the system. Consequently, P700 does not receive any electrons that can be activated for $NADP^+$ reduction. However, O_2 is evolved because all components of Photosystem II are oxidized (see Fig. 19–46).

30. How Often Does a Chlorophyll Molecule Absorb a Photon? The amount of chlorophyll *a* (M_r 892) in a spinach leaf is about 20 μg/cm^2 of leaf. In noonday sunlight (average energy 5.4 J/cm^2·min), the leaf absorbs about 50% of the radiation. How often does a single chlorophyll molecule absorb a photon? Given that the average lifetime of an excited chlorophyll molecule in vivo is 1 ns, what fraction of chlorophyll molecules is excited at any one time?

Answer The leaf absorbs light in units of photons that vary in energy between 170 and 300 kJ/mol, depending on wavelength (see p. 693). The leaf absorbs light energy at the rate of 0.5(5.4 J/cm^2·min) = 2.7 J/cm^2·min. Assuming an average energy of 270 kJ/mol of photons, this rate of light absorption is

$$(2.7 \times 10^{-3} \text{ kJ/cm}^2\cdot\text{min})/(270 \text{ kJ/mol photons}) = 1 \times 10^{-5} \text{ mol/cm}^2\cdot\text{min}$$

The concentration of chlorophyll in the leaf is

$$(20 \times 10^{-6} \text{ g/cm}^2)/(892 \text{ g/mol}) = 2.24 \times 10^{-8} \text{ mol/cm}^2$$

Thus, 1 mol or 1 molecule of chlorophyll absorbs 1 mol of photons every

$$\begin{aligned}(2.24 \times 10^{-8} \text{ mol/cm}^2)/(1 \times 10^{-5} \text{ mol/cm}^2\cdot\text{min}) &= 2.24 \times 10^{-3} \text{ min}\\ &= 0.13 \text{ s}\\ &\approx 1 \times 10^{-1} \text{ s or 100 ms}\end{aligned}$$

Since excitation lasts about 1×10^{-9} s, the fraction of chlorophylls excited at any one time is $(1 \times 10^{-9} \text{ s})/(0.1 \text{ s}) = 1 \times 10^{-8}$, or one in every 10^8 molecules.

31. Effect of Monochromatic Light on Electron Flow The extent to which an electron carrier is oxidized or reduced during photosynthetic electron transfer can sometimes be observed directly with a spectrophotometer. When chloroplasts are illuminated with 700 nm light, cytochrome *f*, plastocyanin, and plastoquinone are oxidized. When chloroplasts are illuminated with 680 nm light, however, these electron carriers are reduced. Explain.

Answer Light at 700 nm activates electrons in P700 and $NADP^+$ is reduced (see Fig. 19–46). This drains all of the electrons from the electron transport system between Photosystems II and I because light at 680 nm is not available to replace electrons by activating Photosystem II. When light at 680 nm activates Photosystem II (but not Photosystem I), all of the carriers between the two systems become reduced because no electrons can be excited in Photosystem I.

32. Function of Cyclic Photophosphorylation When the [NADPH]/[$NADP^+$] ratio in chloroplasts is high, photophosphorylation is predominantly cyclic (see Fig. 19–46). Is O_2 evolved during cyclic photophosphorylation? Is NADPH produced? Explain. What is the main function of cyclic photophosphorylation?

Answer Neither O_2 or NADPH is produced. At high [NADPH]/[$NADP^+$] ratios, electron transport from reduced ferredoxin to $NADP^+$ is inhibited and the electrons are diverted into the cytochrome b_6f complex. These electrons return to P700 and ATP is synthesized by photophosphorylation. Since electrons are not lost from P700, none are needed from Photosystem II. Thus H_2O is not split and no O_2 is produced. In addition, NADPH is not formed because the electrons return to P700. The function of cyclic photophosphorylation is to produce ATP.

Carbohydrate Biosynthesis

1. **Role of Oxidative Phosphorylation in Gluconeogenesis** Is it possible to obtain a net synthesis of glucose from pyruvate if the citric acid cycle and oxidative phosphorylation are totally inhibited?

   Answer No; the transformation of two molecules of pyruvate to one molecule of glucose requires an input of energy (4 ATP + 2 GTP) and reducing power (2 NADPH), which are obtained only through the citric acid cycle and oxidative phosphorylation pathway by catabolism of amino acids, fatty acids, or other carbohydrates.

2. **Pathway of Atoms in Gluconeogenesis** A liver extract capable of carrying out all the normal metabolic reactions of the liver is briefly incubated in separate experiments with the following ^{14}C-labeled precursors:

   (a) [^{14}C]Bicarbonate, $HO{-}^{14}C(=O){-}O^-$

   (b) [1-^{14}C]Pyruvate, $CH_3{-}C(=O){-}^{14}COO^-$

   Trace the pathway of each precursor through gluconeogenesis. Indicate the location of ^{14}C in all intermediates and in the product, glucose.

   Answer

   (a) In the pyruvate carboxylase reaction, $^{14}CO_2$ is added to pyruvate to form [4-^{14}C] oxaloacetate, but the phosphoenolpyruvate carboxykinase reaction removes the *same* CO_2 in the next step. Thus ^{14}C is not (initially) incorporated into glucose.

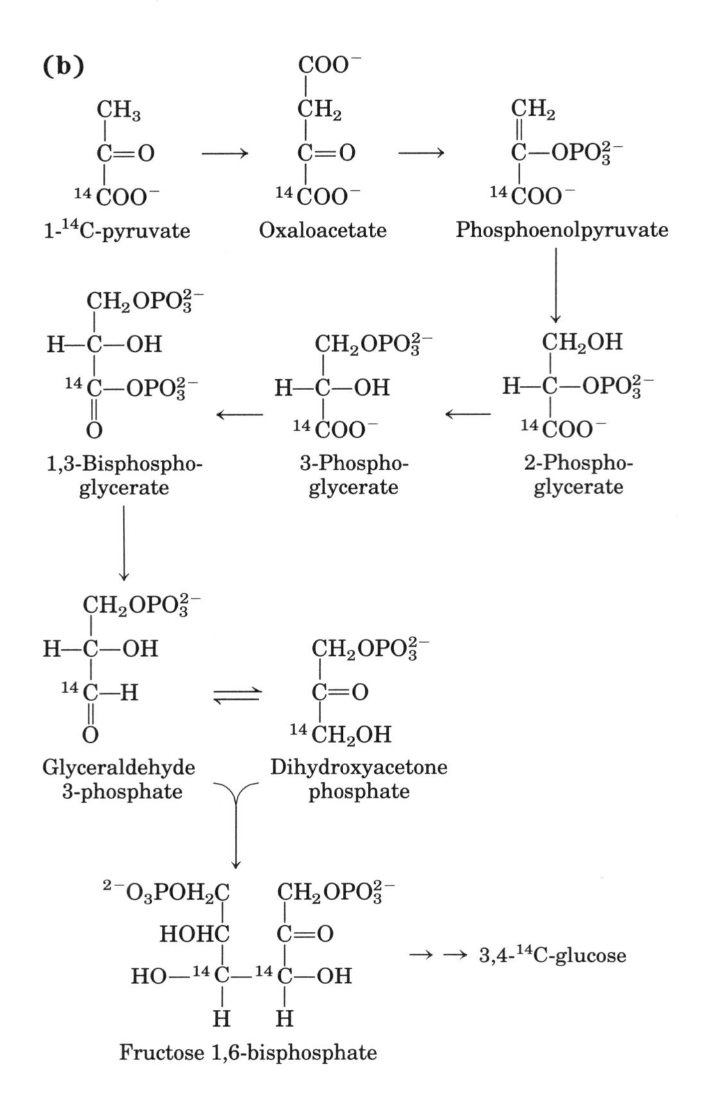

3. Pathway of CO_2 in Gluconeogenesis In the first bypass step of gluconeogenesis, the conversion of pyruvate to phosphoenolpyruvate, pyruvate is carboxylated by pyruvate carboxylase to oxaloacetate, which is subsequently decarboxylated by PEP carboxykinase to yield phosphoenolpyruvate. The observation that the addition of CO_2 is directly followed by the loss of CO_2 suggests that ^{14}C of $^{14}CO_2$ would not be incorporated into PEP, glucose, or any intermediates in gluconeogenesis. However, when a rat liver preparation synthesizes glucose in the presence of $^{14}CO_2$, ^{14}C slowly appears in PEP and eventually at C-3 and C-4 of glucose. How does the ^{14}C label get into PEP and glucose? (Hint: During gluconeogenesis in the presence of $^{14}CO_2$, several of the four-carbon citric acid cycle intermediates also become labeled.)

Answer Because pyruvate carboxylase is a mitochondrial enzyme, the [^{14}C]oxaloacetate (OAA) formed by this reaction mixes with the OAA pool of the citric acid cycle. A mixture of [1-^{14}C] and [4-^{14}C]OAA forms by randomization of the C-1 and C-4 positions in the reversible conversions of OAA to malate to succinate. [1-^{14}C]OAA leads to formation of [3,4-^{14}C]glucose.

4. **Energy Cost of a Cycle of Glycolysis and Gluconeogenesis** What is the cost (in ATP equivalents) of transforming glucose to pyruvate via glycolysis and back again to glucose via gluconeogenesis?

Answer The overall reaction of glycolysis is

$$\text{Glucose} + 2\text{ADP} + 2\text{P}_i + 2\text{NAD}^+ \longrightarrow 2\text{ pyruvate} + 2\text{ATP} + 2\text{NADH} + 2\text{H}^+ + 2\text{H}_2\text{O}$$

The overall reaction of gluconeogenesis is

$$2\text{ Pyruvate} + 4\text{ATP} + 2\text{GTP} + 2\text{NADH} + 2\text{H}^+ + 4\text{H}_2\text{O} \longrightarrow \text{glucose} + 2\text{NAD}^+ + 4\text{ADP} + 4\text{GDP} + 6\text{P}_i$$

The cost of transforming glucose to pyruvate and back to glucose is given by the difference between these two equations:

$$2\text{ATP} + 2\text{GTP} + 2\text{H}_2\text{O} \longrightarrow 2\text{ADP} + 2\text{GDP} + 4\text{P}_i$$

The energy cost is four ATP equivalents per glucose molecule.

5. **Regulation of Glycolysis and Gluconeogenesis** Because both glycolysis and gluconeogenesis are irreversible, there is no thermodynamic barrier to their simultaneous operation. What would the result be if both pathways were operating at the same time and at the same rate? What prevents such simultaneous operation in cells? What considerations govern which pathway is operating at any time?

Answer If both pathways were operating at the same time and at the same rate, the flow of glucose to pyruvate and back to glucose would result in the net hydrolysis of four high-energy phosphate groups (see Problem 4). Under physiological conditions, however, the flow of metabolites through the two pathways is not the same. The pathways are reciprocally regulated—as the flow through one increases, the flow through the other decreases. Glycolysis is stimulated and gluconeogenesis inhibited when the energy charge of the cell, as signaled by high [AMP], is low; glycolysis is inhibited and gluconeogenesis stimulated when the energy charge is high.

6. **Regulation of Fructose 1,6-Bisphosphatase and Phosphofructokinase-1** What are the effects of increasing concentrations of ATP and AMP on the catalytic activities of fructose 1,6-bisphosphatase and phosphofructokinase-1? What are the consequences of these effects on the relative flow of metabolites through gluconeogenesis and glycolysis?

Answer The two enzymes are regulated by adenine nucleotide modifiers in a reciprocal manner: what inhibits one activates the other. PFK-1, which catalyzes the rate-limiting step in glycolysis, is activated by AMP but inhibited by ATP. FBPase-1, which catalyzes the reverse reaction by hydrolysis, is activated by ATP and inhibited by AMP. This relates to "adenylate energy charge": when there is a need for energy (ATP), the flow of metabolites through glycolysis is stimulated and gluconeogenesis is inhibited. Conversely, when there is an abundance of energy, glycolysis is inhibited and the flow in the direction of gluconeogenesis is stimulated.

7. **Glucogenic Substrates** A common procedure for determining the effectiveness of compounds as precursors of glucose in mammals is to starve the animal until the liver glycogen stores are depleted and then administer the compound in question. A substrate that leads to a *net* increase in liver glycogen is termed glucogenic because it must first be converted to glucose 6-phosphate. Show by means of known enzymatic reactions which of the following substances are glucogenic:

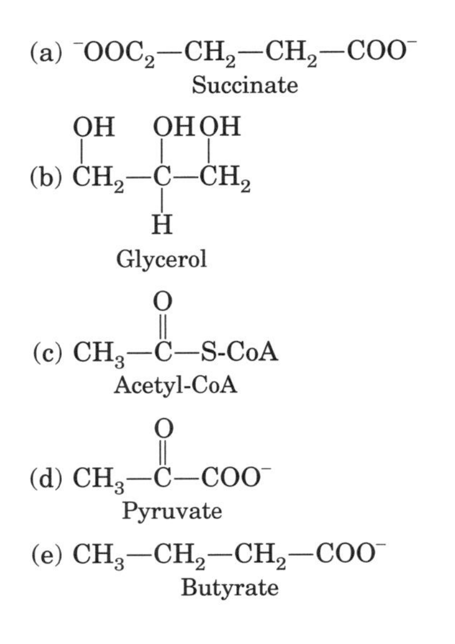

Answer

(a) Succinate is converted by the citric acid cycle to fumarate by succinate dehydrogenase, then to malate by fumarase, then to oxaloacetate by malate dehydrogenase. OAA can then leave the mitochondria via the malate-aspartate shuttle, and is converted to PEP, which is glucogenic in the cytosol.

(b) Glycerol is converted to glycerol 1-phosphate by glycerol kinase, then by a dehydrogenase (using NAD^+) to dihydroxyacetone phosphate, which is glucogenic.

(c) Acetyl-CoA is not glucogenic. Higher animals do not have the enzymes to convert it to pyruvate.

(d) Pyruvate is converted to oxaloacetate by pyruvate carboxylase, which is used for gluconeogenesis as in **(a).**

(e) Butyrate is converted to butyryl-CoA by an acyl-CoA synthetase, and a single turn of the β-oxidation pathway converts butyryl-CoA to two molecules of acetyl-CoA, which is not glucogenic.

8. Blood Lactate Levels during Vigorous Exercise The concentration of lactate in blood plasma before, during, and after a 400 m sprint are shown in the graph

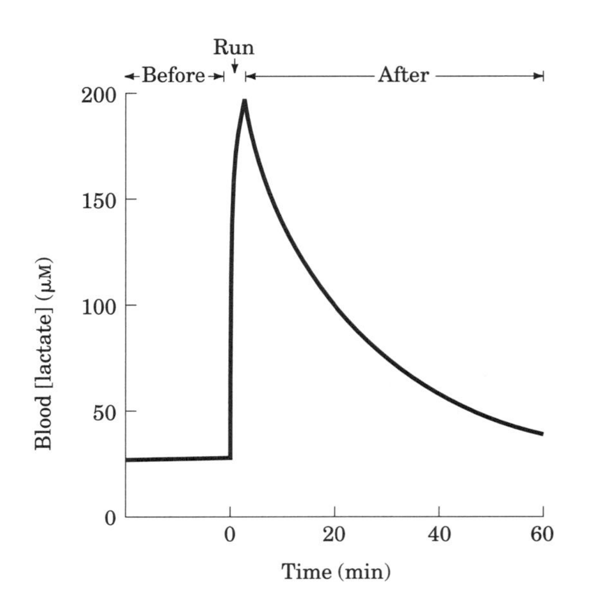

(a) What causes the rapid rise in lactate concentration?

(b) What causes the decline in lactate concentration after completion of the sprint? Why does the decline occur more slowly than the increase?

(c) Why is the concentration of lactate not zero during the resting state?

Answer

(a) Due to rapid depletion of ATP during strenuous muscular exertion, the rate of glycolysis increases dramatically, producing increased cytosolic concentrations of pyruvate and NADH, which are converted to lactate and NAD^+ by lactate dehydrogenase (lactic acid fermentation).

(b) When energy demands are reduced, the oxidative capacity of the mitochondria is again adequate, and lactate is transformed to pyruvate by lactase dehydrogenase and thus to glucose. The rate of the dehydrogenase reaction is slower in this direction because of the limited availability of NAD^+ and because the equilibrium of the reaction is strongly in favor of lactate (conversion of lactate to pyruvate is energy-requiring).

(c) The equilibrium of the lactate dehydrogenase reaction

$$\text{Pyruvate} + \text{NADH} + \text{H}^+ \longrightarrow \text{lactate} + \text{NAD}^+$$

is *strongly* in favor of lactate. Thus, even at very low concentrations of NADH and pyruvate, there is a significant concentration of lactate.

9. Relationship between Fructose 1,6-Bisphosphatase and Blood Lactate Levels A congenital defect in the liver enzyme fructose 1,6-bisphosphatase results in abnormally high levels of lactate in the blood plasma. Explain.

Answer Lactate is transformed to glucose in the liver by gluconeogenesis (see Fig. 20–2). This pathway includes the glycolytic bypass step catalyzed by fructose 1,6-bisphosphatase. A defect in this enzyme would prevent the entry of lactate into the gluconeogenic pathway in hepatocytes, causing lactate to accumulate in the blood.

10. Ethanol Affects Blood Glucose Levels The consumption of alcohol (ethanol), especially after periods of strenuous activity or after not eating for several hours, results in a deficiency of glucose in the blood, a condition known as hypoglycemia. The first step in the metabolism of ethanol by the liver is oxidation to acetaldehyde, catalyzed by liver alcohol dehydrogenase:

$$CH_3CH_2OH + NAD^+ \longrightarrow CH_3CHO + NADH + H^+$$

Explain how this reaction inhibits the transformation of lactate to pyruvate. Why does this lead to hypoglycemia?

Answer The first step in the synthesis of glucose from lactate in the liver is oxidation of the lactate pyruvate and, like the oxidation of ethanol to acetaldehyde, this requires NAD^+. Consumption of alcohol forces a competition for NAD^+ between ethanol metabolism and gluconeogenesis. This reduces the conversion of lactate to glucose, and the result is hypoglycemia. The problem is compounded by strenuous exercise and lack of food because at these times the level of blood glucose is already low.

11. Effect of Phloridzin on Carbohydrate Metabolism Phloridzin, a toxic glycoside from the bark of the pear tree, blocks the normal reabsorption of glucose from the kidney tubule, thus causing blood glucose to be almost completely excreted in the urine. Rats fed phloridzin and sodium succinate excreted about 0.5 moles of glucose (made by gluconeogenesis) for every mole of sodium succinate ingested. How is the succinate transformed to glucose? Explain the stoichiometry.

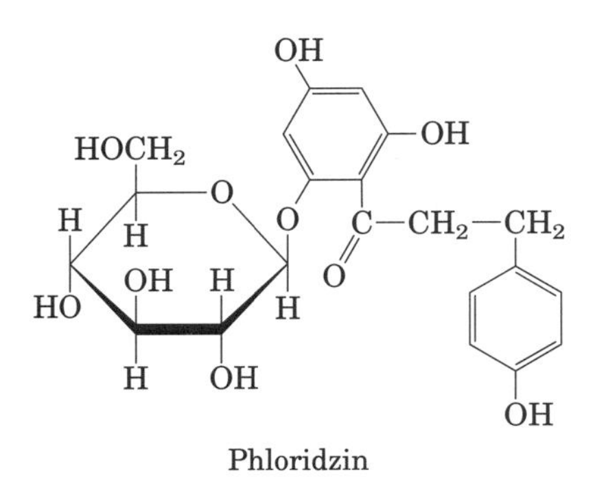

Phloridzin

Answer Excretion of glucose promoted by phloridzin causes a drop in blood glucose, which stimulates gluconeogenesis. The ingested succinate enters the mitochondrion via the dicarboxylate transport system and is transformed to oxaloacetate by enzymes of the citric acid cycle. The oxaloacetate passes into the cytosol and is transformed to phosphoenolpyruvate by PEP carboxykinase. Two moles of PEP are then required to produce a mole of glucose by the route outlined in Figure 20–2, consistent with the observed stoichiometry. Note that the rate of glucose production must be much higher than the rate of utilization by tissues, because almost 100% of the glucose is excreted.

12. Excess O_2 Uptake during Gluconeogenesis Lactate absorbed by the liver is converted to glucose, with the input of 6 mol of ATP for every mole of glucose produced. The extent of this process in a rat liver preparation can be monitored by administering [^{14}C]lactate and measuring the amount of [^{14}C]glucose produced. Because the stoichiometry of O_2 consumption and ATP production is known (Chapter 19), we can predict the extra O_2 consumption above the normal rate when a given amount of lactate is administered. However, when the extra O_2 used in the synthesis of glucose from lactate is actually measured, it is always higher than predicted by known stoichiometric relationships. Suggest a possible explanation for this observation.

Answer If the catabolic and biosynthetic pathways are operating simultaneously, a certain amount of ATP is consumed via futile cycles, in which no useful work is done. Examples of such futile cycles (which are usually blocked by reciprocal regulation of kinases and hydrolases) are between glucose and glucose 6-phosphate or between fructose 6-phosphate and fructose 1,6-bisphosphate. The net hydrolysis of ATP to ADP and P_i increases the consumption of oxygen, the terminal electron acceptor in oxidative phosphorylation.

13. At What Point Is Glycogen Synthesis Regulated? Explain how the following observations identify the point of regulation in the synthesis of glycogen in skeletal muscle.

(a) The measured activity of glycogen synthase in resting muscle, expressed in micromoles of UDP-glucose used per gram per minute, is lower than the activity of phosphoglucomutase or UDP-glucose pyrophosphorylase, each measured in terms of micromoles of substrate transformed per gram per minute.

(b) Stimulation of glycogen synthesis leads to a small decrease in the concentrations of glucose 6-phosphate and glucose 1-phosphate, a large decrease in the concentration of UDP-glucose, but a substantial increase in the concentration of UDP.

Answer The observation **(a)** that glycogen synthase has the lowest measured activity of the enzymes involved in glycogen synthesis suggests that this step is the kinetic bottleneck in the flow of metabolites and is thus a prime candidate for regulation. This is confirmed by the observation **(b)** that *stimulation* of glycogen synthesis by activation of the regulatory enzyme leads to a *decrease* in the steady-state concentrations of the intermediates *before* the regulation point (especially UDP-glucose) and an *increase* in the concentration of metabolites *after* the regulation point (UDP).

14. What Is the Cost of Storing Glucose as Glycogen? Write the sequence of steps and the net reaction required to calculate the cost in ATP molecules of converting a molecule of cytosolic glucose 6-phosphate to glycogen and back to glucose 6-phosphate. What fraction of the maximum number of ATP molecules available from complete catabolism of glucose 6-phosphate to CO_2 and H_2O does this cost represent?

Answer For synthesis of glycogen:

$$\text{Glucose 6-phosphate} \longrightarrow \text{glucose 1-phosphate}$$

$$\text{Glucose 1-phosphate} + \text{UTP} \longrightarrow \text{UDP-glucose} + \text{PP}_i$$

$$\text{PP}_i + \text{H}_2\text{O} \longrightarrow 2\text{P}_i$$

$$\text{UDP-glucose} + \text{glycogen}_n \longrightarrow \text{UDP} + \text{glycogen}_{n+1}$$

$$\text{UDP} + \text{ATP} \longrightarrow \text{UTP} + \text{ADP}$$

Thus, the net reaction for synthesis is:

$$\text{Glucose 6-phosphate} + \text{ATP} + \text{H}_2\text{O} + \text{glycogen}_n \longrightarrow \text{glycogen}_{n+1} + \text{ADP} + 2\text{P}_i$$

For breakdown of glycogen:

$$\text{Glycogen}_{n+1} + \text{P}_i \longrightarrow \text{glucose 1-phosphate} + \text{glycogen}_n$$

$$\text{Glucose 1-phosphate} \longrightarrow \text{glucose 6-phosphate}$$

Thus, the net reaction for breakdown is:

$$\text{Glycogen}_{n+1} + \text{P}_i \longrightarrow \text{glucose 6-phosphate} + \text{glycogen}_n$$

The net reaction for synthesis and breakdown:

$$\text{ATP} + \text{H}_2\text{O} \longrightarrow \text{ADP} + \text{P}_i$$

Each mole of glucose 6-phosphate converted to glycogen and back to glucose 6-phosphate costs 1 ATP. Each mole of glucose 6-phosphate completely catabolized yields 33 mol of ATP (see Table 16–1). Thus, the fraction is 1/33 = 0.030, or 3%. The efficiency of energy storage is 97%.

15. Identification of a Defective Enzyme in Carbohydrate Metabolism A sample of liver tissue was obtained post mortem from the body of a patient believed to be genetically deficient in one of the enzymes of carbohydrate metabolism. A homogenate of the liver sample had the following characteristics: (1) it degraded glycogen to glucose 6-phosphate, (2) it did not synthesize glycogen from any sugar nor did it use galactose as an energy source, and (3) it synthesized glucose 6-phosphate from lactate. Which of the following enzymes was deficient? Give reasons for your choice.

(a) Glycogen phosphorylase
(b) Fructose 1,6-bisphosphatase
(c) UDP-glucose pyrophosphorylase

Answer Characteristic (1) indicates that glycogen phosphorylase is functioning, and (2) indicates that fructose 1,6-bisphosphatase is operative. Thus the deficiency is in **(c)** UDP-glucose pyrophosphorylase. This enzyme is involved in the synthesis of UDP-glucose, a key intermediate in glycogen synthesis and in conversion of galactose (by epimerization) to glucose, a reaction required for the use of galactose as an energy source. Without this enzyme, starting from lactate the reversal of glycolysis would stop at formation of glucose 6-phosphate.

16. Ketosis in Sheep After the birth of a lamb, the udder of a ewe uses almost 80% of the total glucose synthesized by the animal. The glucose is used for milk production, principally in the synthesis of lactose and of glycerol 3-phosphate used in the formation of milk triacylglycerols. During the winter when food quality is poor, milk production decreases and the ewes sometimes develop ketosis: increased levels of plasma ketone bodies. Why do these changes occur? A standard treatment for sheep ketosis is the administration of large doses of propionate (readily converted to succinyl-CoA in ruminants). How does this treatment work?

Answer In ruminants such as sheep, ingested food is fermented by bacteria to yield acetate, lactate, and propionate. In the ewe, the latter two substances are transformed to glucose via gluconeogenesis and are subsequently used for milk production. In the absence of an adequate food supply, the ewe must use glucogenic amino acids from body protein to produce glucose for limited milk production. The energy needs of the ewe are met by the catabolism of body fat and ketogenic amino acids. The diversion of glucose and its precursor oxaloacetate to milk production under conditions of extensive fatty acid catabolism results in ketosis and associated acidosis. This situation is analogous to that in a diabetic. Ruminants can readily transform propionate to succinyl-CoA (via the intermediates propionyl-CoA, D-methylmalonyl-CoA, and L-methylmalonyl-CoA) and thus to oxaloacetate. In the presence of propionate, the activity of the citric acid cycle is therefore increased, and ketosis is averted.

17. Phases of Photosynthesis When a suspension of green algae is illuminated in the absence of CO_2 and then incubated with $^{14}CO_2$ in the dark, $^{14}CO_2$ is converted to [^{14}C]glucose for a brief time. What is the significance of this observation with regard to the CO_2 assimilation process, and how is it related to the light reactions of photosynthesis? Why does the conversion of $^{14}CO_2$ to [^{14}C]glucose stop after a brief time?

Answer This observation suggests that photosynthesis occurs in two phases: (1) a light-dependent phase that generates ATP and NADPH, which are essential for CO_2 fixation, and (2) a light-independent (dark) phase, in which these energy-rich components are used for synthesis of glucose. In the absence of additional illumination, the supplies of NADPH and ATP become exhausted and CO_2 fixation ceases.

18. Identification of Key Intermediates in CO_2 Assimilation Calvin and his colleagues used the unicellular green alga *Chlorella* to study the carbon assimilation reactions of photosynthesis. They incubated $^{14}CO_2$ with illuminated suspensions of algae and followed the time course of appearance of ^{14}C in two compounds, X and Y, under two sets of conditions. Suggest the identities of X and Y based on your understanding of the Calvin cycle.

(a) Illuminated *Chlorella* were grown with unlabeled CO_2, then the lights were turned off and $^{14}CO_2$ was added (vertical dashed line in graph **a**). Under these conditions, X was the first compound to become labeled with ^{14}C; Y was unlabeled.

(b) Illuminated *Chlorella* cells were grown with $^{14}CO_2$. Illumination was continued until all the $^{14}CO_2$ had disappeared (vertical dashed line in graph **b**). Under these conditions, X became labeled quickly but lost its radioactivity with time, whereas Y became more radioactive with time.

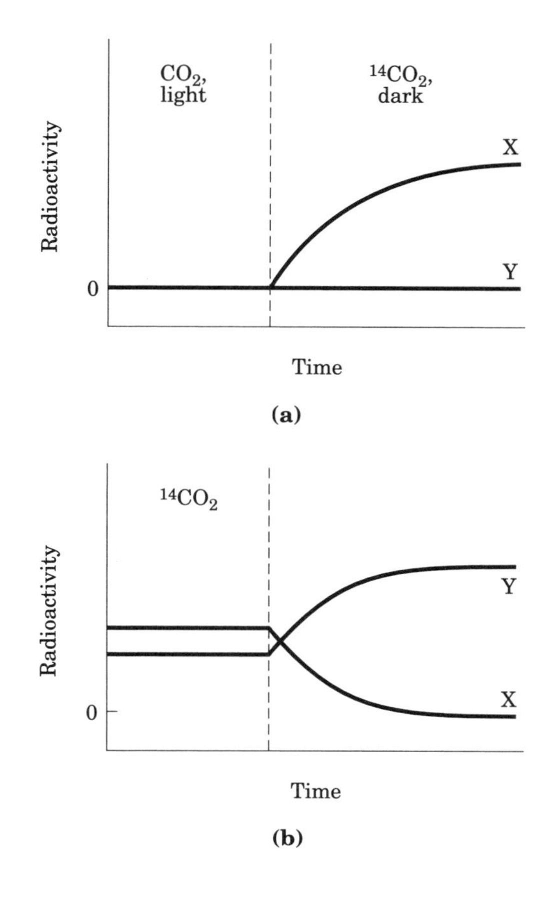

Answer Compound X is 3-phosphoglycerate and compound Y is ribulose 1,5-bisphosphate (see Fig. 20–22).

(a) Illumination of *Chlorella* in the presence of unlabeled CO_2 gives rise to steady-state levels of ribulose 1,5-bisphosphate, 3-phosphoglycerate, ATP, and NADPH. When the light is turned off, the production of ATP and NADPH ceases, but the Calvin cycle continues briefly until the residual ATP and NADPH are exhausted. Once this occurs, the conversion of 3-phosphoglycerate (Stage 2 in Fig. 20–22) to hexoses, which depends on ATP and NADPH, is blocked. Thus the $^{14}CO_2$ added at the time the light is turned off is transformed primarily to 3-phosphoglycerate, but not to other intermediates of the cycle, such as ribulose 1,5-bisphosphate.

(b) Illumination of *Chlorella* in the presence of $^{14}CO_2$ gives rise to steady-state levels of ^{14}C-labeled 3-phosphoglycerate and ribulose 1,5-bisphosphate. If the concentration of CO_2 is rapidly decreased, none is available for the ribulose 1,5-bisphosphate carboxylase reaction, which constitutes a block at the fixation step (Stage 1 in Fig. 20–22). Since this experiment is carried out under conditions of constant illumination, the steps requiring ATP and NADPH are not blocked and all labeled 3-phosphoglycerate (X) can be converted to labeled ribulose 1,5-bisphosphate (Y). This results in a decrease in labeled X and a commensurate increase in labeled Y.

19. Pathway of CO_2 Assimilation in Maize If a maize (corn) plant is illuminated in the presence of $^{14}CO_2$, after about 1 s more than 90% of all the radioactivity incorporated in the leaves is found at C-4 of malate, aspartate, and oxaloacetate. Only after 60 s does ^{14}C appear at C-1 of 3-phosphoglycerate. Explain.

Answer In maize, CO_2 is fixed by the C_4 pathway worked out by Hatch and Slack. Phosphoenolpyruvate is rapidly carboxylated to oxaloacetate, some of which undergoes transamination to aspartate but most of which is reduced to malate in the mesophyll cells. Only after subsequent decarboxylation of labeled malate does $^{14}CO_2$ enter the Calvin cycle for conversion to glucose. The rate of entry into the cycle is limited by the rate of the rubisco-catalyzed reaction.

20. Chemistry of Malic Enzyme: Variation on a Theme Malic enzyme, found in the bundle-sheath cells of C_4 plants, carries out a reaction that has a counterpart in the citric acid cycle. What is the analogous reaction? Explain.

Answer Malic enzyme catalyzes an oxidative decarboxylation of a hydroxycarboxylic acid in the C_4 pathway:

$$\underset{\text{Malate}}{^{-}\text{OOC—CH(OH)—CH}_2\text{—COO}^{-}} + \text{NADP}^{+} \longrightarrow \underset{\text{Pyruvate}}{^{-}\text{OOC—CO—CH}_3} + \text{CO}_2 + \text{NADPH} + \text{H}^{+}$$

which is analogous to the reaction catalyzed in the citric acid cycle by the enzyme isocitrate dehydrogenase:

$$\text{Isocitrate} + \text{NAD}^{+} \longrightarrow \alpha\text{-ketoglutarate} + \text{NADH} + \text{H}^{+}$$

21. Sucrose and Dental Caries The most prevalent infection in humans worldwide is dental caries, which stems from the colonization and destruction of tooth enamel by a variety of acidifying microorganisms. These organisms synthesize and live within a water-insoluble network of dextrans, called dental plaque, composed of ($\alpha1 \rightarrow 6$)-linked polymers of glucose with many ($\alpha1 \rightarrow 3$) branch points. Polymerization of dextran requires dietary sucrose, and the reaction is catalyzed by a bacterial enzyme, dextran-sucrose glucosyltransferase.

(a) Write the overall reaction for dextran polymerization.

(b) In addition to providing a substrate for the formation of dental plaque, how does dietary sucrose also provide oral bacteria with an abundant source of metabolic energy?

Answer

(a) $$\text{Sucrose} + \underset{\text{dextran}_n}{(\text{glucose})_n} \longrightarrow \underset{\text{dextran}_{n+1}}{(\text{glucose})_{n+1}} + \text{fructose}$$

(b) The fructose generated in the synthesis of dextran is readily taken up by the bacteria and metabolized to acidic compounds.

22. Regulation of Carbohydrate Synthesis in Plants Sucrose synthesis occurs in the cytosol and starch synthesis in the chloroplast stroma, yet the two processes are intricately balanced.

(a) What factors shift the reactions in favor of starch synthesis?

(b) What factors shift the reactions to favor sucrose synthesis?

(c) Given that these two synthetic pathways occur in separate cellular compartments, what enables the two processes to influence each other?

Answer

(a) Low levels of P_i in the cytosol and high levels of triose phosphate in the chloroplast favor formation of starch.

(b) High levels of triose phosphate in the cytosol favor formation of sucrose.

(c) The P_i-triose phosphate antiport system.

chapter

21

Lipid Biosynthesis

1. **Pathway of Carbon in Fatty Acid Synthesis** Using your knowledge of fatty acid biosynthesis, provide an explanation for the following experimental observations:
   (a) The addition of uniformly labeled [^{14}C]acetyl-CoA to a soluble liver fraction yields palmitate uniformly labeled with ^{14}C.
   (b) However, the addition of a *trace* of uniformly labeled [^{14}C]acetyl-CoA in the presence of an excess of unlabeled malonyl-CoA to a soluble liver fraction yields palmitate labeled with ^{14}C only in C-15 and C-16.

   Answer Recall that the "loading" of the acyl carrier protein requires an initial addition of acetyl-CoA followed by the addition of malonyl-CoA. Malonyl-CoA is normally produced by the addition of CO_2 to acetyl-CoA by acetyl-CoA carboxylase.
   (a) Consequently, in the presence of an excess of [^{14}C]acetyl-CoA, the metabolic pool of malonyl-CoA becomes labeled at C-1 and C-2. This results in the formation of uniformly labeled palmitate.
   (b) If a *trace* of [^{14}C]acetyl-CoA in the presence of a large excess of unlabeled malonyl-CoA is introduced, however, the metabolic pool of malonyl-CoA does not become labeled because the trace of [^{14}C]acetyl-CoA is loaded onto the acyl carrier protein (to become C-15 and C-16 of palmitate) rather than transformed into malonyl-CoA (a slow, rate-controlling process). In addition, any labeled malonyl-CoA is diluted by the presence of excess unlabeled malonyl-CoA.

2. **Synthesis of Fatty Acids from Glucose** After a person has ingested large amounts of sucrose, the glucose and fructose that exceed caloric requirements are transformed to fatty acids for triacylglycerol synthesis. This fatty acid synthesis consumes acetyl-CoA, ATP, and NADPH. How are these substances produced from glucose?

   Answer Glucose and fructose are degraded to pyruvate via glycolysis in the cytosol. The pyruvate enters the mitochondrion and is oxidatively decarboxylated to acetyl-CoA, some of which enters the citric acid cycle. The reducing equivalents (NADH, $FADH_2$) produced in the citric acid cycle are used for ATP production by oxidative phosphorylation. The remainder of the acetyl-CoA is exported to the cytosol via the acetyl group shuttle for fatty acid synthesis (see Fig. 21–11). Some NADPH is produced in the pentose phosphate pathway (Fig. 15–20) in the cytosol using glucose as a substrate, and some is produced by the action of malic enzyme on cytoplasmic malate.

3. Net Equation of Fatty Acid Synthesis Write the net equation for the biosynthesis of palmitate in rat liver, starting from mitochondrial acetyl-CoA and cytosolic NADPH, ATP, and CO_2.

Answer The majority of acetyl-CoA used in fatty acid synthesis is formed from the oxidation of pyruvate in the mitochondria. Since the mitochondrial membrane is impermeable to acetyl-CoA, transfer across the membrane occurs via the acetyl group shuttle (see Fig. 21–11). This process requires the input of 1 ATP per acetyl-CoA

$$\text{Acetyl-CoA}_{(mit)} + \text{ATP} + \text{H}_2\text{O} \longrightarrow \text{acetyl-CoA}_{(cyt)} + \text{ADP} + \text{P}_i + \text{H}^+ \qquad \text{or}$$

$$8\ \text{Acetyl-CoA}_{(mit)} + 8\text{ATP} + 8\text{H}_2\text{O} \longrightarrow 8\ \text{acetyl-CoA}_{(cyt)} + 8\text{ADP} + 8\text{P}_i + 8\text{H}^+$$

In the synthesis of palmitate, cytosolic acetyl-CoA ("the primer") is condensed with malonyl-CoA, reduced, hydrated, and reduced again. This process is repeated seven times, each time adding an additional acetyl-CoA to the fatty acyl-CoA. The equation for this process is

$$8\ \text{Acetyl-CoA} + 14\text{NADPH} + 6\text{H}^+ + 7\text{ATP} + \text{H}_2\text{O} \longrightarrow \text{palmitate} + 8\text{CoA} + 14\text{NADP}^+ + 7\text{ADP} + 7\text{P}_i$$

(Note that this differs from Eqn 21–3 in the text because the 7 H_2O for hydrolysis of 7 ATP were omitted from Eqn 21–1, as were 7 H^+ in the products of that reaction.)

The net equation for the overall process in the cytosol:

$$8\ \text{Acetyl-CoA}_{(mit)} + 15\text{ATP} + 14\text{NADPH} + 9\text{H}_2\text{O} \longrightarrow \text{palmitate} + 8\text{CoA} + 15\text{ADP} + 15\text{P}_i + 14\text{NADP}^+ + 2\text{H}^+$$

This may also be viewed as the sum of the following two reactions:

$$8\ \text{Acetyl-CoA} + 14\text{NADPH} + 13\text{H}^+ \longrightarrow \text{CH}_3(\text{CH}_2)_{14}\text{COO}^- + 8\text{CoA} + 14\text{NADP}^+ + 6\text{H}_2\text{O} \qquad \text{and}$$

$$15\text{ATP} + 15\text{H}_2\text{O} \longrightarrow 15\text{ADP} + 15\text{P}_i + 15\text{H}^+$$

4. Pathway of Hydrogen in Fatty Acid Synthesis Consider a preparation that contains all the enzymes and cofactors necessary for fatty acid biosynthesis from added acetyl-CoA and malonyl-CoA.

(a) If [2-^{2}H]acetyl-CoA (labeled with deuterium, the heavy isotope of hydrogen):

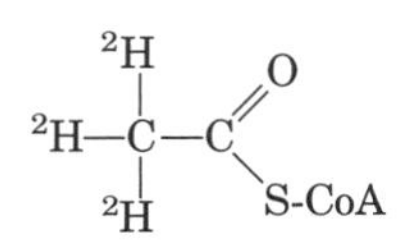

and an excess of unlabeled malonyl-CoA are added as substrates, how many deuterium atoms are incorporated into every molecule of palmitate? What are their locations? Explain.

(b) If unlabeled acetyl-CoA and [2-^{2}H]malonyl-CoA:

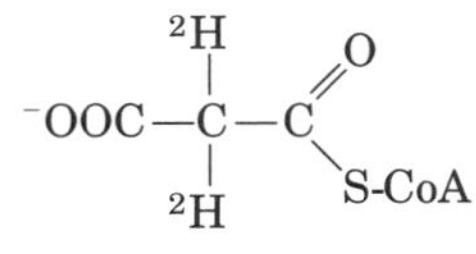

are added as substrates, how many deuterium atoms are incorporated into every molecule of palmitate? What are their locations? Explain.

Answer

(a) Three deuteriums per palmitate; all located on C-16; all other two-carbon units are derived from unlabeled malonyl-CoA. (Note that the unlabeled malonyl-CoA is in excess.)

(b) Seven deuteriums per palmitate; one on each even-numbered carbon except C-16. One deuterium is lost from each labeled even-numbered carbon at the dehydration step (see Fig. 21–5).

5. Energetics of β-Ketoacyl–ACP Synthase In the condensation reaction catalyzed by β-ketoacyl–ACP synthase (Fig. 21–5), a four-carbon unit is synthesized by the combination of a two-carbon unit and a three-carbon unit, with the release of CO_2. What is the thermodynamic advantage of this process over one that simply combines two two-carbon units?

Answer If a four-carbon unit were synthesized by combining two two-carbon units, the reaction would be reversible—for example, the reaction catalyzed by thiolase in the oxidation of fatty acids (see Fig. 17–8).

$$\text{CoA} + \text{CH}_3\text{—}\overset{\overset{\text{O}}{\|}}{\text{C}}\text{—CH}_2\text{—}\overset{\overset{\text{O}}{\|}}{\text{C}}\text{—CoA} \rightleftharpoons \text{CH}_3\text{—}\overset{\overset{\text{O}}{\|}}{\text{C}}\text{—CoA} + \text{CH}_3\text{—}\overset{\overset{\text{O}}{\|}}{\text{C}}\text{—CoA}$$

By using the three-carbon unit malonyl-CoA, the activated form of acetyl-CoA (recall that malonyl-CoA synthesis requires the input of energy from ATP), metabolite flow is driven in the direction of fatty acid synthesis by the exergonic release of CO_2.

6. Modulation of Acetyl-CoA Carboxylase Acetyl-CoA carboxylase is the principal regulation point in the biosynthesis of fatty acids. Some of the properties of the enzyme are described below:

(a) The addition of citrate or isocitrate raises the V_{max} of the enzyme by as much as a factor of 10.

(b) The enzyme exists in two interconvertible forms that differ markedly in their activities:

$$\text{Protomer (inactive)} \rightleftharpoons \text{filamentous polymer (active)}$$

Citrate and isocitrate bind preferentially to the filamentous form, and palmitoyl-CoA binds preferentially to the protomer.

Explain how these properties are consistent with the regulatory role of acetyl-CoA carboxylase in the biosynthesis of fatty acids.

Answer The rate-limiting step in the biosynthesis of fatty acids is the carboxylation of acetyl-CoA catalyzed by acetyl-CoA carboxylase. High levels of citrate and isocitrate indicate that conditions are favorable for fatty acid synthesis: an active citric acid cycle is providing a plentiful supply of ATP, reducing equivalents, and acetyl-CoA. Citrate or isocitrate stimulate (increase the V_{max} of) the rate-limiting enzymatic step in fatty acid biosynthesis **(a).** Furthermore, since citrate and isocitrate bind more tightly to the filamentous form of the enzyme (the active form), the presence of citrate or isocitrate drives the protomer $\rightleftharpoons$ polymer equilibrium in the direction of the active (polymer) form **(b).** In contrast, palmitoyl-CoA (the end product of fatty acid biosynthesis) drives the equilibrium in the direction of the inactive (protomer) form. Hence, when the end product of fatty acid biosynthesis builds up, fatty acid biosynthesis is slowed down.

7. Shuttling of Acetyl Groups across the Inner Mitochondrial Membrane The acetyl group of acetyl-CoA, produced by the oxidative decarboxylation of pyruvate in the mitochondrion, is transferred to the cytosol by the acetyl group shuttle outlined in Figure 21–11.

(a) Write the overall equation for the transfer of one acetyl group from the mitochondrion to the cytosol.

(b) What is the cost of this process in ATPs per acetyl group?

(c) In Chapter 17 we encountered an acyl group shuttle in the transfer of fatty acyl–CoA from the cytosol to the mitochondrion in preparation for β oxidation (see Fig. 17–6). One result of that shuttle was separation of the mitochondrial and cytosolic pools of CoA. Does the acetyl group shuttle also accomplish this?

Answer

(a) The reactions involved in the transfer of one acetyl group:

$$\text{Acetyl-CoA}_{(mit)} + \text{oxaloacetate}_{(mit)} \longrightarrow \text{citrate}_{(mit)} + \text{CoA}_{(mit)}$$

$$\text{Citrate}_{(mit)} \longrightarrow \text{citrate}_{(cyt)}$$

$$\text{Citrate}_{(cyt)} + \text{ATP} + \text{CoA}_{(cyt)} \longrightarrow \text{oxaloacetate}_{(cyt)} + \text{ADP} + \text{P}_i + \text{acetyl-CoA}_{(cyt)}$$

$$\text{Oxaloacetate}_{(cyt)} + \text{NADH} + \text{H}^+ \longrightarrow \text{malate}_{(cyt)} + \text{NAD}^+$$

$$\text{Malate}_{(cyt)} \longrightarrow \text{malate}_{(mit)}$$

$$\text{Malate}_{(mit)} + \text{NAD}^+ \longrightarrow \text{oxaloacetate}_{(mit)} + \text{NADH} + \text{H}^+$$

The overall equation:

$$\text{Acetyl-CoA}_{(mit)} + \text{ATP} + \text{CoA}_{(cyt)} \longrightarrow \text{acetyl-CoA}_{(cyt)} + \text{ADP} + \text{P}_i + \text{CoA}_{(mit)}$$

(b) The transfer of one acetyl group from the mitochondrial matrix to the cytosol (one turn of the acetyl group shuttle) is accompanied by the hydrolysis of ATP.

(c) The transfer requires the input of one cytosolic CoA, and one mitochondrial CoA is released. Thus, the shuttle keeps the cytosolic and mitochondrial pools of CoA separate.

8. Oxygen Requirement for Desaturases The biosynthesis of palmitoleate (Fig. 21–13), a common unsaturated fatty acid with a cis double bond in the Δ^9 position, uses palmitate as a precursor. Can this be carried out under strictly anaerobic conditions? Explain.

Answer No; oxygen is required. The double bond in palmitoleic acid is introduced by an oxidation catalyzed by fatty acyl–CoA oxygenase (or fatty acyl–CoA desaturase; see Fig. 21–14), a mixed-function oxidase that requires molecular oxygen as a cosubstrate.

9. Energy Cost of Triacylglycerol Synthesis Use a net equation for the biosynthesis of tripalmitoylglycerol (tripalmitin) from glycerol and palmitate to show how many ATPs are required per molecule of tripalmitin formed.

Answer The reactions involved in conversion of glycerol and palmitate to tripalmitin:

$$\text{ATP} + \text{glycerol} \longrightarrow \text{glycerol 3-phosphate} + \text{ADP} + \text{H}^+$$

$$3\ \text{Palmitate} + 3\text{ATP} + 3\text{CoA} \longrightarrow 3\ \text{palmitoyl-CoA} + 3\text{AMP} + 3\text{PP}_i + 3\text{H}^+$$

$$3\ \text{H}_2\text{O} + 3\text{PP}_i \longrightarrow 6\text{P}_i + 3\text{H}^+$$

$$2\ \text{Palmitoyl-CoA} + \text{glycerol 3-phosphate} \longrightarrow \text{dipalmitoylglycerol 3-phosphate} + \text{CoA}$$

$$\text{Dipalmitoyl 3-phosphate} + \text{H}_2\text{O} \longrightarrow \text{dipalmitoylglycerol} + \text{P}_i$$

$$\text{Dipalmitoylglycerol} + \text{palmitoyl-CoA} \longrightarrow \text{tripalmitin} + \text{CoA}$$

The overall reaction:

$$3\ \text{Palmitate} + \text{glycerol} + 4\text{ATP} + 4\text{H}_2\text{O} \longrightarrow \text{tripalmitin} + 3\text{AMP} + \text{ADP} + 7\text{P}_i + 7\text{H}^+$$

Including $3\text{AMP} + 3\text{ATP} \longrightarrow 6\text{ADP}$ the overall reaction becomes

$$3\ \text{Palmitate} + \text{glycerol} + 7\text{ATP} + 4\text{H}_2\text{O} \longrightarrow \text{tripalmitin} + 7\text{ADP} + 7\text{P}_i + 7\text{H}^+$$

10. Turnover of Triacylglycerols in Adipose Tissue When [^{14}C]glucose is added to the balanced diet of adult rats, there is no increase in the total amount of stored triacylglycerols, but the triacylglycerols become labeled with ^{14}C. Explain.

Answer In adult rats, stored triacylglycerols are maintained at a steady-state level through a balance of the rates of degradation and biosynthesis. Hence, the depots of fats are continually turned over, explaining the incorporation of ^{14}C from dietary [^{14}C] glucose.

11. Energy Cost of Phosphatidylcholine Synthesis Write the sequence of steps and the net reaction for the biosynthesis of phosphatidylcholine by the salvage pathway from oleate, palmitate, dihydroxyacetone phosphate, and choline. Starting from these precursors, what is the cost (in number of ATPs) of the synthesis of phosphatidylcholine by the salvage pathway?

Answer The sequence of steps required to carry out phosphatidylcholine biosynthesis:

$$\text{Dihydroxyacetone phosphate} + \text{NADH} + \text{H}^+ \longrightarrow \text{glycerol 3-phosphate} + \text{NAD}^+$$

$$\text{Palmitate} + \text{ATP} + \text{CoA} \longrightarrow \text{palmitoyl-CoA} + \text{AMP} + \text{PP}_i$$

$$\text{Oleic acid} + \text{ATP} + \text{CoA} \longrightarrow \text{oleoyl-CoA} + \text{AMP} + \text{PP}_i$$

$$2\text{PP}_i + 2\text{H}_2\text{O} \longrightarrow 2\text{P}_i$$

$$\text{Glycerol 3-phosphate} + \text{palmitoyl-CoA} + \text{oleoyl-CoA} \longrightarrow \text{L-phosphatidate} + 2\text{CoA}$$

$$\text{L-Phosphatidate} + \text{H}_2\text{O} \longrightarrow \text{1,2-diacylglycerol} + \text{P}_i$$

$$\text{ATP} + \text{choline} \longrightarrow \text{ADP} + \text{phosphocholine}$$

$$\text{CTP} + \text{phosphocholine} \longrightarrow \text{CDP-choline} + \text{PP}_i$$

$$\text{PP}_i + \text{H}_2\text{O} \longrightarrow 2\text{P}_i$$

$$\text{CDP-choline} + \text{1,2-diacylglycerol} \longrightarrow \text{CMP} + \text{phosphatidylcholine}$$

The net reaction:

$$\text{Dihydroxyacetone phosphate} + \text{NADH} + \text{palmitate} + \text{oleate} + 3\text{ATP} + \text{CTP} + \text{choline} + 4\text{H}_2\text{O} \longrightarrow \text{phosphatidylcholine} + \text{NAD}^+ + 2\text{AMP} + \text{ADP} + \text{CMP} + 5\text{P}_i + \text{H}^+$$

This can also be viewed as the sum of the following two equations:

$$\text{Dihydroxyacetone phosphate} + \text{palmitate} + \text{oleate} + \text{choline} + \text{NADH} + 2\text{H}^+ \longrightarrow \text{phosphatidylcholine} + \text{NAD}^+ + 3\text{H}_2\text{O}$$

and

$$3\text{ATP} + \text{CTP} + 7\text{H}_2\text{O} \longrightarrow 2\text{AMP} + \text{ADP} + \text{CMP} + 5\text{P}_i + 3\text{H}^+$$

Note that the net equation shows the production of three nucleoside monophosphates (2 AMP + CMP). To recycle these monophosphates to nucleoside diphosphates requires the input of 3 ATP.

$$2\text{AMP} + 2\text{ATP} \rightleftharpoons 4\text{ADP}$$

$$\text{CMP} + \text{ATP} \rightleftharpoons \text{CDP} + \text{ADP}$$

Furthermore, if we consider that CTP is energetically equivalent to ATP, then the synthesis of one phosphatidylcholine requires a total of seven ATP.

12. Salvage Pathway for Synthesis of Phosphatidylcholine A young rat maintained on a diet deficient in methionine fails to thrive unless choline is included in the diet. Explain.

Answer The rat has two pathways for synthesizing phosphatidylcholine: the de novo pathway and the salvage pathway. The de novo pathway requires the transfer of a methyl group from *S*-adenosylmethionine (adoMet; see Fig. 21–25) to phosphatidylethanolamine. When the diet is deficient in methionine (an essential amino acid), the biosynthesis of adoMet and phosphatidylcholine is severely impaired. The salvage pathway, on the other hand, does not require adoMet but utilizes available choline. Thus, phosphatidylcholine can be synthesized when the diet is deficient in methionine as long as choline is available.

13. Synthesis of Isopentenyl Pyrophosphate If 2-[^{14}C]acetyl-CoA is added to a rat liver homogenate that is synthesizing cholesterol, where will the ^{14}C label appear in Δ^3-isopentenyl pyrophosphate, the activated form of an isoprene unit?

Answer Three acetyl units are required for the synthesis of an isoprene unit (see Fig. 21–31). The ^{14}C label at C-2 of acetyl-CoA ends up in three places in the activated isoprene unit:

$$^{14}CH_2{=}C({-}^{14}CH_3){-}^{14}CH_2{-}CH_2{-}$$

14. Activated Donors in Lipid Synthesis In the biosynthesis of complex lipids, components are assembled by transfer of the appropriate group from an activated donor. For example, the activated donor of acetyl groups is acetyl-CoA. For each of the following groups, give the form of the activated donor: **(a)** phosphate; **(b)** D-glucosyl; **(c)** phosphoethanolamine; **(d)** D-galactosyl; **(e)** fatty acyl; **(f)** methyl; **(g)** the two-carbon group in fatty acid biosynthesis; **(h)** Δ^3-isopentenyl.

Answer **(a)** ATP; **(b)** UDP-glucose; **(c)** CDP-ethanolamine; **(d)** UDP-galactose; **(e)** fatty acyl–CoA; **(f)** *S*-adenosylmethionine; **(g)** malonyl-CoA; **(h)** Δ^3-isopentenyl pyrophosphate.

15. Importance of Fats in the Diet When young rats are placed on a totally fat-free diet, they grow poorly, develop a scaly dermatitis, lose hair, and soon die—symptoms that can be prevented if linoleate or plant material is included in the diet. What makes linoleate an essential fatty acid? Why can plant material be substituted?

Answer Linoleate is an unsaturated fatty acid required in the synthesis of prostaglandins. Mammals lack the enzymes required to introduce double bonds into fatty acids beyond the Δ^9 position and therefore are unable to transform oleate to linoleate. Accordingly, linoleate is an essential fatty acid in animals. However, plants can transform oleate to linoleate, and thus they provide animals with the required linoleate (Fig. 21–13).

16. Regulation of Cholesterol Biosynthesis Cholesterol in humans can be obtained from the diet or synthesized de novo. An adult human on a low-cholesterol diet typically synthesizes 600 mg of cholesterol per day in the liver. If the amount of cholesterol in the diet is large, de novo synthesis of cholesterol is drastically reduced. How is this regulation brought about?

Answer The rate-determining step in the biosynthesis of cholesterol is the synthesis of mevalonate catalyzed by hydroxymethylglutaryl-CoA reductase. This enzyme is subject to feedback regulation by the negative modulator cholesterol.

chapter

22 Biosynthesis of Amino Acids, Nucleotides, and Related Molecules

1. **ATP Consumption by Root Nodules in Legumes** Bacteria presiding in the root nodules of the pea plant consume more than 20% of the ATP produced by the plant. Suggest why these bacteria consume so much ATP.

   Answer The bacteria in the root nodules maintain a symbiotic relationship with the plant: the plant supplies ATP and reducing power, and the bacteria supply ammonium ion by reducing atmospheric nitrogen. This reduction requires large quantities of ATP.

2. **Glutamate Dehydrogenase and Protein Synthesis** The bacterium *Methylophilus methylotrophus* can synthesize protein from methanol and ammonia. Recombinant DNA techniques have improved the yield of protein by introducing into *M. methylotrophus* the glutamate dehydrogenase gene from *E.coli.* Why does this genetic manipulation increase the protein yield?

   Answer The synthesis of protein requires the synthesis of amino acids. The transfer of nitrogen from an ammonium ion to carbon skeletons—that is, amino acid synthesis—can be carried out two ways: (1) combination of the ammonium ion with glutamate to form glutamine, catalyzed by glutamine synthetase and (2) reductive amination of α-ketoglutarate to form glutamate, catalyzed by glutamate dehydrogenase. The latter process, which is promoted by the introduction of the *E. coli* enzyme, is especially important because glutamate is the amino group donor in all transamination reactions.

3. **Transformation of Aspartate to Asparagine** There are two routes for transforming aspartate to asparagine at the expense of ATP. Many bacteria have an asparagine synthetase that uses ammonium ion as the nitrogen donor. Mammals have an asparagine synthetase that uses glutamine as the nitrogen donor. Given that the latter requires an extra ATP (for the synthesis of glutamine), why do mammals use this route?

   Answer Recall that the ammonium ion is highly toxic to higher animals, especially to brain tissue. Because ammonium ions are transformed to glutamine in the mammalian route, circulating ammonium ion levels in the organism are reduced and toxic levels are avoided.

4. **Equation for the Synthesis of Aspartate from Glucose** Write the net equation for the synthesis of aspartate (a nonessential amino acid) from glucose, carbon dioxide, and ammonia.

   Answer We can approach this problem by working "backwards" from aspartate to glucose as follows. Aspartate is synthesized from oxaloacetate by transamination from glutamate; the glutamate is synthesized from α-ketoglutarate by glutamate dehydrogenase:

$$\text{Oxaloacetate} + \text{glutamate} \longrightarrow \text{aspartate} + \alpha\text{-ketoglutarate}$$

$$\alpha\text{-Ketoglutarate} + NH_3 + 2H^+ + \text{NADH} \longrightarrow \text{glutamate} + NAD^+ + H_2O$$

The sum of these reactions is

$$\text{Oxaloacetate} + NH_3 + 2H^+ + \text{NADH} \longrightarrow \text{aspartate} + NAD^+ + H_2O$$

Recall from Chapter 16 that oxaloacetate is synthesized from pyruvate by pyruvate carboxylase, and from Chapter 15 that pyruvate is synthesized from glucose via glycolysis:

$$\text{Pyruvate} + CO_2 + \text{ATP} + H_2O \longrightarrow \text{oxaloacetate} + \text{ADP} + P_i + 2H^+$$

$$\text{Glucose} + 2NAD^+ + 2\text{ADP} + 2P_i \longrightarrow 2\ \text{pyruvate} + 2\text{NADH} + 2H^+ + 2\text{ATP} + 2H_2O$$

Thus we can write the net equation for aspartate synthesis:

$$\text{Glucose} + 2CO_2 + 2NH_3 \longrightarrow 2\ \text{aspartate} + 2H^+ + 2H_2O$$

5. **Phenylalanine Hydroxylase Deficiency and Diet** Tyrosine is normally a nonessential amino acid, but individuals with a genetic defect in phenylalanine hydroxylase require tyrosine in their diet for normal growth. Explain.

Answer In animals, tyrosine is synthesized from phenylalanine by phenylalanine hydroxylase. If this enzyme is defective, the biosynthetic route to Tyr is blocked and Tyr must be obtained from the diet.

6. **Cofactors for One-Carbon Transfer Reactions** Most one-carbon transfers are promoted by one of three cofactors: biotin, tetrahydrofolate, or *S*-adenosylmethionine (Chapter 18). *S*-Adenosylmethionine is generally used as a methyl group donor; the transfer potential of the methyl group in N^5-methyltetrahydrofolate is insufficient for most biosynthetic reactions. However, one example of the use of N^5-methyltetrahydrofolate in methyl group transfer is in methionine formation by the methionine synthase reaction (step 9 of Fig. 22–15); methionine is the immediate precursor of *S*-adenosylmethionine (see Fig. 18–17). Explain how the methyl group of *S*-adenosylmethionine can be derived from N^5-methyltetrahydrofolate, even though the transfer potential of the methyl group in N^5-methyltetrahydrofolate is 10^3 times *lower* than that in *S*-adenosylmethionine.

Answer The transfer potential of the methyl group of N^5-methyltetrahydrofolate is quite sufficient for the synthesis of methionine, which has an even lower methyl group transfer potential. The methyl group of methionine is activated by addition of the adenosyl group from ATP when methionine is converted to *S*-adenosylmethionine (see Fig. 18–17). Recall that *S*-adenosylmethionine synthesis is one of only two known biochemical reactions in which triphosphate is released from ATP. Hydrolysis of the triphosphate renders the reaction thermodynamically more favorable.

7. **Concerted Regulation in Amino Acid Biosynthesis** The glutamine synthetase of *E. coli* is independently modulated by various products of glutamine metabolism (see Fig. 22–6). In this concerted inhibition, the extent of enzyme inhibition is greater than the sum of the separate inhibitions caused by each product. For *E. coli* grown in a medium rich in histidine, what would be the advantage of concerted inhibition?

Answer Because the regulatory mechanism is concerted, the amount of inhibition caused by saturating concentrations of histidine is limited—that is, a large excess of one amino acid does not shut down the flow of glutamate to glutamine. Metabolite flow continues, albeit at a reduced rate. If the inhibition of glutamine synthase were not concerted, saturating concentrations of histidine would shut down the enzyme and cut off production of glutamine, which the bacterium needs to synthesize other products.

8. Relationship between Folic Acid Deficiency and Anemia Folic acid deficiency, believed to be the most common vitamin deficiency, causes a type of anemia in which hemoglobin synthesis is impaired and erythrocytes do not mature properly. What is the metabolic relationship between hemoglobin synthesis and folic acid deficiency?

Answer Folic acid is a precursor of the coenzyme tetrahydrofolate (Fig. 18–15), required in the biosynthesis of glycine (Fig. 22–12). Since glycine is a precursor of porphyrins, the heme component of hemoglobin, a folic acid deficiency results in an impairment of hemoglobin synthesis, especially if the diet is also low in glycine.

9. Nucleotide Biosynthesis in Amino Acid Auxotrophic Bacteria Normal *E. coli* cells can synthesize all the standard amino acids, but some mutants, called amino acid auxotrophs, are unable to synthesize a specific amino acid and require its addition to the culture medium for optimal growth. Besides their role in protein synthesis, some amino acids are also precursors for other nitrogenous cell products. Consider the three amino acid auxotrophs that are unable to synthesize glycine, glutamine, and aspartate, respectively. For each mutant, what nitrogenous products other than proteins would the cell fail to synthesize?

Answer Glycine, glutamine, and aspartate are required for the de novo synthesis of purine nucleotides; aspartate is required for the de novo synthesis of UMP; aspartate and glutamine are required for the de novo synthesis of CTP. Thus, glycine auxotrophs would fail to synthesize adenine and guanine nucleotides. Glutamine auxotrophs would fail to synthesize adenine, guanine, and cytosine nucleotides. Aspartate auxotrophs would fail to synthesize adenine, guanine, cytosine, and uridine nucleotides.

10. Inhibitors of Nucleotide Biosynthesis Suggest mechanisms for the inhibition of (a) alanine racemase by L-fluoroalanine and (b) glutamine amidotransferases by azaserine.

Answer

(a) See Figure 18–6 (step 2) for the reaction mechanism of amino acid racemization. The F atom of fluoroalanine is an excellent leaving group. Fluoroalanine causes irreversible (covalent) inhibition of alanine racemase. One plausible mechanism is:

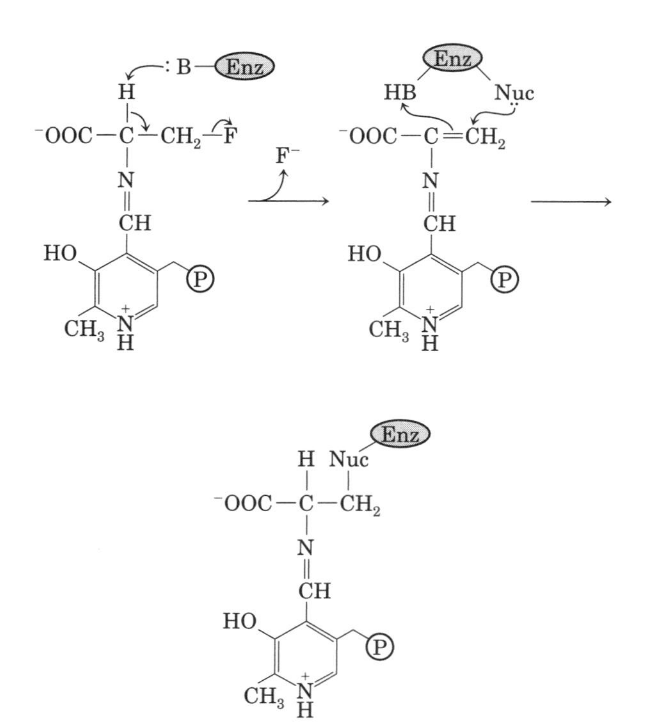

Nuc denotes any nucleophilic amino acid side chain in the enzyme active site.

(b) Azaserine (see Fig. 22–46) is an analog of glutamine. The diazoacetyl group is highly reactive and forms covalent bonds with nucleophiles at the active site of glutamine amidotransferases.

11. Mode of Action of Sulfa Drugs Some bacteria require *p*-aminobenzoate in the culture medium for normal growth, and their growth is severely inhibited by the addition of sulfanilamide, one of the earliest sulfa drugs. Moreover, in the presence of this drug, 5-aminoimidazole-4-carboxamide ribonucleotide (AICAR; see Fig. 22–31) accumulates in the culture medium. These effects are reversed by addition of excess *p*-aminobenzoate.

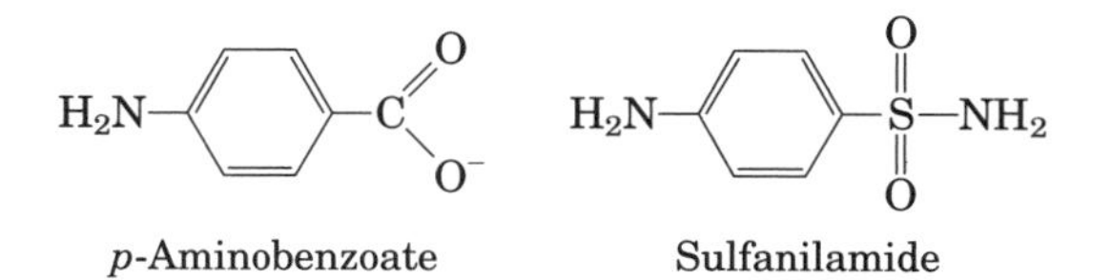

(a) What is the role of *p*-aminobenzoate in these bacteria? (Hint: See Fig. 18–15).
(b) Why does AICAR accumulate in the presence of sulfanilamide?
(c) Why are the inhibition and accumulation reversed by addition of excess *p*-aminobenzoate?

Answer

(a) *p*-Aminobenzoate is a component of tetrahydrofolate (see Fig. 18–15) and its derivative, N^5N^{10}-methylenetetrahydrofolate, the cofactor involved in the transfer of one-carbon units.

(b) Sulfanilamide is a structural analog of *p*-aminobenzoate. In the presence of sulfanilamide, bacteria are unable to synthesize tetrahydrofolate, a cofactor necessary for the transformation of AICAR to *N*-formylaminoimidazole-4-carboxamide ribonucleotide (FAICAR) by the addition of —CHO; thus, AICAR accumulates.

(c) Excess *p*-aminobenzoate reverses the growth inhibition and ribonucleotide accumulation by competing with sulfanilamide for the active site of the enzyme involved in tetrahydrofolate biosynthesis. The competitive inhibition by sulfanilamide is overcome by the addition of excess substrate (*p*-aminobenzoate).

12. Pathway of Carbon in Pyrimidine Biosynthesis Predict the locations of ^{14}C in orotate isolated from cells grown on a small amount of uniformly labeled [^{14}C]succinate. Justify your prediction.

Answer

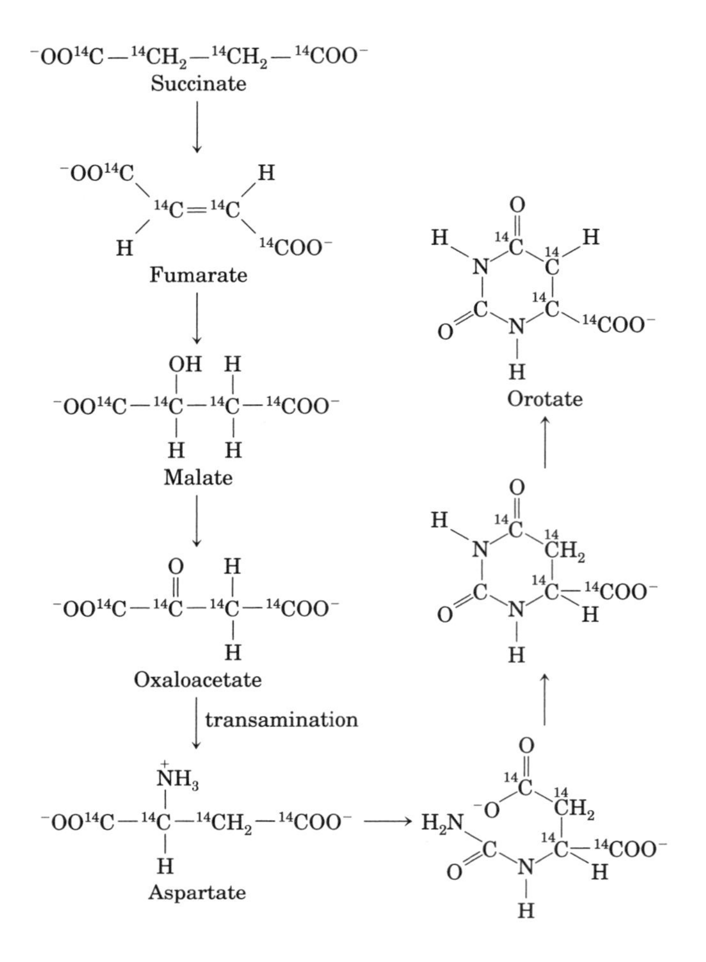

13. Nucleotides as Poor Sourcs of Energy Under starvation conditions, organisms can use proteins and amino acids as sources of energy. Deamination of amino acids produces carbon skeletons that can enter the glycolytic pathway and the citric acid cycle to produce energy in the form of ATP. Nucleotides, on the other hand, are not similarly degraded for use as energy-yielding fuels. What observations about cellular physiology support this statement? What aspect of the structure of nucleotides makes them a relatively poor source of energy?

Answer Organisms do not store nucleotides to be used as fuel and do not completely degrade them, but rather hydrolyze them to release the bases, which can be recovered in salvage pathways. The low C:N ratio of nucleotides makes them poor sources of energy.

14. Treatment of Gout Allopurinol (see Fig. 22–45), an inhibitor of xanthine oxidase, is used to treat chronic gout. Explain the biochemical basis for this treatment. Patients treated with allopurinol sometimes develop xanthine stones in the kidneys, although the incidence of kidney damage is much lower than in untreated gout. Explain this observation in the light of the following solubilities in urine: uric acid, 0.15 g/L; xanthine, 0.05 g/L; and hypoxanthine, 1.4 g/L.

Answer Treatment with allopurinol has two biochemical consequences. (1) Conversion of hypoxanthine to uric acid is inhibited, causing accumulation of hypoxanthine, which is more soluble than uric acid and more readily excreted. This alleviates the clinical problems associated with AMP degradation. (2) Conversion of guanine to uric acid is also inhibited, causing accumulation of xanthine, which is, unfortunately, even less soluble than uric acid. This is the source of xanthine stones. Because less GMP than AMP is degraded, kidney damage caused by xanthine stones is less than that caused by untreated gout.

15. Inhibition of Nucleotide Synthesis by Azaserine The diazo compound *O*-(2-diazoacetyl)-L-serine, known also as azaserine (see Fig. 22–46), is a powerful inhibitor of glutamine amidotransferases. If growing cells are treated with azaserine, what intermediates of nucleotide biosynthesis would accumulate? Explain.

Answer In the de novo pathway of purine biosynthesis, the first step that requires glutamine is the conversion of 5-phosphoribosyl-1-pyrophosphate (PRPP) to 5-phospho-β-D-ribosylamine. In the presence of azaserine, which inhibits this conversion, PRPP will accumulate.

chapter

23 Integration and Hormonal Regulation of Mammalian Metabolism

1. **ATP and Phosphocreatine as Sources of Energy for Muscle** In contracting skeletal muscle, the concentration of phosphocreatine drops while the concentration of ATP remains fairly constant. However, in a classic experiment, Robert Davies found that if he first treated muscle with 1-fluoro-2,4-dinitrobenzene (see p. 141), the concentration of ATP declined rapidly, while the concentration of phosphocreatine remained unchanged during a series of contractions. Suggest an explanation.

Answer Muscle contraction leads to the net hydrolysis of ATP. Although the amount of ATP in muscle is very small, the supply can be rapidly replenished by phosphoryl transfer from the phosphocreatine reservoir, catalyzed by creatine kinase

$$\text{Phosphocreatine} + \text{ADP} \rightleftharpoons \text{creatine} + \text{ATP}$$

Since this reaction is rapid relative to the use of ATP by muscle, the ATP concentration remains in a steady state. The effects of pretreatment with fluoro-2,4-dinitrobenzene suggest that this is an effective inhibitor of creatine kinase. Under working conditions, the small amount of muscle ATP is quickly depleted and cannot be replenished.

2. **Metabolism of Glutamate in the Brain** Glutamate that moves from the blood into the brain is transformed there into glutamine, which is then released into the blood. What is accomplished by this metabolic conversion? How does it take place? Actually, the brain can generate more glutamine than can be made from the glutamate entering from the blood. How does this extra glutamine arise? (Hint: You may want to review amino acid catabolism in Chapter 18; recall that NH_3 is very toxic to the brain.)

Answer Ammonia is very toxic to nervous tissue, especially in the brain. Excess NH_3 in brain cells is removed by the transformation of glutamate to glutamine, catalyzed by glutamine synthetase. Glutamine is then exported to the blood and travels to the liver, where it is subsequently transformed to urea.

The additional glutamate arises from the action of aminotransferases that transfer amino groups from amino acids to α-ketoglutarate (a citric acid cycle intermediate), forming glutamate, which is subsequently converted to glutamine.

3. **Absence of Glycerol Kinase in Adipose Tissue** Glycerol 3-phosphate is a key intermediate in the biosynthesis of triacylglycerols. Adipocytes, specialized for the synthesis and degradation of triacylglycerols, cannot use glycerol directly because they lack glycerol kinase, which catalyzes the reaction

$$\text{Glycerol} + \text{ATP} \longrightarrow \text{glycerol 3-phosphate} + \text{ADP}$$

How does adipose tissue obtain the glycerol 3-phosphate necessary for triacylglycerol synthesis? Explain.

Answer The adipose tissue obtains glycerol 3-phosphate from glucose, by the following route:

$$\text{Glucose} \xrightarrow{\text{glycolysis}} \text{dihydroxyacetone phosphate}$$

$$\text{Dihydroxyacetone phosphate} + \text{NADH} + \text{H}^+ \rightleftharpoons \text{glycerol 3-phosphate} + \text{NAD}^+$$

The glycolytic intermediate dihydroxyacetone phosphate is reduced to glycerol 3-phosphate by the NADH-requiring enzyme glycerol phosphate dehydrogenase.

4. Oxygen Consumption during Exercise A sedentary adult consumes about 0.05 L of O_2 during a 10 s period. A sprinter, running a 100 m race, consumes about 1 L of O_2 during the same time period. After finishing the race, the sprinter will continue to breathe at an elevated but declining rate for some minutes, consuming an extra 4 L of O_2 above the amount consumed by the sedentary individual.

(a) Why do the O_2 needs increase dramatically during the sprint?

(b) Why do the O_2 demands remain high after the sprint has been completed?

Answer

(a) Increased muscular activity increases the demand for ATP, which is met by increased activity of the citric acid cycle enzymes and increased flow of electrons through the electron transfer chain. This results in increased O_2 consumption.

(b) During the sprint, muscle transforms some glycogen to lactate (anaerobic glycolysis). After the sprint, lactate is transported to the liver where it is converted back to glucose and glycogen. This process requires ATP and thus requires O_2 consumption above the resting rate.

5. Thiamine Deficiency and Brain Function Individuals with thiamine deficiency display a number of characteristic neurological signs: loss of reflexes, anxiety, and mental confusion. Why might thiamine deficiency be manifested by changes in brain function?

Answer Glucose is the primary fuel of the brain, and the brain is particularly sensitive to any change in the availability of glucose for energy production. A key reaction in glucose catabolism is the thiamine pyrophosphate—dependent oxidative decarboxylation of pyruvate to acetyl-CoA. Thus a thiamine deficiency reduces the rate of glucose catabolism.

6. Significance of Hormone Concentration Under normal conditions, the human adrenal medulla secretes epinephrine ($C_9H_{13}NO_3$) at a rate sufficient to maintain a concentration of 10^{-10} M in the circulating blood. To appreciate what that concentration means, calculate the diameter of a round swimming pool, with a water depth of 2 m, that would be needed to dissolve 1 g (about 1 teaspoon) of epinephrine to a concentration equal to that in blood.

Answer

Concentration of epinephrine in blood: 10^{-10} M = 10^{-10} mol/L.

Molecular weight of epinephrine: 183.21 g/mol.

Mass of epinephrine (in grams) needed to get 10^{-10} mol/L = (10^{-10} mol)(183.21 g/mol) = 1.83×10^{-8} g

If v = volume in L needed to dissolve 1 g epinephrine to this concentration:

$$1.83 \times 10^{-8} \text{ g/L} = 1 \text{ g}/v$$

$$v = 5.46 \times 10^7 \text{ L} = 5.46 \times 10^{10} \text{ cm}^3$$

This volume, v, is to be contained in a round pool of diameter $2r$ according to this equation:
$v = \pi r^2 h$
Rearranging to solve for r: $r^2 = v/\pi h$
Substituting for v, π, and h:

$$r^2 = (5.46 \times 10^{10}\ \text{cm}^3)/3.14(200\ \text{cm}) = 8.69 \times 10^7\ \text{cm}^2$$

$$r = 9{,}322\ \text{cm} = 93.2\ \text{m}$$

Therefore, the diameter is ($2r$) 186.4 m.
For comparison, an Olympic-size swimming pool is 50 meters long.

7. Regulation of Hormone Levels in the Blood The half-life of most hormones in the blood is relatively short. For example, if radioactively labeled insulin is injected into an animal, half the hormone has disappeared from the blood within 30 min.

(a) What is the importance of the relatively rapid inactivation of circulating hormones?

(b) In view of this rapid inactivation, how can the circulating hormone level be kept constant under normal conditions?

(c) In what ways can the organism make possible rapid changes in the level of circulating hormones?

Answer

(a) Inactivation provides a rapid means to change hormone concentrations.

(b) A constant insulin level is maintained by equal rates of synthesis and degradation.

(c) Other means of varying hormone concentration include changes in the rates of release from storage, of transport, and of conversion from prohormone to active hormone.

8. Water-Soluble versus Lipid-Soluble Hormones On the basis of their physical properties, hormones fall into one of two categories: those that are very soluble in water but relatively insoluble in lipids (e.g., epinephrine) and those that are relatively insoluble in water but highly soluble in lipids (e.g., steroid hormones). In their role as regulators of cellular activity, most water-soluble hormones do not penetrate into the interior of their target cells. The lipid-soluble hormones, by contrast, do penetrate into their target cells and ultimately act in the nucleus. What is the correlation between solubility, the location of receptors, and the mode of action of the two classes of hormones?

Answer Because of their low solubility in lipids, water-soluble hormones cannot penetrate the plasma membrane; they bind to receptors on the outer surface of the cell. In the case of epinephrine, this receptor is an enzyme that catalyzes the formation of a second messenger (cAMP) *inside* the cell. In contrast, lipid-soluble hormones readily penetrate the hydrophobic core of the plasma membrane. Once inside the cell they can act on their target molecules or receptors directly.

9. Metabolic Differences in Muscle and Liver in a "Fight or Flight" Situation During a "fight or flight" situation, the release of epinephrine promotes glycogen breakdown in the liver, heart, and skeletal muscle. The end product of glycogen breakdown in the liver is glucose. In contrast, the end product in skeletal muscle is pyruvate.

(a) Why are different products of glycogen breakdown observed in the two tissues?

(b) What is the advantage to the organism during a "fight or flight" condition of having these specific glycogen breakdown routes?

Answer

(a) Glycogen breakdown in hepatocytes in response to epinephrine produces glucose 6-phosphate, which can be exported to the blood following dephosphorylation to glucose by glucose 6-phosphate phosphatase. Heart and skeletal muscle lack this phosphatase. Any glucose 6-phosphate that is produced in these cells enters the glycolytic pathway and is metabolized to pyruvate. Under O_2-deficient conditions (actively contracting muscle), pyruvate is converted into lactate to regenerate NAD^+ for continued glycolysis.

(b) In a "fight or flight" situation, the concentration of glycolytic precursors in myocytes needs to be high in preparation for muscular activity. Glucose 6-phosphate cannot escape from myocytes because the glucose transporter does not recognize glucose 6-phosphate. The liver, on the other hand, must release the glucose necessary to maintain the blood glucose level. The glucose formed from glucose 6-phosphate is transported from the hepatocytes to the bloodstream, thereby increasing blood glucose levels for use by muscle.

10. Excessive Amounts of Insulin Secretion: Hyperinsulinism Certain malignant tumors of the pancreas cause excessive production of insulin by the β cells. Affected individuals exhibit shaking and trembling, weakness and fatigue, sweating, and hunger.

(a) What is the effect of hyperinsulinism on the metabolism of carbohydrate, amino acids, and lipids by the liver?

(b) What are the causes of the observed symptoms? Suggest why this condition, if prolonged, leads to brain damage.

Answer

(a) Overproduction of insulin leads to excessive uptake of blood glucose by the liver to make glycogen, leading to hypoglycemia. In addition, there would be a general shutdown of both amino acid and fatty acid catabolism.

(b) Under conditions of hyperinsulinism, little circulating fuel is available to meet the requirements for ATP production, so the function of energy-demanding tissues such as brain and muscle is compromised. This accounts for the shaking and trembling, weakness and fatigue. Excess insulin affects the mechanisms that regulate the autonomic nervous system, which in turn controls sweating. Finally, brain damage can result because glucose is the main source of fuel for the brain.

11. Thermogenesis Caused by Thyroid Hormones Thyroid hormones are intimately involved in regulating the basal metabolic rate. Liver tissue of animals given excess thyroxine shows an increased rate of O_2 consumption and increased heat output (thermogenesis), but the ATP concentration in the tissue is normal. Different explanations have been offered for the thermogenic effect of thyroxine. One is that excess thyroid hormone causes uncoupling of oxidative phosphorylation in mitochondria. How could such an effect account for the observations? Another explanation suggests that the thermogenesis is due to an increased rate of ATP utilization by the thyroxine-stimulated tissue. Is this a reasonable explanation? Why?

Answer The observations are consistent with the thesis that thyroxine acts as an uncoupler of oxidative phosphorylation. Uncouplers lower the P/O ratio of tissues, and thus the tissue must increase respiration to meet the normal ATP demands. The observed thermogenesis could also be due to the increased rate of ATP utilization by the thyroxine-stimulated tissue because the increased ATP demands are met by increased oxidative phosphorylation and thus respiration.

12. Function of Prohormones What are the possible advantages in the synthesis of hormones as prohormones?

Answer Because prohormones and preprohormones are inactive, they can be stored in quantity in secretory granules. Rapid activation is achieved by enzymatic cleavage in response to an appropriate signal.

13. Sources of Glucose in the Human Body The typical human adult uses about 160 g of glucose per day, with 120 g being used by the brain. The available reserve of glucose is adequate for about one day (about 20 g of circulating glucose and 190 g of glycogen). After the reserve becomes depleted, how would a starved body obtain more glucose?

Answer In animals, many precursors lead to the synthesis of glucose by gluconeogenesis (Fig. 20–1). In humans, the principal precursors (obtained from storage depots) are glycerol from triacylglycerols and glucogenic amino acids from protein. Oxaloacetate formed from CO_2 and pyruvate by pyruvate carboxykinase is also a potential source.

14. Parabiotic *ob/ob* mice By careful surgery, it is possible to connect the circulatory systems of two mice so that the same blood circulates through both animals. In these **parabiotic** mice, products released into the blood by one animal reaches the other animal via the shared circulation. Both animals are free to eat independently. When an *ob/ob* mouse (both copies of *OB* gene are defective) and a normal *OB/OB* mouse (two good copies of the *OB* gene) are made parabiotic, what will happen to the weight of each mouse?

Answer The mutant *(ob/ob)* mouse is obese because it does not produce functional leptin, the signal to eat less and be more physically active. When the circulatory system of this mouse is joined with that of the normal mouse, leptin produced in the normal mouse reaches the hypothalamus of the obese mouse, triggering the "eat less" response. The obese mouse therefore loses weight. The normal *OB/OB* mouse continues to have enough leptin in its blood to maintain the signal to eat moderately, and it retains its normal body weight.

Genes and Chromosomes

1. **Packaging of DNA in a Virus** Bacteriophage T2 has a DNA of molecular weight 120×10^6 contained in a head about 210 nm long. Calculate the length of the DNA (assume the molecular weight of a nucleotide pair is 650) and compare it with the length of the T2 head.

   Answer The T2 DNA molecule has $120 \times 10^6/650 = 184{,}615$ nucleotide pairs. Recall from Chapter 10 that a nucleotide pair occupies 3.4 Å, so the length of the T2 DNA molecule is (184,615 nucleotide pairs) (3.4 Å/base pair) = 627,691 Å = 62,769 nm.

   The length of the DNA molecule compared to the length of the viral head is 62,769 nm/ 210 nm = 299 or approximately 300 times the length of the compartment it must fit into. This nicely illustrates the necessity of compact packaging of DNA in viruses (see Fig. 24–1 and Table 24–1 for the relative lengths of DNA and the viral head for other bacteriophages).

2. **The DNA of Phage M13** The base composition of phage M13 DNA is A, 23%; T, 36%; G, 21%; C, 20%. What does this tell you about the DNA of phage M13?

   Answer The complementarity between A and T, and between G and C, in the two strands of duplex DNA explains Chargaff's rules that the sum of pyrimidine nucleotides equals that of purine nucleotides in DNAs from (virtually) all species: A = T, G = C, and A + G = C + T for duplex DNA. In M13 DNA, the percentage of A (23%) does not equal that of T (36%), nor does that of G (21%) equal that of C (20%). A + G = 44%, whereas C + T = 56%. This lack of equality between purine and pyrimidine nucleotides shows that M13 DNA is single-stranded, *not* double stranded, because the relationships expected from complementarity between the two strands of duplex DNA are not seen. The DNA is double-stranded only when it is replicating in the host cell.

3. **The *Mycoplasma* Genome** The complete genome of the simplest bacterium known, *Mycoplasma genitalium,* is a circular DNA molecule with 580,070 base pairs. Calculate the molecular weight and contour length (when relaxed) of this molecule. What is Lk_0 for the *Mycoplasma* chromosome? If $\sigma = -0.06$, what is Lk?

   Answer The molecular weight of a single nucleotide pair is 650 (from question 1 above) so the approximate molecular weight of the DNA molecule is $580{,}070 \times 650 = 3.77 \times 10^8$. The contour length is 580,070 bp × 3.4 Å per bp = 1,972,238 Å or approximately 1.97×10^6 Å = 197 μm.

   For relaxed circular DNA the linking number is the number of base pairs divided by 10.5. For *M. genitalium,* $Lk_0 = 580{,}070/10.5 = 55{,}245$. If $\sigma = -0.06$ then $\Delta Lk = \sigma(Lk_0) = (-0.06)(55{,}245) = -3{,}315$. This result means that three helical turns present in the relaxed circular DNA molecule have been removed. Therefore the $Lk = 55{,}245 - 3{,}315 = 51{,}930$.

4. Size of Eukaryotic Genes An enzyme isolated from rat liver has 192 amino acid residues and is coded for by a gene with 1,440 base pairs. Explain the relationship between the number of amino acid residues in the enzyme and the number of nucleotide pairs in its gene.

Answer Each amino acid is encoded by a triplet of three nucleotide pairs, so the 192 amino acids are encoded by 576 nucleotide pairs. The gene is in fact longer (1,440 nucleotide pairs). The additional 864 nucleotide pairs could be in introns (noncoding DNA, interrupting a coding segment) or they could code for a signal sequence (or leader peptide). In addition, as will be discussed in Chapter 26, eukaryotic mRNAs have untranslated segments before and after the portion coding for the polypeptide chain; these also contribute to the "extra" size of genes.

5. Linking Number A closed-circular DNA molecule in its relaxed form has an *Lk* of 500. Approximately how many base pairs are in this DNA? How is the linking number altered (increases, decreases, doesn't change, becomes undefined) when (a) a protein complex is bound to form a nucleosome, (b) one DNA strand is broken, (c) DNA gyrase and ATP are added to the DNA solution, or (d) the double helix is denatured by heat?

Answer
In relaxed DNA, the linking number (*Lk*) is equivalent to the number of turns in the DNA helix. *Lk* is a topological property, and thus does not vary when duplex DNA is twisted or deformed, as long as both DNA strands remain intact. It can change only if one or both strands are broken and rejoined. If a DNA strand remains broken, the molecule is no longer topologically constrained (the strands can unravel) and *Lk* is undefined.

The *Lk* of relaxed DNA is equivalent to the number of turns of DNA, and there are about 10.5 base pairs per turn of relaxed B-form DNA. Thus the DNA has approximately (500 turns)(10.5 bp/turn) = 5,250 base pairs.

(a) Doesn't change; the DNA strands are not cleaved and rejoined.

(b) Becomes undefined; one of the strands remains broken.

(c) Decreases; in the presence of ATP, gyrase (a type 2 topoisomerase that introduces negative supercoils) will underwind the DNA.

(d) Doesn't change; again, the DNA strands are not broken and rejoined.

6. Superhelical Density Bacteriophage λ infects *E. coli* by integrating its DNA into the bacterial chromosome. The success of this recombination depends on the topology of the *E. coli* DNA. When the superhelical density (σ) of the *E. coli* DNA is greater than -0.045, the probability of integration is $< 20\%$; when σ is less than -0.06, the probability is $> 70\%$.

DNA isolated from an *E. coli* culture is found to have a length of 13,800 base pairs and an *Lk* of 1,222. Calculate σ for this DNA and predict the likelihood that bacteriophage λ will be able to infect this culture.

Answer Superhelical density, also known as the specific linking difference, is given by

$$\begin{aligned}\sigma &= (Lk - Lk_0)/Lk_0 \\ &= (1{,}222 - 1{,}314)/1{,}314 \\ &= -92/1{,}314 = -0.07\end{aligned}$$

Since -0.07 is less than -0.06, there is a greater than 70% probability that the DNA phage will be incorporated into the *E.coli* DNA. For more information about the relationship between the probability of recombination and the extent of DNA supercoiling, see "Topoisomerase IV, not gyrase, decantenates products of site-specific recombination in *Escherichia coli*" by Zechiedrich, E.L., Khodursky, A.B., Cozzarelli, N.R. (1997) *Genes & Development* *11*:2580–2592, specifically, Fig. 4 on p. 2585.

7. DNA Structure Explain how the underwinding of a B-DNA helix might facilitate or stabilize the formation of Z-DNA.

Answer The shift from B- to Z-DNA requires a shift from a right-handed helix to a left-handed helix. You can visualize the underwinding as aiding in the transition by providing a local region of unwinding that allows it to shift to Z-DNA. Second, the left-handed helix of Z-DNA has a negative *Lk*, and underwinding the DNA lowers the *Lk*. Third, the underwinding essentially stores some free energy within the DNA that can be used to facilitate the B to Z transition.

8. Chromatin Early evidence that helped researchers define nucleosome structure is illustrated by the agarose gel shown on p. 930 of the text, in which the thick bands represent DNA. It was generated by briefly treating chromatin with an enzyme that degrades DNA, then removing all protein and subjecting the purified DNA to electrophoresis. Numbers at the side of the gel denote the position to which a linear DNA of the indicated size (in base pairs) would migrate. What does this gel tell you about chromatin structure? Why are the DNA bands thick and spread out rather than sharply defined?

Answer The bands have a periodicity of about 200 nucleotide pairs, or base pairs (bp 200, 400, 600, etc.), showing that the chromatin is protected from nuclease digestion at regular intervals of 200 bp. This suggests that the nucleosomal cores (146 bp) were providing the protection, which was verified in numerous subsequent investigations. Thus the nucleosomes themselves are in a fairly regular array, occurring about once every 200 bp. The nuclease is cutting the regions of doublestranded DNA that link but do not bind to the nucleosome cores—specifically, the spacer regions of about 60 bp. These regions are not always digested to completion; consequently, some bands correspond to the DNA from single nucleosomes (200 bp), others to two nucleosomes (400 bp), and so forth. If the nucleosomes had been randomly distributed in the chromatin, a large number of differently sized DNA fragments would have been generated by the nuclease cleavage, and a heterogeneous population of DNA fragments would have smeared through the gel.

The bands are thick because the spacer is fairly long (60 bp in this example) relative to the size of the nucleosomal core (146 bp). The nuclease can cut essentially anywhere in the spacer, so the band corresponding to, for example, mononucleosomes, has DNAs ranging from 146 to 206 bp.

9. Yeast Artificial Chromosomes (YACs) YACs are used to clone large pieces of DNA in yeast cells. What three types of DNA sequences are required to ensure proper replication and propagation of a YAC in a yeast cell?

Answer Stable artificial chromosomes require only three components: a centromere, telomeres at the ends, and an autonomous replicating sequence (replication origin).

DNA Metabolism

1. **Conclusions from the Meselson-Stahl Experiment** The Meselson-Stahl experiment (see Fig. 25–2) proved that DNA undergoes semiconservative replication in *E. coli*. In the "dispersive" model of DNA replication, the parental DNA strands are cleaved into pieces of random size, then joined with pieces of newly replicated DNA to yield daughter duplexes. In the Meselson-Stahl experiment, each strand would contain random segments of heavy and light DNA. Explain how the results of Meselson and Stahl's experiment ruled out such a model.

   Answer If random, dispersive replication had taken place, the density of the first-generation DNA would have been the same as actually observed, a single band midway between heavy and light DNA. In the second generation, all of the DNA would again have had the same density and would have appeared as a single band, midway between that observed in the first generation and that of light DNA. In fact, two bands were observed in the Meselson-Stahl experiment.

2. **Heavy Isotope Analysis of DNA Replication** A culture of *E. coli* growing in a medium containing $^{15}NH_4Cl$ is switched to a medium containing $^{14}NH_4Cl$ for three generations (an eightfold increase in population). What is the molar ratio of hybrid DNA ($^{15}N–^{14}N$) to light DNA ($^{14}N–^{14}N$) at this point?

   Answer This experiment is an extension of the Meselson-Stahl experiment, which demonstrates that replication in *E. coli* is semiconservative. Hence, after three generations the molar ratio of $^{15}N–^{14}N$ DNA to $^{14}N–^{14}N$ DNA is 2/6 = 0.33.

3. **Replication of the *E. coli* Chromosome** The *E. coli* chromosome contains 4,639,221 base pairs.
   (a) How many turns of the double helix must be unwound during replication of the *E. coli* chromosome?
   (b) From the data in this chapter, how long would it take to replicate the *E. coli* chromosome at 37 °C if two replication forks proceed from the origin? Assume replication occurs at a rate of 1,000 base pairs per second. Under some conditions *E. coli* cells can divide every 20 min. How might this be possible?
   (c) In the replication of the *E. coli* chromosome, about how many Okazaki fragments would be formed? What factors guarantee that the numerous Okazaki fragments are assembled in the correct order in the new DNA?

   Answer
   (a) During DNA replication, the complementary strands must unwind completely to allow the synthesis of a new strand on each template. Since there are 10.5 bp/turn in B-DNA,

$$\text{the number of helical turns} = \frac{\text{number of base pairs}}{\text{number of base pairs per helical turn}}$$

$$\frac{4.64 \times 10^6 \text{ bp}}{10.5 \text{ bp/turn}} = 4.42 \times 10^5 \text{ turns}$$

(b) Chromosomal DNA replication in *E. coli* starts at a fixed origin and proceeds bidirectionally. Each replication fork travels $(4.64 \times 10^6)/2 = 2.32 \times 10^6$ nucleotide pairs during replication. If we assume a replication rate of 1,000 bp/sec, the time required for the completion of DNA synthesis in each replication fork is $(2.32 \times 10^6)/1000 = 2320$ s = 39 min.

One possible explanation is that replication of an *E. coli* chromosome starts from two origins, each proceeding bidirectionally to yield four replication forks. In this mode, it would take 19.5 min to complete the replication of the chromosome. There is, however, only one replication origin in the *E. coli* chromosome. Thus, an alternative possibility is that a new round of replication begins before the previous one is completed: for cells dividing every 20 min, a replicative cycle is initiated every 20 min, and each daughter cell would receive a chromosome that is half-replicated. This latter mode has in fact been experimentally verified.

(c) About 2,350 to 4,700 Okazaki fragments (4.7×10^6 bp/2000 bp and 4.7×10^6 bp/1000 bp, respectively) are formed by DNA polymerase III from an RNA primer and a DNA template. Because the Okazaki fragments in *E. coli* are 1,000 to 2,000 bases long (see Fig. 25–4), they are firmly bound to the template strand by base pairing. Each fragment is quickly joined to the lagging strand by the successive action of DNA polymerase I and DNA ligase, thus preserving the correct order of the fragments. A mixed pool of different Okazaki fragments, detached from their template, does *not* form during normal replication.

4. Base Composition of DNAs Made from Single-Stranded Templates Predict the base composition of the total DNA synthesized by DNA polymerase on templates provided by an equimolar mixture of the two complementary strands of bacteriophage ϕX174 DNA (a circular DNA molecule). The base composition of one strand is A, 24.7%; G, 24.1%; C, 18.5%; and T, 32.7%. What assumption is necessary to answer this problem?

Answer The sequence of one strand of duplex DNA is complementary to that of the other strand, as determined by Watson-Crick base pairing (A with T, and G with C). The DNA strand made from the given template strand has A, 32.7%; G, 18.5%; C, 24.1%; T, 24.7%. The DNA strand made from the complementary template strand has A, 24.7%; G, 24.1%; C, 18.5%; T, 32.7%. Thus the composition of the *total* DNA synthesized is A, 28.7%; G, 21.3%; C, 21.3%; T, 28.7%. It is assumed that both template strands are completely replicated.

5. DNA Replication Kornberg and his colleagues incubated soluble extracts of *E. coli* with a mixture of dATP, dTTP, dGTP, and dCTP, all labeled with ^{32}P in the α-phosphate group. After a time, the incubation mixture was treated with trichloroacetic acid, which precipitates the DNA but not the nucleotide precursors. The precipitate was collected, and the extent of precursor incorporation into DNA was determined from the amount of radioactivity present in the precipitate.

(a) If any one of the four nucleotide precursors were omitted from the incubation mixture, would radioactivity be found in the precipitate? Explain.

(b) Would ^{32}P be incorporated into the DNA if only dTTP were labeled? Explain.

(c) Would radioactivity be found in the precipitate if ^{32}P labeled the β or γ phosphate rather than the α phosphate of the deoxyribonucleotides? Explain.

Answer

(a) The incorporation of ^{32}P label into DNA results from the synthesis of new DNA. The synthesis of DNA requires the presence of all four nucleotide precursors.

(b) Yes. Although all four nucleotide precursors must be present for DNA synthesis to occur, only one of them has to be radioactive in order for radioactivity to be observed in the new DNA.

(c) No radioactivity would be incorporated if the ^{32}P label were not in the α phosphate because DNA polymerase, which catalyzes this reaction, cleaves off pyrophosphate, that is, the β and γ phosphate groups.

6. Leading and Lagging Strands Prepare a table that lists the names and compares the functions of the precursors, enzymes, and other proteins needed to make the leading versus lagging strands during DNA replication in *E. coli.*

Answer In DNA replication, the *leading strand* is produced by continuous replication of the DNA template strand in the 5 → 3 direction. The *lagging strand* is synthesized in the form of short Okazaki fragments, which are then spliced together. Below is a table that is specific for DNA replication in *E. coli.*

Component	Function	Leading Strand	Lagging Strand
DNTPs (dATP, dGTP, dGTP, dTTP)	Source of new nucleotides for new DNA strand; ATP energy source	X	X
Template	"Parent" strand provides information for identities of incoming nucleotides; stabilizes association of nucleotides in growing strand by base pair reactions.	X	X
DNA primer	Provides a free 3′ —OH, which is the point of attachment for incoming nucleotides	X	
DNA helicase	Unwinds double-stranded DNA just ahead of the replication fork; requires ATP	X	X
DNA gyrase	Topoisomerase that favors unwinding of the DNA at the replication fork by twisting the DNA; requires ATP	X	X
Single-strand DNA-binding proteins	Prevents base-pairing of unwound DNA strands; stabilizes single-stranded DNA	X	X
DNA polymerase III	Enzyme that elongates the new DNA strand by adding nucleotides; requires the cofactors Mg^{2+} and Zn^{2+}	X	X
Pyrophosphatase	Hydrolyzes the PP_i released by polymerase activity; helps "pull" the reaction in the forward direction.	X	X
RNA Primer	Same as DNA primer; used to start each Okazaki fragment		X
Primase	Enzyme that synthesizes RNA primer		X

Component	Function	Leading Strand	Lagging Strand
NTPs (ATP, GTP, CTP, UTP)	Used in synthesis of RNA primer		X
DNA polymerase I	Exonuclease that removes RNA primer by replacing NMPs with dNMPs in Okazaki fragments; requires the cofactors Mg^{2+} and Zn^{2+}		X
DNA ligase	Joins the Okazaki fragments in the lagging strand by catalyzing the formation of a phosphodiester bond; requires NAD^+ as energy source		X

7. Function of DNA Ligase Some *E. coli* mutants contain defective DNA ligase. When these mutants are exposed to ^{3}H-labeled thymine and the DNA produced is sedimented on an alkaline sucrose density gradient, two radioactive bands appear. One corresponds to a high molecular weight fraction, the other to a low molecular weight fraction. Explain.

Answer During replication of the DNA duplex, the leading strand is replicated continuously and the lagging strand is replicated in short fragments (Okazaki fragments), which are then spliced together by means of DNA ligase. Mutants with defective DNA ligase produce DNA duplex in which one of the strands remains fragmented. Consequently, when this duplex is denatured by the alkaline conditions of the sucrose gradient, sedimentation results in one fraction containing the intact single strand (the high molecular weight band) and in one fraction containing the unspliced fragments (the low molecular weight band).

8. Fidelity of Replication of DNA What factors promote the fidelity of replication during the synthesis of the leading strand of DNA? Would you expect the lagging strand to be made with the same fidelity? Give reasons for your answers.

Answer Fidelity of replication is ensured by Watson-Crick base pairing between the template and leading strand, and proofreading and removal of wrongly inserted nucleotides by the 3′-exonuclease activity of DNA polymerase III. The same fidelity would be expected in the lagging strand—maybe. The factors ensuring fidelity of replication are operative in both the leading and the lagging strands. However, the greater number of distinct chemical operations involved in making the lagging strand might provide a greater opportunity for errors to arise.

9. Importance of DNA Topoisomerases in DNA Replication DNA unwinding, such as that occurring in replication, affects the superhelical density of DNA. In the absence of topoisomerases, the DNA would become overwound ahead of a replication fork as the DNA is unwound behind it. A bacterial replication fork will stall when the superhelical density (σ) of the DNA ahead of the fork reaches $+0.14$, (see Chapter 24).

Bidirectional replication is initiated at the origin of a 6,000 base pair plasmid in vitro, in the absence of topoisomerases. The plasmid initially has a σ of -0.06. How many base pairs will be unwound and replicated by each replication fork before the forks stall? Assume that each fork travels at the same rate and that each includes all components necessary for elongation except topoisomerase.

Answer About 1,200 base pairs will be unwound, or about 600 in each direction. The DNA is initially negatively supercoiled, with an *Lk* of about 537 and an Lk_0 of about 571 (see Chapter 24). Unwinding 34 turns of DNA will relax the DNA, and further unwinding will add positive supercoils. Unwinding another 79 turns (or 113 total) will bring the superhelical density in the remaining wound DNA to about 0.14. These 113 turns are equivalent to just under 1,200 bp of DNA.

10. The Ames Test In a nutrient medium that lacks histidine, a thin layer of agar containing about 10^9 *Salmonella typhimurium* histidine auxotrophs (mutant cells that require histidine to survive) produces about 13 colonies over a 2-day incubation period at 37 °C (see Fig. 25–18). How do these colonies arise in the absence of histidine? The experiment is repeated in the presence of 0.4 μg of 2-aminoanthracene. The number of colonies produced over 2 days exceeds 10,000. What does this indicate about 2-aminoanthracene? What can you surmise about its carcinogenicity?

Answer Occasionally, some of the histidine-requiring mutants spontaneously undergo a back-mutation and regain their capacity to synthesize histidine and grow in a medium lacking histidine. The observation that only 13 bacteria out of about 10^9 produce colonies indicates that the rate of back-mutation is quite low. In contrast, the addition of 2-aminoanthracene increases the rate of back-mutations about 1,000-fold. This indicates that 2-aminoanthracene is mutagenic. Since about 90% of 300 known carcinogens are mutagenic, these observations suggest that it is very likely that 2-aminoanthracene is carcinogenic.

11. DNA Repair Mechanisms Vertebrate and plant cells often methylate cytosine in DNA to form 5-methylcytosine (see Fig. 10–5a). In these same cells, a specialized repair system recognizes G–T mismatches and repairs them to G≡C base pairs. How might this repair system be advantageous to the cell? (Explain in terms of the presence of 5-methylcytosine in the DNA.)

Answer This problem is very similar to Problem 7 of Chapter 10. Spontaneous deamination of 5-methylcytosine produces thymine, and thus a G–T mismatched pair. Such G–T mismatches are among the most common mismatches in the DNA of eukaryotes. The specialized repair system restores the G≡C pair.

12. DNA Repair in People with Xeroderma Pigmentosum The condition known as xeroderma pigmentosum (XP) arises from mutations in at least seven different human genes. The deficiencies are generally in genes encoding enzymes involved in some part of the pathway for human nucleotide excision repair. The various types of XP are labeled A–G (XP-A, XP-B, etc.), with a few additional variants lumped under the label XP-V.

Cultures of cells from normal individuals and from some patients with XP-G are irradiated with ultraviolet light. The DNA is isolated and denatured, and the resulting single-stranded DNA is characterized by analytical ultracentrifugation.

(a) Samples from the normal fibroblasts show a significant reduction in the average molecular weight of the single-stranded DNA after irradiation, but samples from the XP-G fibroblasts show no such reduction. Why might this be?

(b) If you assume that a nucleotide-excision repair system is operative, which step might be defective in the fibroblasts from the patients with XP-G? Explain.

Answer

(a) Ultraviolet irradiation of skin fibroblast DNA results in the formation of pyrimidine dimers. In normal fibroblasts, the damaged DNA is repaired by excision of the pyrimidine dimer. One step in this process is cleavage of the damaged strand by a special excinuclease. Thus the denatured single-stranded DNA from normal cells isolated after irradiation contains the many fragments caused by the cleavage, and the average molecular weight is lowered.

(b) The absence of fragments in the single-stranded DNA from the XP-G cells (no change in average molecular weight) after irradiation suggests that the special repair excinuclease is defective or missing in these cells.

13. Holliday Intermediates How does the formation of Holliday intermediates in homologous genetic recombination differ from their formation in site-specific recombination?

Answer During homologous genetic recombination, a Holliday intermediate may be formed almost anywhere within the two paired, homologous chromosomes. Once formed, the branch point of the intermediate may move extensively by branch migration. In site-specific recombination, the Holliday intermediate is formed between two specific sites, and branch migration is generally restricted by heterologous sequences on either side of the recombination sites.

RNA Metabolism

1. RNA Polymerase

(a) How long would it take for the *E. coli* RNA polymerase to synthesize the primary transcript for the *E. coli* genes encoding the enzymes for lactose metabolism (the 5,300 base pair *lac* operon, considered in Chapter 28)?

(b) How far along the DNA would the transcription "bubble" formed by RNA polymerase move in 10 s?

Answer

(a) Elongation of the RNA transcript in *E. coli* proceeds at about 50–90 nucleotides per second. Thus the time required to produce the primary transcript is

$$\frac{5{,}300 \text{ nucleotides}}{50\text{–}90 \text{ nucleotides/s}} = [59\text{–}106 \text{ s}]$$

(b) Elongation of the RNA transcript proceeds at about 50–90 nucleotides per second. Thus in 10 s the bubble travels

$$(10 \text{ s})(50\text{–}90 \text{ nucleotides/s}) = 500\text{–}900 \text{ nucleotides}$$

2. Error Correction by RNA Polymerases DNA polymerases are capable of editing and error correction, but RNA polymerases do not appear to have this capacity. Given that a single base error in either replication or transcription can lead to an error in protein synthesis, suggest a possible biological explanation for this striking difference.

Answer Since many RNA molecules are made from each gene, an error in any one molecule will lead to only a small fraction of protein with an incorrect amino acid. The incorrect protein will probably be degraded fairly quickly. The error in the mRNA will not be propagated in subsequent generations of cells because the mRNA itself is degraded. For DNA replication, however, errors would be transmitted to the next generation of cells.

3. RNA Posttranscriptional Processing Predict the likely effects of a mutation in the sequence (5′)AAUAAA in a eukaryotic mRNA transcript.

Answer One of the key signals for cleavage and 3′ polyadenylation is the sequence AAUAAA. After RNA polymerase II has transcribed beyond this sequence, an endonuclease (uncharacterized at this time) cleaves the primary transcript at a position about 25 to 30 nucleotides 3′ to the AAUAAA. The enzyme polyadenylate polymerase then adds a string of 20 to 250 A residues to the free 3′ end, generating the 3′ poly(A) tail. The mutation would prevent cleavage and polyadenylation at the usual site. If the transcript is not polyadenylated it will be quite unstable, the steady-state levels of mRNA will be very low, and little or no protein product will be made.

4. Coding vs. Template Strands The RNA genome of phage Qβ is the nontemplate or coding strand, and when introduced into the cell it functions as an mRNA. Suppose the RNA replicase of phage Qβ synthesized primarily template-strand RNA and uniquely incorporated this, rather than nontemplate strands, into the viral particles. What would be the fate of the template strands when they entered a new cell? What enzyme would such a template-strand virus need to include in the viral particles for successful invasion of a host cell?

Answer A template-strand RNA does not encode proteins, and by itself would not cause a productive infection. However, if an RNA-dependent RNA polymerase were included in the viral particle, it could copy the template strand to form a nontemplate strand after entry into the host cell. The nontemplate strand would serve as mRNA for synthesis of viral proteins, leading to a productive infection.

5. The Chemistry of Nucleic Acid Biosynthesis Describe three properties common to the reactions catalyzed by DNA polymerase, RNA polymerase, reverse transcriptase, and RNA replicase. How is the enzyme polynucleotide phosphorylase similar to and different from these three enzymes?

Answer These enzymes have at least four properties in common.

(1) All are template directed, synthesizing a sequence complementary to the template.

(2) Synthesis occurs in a $5' \rightarrow 3'$ direction.

(3) All catalyze the addition of a nucleotide by the formation of a phosphodiester bond.

(4) All use (deoxy)ribonucleoside triphosphates as substrate, and release pyrophosphate as a product.

The enzyme polynucleotide phosphorylase differs from these polymerase in points 1 and 4 (p. 1006). Polynucleotide phosphorylase does not use a template, but rather adds ribonucleotides to an RNA in a highly reversible reaction. The substrates (in the direction of synthesis) are ribonucleoside diphosphates, which are added with the release of phosphate as a product. In the cell, this enzyme probably catalyzes the reverse reaction to degrade RNAs.

6. RNA Splicing What is the minimum number of transesterification reactions needed to splice an intron from an mRNA transcript? Why?

Answer A minimum of two transesterification steps are required. The mechanism for removal of introns from pre-mRNAs involves the formation of a lariat intermediate after the reaction is initiated. Each of the cleavage and rejoining reactions is a transesterification, in which a new phosphodiester bond is formed for every one that is broken. The first step is initiated by the attack of the 2′-hydroxyl group of an A residue within the intron on the bond linking the 3′ end of the first exon with the 5′ end of the intron. This generates a 3′-hydroxyl on the nucleotide at the 3′ end of the first exon, and effectively takes the intron out of the series of transesterifications by forming a lariat structure (see Fig. 26–15). The 3′ nucleotide of the first intron can then link to the first nucleotide of the second exon, again by a transesterification. The result of this second step is the union of the first and second exons, with the intron released as a lariat intermediate.

7. RNA Genomes The RNA viruses have relatively small genomes. For example, the single-stranded RNAs of retroviruses have about 10,000 nucleotides and the Qβ RNA is only 4,220 nucleotides long. Given the properties of reverse transcriptase and RNA replicase described in this chapter, can you suggest a reason for the small size of these viral genomes?

Answer Neither reverse transcriptase nor RNA replicase has a proofreading function, and hence these enzymes are much more error-prone than DNA polymerases. The smaller the genomes, the fewer are the mutable sites, and thus it is less likely that a lethal mutational load will accumulate (i.e., that the RNA will sustain one or a group of mutations that inactivates the virus).

8. Screening RNAs by SELEX The practical limit for the number of different RNA sequences that can be screened in a SELEX experiment is 10^{15}.

(a) Suppose you are working with oligonucleotides 32 nucleotides in length. How many sequences exist in a randomized pool containing every sequence possible?

(b) What percentage of these can be screened in a SELEX experiment?

(c) Suppose you wish to select an RNA molecule that catalyzes the hydrolysis of a particular ester. From what you know about catalysis (Chapter 8), propose a SELEX strategy that might allow you to select the appropriate catalyst.

Answer

(a) If there are 4 different nucleotides, the number of possible sequences in any nucleic acid of length n residues is 4^n. Thus, for an oligonucleotide with 32 residues, the number of possible sequences is 4^{32} or 1.8×10^{19}.

(b) Only 0.005% of these could be sampled in an experiment limited to a sampling of 10^{15} sequences.

(c) For the "unnatural selection" step, use a chromatographic resin to which is bound a molecule that is a transition-state analog of the ester hydrolysis reaction (e.g., an appropriate phosphonate compound; see Box 8–3). RNAs that bind tightly to the transition-state analog may catalyze the corresponding reaction, much as catalytic antibodies do.

9. Slow Death The death cap mushroom, *Amanita phalloides,* contains several dangerous substances including the lethal toxin α-amanitin. This toxin blocks RNA elongation in consumers of the mushroom by binding to eukaryotic RNA polymerase II with very high affinity; it is deadly in concentrations as low as 10^{-8} M. The initial reaction to ingestion of the mushroom is gastrointestinal distress (caused by some of the other toxins). These symptoms disappear, but about 48 hours later, the mushroom-eater dies, usually from liver dysfunction. Speculate on why it takes this long for α-amanitin to kill.

Answer The turnover rates of mRNAs and their protein products vary, based on the rates of synthesis and of degradation. Though RNA synthesis is quickly halted by the α-amanitin toxin, it takes several days for the critical mRNAs and proteins in the liver to degrade; once degraded, though, they are not replenished, and the system fails.

10. Determination of Rifampicin-Resistant Strains of TB Rifampicin is an important antibiotic used to treat tuberculosis (TB), as well as other mycobacterial diseases. Some strains of *Mycobacterium tuberculosis,* the causative agent of TB, are resistant to rifampicin. These strains become resistant due to mutations that alter the *rpo*B gene, which encodes the β-subunit of the RNA polymerase. Rifampicin cannot bind to the mutant RNA polymerase and so is unable to block the initiation of transcription. DNA sequences from a large number of rifampicin-resistant *M. tuberculosis* strains have been found to have mutations in a specific 69 base-pair region of *rpo*B. One well-characterized strain with rifampicin resistance has a single base-pair alteration in *rpo*B that results in a single amino acid substitution in the β-subunit protein: a His residue is replaced by an Asp residue.

(a) Based on your knowledge of protein chemistry (Chapters 5 and 6), suggest a technique that would allow detection of the rifampicin-resistant strain containing this particular mutant protein.

(b) Based on your knowledge of nucleic acid chemistry (Chapter 10), suggest a technique to identify the mutant form of *rpo*B.

Answer

(a) After lysis of the cells and partial purification, the protein extract could be subjected to isoelectric focusing. The β-subunit protein could be detected by an antibody-based assay. The difference in amino acid residues between the normal β subunit and the mutated form (i.e., the different charges on the amino acids) would alter the electrophoretic mobility of the mutant β-subunit protein in an isoelectric focusing gel relative to the protein from a non-resistant strain.

(b) Direct DNA sequencing (by the Sanger method) of the *rpo*B gene would allow detection of this particular mutation.

Biochemistry on the Internet

11. The Ribonuclease Gene Human pancreatic ribonuclease has 128 amino acid residues.

(a) What is the minimum number of nucleotide pairs required to code for this protein?

(b) The mRNA expressed in human pancreas cells was copied with reverse transcriptase to create a "library" of human DNA. The sequence of the mRNA coding for human pancreatic ribonuclease was determined by sequencing the "complementary DNA" or cDNA from this library that included an open reading frame for the protein. Use the Entrez database system to find the published sequence of this mRNA (accession number D26129). See the *Principles of Biochemistry* Web site for the current URL and more detailed instructions. What is the length of this mRNA?

(c) How can you account for the discrepancy between the size you calculated in (a) and the actual length of the mRNA?

Answer

(a) Individual amino acids in proteins are coded for by a triplet of nucleotides. Therefore, encoding a protein with 128 amino acid residues will require $128 \times 3 = 384$ nucleotide pairs in the DNA.

(b) The GenBank report indicates that the cDNA derived from the mature mRNA contains 1,620 nucleotide pairs.

(c) The mRNA for pancreatic ribonuclease is clearly much larger than is necessary to code for the protein itself. (To see this rather dramatically, display the information graphically by clicking on the Graphic option at the top of the page.) Most of the nucleotides are untranslated regions at the 3′ and 5′ ends of the mRNA. The mRNA encodes a primary transcript longer than the mature ribonuclease; a signal sequence (see Chapter 27) is cleaved off to produce the mature and functional protein.

chapter

27 Protein Metabolism

1. **Messenger RNA Translation** Predict the amino acid sequences of peptides formed by ribosomes in response to the following mRNA, assuming that the reading frame begins with the first sequences, three bases in each sequence.
   (a) GGUCAGUCGCUCCUGAUU
   (b) UUGGAUGCGCCAUAAUUUGCU
   (c) CAUGAUGCCUGUUGCUAC
   (d) AUGGACGAA

   Answer The genetic code is nonoverlapping, unpunctuated, and in triplet (see Fig. 27–7).
   (a) Gly-Gln-Ser-Leu-Leu-Ile
   (b) Leu-Asp-Ala-Pro
   (c) His-Asp-Ala-Cys-Cys-Tyr
   (d) Met-Asp-Glu in eukaryotes; fMet-Asp-Glu in prokaryotes

2. **How Many Different mRNA Sequences Can Specify One Amino Acid Sequence?** Write all the possible mRNA sequences that can code for the simple tripeptide segment Leu–Met–Tyr. Your answer will give you some idea about the number of possible mRNAs that can code for one polypeptide.

   Answer The genetic code is degenerate, meaning that a given amino acid may be specified by more than one codon (see Table 27–4, Fig. 27–7). Leu is specified by six different codons: UUA, UUG, CUU, CUC, CUA, CUG. Met, when not used as an initiation codon, is specified by AUG; Tyr is specified by two codons: UAC, UAU. Thus, there are $6 \times 1 \times 2 = 12$ possible mRNA sequences that can code for a tripeptide segment Leu–Met–Tyr:

   UUA AUG UAU, UUA AUG UAC,
   UUG AUG UAU, UUG AUG UAC,
   CUU AUG UAU, CUU AUG UAC,
   CUC AUG UAU, CUC AUG UAC,
   CUA AUG UAU, CUA AUG UAC,
   CUG AUG UAU, CUG AUG UAC

3. **Can the Base Sequence of an mRNA Be Predicted from the Amino Acid Sequence of Its Polypeptide Product?** A given sequence of bases in an mRNA will code for one and only one sequence of amino acids in a polypeptide, if the reading frame is specified. From a given sequence of amino acid residues in a protein such as cytochrome *c,* can we predict the base sequence of the unique mRNA that coded it? Give reasons for your answer.

Answer No; because nearly all the amino acids have more than one codon, any given polypeptide can be coded for by a number of different base sequences (see Problem 2). However, because some amino acids are encoded by only one codon and those with multiple codons often share the same nucleotide at two of the three positions, *certain parts* of the mRNA sequence encoding a protein of known amino acid sequence can be predicted with high certainty.

4. Coding of a Polypeptide by Duplex DNA The template strand of a segment of double-helical DNA contains the sequence

(5′)CTTAACACCCCTGACTTCGCGCCGTCG

(a) What is the base sequence of the mRNA that can be transcribed from this strand?

(b) What amino acid sequence could be coded by the mRNA in (a), starting from the 5′ end?

(c) If the complementary strand of this DNA were transcribed and translated, would the resulting amino acid sequence be the same as in (b)? Explain the biological significance of your answer.

Answer The template strand serves as the template for RNA synthesis; the nontemplate strand is identical in sequence to the RNA transcribed from the gene, with U in place of T.

(a) (5′)CGACGGCGCGAAGUCAGGGGUGUUAAG(3′)

(b) Arg-Arg-Arg-Glu-Val-Arg-Gly-Val-Lys

(c) No; the base sequence of mRNA transcribed from the nontemplate strand would be (5′)CUUAACACCCCUGACUUCGCGCCGUCG. This mRNA, when translated, would result in a different peptide from (b) and would have the amino acid sequence Leu-Asn-Thr-Pro-Asp-Phe-Ala-Pro-Ser. The complementary antiparallel strands in double-helical DNA do not have the same base sequence in the 5′ → 3′ direction. RNA is transcribed from only one specific strand of duplex DNA. The RNA polymerase must therefore recognize and bind to the correct strand.

5. Methionine Has Only One Codon Methionine is one of two amino acids with only one codon. How does the single codon for methionine specify both the initiating residue and interior Met residues of polypeptides synthesized by *E. coli*?

Answer There are two tRNAs for methionine: $tRNA^{fMet}$, the initiating tRNA, and $tRNA^{Met}$, which can insert Met in interior positions in a polypeptide. The $tRNA^{fMet}$ reacts with Met to yield Met-$tRNA^{fMet}$, promoted by methionine aminoacyl-tRNA synthetase. The amino group of its Met residue is then formylated by N^{10}-formyltetrahydrofolate to yield fMet-$tRNA^{fMet}$. Free Met or Met-$tRNA^{Met}$ cannot be formylated. Only fMet-$tRNA^{fMet}$ is recognized by the initiation factor IF-2 and is aligned with the initiating AUG positioned at the ribosomal P site in the initiation complex. AUG codons in the interior of the mRNA are eventually positioned at the ribosomal A site and can bind and incorporate only Met-$tRNA^{Met}$.

6. Synthetic mRNAs The genetic code was elucidated with polyribonucleotides synthesized either enzymatically or chemically in the laboratory. Given what we now know about the genetic code, how would you make a polyribonucleotide that could serve as an mRNA coding predominantly for many Phe residues and a small number of Leu and Ser residues? What other amino acid(s) would be coded for by this polyribonucleotide, but in smaller amounts?

Answer Polynucleotide phosphorylase is template-independent and does not require a primer. The base composition of the RNA formed by this enzyme reflects the relative concentrations of the nucleoside 5′-diphosphates in the reaction mixture. To prepare the required polyribonucleotide, allow polynucleotide phosphorylase to act on a mixture of UDP and CDP in which UDP has, say, five times the concentration of CDP. The result would be a synthetic RNA polymer with many UUU triplets (coding for Phe), a smaller number of UUC (also Phe), UCU (Ser), and CUU (Leu), and a yet smaller number of UCC (also Ser), CUC (also Leu), and CCU (Pro).

7. **Energy Cost of Protein Biosynthesis** Determine the minimum energy cost, in terms of high-energy phosphate groups expended, required for the biosynthesis of the β-globin chain of hemoglobin (146 residues), starting from a pool including all necessary amino acids, ATP, and GTP. Compare your answer with the direct energy cost of the biosynthesis of a linear glycogen chain of 146 glucose residues in ($\alpha 1 \rightarrow 4$) linkage, starting from a pool including glucose, UTP, and ATP (Chapter 20). From your data, what is the *extra* energy cost of making a protein, in which all the residues are ordered in a specific sequence, compared to the cost of making a polysaccharide containing the same number of residues, but lacking the informational content of the protein?

In addition to the direct energy cost for the synthesis of a protein, there are indirect energy costs—those required for the cell to make the necessary enzymes for protein synthesis. Compare the magnitude of the indirect costs to a eukaryotic cell of the biosynthesis of linear ($\alpha 1 \rightarrow 4$) glycogen chains and the biosynthesis of polypeptides, in terms of the enzymatic machinery involved.

Answer The number of high-energy phosphate groups required for the synthesis of a *polypeptide* with n residues is:

$2n$ for the charging of tRNA (ATP $\longrightarrow$ AMP + PP_i; $PP_i \longrightarrow 2P_i$)

1 for initiation (GTP $\longrightarrow$ GDP + P_i)

$n - 1$ for the formation of $n - 1$ peptide bonds (GTP $\longrightarrow$ GDP + P_i)

$n - 1$ for the $n - 1$ translocation steps (GTP $\longrightarrow$ GDP + P_i)

The total number for the polypeptide $= 4n - 1$.

The number of high-energy phosphate groups required for the synthesis of a linear *glycogen* chain of n glucose residues is:

n for the phosphorylation of glucose (Glucose + ATP $\longrightarrow$ ADP + glucose 6-phosphate)

None for the conversion of glucose 6-phosphate to glucose 1-phosphate

$2n$ for the activation of glucose 1-phosphate to UDP-glucose (Glucose 1-phosphate + UTP $\longrightarrow$ UDP-glucose + PP_i; $PP_i \rightarrow 2P_i$)

$-n$ (i.e., n generated) for the formation of the polymer from UDP-glucose

(UDP-glucose + glycogen $\longrightarrow$ UDP + glycogen-glucose)

The total number for the glycogen chain $= 2n$.

Thus the total number of high-energy phosphate groups required for the synthesis of one molecule of β-globin is $(4 \times 146) - 1 = 583$. The total number for the synthesis of a linear chain of 146 glucose residues is $2 \times 146 = 292$. The extra energy cost for the synthesis of β-globin is $583 - 292 = 291$ high-energy phosphate groups; this reflects the cost of the information contained in the protein.

In order to synthesize a protein from amino acids at least 20 aminoacyl-tRNA synthetases (activating enzymes), 70 ribosomal proteins, 4 rRNAs, 32 or more tRNAs, an mRNA, and 10 or more auxiliary enzymes must be made by the eukaryotic cell. Synthesis of these proteins and RNA molecules is energetically expensive. In contrast, the synthesis of an ($\alpha 1 \rightarrow 4$) chain of glycogen from glucose requires only four or five enzymes (see Chapter 20).

8. Predicting Anticodons from Codons Most amino acids have more than one codon and attach to more than one tRNA, each with a different anticodon. Write all possible anticodons for the four codons of glycine: (5′)GGU, GGC, GGA, and GGG.

(a) From your answer, which of the positions in the anticodons are primary determinants of their codon specificity in the case of glycine?

(b) Which of these anticodon-codon pairings has/have a wobbly base pair?

(c) In which of the anticodon-codon pairings do all three positions exhibit strong Watson-Crick hydrogen bonding?

Answer All the anticodons for the four Gly codons have the sequence (5′)XCC. The first position of each anticodon is determined by the specific codon it interacts with, and by the wobble hypothesis (see Table 27–5). For example, the wobble position of the codon GGU is U, which can be recognized by either A, G, or I. Thus, this codon has three possible anticodons: ACC, GCC, and ICC. By the same token, the anticodons for the GGC codon are GCC and ICC; the anticodons for the GGA codon are UCC and ICC; and the anticodons for the GGG codon are CCC and UCC.

(a) The 3′ and the middle position. The 5′ position is the wobble position.

(b) Anticodons GCC, ICC, and UCC each recognize more than one codon by virtue of wobble base pairing. Anticodons ACC and CCC each only recognize one codon and thus are not involved in wobble base pairing.

(c) The pairing of anticodons ACC and CCC with their respective codons involves Watson-Crick base pairing in all three positions, A pairing with U, and C pairing with G.

9. Effect of Single-Base Changes on Amino Acid Sequence Much important confirmatory evidence on the genetic code has come from the nature of single-residue changes in the amino acid sequence of mutant proteins. Which of the following single-residue replacements would be consistent with the genetic code? Which cannot be the result of a single-base mutation? Why?

(a) Phe → Leu
(b) Lys → Ala
(c) Ala → Thr
(d) Phe → Lys
(e) Ile → Leu
(f) His → Glu
(g) Pro → Ser

Answer For each part of this problem "yes" indicates that the replacement is consistent with a single-base change. The various codons for each amino acid are listed. The single-base changes (where they exist) that would cause the mutation are underlined.

(a) Yes.
Phe: UUU UUC UUA UUG
Leu: CUU CUC CUA CUG

(b) No single-base mutation could convert the codons for Lys to the codons for Ala.
Lys: AAA AAG
Ala: GCU GCC GCA GCG

(c) Yes.
Ala: GCU GCC GCA GCG
Thr: ACU ACC ACA ACG

(d) No single-base mutation could convert the codons for Phe to the codons for Lys.
Phe: UUU UUC UUA UUG
Lys: AAA AAG

(e) Yes.
Ile: AUU AUC AUA
Leu: CUU CUC CUA

(f) No single-base mutation could convert the codons for His to the codons for Glu.
His: CAU CAC
Glu: GAA GAG

(g) Yes.
Pro: CCU CCC CCA CCG
Ser: UCU UCC UCA UCG

10. Basis of the Sickle-Cell Mutation Sickle-cell hemoglobin has a Val residue at position 6 of the β-globin chain, instead of the Glu residue found in normal hemoglobin A. Can you predict what change took place in the DNA codon for glutamate to account for its replacement by valine?

Answer The two DNA codons for Glu are GAA and GAG, and the four DNA codons for Val are GTT, GTC, GTA, and GTG. A single-base change in GAA to form GTA or in GAG to form GTG could account for the Glu → Val replacement in sickle-cell hemoglobin. Much less likely are two-base changes from GAA to GTG, GTT, or GTC; and from GAG to GTA, GTT, or GTC.

11. Importance of the "Second Genetic Code" Some aminoacyl-tRNA synthetases do not recognize and bind the anticodon of their cognate tRNAs but instead use other structural features of the tRNAs to impart binding specificity. The tRNAs for alanine apparently fall into this category.

(a) What features of tRNA^{Ala} are recognized by Ala-tRNA synthetase?

(b) Describe the consequences of a C → G mutation in the third position of the anticodon of tRNA^{Ala}

(c) What other kinds of mutations might have similar effects?

(d) Mutations of these types are never found in natural populations of organisms. Why? (Hint: Consider what might happen both to individual proteins and to the organism as a whole.)

Answer

(a) The only nucleotides of tRNA^{Ala} required for recognition by Ala-tRNA synthetase are those of the G3-U70 base pair in the amino acid arm (see Fig. 27–20a).

(b) Mutations in the anticodon region would produce a tRNA^{Ala} capable of recognizing and binding to codons for amino acids other than alanine. However, changes in the anticodon of tRNA^{Ala} would not affect the specificity of the charging reaction by Ala-tRNA synthetase.

There are four Ala codons, GCU, GCC, GCA, and GCG. The third position of each tRNA^{Ala} should be a C, because this position interacts with the first position of the Ala codons, which is a G in all four. Thus, changing the C in the third position of tRNA^{Ala} to a G would allow the mutant tRNA^{Ala} to recognize CCU, CCC, CCA, and CCG, all of which specify Pro. Thus Ala residues would be inserted at sites coding for Pro.

(c) Another mutation that would have similar effects would be, for example, a mutation to tRNA^{Ala} synthetase, which now recognizes proline vs. alanine.

(d) The Ala → Pro replacement resulting from these mutations will render most of the proteins in the cell inactive, making these mutations lethal; hence, their effects would not be observed.

12. Maintaining the Fidelity of Protein Synthesis The chemical mechanisms used to avoid errors in protein synthesis are different from those used during DNA replication. DNA polymerases use a 3′ → 5′ exonuclease proofreading activity to remove mispaired nucleotides incorrectly inserted into a growing DNA strand. There is no analogous proofreading function on ribosomes; and, in fact, the identity of an amino acid attached to an incoming tRNA and added to the growing polypeptide is never checked. A proofreading step that hydrolyzed the previously formed peptide bond after an incorrect amino acid had been inserted into a growing polypeptide (analogous to the proofreading step of DNA polymerases) would be impractical. Why? (Hint: Consider how the link between the growing polypeptide and the mRNA is maintained during elongation; see Figs. 27–26 and 27–27.)

Answer The amino acid most recently added to a growing polypeptide chain is the only one covalently attached to a tRNA and hence is the only link between the polypeptide and the mRNA that is encoding it. A proofreading activity that severed this link would halt synthesis of the polypeptide and release it from the mRNA.

13. Predicting the Cellular Location of a Protein The gene for a eukaryotic polypeptide 300 amino acid residues long is altered so that a signal sequence recognized by SRP occurs at the polypeptide's amino terminus, and a nuclear localization signal (NLS) occurs internally beginning at residue 150. Where is the protein likely to be found in the cell?

Answer The protein will be directed into the endoplasmic reticulum, and from there the targeting will depend upon additional signals. The SRP will bind the amino-terminal signal early in protein synthesis and direct the nascent polypeptide and ribosome to receptors in the endoplasmic reticulum. Because the protein is translocated into the lumen of the endoplasmic reticulum as it is synthesized, the NLS will never be accessible to the proteins involved in nuclear targeting.

14. Requirements for Protein Translocation across a Membrane The secreted bacterial protein OmpA has a precursor ProOmpA, which has the amino-terminal signal sequence required for secretion. If purified ProOmpA is denatured with 8 M urea and the urea is then removed (e.g., by running the protein solution rapidly through a gel filtration column), the protein can be translocated across isolated bacterial inner membranes in vitro. However, translocation becomes impossible if ProOmpA is allowed to incubate for a few hours in the absence of urea. Furthermore, the capacity for translocation is maintained for an extended period if ProOmpA is incubated in the presence of another bacterial protein called trigger factor. Describe the probable function of this factor.

Answer Trigger factor is a molecular chaperone that stabilizes an unfolded and translocation-competent conformation of ProOmpA.

15. Protein Coding Capacity of a Viral DNA The 5,386 base pair genome of bacteriophage ϕX174 (see Fig. 27–10) includes genes for 10 proteins, designated A to K, with sizes given in the table below. How much DNA would be required to encode these 10 proteins? How can you reconcile the size of the ϕX174 genome with its protein-coding capacity?

Protein	Number of amino acid residues
A	455
B	120
C	86
D	152
E	91
F	427
G	175
H	328
J	38
K	56

Answer

Protein	Number of amino acid residues	Minimum base pairs
A	455	1,365
B	120	360
C	86	258
D	152	456
E	91	273
F	427	1,281
G	175	525
H	328	984
J	38	114
K	56	168
	Minimum base pairs required = 5,784 bp	

The 5,386 base pair genome of bacteriophage ϕX174 is not large enough even to code for the ten amino acid sequences, not to mention the necessary regulatory elements and promoters. Therefore, some of the coding sequences for these viral proteins must be nested or must overlap.

Regulation of Gene Expression

1. **Effect of mRNA and Protein Stability on Regulation** *E. coli* cells are growing in a medium with glucose as the sole carbon source. Tryptophan is suddenly added. The cells continue to grow, and divide every 30 min. Describe (qualitatively) how the amount of tryptophan synthase activity in the cells changes under the following conditions:
   (a) The *trp* mRNA is stable (degraded slowly over many hours).
   (b) The *trp* mRNA is degraded rapidly, but tryptophan synthase is stable.
   (c) The *trp* mRNA and tryptophan synthase are both degraded more rapidly than normal.

   Answer The mRNA from the *trp*EDCBA operon encodes several enzymes for Trp biosynthesis; the *trp*B and *trp*A genes encode tryptophan synthase. The complete *trp* mRNA is synthesized only when the concentration of Trp (actually, that of charged Trp-tRNA) is low. This is the result of both repression and attenuation (see Problem 6). There is no strong regulation at translation, so when *trp* mRNA is present, tryptophan synthase will be produced. Although the regulation of Trp biosynthesis, like most biosynthetic pathways, is fine-tuned by feedback control, feedback inhibition by Trp is exerted at the branch point anthranilate synthase (the product of the *trp*E and *trp*D genes; see Fig. 28–21). Hence the activity of tryptophan synthase is not strongly affected by [Trp].
   (a) If the *trp* mRNA is stable relative to cell generation time, it will persist in the population of bacteria even after Trp has been added, and the enzyme tryptophan synthase will continue to be synthesized and be active. In a simple model, one would expect to see the enzyme activity decrease roughly twofold per cell for each generation (30 min); that is, the activity would be slowly diluted out by the increasing numbers of cells.
   (b) Again, if the enzyme is stable relative to the generation time, it will persist in the population, even after the addition of Trp, and remain active.
   (c) If both the mRNA and enzyme are unstable (degraded more rapidly than the cells divide), the attenuation of transcription of the *trp* operon caused by the addition of Trp to the medium will lead to an abrupt decrease in levels of *trp* mRNA and tryptophan synthase.

2. **Negative Regulation** Describe the probable effects on gene expression in the *lac* operon of mutations in
   (a) the *lac* operator that deletes most of O_1
   (b) the *lac*I gene that inactivates the repressor
   (c) the promoter that eliminates the region in the *lac* gene around position -10.

Answer The *lac* operon is negatively regulated by a repressor, the product of the *lac*I gene. The Lac repressor binds to specific DNA sequences called the operators (*lac*O_1 and pseudo-operators O_2 and O_3). Binding of the repressor prevents efficient initiation of transcription by RNA polymerase from the promoter (*lac*P). An inducer (allolactose or an analog) binds to the repressor and prevents its binding to the operator, thereby releasing the repression and allowing transcription of the *lac* operon.

(a) Most mutations in the operator, the binding site for repressor, lead to lower affinity for the repressor and hence less binding. Thus these mutations allow continued transcription (and thus expression) of the *lac* operon even in the absence of inducer; this is referred to as constitutive expression.

(b) Mutations in the *lac*I gene that produce a repressor that cannot bind to the operator will lead to constitutive expression (no repression in the absence of inducer). Mutations that prevent binding of the repressor to the inducer without affecting the ability to bind to the operator lead to a noninducible phenotype (constant repression).

(c) The region of the *lac* gene around position -10 is the promoter region. The *lac* promoter is not particularly strong. Mutations have been observed that either increase or decrease its efficiency of initiating transcription. Base substitutions that make the promoter sequence more similar to the consensus generate a stronger promoter (promoter "up" mutations), whereas those that make the promoter less similar to the consensus generate a weaker promoter (promoter "down" mutations). An "up" mutation would make the *lac* operon independent of positive regulation by the cAMP-CAP complex (when the operon is induced). A "down" mutation would not allow expression even in the derepressed state (presence of inducer) and hence would produce a noninducible phenotype.

3. **Specific DNA Binding by Regulatory Proteins** A typical prokaryotic repressor protein discriminates between its specific DNA-binding site (operator) and nonspecific DNA by a factor of 10^5 to 10^7. About ten molecules of repressor per cell are sufficient to ensure a high level of repression. Assume that a very similar repressor existed in a human cell, with a similar specificity for its binding site. How many copies of the repressor would be required to elicit a level of repression similar to that in the prokaryotic cell? (Hint: The *E. coli* genome contains about 4.7 million base pairs; the human haploid genome has about 2.4 billion base pairs.)

Answer The discrete DNA-binding domains of transcriptional regulatory proteins form specific complexes with defined sequences of DNA. Their affinity for these defined sequences is about 10^5 to 10^7 greater than their affinity for other sequences.

Using the example of the Lac repressor, the binding site (operator) is 22 base pairs (bp) long. Ten molecules of the Lac repressor are sufficient to keep this operator in a bound state even in the context of 4.7×10^6 bp of nonspecific DNA (the rest of the *E. coli* genome). This amounts to finding one specific site in a sea of $(4.7 \times 10^6 \text{ bp})/(22 \text{ bp}) = 2.1 \times 10^5$ nonspecific sites. For the hypothetical repressor in a human cell, let's use the same size binding site (22 bp), although this is larger than most sites so far characterized. The human repressor must find its specific site within $(2.4 \times 10^9 \text{ bp})/(22 \text{ bp}) = 1.1 \times 10^8$ nonspecific sites. Thus the ratio of nonspecific to specific sites is $(1.1 \times 10^8)/(2.1 \times 10^5) = 519$ times greater in the human cell. Extrapolating from the Lac repressor information, we estimate that $(519)(10 \text{ molecules}) = 5{,}190 \approx 5{,}000$ molecules of repressor will be needed per cell.

4. **Repressor Concentration in *E. coli*** The dissociation constant for a particular repressor-operator complex is very low, about 10^{-13} M. An *E. coli* cell (volume 2×10^{-12} mL) contains 10 copies of the repressor. Calculate the cellular concentration of the repressor protein. How does this value compare with the dissociation constant of the repressor-operator complex? What is the significance of this result?

Answer

The concentration of the repressor

$$= 10 \text{ molecules} \times \frac{1 \text{ mol}}{6.02 \times 10^{23} \text{ molecules}} \times \frac{10^3 \text{ mL}}{\text{L}} \times \frac{1}{2 \times 10^{-12} \text{ mL}}$$
$$= 8 \times 10^{-9} \text{ M}$$

The repressor concentration is about 10^5 times greater than the dissociation constant. Because the dissociation constant represents the concentration of repressor needed for half the operator sites to be filled, we conclude that, with 10 copies of active repressor in the cell, the operator site is always bound by a repressor molecule.

5. Catabolite Repression *E. coli* cells are growing in a medium containing lactose but no glucose. Indicate whether each of the following changes or conditions would increase, decrease, or not change the expression of the *lac* operon. It may be helpful to draw a model depicting what is happening in each situation.

(a) Addition of a high concentration of glucose
(b) A mutation that prevents dissociation of the Lac repressor from the operator
(c) A mutation that completely inactivates β-galactosidase
(d) A mutation that completely inactivates galactoside permease
(e) A mutation that prevents binding of CRP to its binding site near the *lac* promoter

Answer Each condition would decrease expression of *lac* operon genes.

(a) Increased [glucose] would result in a decrease in [cAMP]. A cAMP-CRP complex is required for strong binding of the RNA polymerase to the *lac* promoter. As [cAMP] declines so will expression of the *lac* operon.

(b) A repressor that binds irreversibly to the *lac* operator will block the binding of RNA polymerase, even if lactose is present, thus preventing expression of the *lac* operon.

(c) If β-galactosidase is not active, there will be no conversion of lactose to allolactose, and hence the normal inducer of the operon will not be present. Expression of the operon will decrease.

(d) A mutation that inactivates the galactoside permease will reduce transport of lactose into the cell, even when extracellular levels of lactose are high. Little or no allolactose will be produced to induce expression of the *lac* operon, so expression will decrease.

(e) In the presence of cAMP, CRP binds to a site near the *lac* promoter and significantly enhances transcription of the operon by stabilizing the open complex of the RNA polymerase. Mutations that prevent CRP binding to the DNA will decrease expression of the *lac* operon.

6. Transcription Attenuation How would transcription of the *E. coli trp* operon be affected by the following manipulations of the leader region of the *trp* mRNA?

(a) Increasing the distance (number of bases) between the leader peptide gene and sequence 2
(b) Increasing the distance between sequences 2 and 3
(c) Removing sequence 4
(d) Changing the two Trp codons in the leader peptide gene to His codons
(e) Eliminating the ribosome-binding site for the gene that encodes the leader peptide
(f) Changing several nucleotides in sequence 3 so that it can base-pair with sequence 4 but not with sequence 2

Answer The *trp* operon is subject to regulation by both repression and attenuation. Attenuation depends on the tight coupling between transcription and translation in bacteria. When [Trp] is high, the *trp* leader is completely translated and the ribosome blocks sequence 2. This allows the transcribed sequences 3 and 4 to form the stem and loop attenuator structure (see Fig. 28–23). Formation of the 3:4 loop, which resembles a ρ-independent transcription terminator, terminates transcription of the *trp* operon before genes E, D, C, B, and A are transcribed, and the enzymes for Trp biosynthesis are not produced. When [Trp] is low, translation of the *trp* leader stalls at two Trp codons. In this position, the ribosome does not cover sequence 2, and sequence 2 is free to base-pair with sequence 3 in an alternative secondary structure. Formation of the 2:3 stem and loop precludes formation of the 3:4 attenuator loop, and transcription proceeds through the *trp*EDCBA genes. Thus when [Trp] is low, the biosynthetic genes are expressed and more Trp is synthesized.

(a) Increasing the distance between sequence 1 (encoding the *trp* leader peptide) and sequence 2 will result in constitutive expression of the *trp* operon. Whether the ribosome stalls at the leader sequence due to lack of Trp or translates through the leader sequence in the presence of Trp, sequence 2 will be free to form the 2:3 stem and loop, preventing formation of the 3:4 attenuator.

(b) A large increase in the distance between sequences 2 and 3 would discourage formation of the 2:3 stem and loop. Thus at low [Trp], even though the ribosome stalls, the 3:4 attenuator will be more likely to form than the 2:3 stem and loop, and expression of the *trp* operon will decrease.

(c) Because sequence 4 is required to form the 3:4 attenuator stem and loop, no attenuation would occur in its absence.

(d) Changing the Trp codons in the leader sequence to His will cause expression of the operon to be regulated by the availability of histidine rather than tryptophan.

(e) Eliminating the ribosome binding site from the leader sequence would prevent the synthesis of the leader protein, which is the tryptophan sensor in this system.

(f) Changing sequence 3 so that it can base-pair with sequence 4, but not sequence 2 would mean that only the 3:4 attenuator structure would form. Thus, there would be an inappropriate decrease in *trp* operon expression when tryptophan levels were low.

7. Repressors and Repression How would the SOS response in *E. coli* be affected by a mutation in the *lex*A gene that prevented autocatalytic cleavage of the LexA protein?

Answer Induction of the SOS response could not occur, making the cells more sensitive to high levels of DNA damage.

8. Regulation by Recombination In the phase variation system of *Salmonella,* what would happen to the cell if the Hin recombinase became more active and promoted recombination (DNA inversion) several times in each cell generation?

Answer *Salmonella* will switch expression between two different flagellin genes, *H1* and *H2*, about once every 1,000 generations in order to evade the immune system of the host organism. This switch in expression is accomplished by a site-specific recombination system, requiring the action of the Hin recombinase on *hix* sites that flank the *H2* gene. *H2* is expressed and *H1* is repressed in one orientation of *hin* (the one in which the *H2* promoter is orientated toward the *H2* gene), whereas *H1* is expressed and *H2* is inactive in the other orientation (see Fig. 28–28).

More rapid switching due to a more active Hin recombinase would lead to a mixed population of *Salmonella,* some with H1 flagellin and others with H2 flagellin. The host immune system, faced with both types of flagella, would mount an attack on both, greatly reducing the numbers of *Salmonella.* In other words, the protective advantage (to *Salmonella*) of phase variation would be lost.

9. Initiation of Transcription in Eukaryotes A new RNA polymerase activity is discovered in crude extracts of cells derived from an exotic fungus. The RNA polymerase initiates transcription only from a single, highly specialized promoter. As the polymerase is purified its activity declines, and the purified enzyme is completely inactive unless crude extract is added to the reaction mixture. Suggest an explanation for these observations.

Answer The observation that the new RNA polymerase loses activity as it is purified, and that this activity is restored by addition of crude extract, suggests that some factor remaining in the crude extract is necessary for activity under the assay conditions. Any of these factors could have been disassociated and eliminated during purification: (1) an activator such as cAMP-CRP; (2) a loosely associated subunit of the protein that is necessary for activity; or (3) a specificity factor (similar to the σ subunit of the *E. coli* RNA polymerase) necessary for efficient transcription from its promoter. Finally, proteolysis could be responsible for loss of activity during successive stages of purification. Small amounts of polymerase remaining in the crude extract then supply the only activity to the fraction. In the absence of quantitative information (a purification table), it is impossible to determine if this explanation is feasible.

10. Functional Domains in Regulatory Proteins A biochemist replaces the DNA-binding domain of the yeast Gal4 protein with the DNA-binding domain from the Lac repressor and finds that the engineered protein no longer regulates transcription of the *GAL* genes in yeast. Draw a diagram of the different functional domains you would expect to find in the Gal4 protein and in the engineered protein. Why does the engineered protein no longer regulate transcription of the *GAL* genes? What might be done to the DNA-binding site recognized by this chimeric protein to make it functional in activating transcription of *GAL* genes?

Answer Transcriptional activators have at least two domains that frequently function separately: the DNA-binding domain and the activation domain. The DNA-binding domain is required for the sequence-specific binding of the protein to DNA. A different portion of the protein is responsible for activation; this domain may directly contact the RNA polymerase or it may facilitate the action of coactivators or other proteins that stimulate transcription.

Gal4 protein

Gal4 DNA-binding domain	Gal4 activator domain

Engineered protein

Lac repressor DNA-binding domain	Gal4 activator domain

The biochemist has done part of a domain-swap experiment; the activation domain of Gal4 is fused to the DNA-binding domain of the Lac repressor. This new hybrid will no longer recognize the Gal4 binding site in DNA (called UAS_G) because that DNA-binding domain is no longer present. However it will be able to bind to the repressor binding site in the operator portion of the *lac* operon. Therefore, replacement of UAS_G with the *lac* operator (binding site for the Lac repressor) in the *GAL* operon should allow the Lac repressor-Gal4 hybrid protein to function as a transcriptional activator of *GAL* genes in yeast.

11. **Inheritance Mechanisms in Development** A *Drosophila* egg that is bcd^-/bcd^- may develop normally but as an adult will not be able to produce viable offspring. Explain.

Answer The *bicoid (bcd)* gene product is a major anterior morphogen. mRNA from the *bicoid* gene is synthesized by nurse cells in female flies and deposited in the unfertilized egg near its anterior pole. The Bicoid protein produced by the maternal mRNA diffuses through the egg cell creating a concentration gradient that establishes the location of the anterior pole of the embryo. The amounts of Bicoid protein present in various parts of the embryo affect the subsequent expression of a number of other genes in a threshold-dependent manner. Lack of Bicoid protein results in the development of an embryo with two abdomens but neither head nor thorax; this embryo dies before it hatches as a larvae.

Consequently, if a female fly that is heterozygous for the *bicoid* gene (bcd^+/bcd^-) is mated with a male that has a *bicoid* deficiency, this female will be able to produce *bicoid* mRNA allowing for the development of phenotypically normal female progeny that lack the bicoid gene (bcd^-/bcd^-). The female progeny from this cross, however, will not be able to produce viable offspring because their nurse cells are incapable of producing *bicoid* mRNA for deposition in their egg cells.

Biochemistry on the Internet

12. **TATA Binding Protein and the TATA Box** To examine the interactions between transcription factors and DNA, go to the Protein Data Bank and download the PDB file 1TGH. This file models the interactions between a human TATA binding protein and a segment of double-stranded DNA. Use the Noncovalent Bond Finder found at the Chime Resources Web site to examine the roles of hydrogen bonds and hydrophobic interactions involved in the binding of this transcription factor to the TATA box in DNA. (For the current URLs of these Web sites and for further instructions on how to use the Noncovalent Bond Finder, go to http://www.worthpublishers.com/lehninger.)

Within the Noncovalent Bond Finder program, load the PDB file and display the protein in Space-fill mode and the DNA in Wireframe mode. Go to the Lehninger Web site for more detailed instructions in order to answer the following questions.

(a) Which of the base pairs in the DNA form hydrogen bonds with the protein? Which of these contribute to the specific recognition of the TATA box by this protein? (Hydrogen bond lengths range from 2.5 to 3.3 Å between the hydrogen donor and hydrogen acceptor.)

(b) Which amino acid residues in the protein interact with these base pairs? On what basis did you make this determination? Do these observations agree with the information presented in the text?

(c) What is the sequence of the DNA in this model and which portions of the sequence are recognized by the TATA binding protein?

(d) Can you identify any hydrophobic interactions in this complex? (Hydrophobic interactions usually occur with interatomic distances of 3.3 to 4.0 Å.)

Answer

(a) There are seven red oxygens and two blue nitrogens in the DNA sequence that are highlighted when you click the **Find** button. Five of these interactions are with backbone oxygens and four are located on the bases themselves. Click on the highlighted atom to identify the specific residue in the DNA to which it belongs. Interactions between the protein and the individual bases (not the backbone) are the ones that contribute directly to DNA sequence recognition.

The nucleotides in the DNA backbone that are involved in hydrogen bonding include Chain B: A106, A110 and Chain C: A118, T119, and A122. The nucleotides in the bases that are involved in hydrogen bonding include Chain B: A106, T107 and Chain C: A118, T119.

(b) The amino acids that are involved in hydrogen bonds with the base pairs in this DNA are shown below. These interactions can be visualized by zooming in on the protein and looking for the residues in the protein that have hydrogens that contact the identified atoms in the DNA. To check whether the amino acid residues are close enough to form hydrogen bonds with the DNA molecules, you can change the Mouse Clicks window to **Distance: Monitor Lines.** By clicking on two individual atoms you will be presented with the distance between them in angstroms. See the Table below for specific interactions.

DNA Chain B		DNA Chain C	
Backbone	**Base pair**	**Backbone**	**Base pair**
A106 → Arg^{290}	A106 → Asn^{253}	A118 → Arg^{199}	A118 → Asn^{163}
A110 → Ser^{212}	T107 → Asn^{253}	T119 → Arg^{204}	T119 → Asn^{163}
		A122 → Ser^{303}	

The amino acid residues whose side chains are most often found hydrogen-bonded to bases in the DNA include Asn, Gln, Glu, Lys, and Arg. The majority of residues in the TATA binding protein that are involved in hydrogen bonds are Arg and Asn.

(c) To better visualize the DNA, click on the MDL logo, select the DNA **(Select → Nucleic → Nucleic),** and then **Display** it as a **Wireframe** or **Ball and Stick** model. By clicking on the individual bases in each strand you should get the following sequence:

Chain B: TATATATA (residues 103 to 110)

Chain C: ATATATAT (residues 122 to 115)

The TATA binding protein makes specific contact with the base pairs A106, T107/T119, A118 in the middle of this sequence and has interactions with the backbone at each end at A104/T121 and T109/A116.

(d) Hydrophobic interactions usually involve carbon atoms and have carbon–carbon distances slightly longer than hydrogen bonds, usually 3.3 to 4.0 Å. Click the **Reset** button to restore the original model and then enter "carbon" in the "Find only" window. Increase the Increment distance to 0.5 Å and **Find** carbon–carbon interactions that occur up to 4 Å. You should see quite a few gray carbons in the bases being highlighted.

chapter

29 Recombinant DNA Technology

1. **Cloning** When joining two or more DNA fragments, a researcher can adjust the sequence at the junction in a variety of subtle ways, as seen in the following exercises.

   (a) Draw the structure of each end of a linear DNA fragment that was produced by an *Eco*RI restriction digest (include those sequences remaining from the *Eco*RI recognition sequence).

   (b) Draw the structure resulting from the reaction of this end sequence with DNA polymerase I and the four deoxynucleoside triphosphates.

   (c) Draw the sequence produced at the junction that arises if two ends with the structure derived in **(b)** are ligated.

   (d) Draw the structure produced if the structure derived in **(a)** is treated with a nuclease that degrades only single-stranded DNA.

   (e) Draw the sequence of the junction produced if an end with structure **(b)** is ligated to an end with structure **(d).**

   (f) Draw the structure of the end of a linear DNA fragment that was produced by a *Pvu*II restriction digest (as in **(a)**).

   (g) Draw the sequence of the junction produced if an end with structure **(b)** is ligated to an end with structure **(f).**

   (h) Suppose you can synthesize a short duplex DNA fragment with any sequence you desire. With such a synthetic fragment and the procedures described in **(a)** through **(g),** design a protocol that will remove an *Eco*RI restriction site from a DNA molecule and incorporate a new *Bam*HI restriction site at approximately the same location. (Hint: See Fig. 29–3.)

   (i) Design four different short synthetic double-stranded DNA fragments that would permit ligation of structure **(a)** with a DNA fragment produced by a *Pst*I restriction digest. In one of these synthetic fragments, design the sequence so that the final junction contains the recognition sequences for both *Eco*RI and *Pst*I. In the second and third synthetic fragments, design the sequence so that the junction contains only the *Eco*RI or the *Pst*I recognition sequence, respectively. Design the sequence of the fourth fragment so that neither the *Eco*RI nor the *Pst*I sequence appears in the junction.

   Answer Type II restriction enzymes cleave double-stranded DNA within recognition sequences to create either blunt-ended DNA or sticky-ended fragments. Blunt-ended DNA fragments can be joined by the action of T4 DNA ligase. Sticky-ended DNA fragments can be joined by either *E. coli* or T4 DNA ligases, provided that the sticky ends are complementary. Sticky-ended DNA fragments without complementary sticky ends can be joined only after the ends are made blunt either by exonucleases or *E. coli* DNA polymerase I.

   (a) The recognition sequence for *Eco*RI is (5′) GAATTC(3′), with the cleavage site between G and A (see Table 29–2). Thus, digestion of a DNA molecule with one *Eco*RI site

   (5′)---GAATTC---(3′)
   ---CTTAAG---

would yield two fragments:

(5′)---G(3′) and (5′)AATTC---(3′)
---CTTAA G---

(b) DNA polymerase I catalyzes the synthesis of DNA in the 5′ → 3′ direction in the presence of four deoxyribonucleoside triphosphates. Therefore, the ends of both fragments generated in **(a)** will be made blunt ended

(5′)---GAATT(3′) and (5′)AATTC---(3′)
---CTTAA TTAAG---

(c) The two fragments generated in **(b)** can be ligated by T4 DNA ligase to form

(5′)---GAATTAATTC---(3′)
---CTTAATTAAG---

(d) The fragments shown in **(a)** have sticky ends with a 5′ protruding single-stranded region. Treatment of these DNA fragments with a single-strand-specific nuclease will yield DNA fragments with blunt ends

(5′)---G(3′) and (5′)C---(3′)
---C G---

(e) The left-hand DNA fragment in **(b)** can be joined with the right-hand DNA fragment in **(d)** to yield

(5′)---GAATTC---(3′)
---CTTAAG---

The same recombinant DNA molecule can be produced by joining the right-hand DNA fragment in **(b)** with the left-hand DNA fragment in **(d).**

(f) The recognition sequence for *Pvu*II is (5′)CAGCTG (3′), with the cleavage site lying between G and C (see Table 29–2). Thus, a DNA molecule with a *Pvu*II site will yield two fragments when digested with *Pvu*II

(5′)---CAG(3′) and (5′)CTG---(3′)
---GTC GAC---

(g) The left-hand DNA fragment in **(b)** can be joined with the right-hand DNA fragment in **(f)** to yield

(5′)---GAATTCTG---(3′)
---CTTAAGAC---

The same DNA fragment can also be produced by joining the right-hand DNA fragment in **(b)** with the left-hand DNA fragment in **(f).**

(5′)---CAGAATTC---(3′)
---GTCTTAAG---

(h) There are two different methods by which one can convert an *Eco*RI restriction site to a *Bam*HI restriction site on a DNA molecule.

Method 1: Digest DNA with *Eco*RI and then create blunt ends by using either DNA polymerase I to fill in the single-stranded region as in **(b),** or a single-strand specific nuclease to remove the single-stranded region as in **(d).** Ligate a synthetic linker that contains the recognition sequence of *Bam*HI,

(5′)GCGGATCCCG(3′)
CGCCTAGGGC

between the two blunt-ended DNA fragments to yield, if the *Eco*RI digested DNA is treated as in **(b),**

(5′)---GAATTGCGGATCCCGAATTC---(3′)
---CTTAACGCCTAGGGCTTAAG---

or if the *Eco*RI digested DNA is treated as in **(d)**

(5′)---GGCGGATCCCG(3′)
---CCGCCTAGGGC

Notice that the *Eco*RI site is not regenerated after the ligation of the linker.

Method 2: This method utilizes a conversion adaptor to introduce a *Bam*HI site into the DNA molecule. A synthetic oligonucleotide with the sequence (5′)AATTGGATCC(3′) is partially self-complementary, and it spontaneously forms the structure

(5′)AATTGGATCC
CCTAGGTTAA

The sticky ends of this adaptor are complementary to the sticky ends generated by *Eco*RI digestion so that this adaptor can be ligated between the two *Eco*RI fragments to form

(5′)---GAATTGGATCCAATT---(3′)
---CTTAACCTAGGTTAA---

Since ligation between DNA molecules with compatible sticky ends is more efficient than ligation between DNA molecules with blunt ends, Method 2 is preferred over Method 1.

(i) In order for the DNA fragments shown in **(a)** to be joined with a DNA fragment generated by *Pst*I digestion, a conversion adaptor has to be used; this adaptor should contain a single-stranded region complementary to the sticky end of an *Eco*RI-generated DNA fragment, and a single-stranded region complementary to the sticky end generated by *Pst*I digestion. The four adaptor sequences that fulfill this requirement are shown below, in order of discussion in the problem (N = any nucleotide):

(5′)AATTCNNNNCTGCA
GNNNNG

(5′)AATTCNNNNGTGCA
GNNNNC

(5′)AATTGNNNNCTGCA
CNNNNG

(5′)AATTGNNNNGTGCA
CNNNNC

Ligation of the first adaptor to the *Eco*RI-digested DNA molecule would yield

(5′)---GAATTCNNNNCTGCA(3′)
---CTTAAGNNNNG

This DNA molecule can now be ligated to a DNA fragment produced by a *Pst*I digest which has the terminal sequence

(5′) G---(3′)
ACGTC---

to yield

(5′)---GAATTCNNNNCTGCAG---(3′)
---CTTAAGNNNNGACGTC---

(Notice that both *Eco*RI and *Pst*I sites are retained.)

In a similar fashion, the other three adaptors can each be ligated to the *Eco*RI-digested DNA molecule, and the ligated DNA molecule can subsequently be joined to a DNA fragment produced by a *Pst*I digest. The final products with the use of the second, third, and fourth adaptor, are

(5′)---GAATTCNNNNGTGCAG---(3′)
---CTTAAGNNNNCACGTC---

(Notice that the *Eco*RI site is retained, but not the *Pst*I site.)

(5′)---GAATTGNNNNCTGCAG---(3′)
---CTTAACNNNNGACGTC---

(Notice that the *Pst*I site is retained, but not the *Eco*RI site.)

(5′)---GAATTGNNNNGTGCAG---(3′)
---CTTAACNNNNCACGTC---

(Notice that neither the *Eco*RI nor the *Pst*I site is retained.)

2. Selecting for Recombinant Plasmids When cloning a foreign DNA fragment into a plasmid, it is often useful to insert the fragment at a site that interrupts a selectable marker (such as the tetracycline-resistance gene of pBR322). The loss of function of the interrupted gene can be used to identify clones containing recombinant plasmids with foreign DNA. With a bacteriophage λ vector it is not necessary to do this, yet one can easily distinguish vectors that incorporate large foreign DNA fragments from those that do not. How are these recombinant vectors identified?

Answer λ phage DNA can be packaged into infectious phage particles only if it is between 40,000 and 53,000 base pairs in length. Since bacteriophage λ vectors generally include about 30,000 base pairs (in two pieces), they will not be packaged into phage particles unless they contain a sufficient length of inserted DNA (10,000 to 23,000 base pairs).

3. DNA Cloning The plasmid cloning vector pBR322 (see Fig. 29–5) is cleaved with the restriction endonuclease *Pst*I. An isolated DNA fragment from a eukaryotic genome (also produced by *Pst*I cleavage) is added to the prepared vector and ligated. The mixture of ligated DNAs is then used to transform bacteria, and plasmid-containing bacteria are selected by growth in the presence of tetracycline.

(a) In addition to the desired recombinant plasmid, what other types of plasmids might be found among the transformed bacteria that are tetracycline resistant? How can they be distinguished?

(b) The cloned DNA fragment is 1,000 bp in length and has an *Eco*RI site 250 bp from one end. Three different recombinant plasmids are cleaved with *Eco*RI and analyzed by gel electrophoresis, giving the patterns shown. What does each pattern say about the cloned DNA? Note that in pBR322, the *Pst*I and *Eco*RI restriction sites are about 750 bp apart. The entire plasmid with no cloned insert is 4,361 bp. Size markers in lane 4 have the number of nucleotides noted.

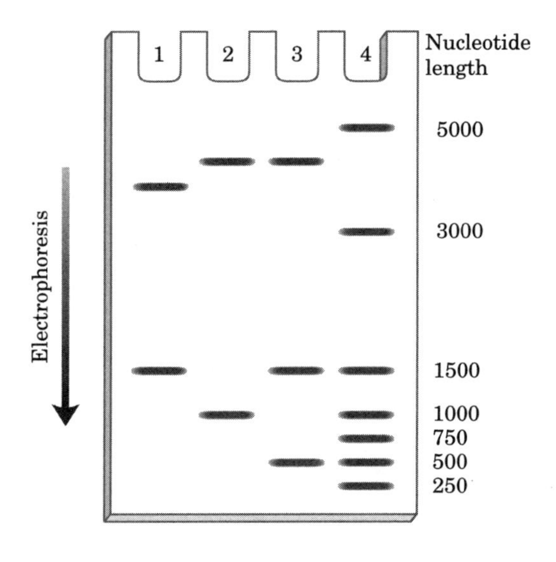

Answer

(a) Ligation of the linear pBR322 to regenerate circular pBR322 is a unimolecular process and thus can occur more efficiently than the ligation of a foreign DNA fragment to the linear pBR322, which is a bimolecular process (assuming equimolar amounts of linear pBR322 and a foreign DNA fragment in the reaction mixture). The tetracycline-resistant bacteria would include recombinant plasmids and plasmids in which the original pBR322 was regenerated without insertion of a foreign DNA fragment. These would retain resistance to ampicillin. Also, two or more molecules of pBR322 might be ligated together with or without insertion of foreign DNA.

(b) The clones in lanes 1 and 2 each have one DNA fragment inserted in different orientations. The clone in lane 3 has two DNA fragments, ligated such that the *Eco*RI proximal ends are joined.

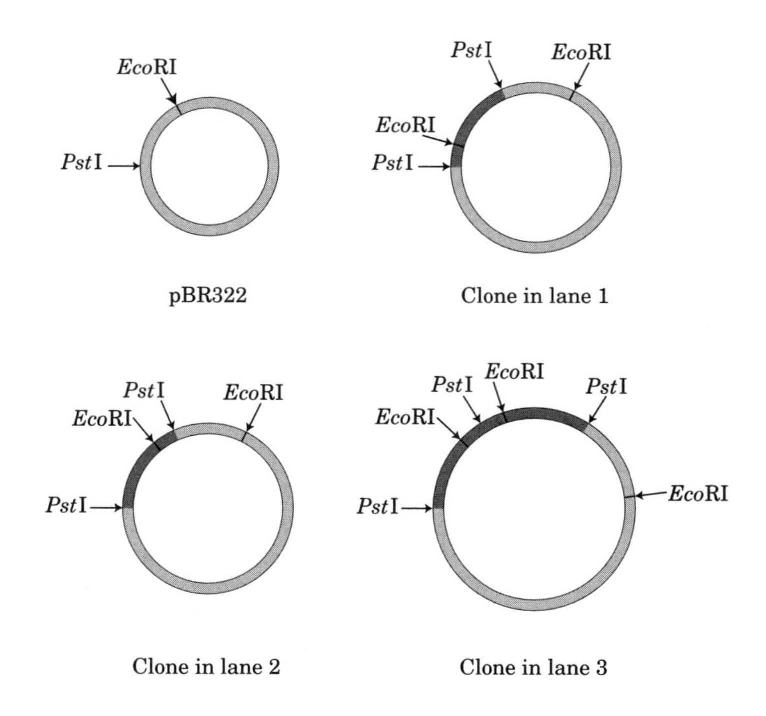

4. Expressing a Cloned Gene You have isolated a plant gene that encodes a protein in which you are interested. On the drawing below, indicate sequences or sites that you will need to get this gene transcribed, translated, and regulated in *E. coli.*

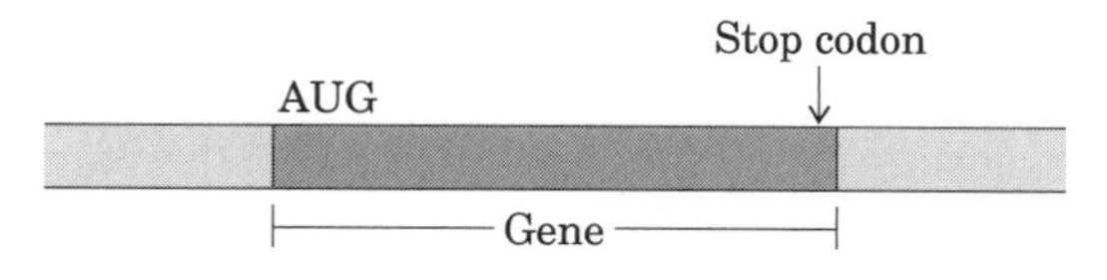

Answer The required sites and sequences are an *E. coli* promoter, because *E. coli* RNA polymerase does not interact with eukaryotic promoters; a ribosome-binding site positioned at an appropriate distance upstream from the ATG codon, because eukaryotic mRNA does not utilize such a site for translation initiation; and an operator site for regulation of transcription in *E. coli*.

5. Identifying the Gene for a Protein with a Known Amino Acid Sequence Design a DNA probe that would allow you to identify the gene for a protein with the following amino-terminal amino acid sequence. The probe should be 18 to 20 nucleotides long, a size that provides adequate specificity if there is sufficient homology between the probe and the gene.

$H_3\overset{+}{N}$—Ala—Pro—Met—Thr—Trp—Tyr—Cys—
Met—Asp—Trp—Ile—Ala—Gly—Gly—Pro—
Trp—Phe—Arg—Lys—Asn—Thr—Lys—

Answer Recall from Chapter 27 that most amino acids are encoded by two or more codons. To minimize the degree of ambiguity in codon assignment for a given peptide sequence, we must select a region of the peptide that contains mostly amino acids specified by a small number of codons. Focus on the amino acids with the fewest codons: Met and Trp (see Fig. 27–7 and Table 27–4). The best possibility is the span of DNA from the codon for the first Trp residue to the first two nucleotides of the codon for Ile. The sequence of the probe would be

(5′)UGGUA(U/C)UG(U/C)AUGGA(U/C)UGGAU

The synthesis would be designed to incorporate either U or C where indicated, producing a mixture of eight 20-nucleotide probes that differ only at one or more of these positions.

6. Cloning in Plants The strategy outlined in Figure 29–19 employs *Agrobacteria* that contain two separate plasmids. Suggest a reason why the sequences on the two plasmids are not combined on one plasmid.

Answer Simply for convenience; the 200,000 base pair Ti plasmid, even when the T DNA is removed, is too large to isolate in quantity and manipulate in vitro. It is also too large to reintroduce into a cell by standard transformation techniques. Single-plasmid systems in which the T DNA of a Ti plasmid has been replaced by foreign DNA (by means of low efficiency recombination in vivo) have been used successfully, but this approach is very laborious. The *vir* genes will facilitate the transfer of any DNA between the T DNA repeats, even if they are on a separate plasmid. The second plasmid in the two-plasmid system, because it requires only the T DNA repeats and a few sequences necessary for plasmid selection and propagation, is relatively small, easily isolated, and easily manipulated (foreign DNA is easily added and/or altered). It can be propagated in either *E. coli* or *Agrobacterium* and readily reintroduced into either bacterium.

7. Cloning in Mammals The retroviral vectors described in Figure 29–23 make it possible to integrate foreign DNA efficiently into a mammalian genome. Explain how these vectors, which lack genes for replication and viral packaging *(gag, pol, env)*, are assembled into infectious viral particles. Suggest why it is important that these vectors lack the replication and packaging genes.

Answer The vectors must be introduced into a cell infected with a helper virus that can provide the necessary replication and packaging functions but cannot itself be packaged. The vectors packaged into infectious viral particles are used to introduce the recombinant DNA into a mammalian cell. Once this DNA is integrated into the chromosome of the target cell, the lack of recombination and packaging functions makes the integration very stable by preventing the deletion or replication of the integrated DNA.

8. Designing a Diagnostic Test for a Genetic Disease Huntington's disease (HD) is an inherited neurodegenerative disorder, characterized by the gradual, irreversible impairment of psychological, motor, and cognitive functions. Symptoms typically appear in middle age, but onset can occur at almost any age. The course of the disease can last 15 to 20 years. The molecular basis of the disease is increasingly well understood. The genetic mutation underlying HD has been traced to a gene encoding a protein (M_r 350,000) of unknown function. In individuals who will not develop HD, a region near the end of the gene that encodes the amino terminus of the protein has a sequence of CAG codons (for glutamine) that is repeated 6 to 39 times in succession. In individuals with adult-onset HD, this codon is typically repeated 40 to 55 times. In individuals with childhood-onset HD, this codon is repeated more than 70 times. The length of this simple trinucleotide repeat indicates whether an individual will develop HD, and at approximately what age the first symptoms will occur.

A small portion of the amino-terminal coding sequence of the 3,143-codon HD gene is given below. The nucleotide sequence of the DNA is shown in black, the amino acid sequence of the open reading frame is below it, and the CAG repeat is shaded. Outline a PCR-based test for HD that could be carried out using a blood sample. Assume the PCR primer must be 25 nucleotides long. You may wish to refer to Chapter 26 (p. 982) for conventions for displaying gene sequences.

```
307 ATGGCGACCCTGGAAAAGCTGATGAAGGCCTTCGAGTCCCTCAAGTCCTTC
1   M  A  T  L  E  K  L  M  K  A  F  E  S  L  K  S  F

358 CAGCAGTTCCAGCAGCAGCAGCAGCAGCAGCAGCAGCAGCAGCAGCAGCAG
18  Q  Q  F  Q  Q  Q  Q  Q  Q  Q  Q  Q  Q  Q  Q  Q  Q

409 CAGCAGCAGCAGCAGCAGCAGCAACAGCCGCCACCGCCGCCGCCGCCGCCG
35  Q  Q  Q  Q  Q  Q  Q  Q  Q  P  P  P  P  P  P  P  P

460 CCGCCTCCTCAGCTTCCTCAGCCGCCGCCG
52  P  P  P  Q  L  P  Q  P  P  P
```

Source: The Huntington's Disease Collaborative Research Group (1993). A novel gene containing a trinucleotide repeat that is expanded and unstable on Huntington's disease chromosomes. *Cell* **72,** 971–983.

Answer Your test would require DNA primers, a heat-stable DNA polymerase, deoxynucleoside triphosphates, and a PCR machine (thermal cycler). The primers would be designed to simplify a DNA segment encompassing the CAG repeat. The DNA strand shown is the coding strand, oriented 5′ → 3′ left to right. The primer targeted to DNA to the left of the repeat would be identical to any 25-nucleotide sequence shown in the region to the left of the CAG repeat. Such a primer will direct synthesis of DNA across the repeat from left to right. The primer on the right side must be *complementary* and *antiparallel* to a 25-nucleotide sequence to the right of the CAG repeat. Such a primer will direct 5′ → 3′ synthesis of DNA across the repeat from left to right. Choosing unique sequences relatively close to the CAG repeat will make the amplified region smaller and the test more sensitive to small changes in size. Using the primers, DNA including the CAG repeat would be amplified by PCR, and its size would be determined by comparison to size markers after electrophoresis. The length of the DNA would reflect the length of the CAG repeat, providing a simple test for the disease. Such a test could be carried out on a blood sample and completed in less than a day.

9. Using PCR to Detect Circular DNA Molecules A segment of genomic DNA in a ciliated protozoan is sometimes deleted. The deletion is a genetically programmed reaction associated with cellular mating. A researcher proposes that the DNA is deleted in a site-specific recombination event (Fig. 25–36), with the deleted DNA ending up as a circular DNA reaction product as shown below. Suggest how the researcher might use the polymerase chain reaction (PCR) to detect the presence of the circular form of the deleted DNA in an extract of the protozoan.

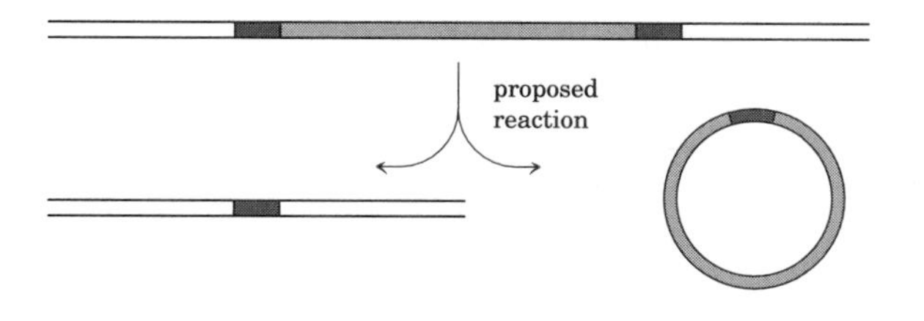

Answer Design PCR primers complementary to DNA in the deleted segment, but which would direct DNA synthesis away from each other. No PCR product will be generated unless the ends of the deleted segment are joined to create a circle.

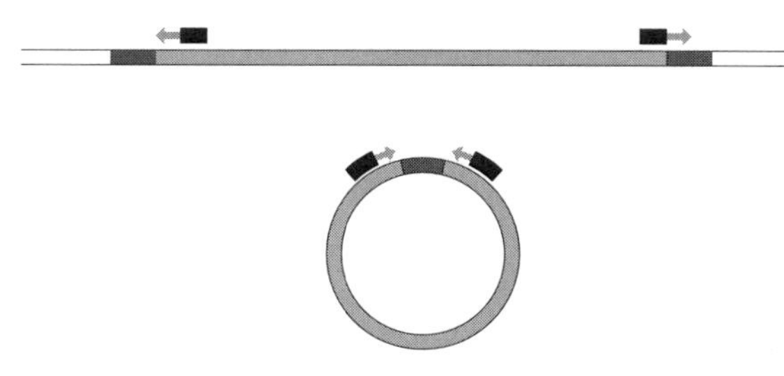

10. DNA Fingerprinting and RFLP Analysis DNA is extracted from the blood cells of two different humans. In separate experiments, the DNA from each individual is cleaved by restriction endonucleases A, B, and C, and the fragments are separated by electrophoresis. A hypothetical map of a 10,000 base pair segment of a human chromosome is shown. Individual #2 has point mutations that eliminate restriction recognition sites B* and C*. You probe the gel with a radioactive oligonucleotide complementary to the indicated sequence and expose a piece of x-ray film to the gel. Indicate where you would expect to see bands on the film. The lanes of the gel are marked on the accompanying figure.

A B C probe B C* B* A C

kbp 0 1 2 3 4 5 6 7 8 9 10

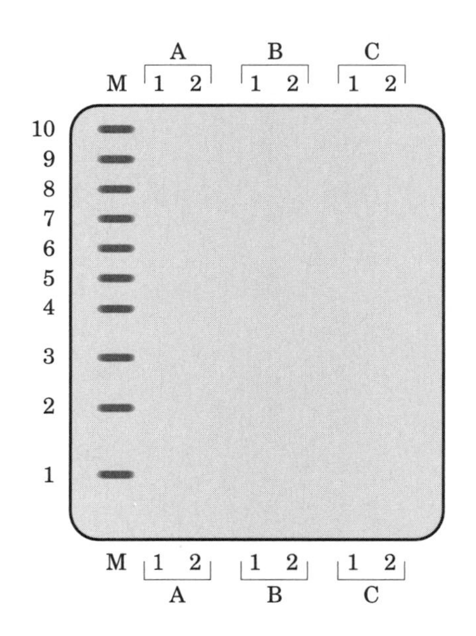

Answer Cleaving DNA with restriction enzyme A will produce identical fragments in both individuals, i.e., 6.5 kbp and 3.5 kbp fragments. The probe will hybridize to each of the 6.5 kbp fragments resulting in 2 identical bands being labeled in column A. Restriction enzyme B will produce different cleavage products. DNA from individual #1 will be cleaved into 3, 2, and 4 kbp fragments while that from individual #2 (who has an altered B recognition sequence) will be cleaved into 3 and 6 kbp fragments. However, the probe will bind to the 3 kbp fragments from both individuals and therefore will produce the same pattern of bands on the gel (shown in column B). Restriction enzyme C will cleave DNA from individual #1 into 2.5 and 4.5 kbp fragments and the probe will hybridize and label the 2.5 kbp piece. DNA from individual #2, however, will be cleaved to produce a single 7 kbp fragment, which will hybridize with the probe. Thus only in column C will a difference in DNA sequence between individual #1 and #2 become apparent. This exercise points out the importance of the choice of restriction enzymes as well as the choice of probes when performing DNA fingerprinting and RFLP analysis.

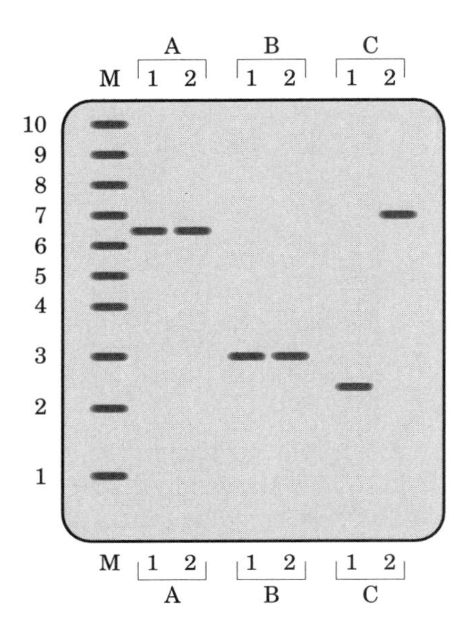

11. RFLP Analysis for Paternity Testing DNA fingerprinting and RFLP analysis are often used to test paternity. A child inherits chromosomes from both mother and father, so DNA from each child displays restriction fragments derived from each parent. In the gel shown here, which child, if any, can be excluded as being the biological offspring of the father? Explain your reasoning. Lane M is the sample from the mother, F from the father, and C1, C2, and C3 from the children.

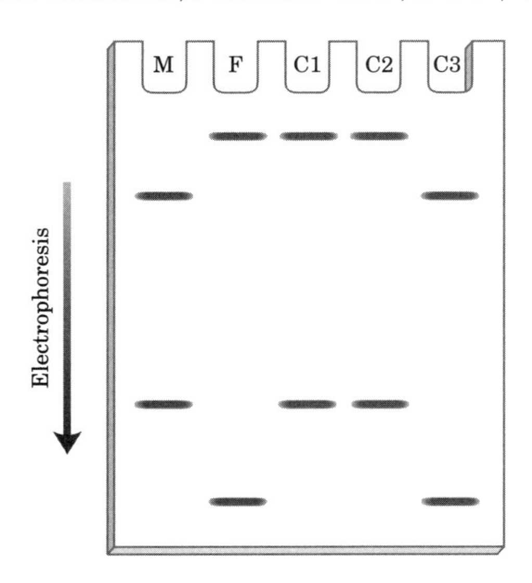

Answer None of the children can be excluded. Each has one band that could be derived from the father.